Structural Analysis with the Finite Element Method

Linear Statics

Volume 1. Basis and Solids

Lecture Notes on Numerical Methods in Engineering and Sciences

Aims and Scope of the Series

This series publishes text books on topics of general interest in the field of computational engineering sciences.

The books will focus on subjects in which numerical methods play a fundamental role for solving problems in engineering and applied sciences. Advances in finite element, finite volume, finite differences, discrete and particle methods and their applications to classical single discipline fields and new multidisciplinary domains are examples of the topics covered by the series.

The main intended audience is the first year graduate student. Some books define the current state of a field to a highly specialised readership; others are accessible to final year undergraduates, but essentially the emphasis is on accessibility and clarity.

The books will be also useful for practising engineers and scientists interested in state of the art information on the theory and application of numerical methods.

Titles:

1. E. Oñate, Structural Analysis with the Finite Element Method.
   Linear Statics. Volume 1. Basis and Solids, 2009

Structural Analysis with the Finite Element Method

Linear Statics

Volume 1. Basis and Solids

Eugenio Oñate

International Center for Numerical Methods in Engineering (CIMNE)
School of Civil Engineering
Universitat Politècnica de Catalunya (UPC)
Barcelona, Spain

ISBN: 978-1-4020-8732-5 (HB)
ISBN:978-1-4020-8733-2 (e-book)

Depósito legal: B-11715-09

A C.I.P. Catalogue record for this book is available from the Library of Congress

Typesetting: **Mª Jesús Samper**, CIMNE, Barcelona, Spain

Lecture Notes Series Manager: **Adriana Hanganu,** CIMNE, Barcelona, Spain

Cover page: **Pallí Disseny i Comunicació,** www.pallidisseny.com

Printed by: **Artes Gráficas Torres S.L.**
Morales 17, 08029 Barcelona, España
www.agraficastorres.es

Printed on elemental chlorine-free paper

Structural Analysis with the Finite Element Method. Linear Statics.
Volume 1. Basis and Solids
Eugenio Oñate

First edition, March 2009

To my family

Foreword

It is just over one-half century since papers on element based approximate solutions to structural problems first appeared in print. The term *Finite Element Method* was introduced in 1960 by Professor R.W. Clough to define this class of solution methods. In 1967, Professor O.C. Zienkiewicz published the first book describing applications of the method. Since these early contributions the finite element method has become indispensable to engineers and scientists involved in the analysis and design of a very wide range of practical structural problems: These include concrete dams, automobiles, aircraft, electronic parts, and medical devices, to name a few. Professor Eugenio Oñate, the author of *Structural Analysis with the Finite Element Method*, is a well recognized educator and research scholar in the area of computational mechanics. He completed his doctoral studies under the supervision of Professor O.C. Zienkiewicz at the University of Wales, Swansea. Professor Oñate is the founder and director of the International Center for Numerical Methods in Engineering (CIMNE) at the Universitat Politècnica de Catalunya in Barcelona, Spain. He has more than thirty years experience in development of finite element methods and related software.

This two volume book presents the results of the author's extensive experience in teaching and research on the finite element method. The content of the book develops the theory and practical implementation of the finite element method for application to linear structural problems. In the first volume, the finite element method is described to solve linear elastic problems for solids. The second volume extends the method to solve beam, plate and shell structures.

The style of presentation allows the reader to fully comprehend the fundamental steps in a finite element solution process. In the first volume, the equations of elasticity are developed explicitly and are combined with the principal of virtual work to describe the matrix problem to be solved. The book starts with one dimensional problems and builds systematically through two and three dimensional applications for solids. The first nine chapters present the theory of finite element analysis in detail – including the required steps to approximate element variables by isoparametric shape functions, to carry out numerical integration, and to perform assembly of final equations. Numerous examples are completely worked out and are complemented by color plates of results from analyses of practical problems. The first volume concludes with a chapter on mesh generation and visualization and a chapter on programming the finite element method. Use of the GiD program permits the reader to rapidly generate a mesh, while the chapter on programming describes how the reader can combine the computational advantages of MATLAB with the graphical capabilities of GiD to solve problems and visualize results. The reader can attain a deeper understanding of the finite element method by studying these chapters in parallel with the earlier theoretical chapters.

The second volume builds on the first to develop finite element formulations for beam, plate and shell problems. The pattern of development is identical with the first volume – namely starting with beam theories and building systematically through the development of various plate and shell finite element forms.

These two volumes enhance the reader's ability to master the basic concepts of the finite element method. Moreover, they provide the necessary background for further study on inelastic material behavior, contact interactions, and large deformation of solids and shells. Thus, the book is an extremely valuable contribution toward practical application of the finite element method in analysis and design of structures.

Robert L. Taylor
University of California, Berkeley, USA
December 2008

Preface

This two-volume book presents an overview of the possibilities of the Finite Element Method (FEM) for linear static analysis of structures. The text is a revised extension of the Spanish version of the book published by the International Center for Numerical Methods in Engineering (CIMNE) in 1992 and 1995 (2nd edition). The content of the book is based on the lectures of the course on Finite Element Structural Analysis taught by the author since 1979 to final year students in the School of Civil Engineering at the Technical University of Catalonia (UPC) in Barcelona, Spain.
Volume 1 of the book presents the basis of the FEM and its application to structures that can be modelled as two-dimensional (2D), axisymmetric and three-dimensional (3D) solids using the assumptions of general linear elasticity theory.
Volume 2 covers the FEM analysis of beam, plate, folded plate, axisymmetric shell and arbitrary shape shell structures. Emphasis is put in the treatment of structures with composite materials.
Each chapter of the book presents the main theoretical concepts on the particular structural model considered, such as the kinematic description, the constitutive relationship between stresses and strains and the equilibrium equations expressed by the Principle of Virtual Work. This is followed by a detailed derivation of the FEM equations and some applications to academic and practical examples of structural analysis. Complementary topics such as error estimation, adaptive mesh refinement, mesh generation and visualization of FEM results and computer programming of the FEM are also covered in the last chapters of Volume 1.
The book is particularly addressed to those interested in the analysis and design of solids and structures, understood here in a broad sense. The FEM concepts explained in the book are therefore applicable to the analysis of structures in civil engineering constructions, buildings and historical constructions, mechanical components and structural parts in automotive, naval and aerospace engineering, among many other applications.
The background knowledge required for study of the book is the standard one on mathematics, numerical analysis, elasticity and strength of

materials, matrix structural analysis and computer programming covered in the first courses of engineering and architecture schools at technical universities. In any case, the key theoretical concepts of each chapter are explained in some detail so as to facilitate its study.

Chapter 1 of Volume 1 presents first the concepts of structural and computational models. Then the basic steps of matrix analysis of bar structures are summarized. This chapter is important as the FEM follows very closely the methodology of matrix structural analysis. Understanding clearly the concept of splitting a structure in different elements, the equilibrium of the individual elements and the assembly of the global equilibrium equations of the structure from the contributions of the different elements is essential in order to follow the rest of the book.

Chapters 2 and 3 introduce the FEM formulation for the analysis of simple axially loaded bars using one-dimensional (1D) bar elements. The key ingredients of the FEM, such as discretization, interpolation, shape functions, numerical integration of the stiffness matrix and the equivalent nodal force vector for the element are explained in detail, as well as other general concepts such as the patch test, the conditions for convergence of the FE solution, the types of errors, etc.

Chapter 4 focuses on the study of structures under the assumption of 2D elasticity. These structures include dams, tunnels, pipes and retaining walls, among many others. The key ideas of 2D elasticity theory are explained, as well as the formulation of the 3-noded triangular element. Details of the explicit form of the element stiffness matrix and the equivalent nodal force vector are given.

Chapter 5 explains the derivation of the shape functions for 2D solid elements of rectangular and triangular shape and different orders of approximation. The resulting expressions for the shape functions are applicable to axisymmetric solid elements, as well as for many plate and shell elements studied in Volume 2.

Chapter 6 focuses on the formulation of 2D solid elements of arbitrary shape (i.e. irregular quadrilateral and triangular elements with straight or curved sides) using the isoparametric formulation and numerical integration. These concepts are essential for the organization of a general FEM computer program applicable to elements of different shape and approximation order. Examples of application to civil engineering constructions are presented.

Chapter 7 describes the formulation of axisymmetric solid elements. Use is made of the concepts explained in the previous two chapters, such as the derivation of the element shape functions, the isoparametric formulation and numerical integration. Applications to the analysis of axisymmetric solids and structures are presented.

Chapter 8 studies 3D solid elements of tetrahedral and hexahedral shapes. 3D solid elements allow the FEM analysis of any structure. Details of the derivation of the stiffness matrix and the equivalent nodal force vector are given for the simple 4-noded tetrahedral element. The formulation of higher order 3D solid elements is explained using the isoparametric formulation and numerical integration. Applications of 3D solid elements to a wide range of structures such as dams, buildings, historical constructions and mechanical parts are presented.

Chapter 9 covers miscellaneous topics of general interest for FEM analysis. These include the treatment of inclined supports, the blending of elements of different types, the study of structures on elastic foundations, the use of substructuring techniques, the procedures for applying constraints on the nodal displacements, the computation of stresses at the nodes and the key concepts of error estimation and adaptive mesh refinement strategies.

Chapter 10 introduces the basic ideas of mesh generation and visualization of the FEM results. The advancing front method and the Delaunay method for generation of unstructured meshes are explained in some detail.

Chapter 11 finally describes the organization of a simple computer program for FEM analysis of 2D structures using the 3-noded triangle and the 4-noded quadrilateral using MATLAB as a programming tool and the GiD pre-postprocessing system.

The four annexes cover the basic concepts of matrix algebra (Annex A), the solution of simultaneous linear algebraic equations (Annex B), the computation of the parameters for adaptive mesh refinement analysis (Annex C) and details of the GiD pre-postprocessing system developed at CIMNE (Annex D).

I want to express my gratitude to Dr. Francisco Zárate who was responsible for writing the computer program Mat-fem explained in Chapter 11 and also undertook the task of the writing this chapter.

Many thanks also to my colleagues in the Department of Continuum Mechanics and Structural Analysis at the Civil Engineering School of UPC for their support and cooperation over many years. Special thanks to Profs. Benjamín Suárez, Miguel Cervera and Juan Miquel and Drs. Francisco Zárate and Daniel di Capua with whom I have shared the teaching of the course on Finite Element Structural Analysis at UPC.

Many examples included in the book are the result of problems solved by academics and research students at UPC and CIMNE in cooperation with companies which are acknowledged in the text. I thank all of them for their contributions. Special thanks to the GiD team at CIMNE for providing the text for Annex D and many pictures shown in the book.
Many thanks also to my colleagues and staff at CIMNE for their cooperation and support during so many years that has made possible the publication of this book.
I am particularly grateful to Prof. O.C. Zienkiewicz from University of Swansea (UK) and Prof. R.L. Taylor from University of California at Berkeley (USA). Their ideas and suggestions during many visits at CIMNE and UPC in the period 1987-2007 have been a source of inspiration for the writing of this book.
Prof. Zienkiewicz, one of the giants in the field of computational mechanics, unfortunately passed away on January 2nd 2009 and has been unable to see the publication of this book. I express my deep sorrow for such a big loss and my recognition and gratitude for his support and friendship throughout my career.
Thanks also to Mrs. Adriana Hanganu from CIMNE for supervising the joint publication of the book by CIMNE and Springer.
Finally, my special thanks to Mrs. María Jesús Samper from CIMNE for her excellent work in the typing and editing of the manuscript.

Eugenio Oñate
Barcelona, January 2009

Contents

1 INTRODUCTION TO THE FINITE ELEMENT METHOD FOR STRUCTURAL ANALYSIS

1.1 WHAT IS THE FINITE ELEMENT METHOD?

The Finite Element Method (FEM) is a procedure for the numerical solution of the equations that govern the problems found in nature. Usually the behaviour of nature can be described by equations expressed in differential or integral form. For this reason the FEM is understood in mathematical circles as a numerical technique for solving partial differential or integral equations. Generally, the FEM allows users to obtain the evolution in space and/or time of one or more variables representing the behaviour of a physical system.

When referred to the analysis of structures the FEM is a powerful method for computing the displacements, stresses and strains in a structure under a set of loads. This is precisely what we aim to study in this book.

1.2 ANALYTICAL AND NUMERICAL METHODS

The conceptual difference between analytical and numerical methods is that the former search for the universal mathematical expressions representing the general and "exact" solution of a problem governed typically by mathematical equations. Unfortunately exact solutions are only possible for a few particular cases which frequently represent coarse simplifications of reality.

On the other hand, numerical methods such as the FEM aim to providing a solution, in the form of a set of numbers, to the mathematical equations governing a problem. The strategy followed by most numerical methods is to transform the mathematical expressions into a set of algebraic equations which depend on a finite set of parameters. For practical problems these equations involve many thousands (or even millions) of unknowns and therefore the final system of algebraic equations can only be solved with the help of computers. This explains why even though many numerical methods were known since the XVIII century, their development and popularity has occurred in tandem to the progress of modern computers in the XX century. The term *numerical method* is synonymous of *computational method* in this text.

Numerical methods represent, in fact, the return of numbers as the true protagonists in the solution of a problem. The loop initiated by Pythagoras some 25 centuries ago has been closed in the last few decades with the evidence that, with the help of numerical methods, we can find precise answers to any problem in science and engineering.

We should keep in mind that numerical methods for structural engineering are inseparable from mathematics, material modelling and computer science. Nowadays it is unthinkable to attempt the development of a numerical method for structural analysis without referring to those disciplines. As an example, any method for solving a large scale structural problem has to take into account the hardware environment where it will be implemented (most frequently using parallel computing facilities). Also a modern computer program for structural analysis should be able to incorporate the continuous advances in the modelling of new materials.

The concept which perhaps best synthesizes the immediate future of numerical methods is "multidisciplinary computations". The solution of problems will not be attempted from the perspective of a single discipline and it will involve all the couplings which characterize the complexity of reality. For instance, the design of a structural component for a vehicle (an automobile, an aeroplane, etc.) will take into account the manufacturing process and the function which the component will play throughout its life time. Structures in civil engineering will be studied considering the surrounding environment (soil, water, air). Similar examples are found in mechanical, naval and aeronautical engineering and indeed in practically all branches of engineering science. Accounting for the non-deterministic character of data will be essential for estimating the probability that the new products and processes conceived by men behave as planned. The huge computational needs resulting from a *stochastic multidisciplinary* viewpoint will demand better numerical methods, new material models and, indeed, faster computers.

It is only through the integration of a deep knowledge of the physical and mathematical basis of a problem and of numerical methods and informatics, that effective solutions will be found for the large-scale multidisciplinary problems in structural engineering of the twenty-first century.

1.3 WHAT IS A FINITE ELEMENT?

A finite element can be visualized as a small portion of a continuum (in this book a solid or a structure). The word "finite" distinguishes such a portion from the "infinitesimal" elements of differential calculus. The geometry of the continuum is considered to be formed by the assembly of a collection of non-overlapping domains with simple geometry termed finite elements. Triangles and quadrilaterals in two dimensions (2D) or tetrahedra and hexahedra in three dimensions (3D) are typically chosen to represent the "elements". It is usually said that a "mesh" of finite elements "discretizes" the continuum (Figure 1.1). The space variation of the problem parameters (i.e. the displacements in a structure) is expressed within each element by means of a polynomial expansion. Since the "exact" analytical variation of such parameters is more complex and generally unknown, the FEM only provides an *approximation* to the exact solution.

1.4 STRUCTURAL MODELLING AND FEM ANALYSIS

1.4.1 Classification of the problem

The first step in the solution of a problem is the identification of the problem itself. Hence, before we can analyze a structure we must ask ourselves the following questions: Which are the more relevant physical phenomena influencing the structure? Is the problem of static or dynamic nature? Are the kinematics or the material properties linear or non-linear? Which are the key results requested? What is the level of accuracy sought? The answers to these questions are essential for selecting a structural model and the adequate computational method.

1.4.2 Conceptual, structural and computational models

Computational methods, such as the FEM, are applied to *conceptual models* of a real problem, and not to the actual problem itself. Even experimental methods in structural laboratories make use of scale reproductions of the conceptual model chosen (also called physical models) unless the

Fig. 1.1 Discretization of different solids and structures with finite elements

actual structure is tested in real size, which rarely occurs. A conceptual model can be developed once the physical nature of a problem is clearly understood. In the derivation of a conceptual model we should aim to exclude superfluous details and include all the relevant features of the problem under consideration so that the model can describe reality with enough accuracy.

A conceptual model for the study of a structure should include all the data necessary for its representation and analysis. Clearly different persons will have different perceptions of reality and, consequently, the conceptual model for the same structure can take a variety of forms.

After selecting a conceptual model of a structure, the next step for the numerical (and analytical) study is the definition of a *structural model* (sometimes called *mathematical model*).

A *structural model* must include three fundamental aspects. The *geometric description* of the structure by means of its geometrical components (points, lines, surfaces, volumes), the *mathematical expression* of the basic physical laws governing the behaviour of the structure (i.e. the force-equilibrium equations and the boundary conditions) usually written in terms of differential and/or integral equations and the specification of the *properties of the materials* and of the *loads* acting on the structure. Clearly the same conceptual model of a structure can be analyzed using different structural models depending on the accuracy and/or simplicity sought in the analysis. As an example, a beam can be modelled using the general 3D elasticity theory, the 2D plane stress theory or the simpler beam theory. Each structural model provides a different set out for the analysis of the actual structure. We should bear in mind that a solution found by starting from an incorrect conceptual or structural model will be a wrong solution, far from correct physical values, even if obtained with the most accurate numerical method.

The next step in the structural analysis sequence is the definition of a *numerical method*, such as the FEM. The application of the FEM invariably requires its implementation in a computer code. The analysis of a structure with the FEM implies feeding the code with quantitative information on the mechanical properties of the materials, the boundary conditions and the applied loads (the physical parameters) as well as the features of the discretization chosen (i.e. element type, mesh size, etc). The outcome of this process is what we call a *computational model* for the analysis of a structure (Figure 1.2).

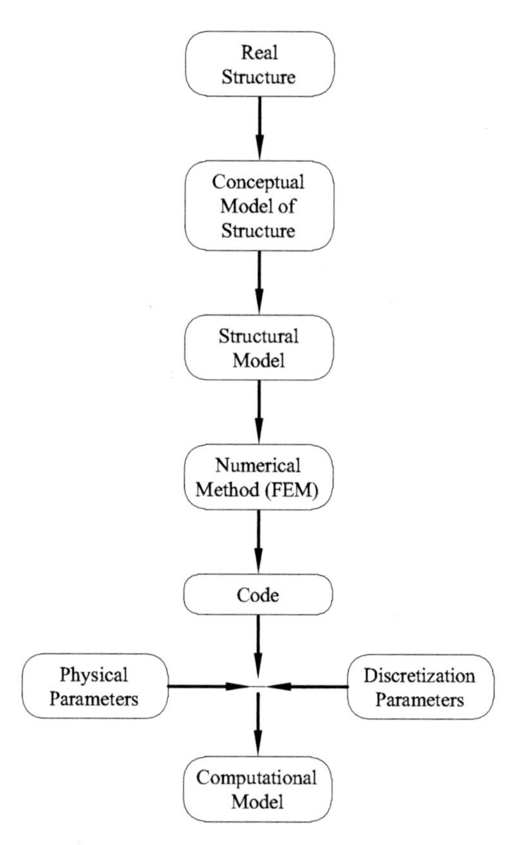

Fig. 1.2 The path from the real structure to the computational model

In this book we will study the application of the FEM to a number of structural models covering most structures found in the engineering practice. The material properties will be considered to be *linear elastic.* Furthermore the analysis will be restricted to *linear kinematics* and to *static loading.* The structures are therefore analyzed under ***linear static conditions***. Despite their simplicity, these assumptions are applicable to most of the situations found in the everyday practice of structural analysis and design.

The structural models considered in this book are classified as *solid models* (2D/3D solids and axisymmetric solids), *beam and plate models* and *shell models* (faceted shells, axisymmetric shells and curved shells). Figure 1.3 shows the general features of a typical member of each structural model family. The structures that can be analyzed with these models

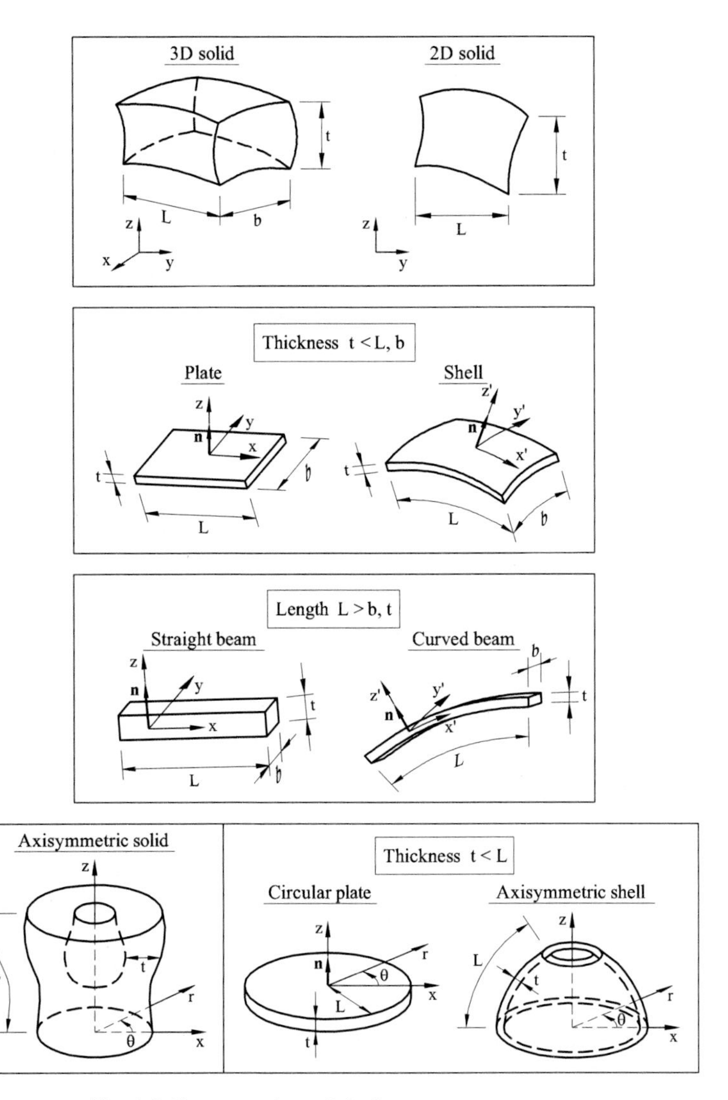

Fig. 1.3 Structural models for some structures

include frames, buildings, slabs, foundations, retaining walls, dams, tunnels, bridges, cylindrical tanks, shell roofs, ship hulls, mechanical parts, airplane fuselages, vehicle components, etc.

Volume 1 of this book studies structures that can be analyzed using solid finite element models. The finite element analysis of beam, plate and shell structures is covered in Volume 2 [On].

1.4.3 Structural analysis by the FEM

The geometry of a structure is *discretized* when it is split into a mesh of finite elements of a certain accuracy. Clearly the discretization introduces another approximation. With respect to reality we have therefore two error sources from the outset: the *modelling error* and the *discretization error*. The former can be reduced by improving the conceptual and structural models which describe the actual behaviour of the structure, as previously explained. The discretization error, on the other hand, can be reduced by using a finer mesh (i.e. more elements), or else by increasing the accuracy of the finite elements chosen using higher order polynomial expansions for approximating the displacement field within each element.

Additionally, the use of computers introduces *numerical errors* associated with their ability to represent data accurately with numbers of finite precision. The numerical error is usually small, although it can be large in some problems, such as when some parts of the structure have very different physical properties. The sum of discretization and numerical errors contribute to the *error of the computational model*. Note that even if we could reduce the computational error to zero, we would not be able to reproduce accurately the actual behaviour of the structure, unless the conceptual and structural models were perfect.

Figure 1.4 shows schematically the discretization of some geometrical models of structures using finite elements. Figure 1.5 shows the actual image of a car panel, the geometrical definition of the panel surface by means of NURBS (non-uniform rational B-splines) patches [PT] using computer-aided design (CAD) tools (see Chapter 10), the discretization of the surface by a mesh of 3-noded shell triangles and some numerical results of the FEM analysis. The differences between the real structure of the panel, the geometrical description and the analysis mesh can be seen clearly. A similar example of the FEM analysis of an office building is shown in Figure 1.6.

1.4.4 Verification and validation of FEM results

Developers of structural finite element computer codes, analysts who use the codes and decision makers who rely on the results of the analysis face a critical question: How should confidence in modelling and computation be critically assessed? *Validation* and *verification* of FEM results are the primary methods for building and quantifying this confidence. In essence, validation is the assessment of the accuracy of the structural and compu-

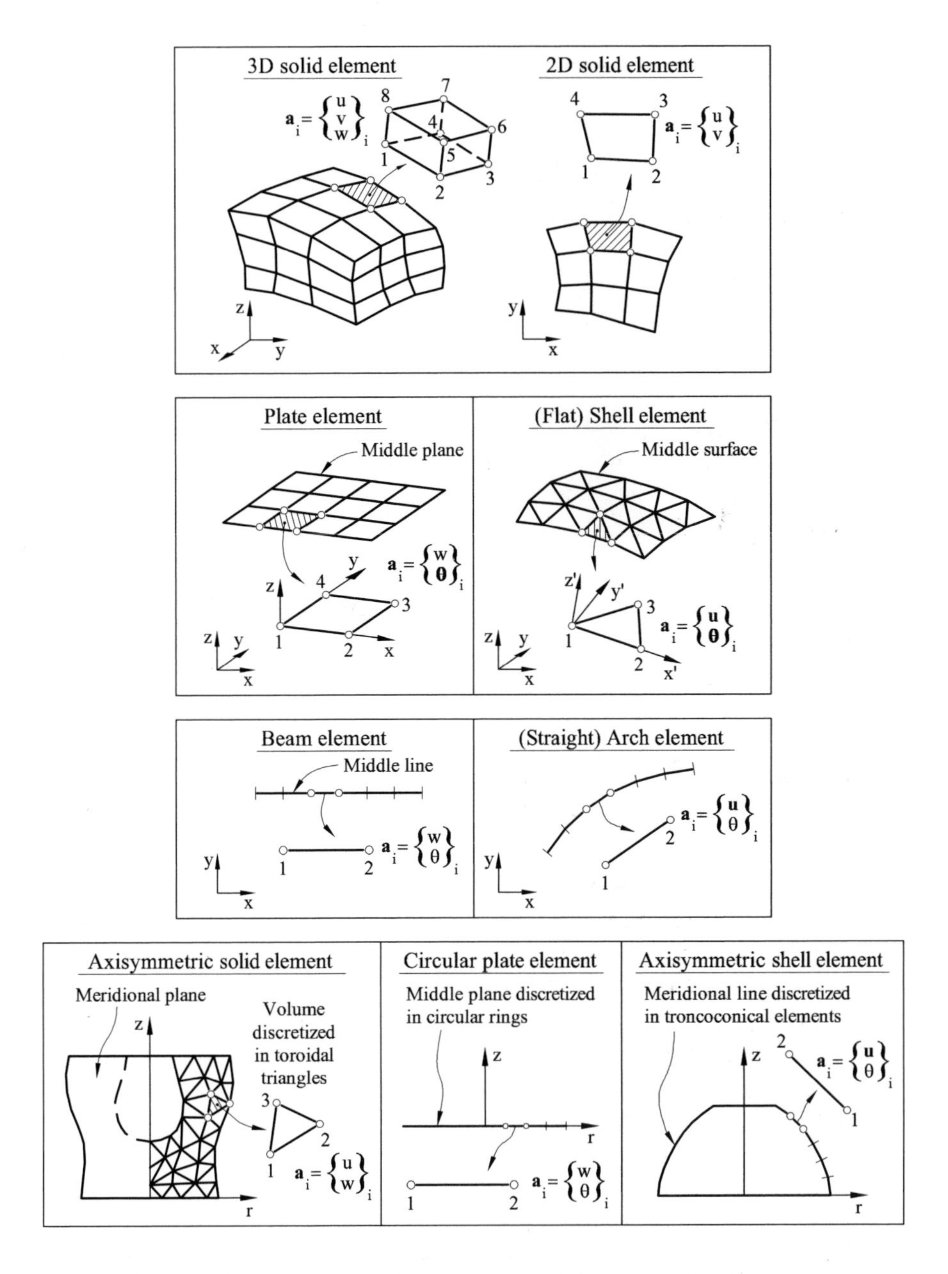

Fig. 1.4 Discretization of structural models into finite elements

tational models by comparison of the numerical results with *experimental data*. Experiments are usually performed in laboratory using scale models

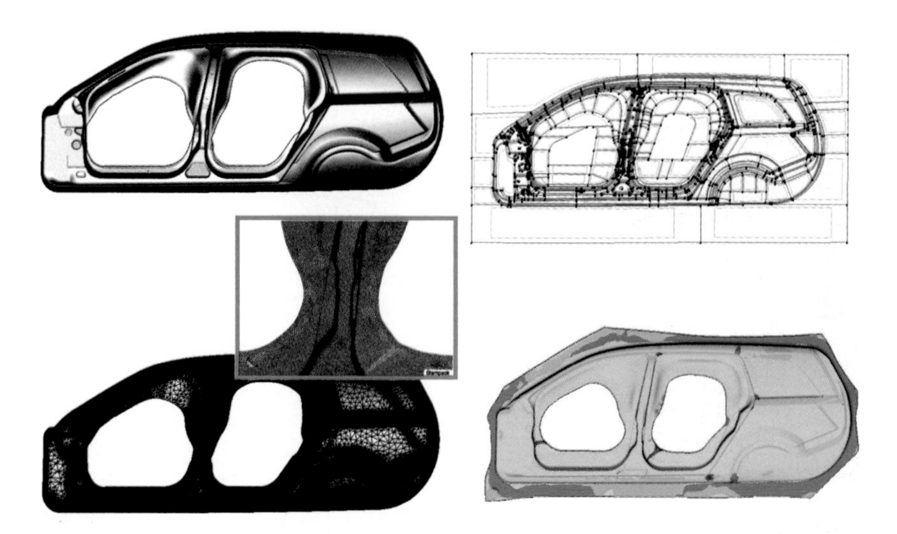

Fig. 1.5 (a) Actual geometry of an automotive panel. (b) CAD geometrical description by NURBS patches. (c) Finite element mesh of 3-noded shell triangles discretizing the panel geometry. (d) FEM numerical results of the structural analysis showing the equivalent strain distribution. Images by courtesy of Quantech ATZ S.A., www.quantech.es

of a structure, and in special occasions on actual structures. The correct definition of the experimental tests and the reliability of the experimental results are crucial issues in the validation process.

Verification, on the other hand, is the process of determining that a computational model accurately represents the underlying structural model and its solution. In verification, therefore, the relationship between the numerical results to the real world is not an issue. The verification of FEM computations is made by comparing the numerical results for simple benchmark problems with "exact" solutions obtained analytically, or using more accurate numerical methods. Figure 1.7 shows an scheme of the verification and validation steps [ASME,Sch].

A careful examination of the verification process indicates that there are two fundamental parts of verification: 1) *code verification*, in order to establish confidence that the mathematical model and the solution algorithms are working correctly, and 2) *calculation verification* aiming to establish confidence that the discrete solution of the mathematical model is accurate.

Among the code verification techniques, the most popular one is to compare code outputs with *analytical solutions*. As the number of such

Fig. 1.6 FEM analysis of the Agbar tower (Barcelona). Actual structure and discretization into shell and 3D beam elements. Deformed mesh (amplified) under wind load. Images are courtesy of Compass Ingeniería y Sistemas SA, www.compassis.com and Robert Brufau i Associats, S.A. www.robertbrufau.com

solutions is very limited, a code verification procedure with the potential to greatly expand is the use of *manufactured solutions.*

The basic concept of a manufactured solution is simple. Given a partial differential equation (PDE) and a code that provides general solutions of that PDE, an arbitrary solution to the PDE is manufactured, i.e. made up, then substituted into the PDE along with associated boundary conditions, also manufactured. The result is a forcing function (right-hand side) that exactly reproduces the originally selected *manufactured* solution. The code is then subjected to this forcing function and the numerical results

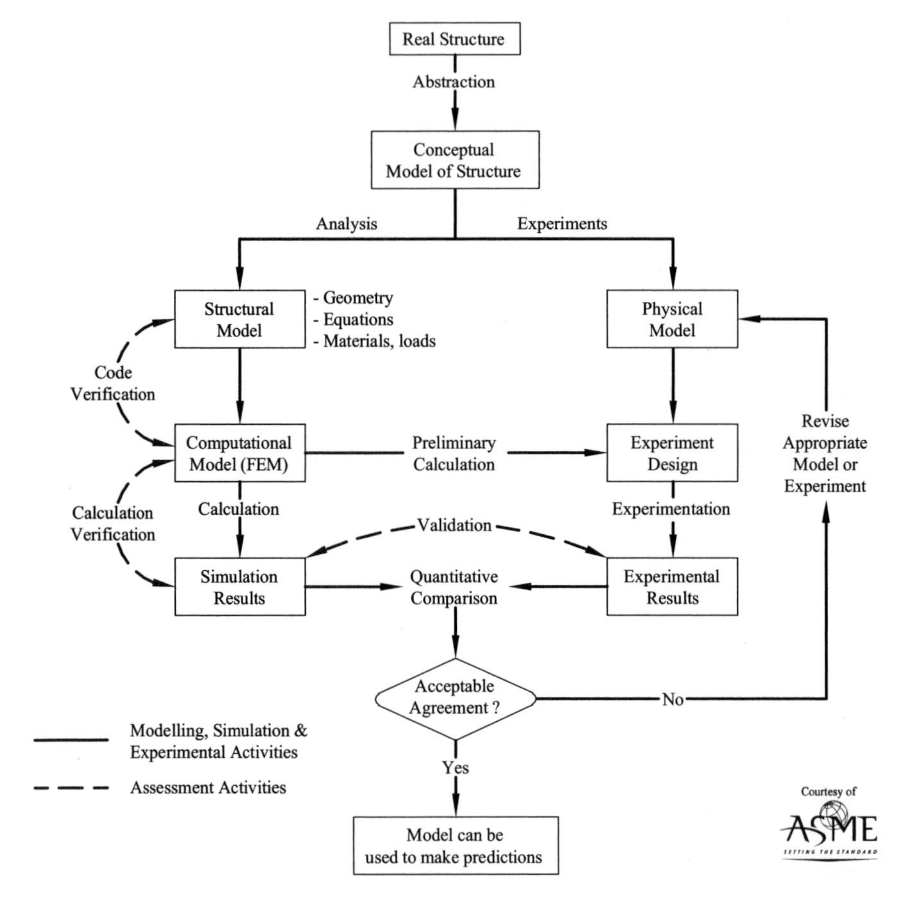

Fig. 1.7 Scheme of the verification and validation processes in the FEM. Flowchart concept taken from [ASME,Sch] and reprinted by permission of the American Society of Mechanical Engineering (ASME)

compared with the manufactured solution. If the code has no errors the two solutions should agree [Sch].

As an illustration of a manufactured solution, let us consider the ordinary differential equations for an Euler-Bernouilli beam of length L with a constant cross section (Chapter 1 of Volume 2 [On] and [Ti])

$$EI\frac{d^4w}{dx^4} = f(x)$$

where w is the beam deflection, E and I are the Young modulus and the inertia of the beam cross section, respectively and $f(x)$ is a uniformly distributed loading. The following manufactured solution is assumed

$$w(x) = A\sin\frac{\alpha x}{L} + Be^{x/L} + C$$

where the four constants A, α, B and C are determined from the boundary conditions. Substitution of the manufactured solution into the beam equation results in the following expression for the loading term

$$f(x) = EI\left[A\left(\frac{\alpha}{L}\right)^4 \sin\frac{\alpha x}{L} + \frac{B}{L^4}e^{x/L}\right]$$

This loading function would be prescribed as input data to the discrete beam finite element code and the code's solution for $w(x)$ is then compared with the selected manufactured solution.

Code verification is only half of the verification effort. The other half is the *calculation verification*, or, in other words, estimating the error in the numerical solution due to discretization. These errors can be appraised using error estimation techniques (Chapter 9). A more accurate numerical solution can be found with a finer discretization or by using higher order elements.

The subsequent validation step (Figure 1.7) has the goal of assessing the predictive capability of the model. This assessment is made by comparing the numerical results with validation experiments performed on physical models in laboratory or in real structures. If these comparisons are satisfactory, the model is deemed validated for its intended use. In summary, the validation exercise provides insight on the capacity of the overall structural model to reproduce the behaviour of a real structure (or the physical model chosen) with enough precision. Although both the accuracy of the structural model and the computational method are assessed in a validation process, a large validation error for an already verified code typically means that the structural model chosen is not adequate and that a better structural model should be used.

In conclusion, verification serves to check that we are solving structural problems accurately, while validation tell us that we are solving the right problem. Simply put, if the model passes the tests in the verification and validation plan, then it can be used to make the desired predictions with confidence. More details on the issue of verification and validation of the FEM in solid mechanics can be found in [ASME,Ro,Sch].

In the following sections we will revisit the basic concepts of the matrix analysis of bar structures, considered here as a particular class of the so-called *discrete systems*. Then we will summarize the general steps in the

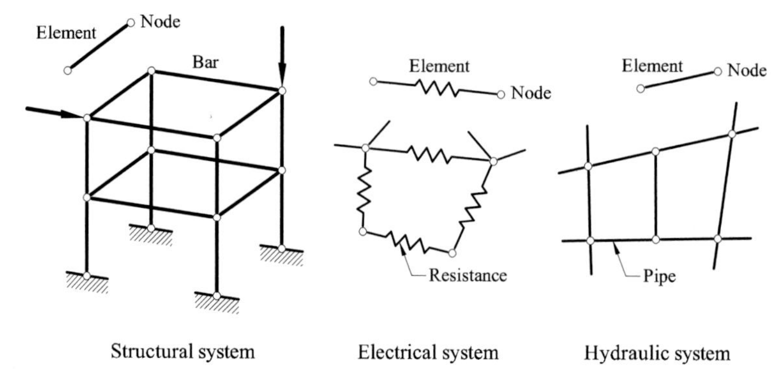

Fig. 1.8 Some discrete systems. Elements and joint points (nodes)

analysis of "continuous" structures by the FEM. The interest of classical matrix structural analysis is that it provides a general solution framework which reassembles very closely that followed in the FEM.

1.5 DISCRETE SYSTEMS. BAR STRUCTURES

The solution of many technical problems requires the analysis of a network system formed by different "elements" connected by their extremities or joints, and subjected to a set of "loads" which are usually external to the system. Examples of such systems, which we will call *discrete systems*, are common in structural engineering (pin-jointed bar structures, frames, grillages, etc.) and in many other different engineering problems, e.g.: hydraulic piping networks, electric networks, transport planning networks, production organization systems (PERT, etc) amongst others. Figure 1.8 shows some of these discrete systems.

Discrete systems can be studied using *matrix analysis* procedures which have a very close resemblance to the FEM. In Appendix A the basic concepts of matrix algebra are summarized. An outline of matrix analysis techniques for bar structures and other discrete systems such as electric and hydraulic networks is presented in the next section.

1.5.1 Basic concepts of matrix analysis of bar structures

Matrix analysis is the most popular technique for the solution of bar structures [Li,Pr]. Matrix analysis also provides a general methodology for the application of the FEM to other structural problems. A good knowledge of matrix analysis is essential for the study of this book.

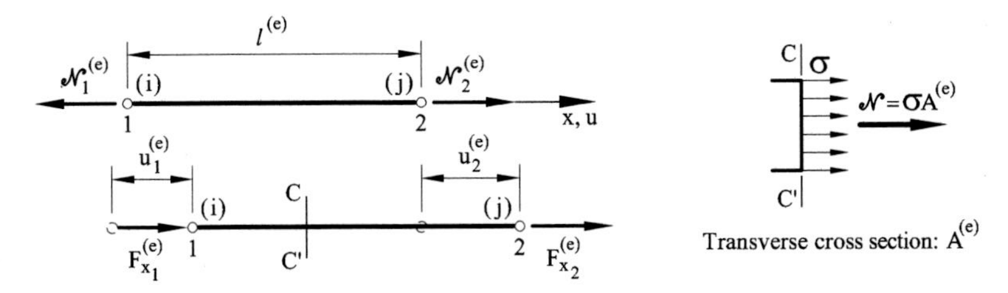

Fig. 1.9 Deformation of a bar subjected to axial end forces. Number in brackets at joints denotes global joint number

The matrix equations for a bar structure are obtained from the equations expressing the equilibrium of forces for each bar and for the structure as a whole. Let us consider an isolated bar, e, of length $l^{(e)}$ subjected to axial forces $F_{x_1}^{(e)}$ and $F_{x_2}^{(e)}$ acting at the beam joints (Figure 1.9). The x axis has the direction of the bar. Strength of Materials defines the strain at any point in the bar by the relative elongation [Ti], i.e.

$$\varepsilon = \frac{\Delta l^{(e)}}{l^{(e)}} = \frac{u_2^{(e)} - u_1^{(e)}}{l^{(e)}} \tag{1.1}$$

where $u_1^{(e)}$ and $u_2^{(e)}$ are the displacements of the joint points 1 and 2 in the x direction, respectively. In Eq.(1.1) and the following the superindex e denotes values associated to an individual bar. Generally indexes 1 and 2 are *local joint numbers* for the bar and correspond to the actual *global numbers* i, j of the joints in the structure. Hence $u_1^{(e)} = u_i$ and $u_2^{(e)} = u_j$ (Figure 1.9 and Example 1.1).

The axial stress σ is related to the strain ε by Hooke law [Ti] as

$$\sigma = E\varepsilon = E\frac{u_2^{(e)} - u_1^{(e)}}{l^{(e)}} \tag{1.2}$$

where E is the Young modulus of the material. The *axial force* $\mathcal{N}$ at each section is obtained by integrating the stress over the cross sectional area. The axial force $\mathcal{N}$ is transmitted to the adjacent bars through the joints. For homogeneous material we have (Figure 1.9)

$$\mathcal{N}_2^{(e)} = A^{(e)}\sigma = (EA)^{(e)}\frac{u_2^{(e)} - u_1^{(e)}}{l^{(e)}} = \mathcal{N}_1^{(e)} \tag{1.3}$$

The force equilibrium equation for the bar of Figure 1.9 is simply

$$F_{x_1}^{(e)} + F_{x_2}^{(e)} = 0 \tag{1.4a}$$

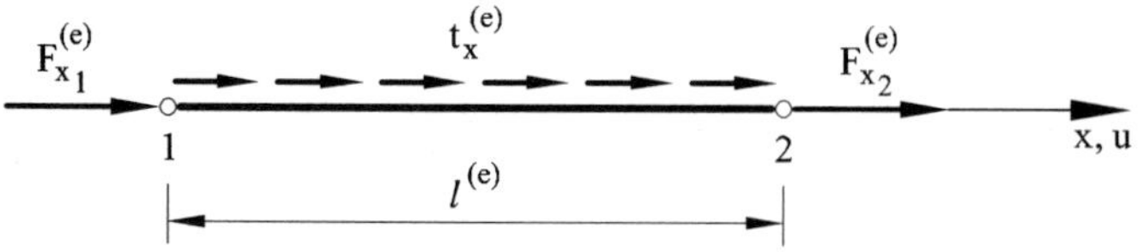

$$F_{x_1}^{(e)} + \frac{(lt_x)^{(e)}}{2} = k^{(e)}(u_1^{(e)} - u_2^{(e)})$$

$$F_{x_2}^{(e)} + \frac{(lt_x)^{(e)}}{2} = k^{(e)}(u_2^{(e)} - u_1^{(e)}) \quad , \quad k^{(e)} = \left(\frac{EA}{l}\right)^{(e)}$$

Fig. 1.10 Equilibrium equations for a bar subjected to axial joint forces and a uniformly distributed axial load $t_x^{(e)}$

with

$$F_{x_2}^{(e)} = \mathcal{N}_2^{(e)} = (EA)^{(e)} \frac{u_2^{(e)} - u_1^{(e)}}{l^{(e)}} = k^{(e)}(u_2^{(e)} - u_1^{(e)})$$

and (1.4b)

$$F_{x_1}^{(e)} = -F_{x_2}^{(e)} = k^{(e)}(u_1^{(e)} - u_2^{(e)}) = -\mathcal{N}_1^{(e)}$$

where $k^{(e)} = \left(\frac{EA}{l}\right)^{(e)}$. Eqs.(1.4b) can be written in matrix form as

$$\mathbf{q}^{(e)} = \begin{Bmatrix} F_{x_1}^{(e)} \\ F_{x_2}^{(e)} \end{Bmatrix} = k^{(e)} \begin{bmatrix} 1 & -1 \\ -1 & 1 \end{bmatrix} \begin{Bmatrix} u_1^{(e)} \\ u_2^{(e)} \end{Bmatrix} = \mathbf{K}^{(e)}\mathbf{a}^{(e)} \tag{1.5a}$$

where

$$\mathbf{K}^{(e)} = k^{(e)} \begin{bmatrix} 1 & -1 \\ -1 & 1 \end{bmatrix} \tag{1.5b}$$

is the *stiffness matrix* of the bar, which depends on the geometry of the bar ($l^{(e)}, A^{(e)}$) and its mechanical properties ($E^{(e)}$) only; $\mathbf{a}^{(e)} = [u_1^{(e)}, u_2^{(e)}]^T$ and $\mathbf{q}^{(e)} = [F_{x_1}^{(e)}, F_{x_2}^{(e)}]^T$ are the joint displacement vector and the joint equilibrating force vector for the bar, respectively.

A uniformly distributed external axial load of intensity $t_x^{(e)}$ can easily be taken into account by adding one half of the total external load to each axial force at the bar joints. The equilibrium equations now read (Figure 1.10)

$$\mathbf{q}^{(e)} = \begin{Bmatrix} F_{x_1}^{(e)} \\ F_{x_2}^{(e)} \end{Bmatrix} = k^{(e)} \begin{bmatrix} 1 & -1 \\ -1 & 1 \end{bmatrix} \begin{Bmatrix} u_1^{(e)} \\ u_2^{(e)} \end{Bmatrix} - \frac{(lt_x)^{(e)}}{2} \begin{Bmatrix} 1 \\ 1 \end{Bmatrix} = \mathbf{K}^{(e)}\mathbf{a}^{(e)} - \mathbf{f}^{(e)} \tag{1.6a}$$

where

$$\mathbf{f}^{(e)} = \begin{Bmatrix} f_{x_1}^{(e)} \\ f_{x_2}^{(e)} \end{Bmatrix} = \frac{(lt_x)^{(e)}}{2} \begin{Bmatrix} 1 \\ 1 \end{Bmatrix} \tag{1.6b}$$

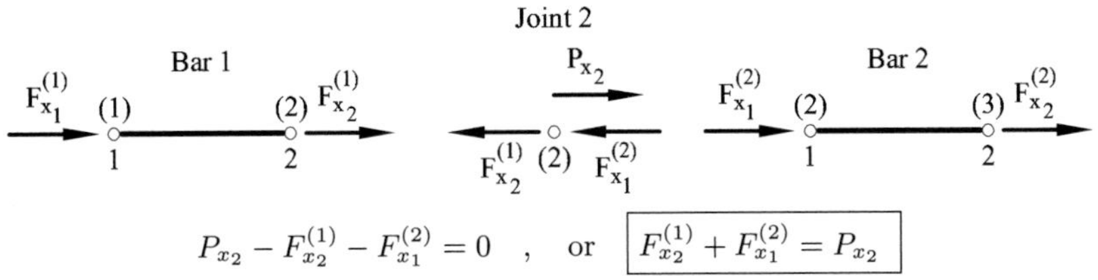
Joint 2
Bar 1
Bar 2
P_{x_2}
$F_{x_1}^{(1)}$ (1)
(2) $F_{x_2}^{(1)}$
1
2
$F_{x_2}^{(1)}$ (2) $F_{x_1}^{(2)}$
$F_{x_1}^{(2)}$ (2)
(3) $F_{x_2}^{(2)}$
1
2
$P_{x2} - F_{x2}^{(1)} - F_{x1}^{(2)} = 0$, or $F_{x2}^{(1)} + F_{x1}^{(2)} = P_{x2}$

Fig. 1.11 Equilibrium of axial forces $F_{x_2}^{(1)}$ and $F_{x_1}^{(2)}$ and external force P_{x_2} at joint 2 connecting bars 1 and 2. Number in brackets at joint denotes global joint number

is the vector of forces at the beam joints due to the distributed loading.

The equilibrium equations for the whole structure are obtained by imposing the equilibrium of axial and external forces at each of the N joints. This condition can be written as [Li,Pr]

$$\sum_{e=1}^{n_e} F_{x_i}^{(e)} = P_{x_j} \quad , \quad j = 1, N \tag{1.7}$$

The sum on the left hand side (l.h.s.) of Eq.(1.7) extends over all bars n_e sharing the joint with global number j and P_{x_j} represents the external point load acting at that joint (Figure 1.11). The joint forces $F_{x_i}^{(e)}$ for each bar are expressed in terms of the joint displacements using Eq.(1.6). This process leads to the system of global equilibrium equations. In matrix form

$$\begin{bmatrix} K_{11} & K_{12} & \cdots\cdots & K_{1N} \\ K_{21} & K_{22} & \cdots\cdots & K_{2N} \\ \vdots & & & \\ K_{N1} & K_{N2} & \cdots\cdots & K_{NN} \end{bmatrix} \begin{Bmatrix} u_1 \\ u_2 \\ \vdots \\ u_N \end{Bmatrix} = \begin{Bmatrix} f_1 \\ f_2 \\ \vdots \\ f_N \end{Bmatrix}$$

or

$$\mathbf{K}\mathbf{a} = \mathbf{f} \tag{1.8a}$$

where $\mathbf{K}$ is the global stiffness matrix of the structure and $\mathbf{a}$ and $\mathbf{f}$ are the global joint displacement vector and the global joint force vector, respectively. The derivation of Eq.(1.8a) is termed the *assembly process*. Solution of Eq.(1.8a) yields the displacements at all joint points from which the value of the axial force in each bar can be computed as

$$\mathcal{N}^{(e)} = (EA)^{(e)} \frac{u_2^{(e)} - u_1^{(e)}}{l^{(e)}} \tag{1.8b}$$

The axial forces at the joints can be computed from Eqs.(1.4b) and (1.6a) as

$$\mathbf{q}^{(e)} = \begin{Bmatrix} -\mathcal{N}_1^{(e)} \\ \mathcal{N}_2^{(e)} \end{Bmatrix} = \mathbf{K}^{(e)}\mathbf{a}^{(e)} - \mathbf{f}^{(e)} \tag{1.9}$$

Note that $\mathcal{N}_2^{(e)} = -\mathcal{N}_1^{(e)} = \mathcal{N}^{(e)}$.

The components of $\mathbf{q}^{(e)}$ can therefore be interpreted as the joint equilibrating forces for each bar necessary for imposing global equilibrium of forces at the joints (Eq.(1.6a)), or as the axial forces at the bar joints (Eq.(1.9)) which are useful for design purposes. This coincidence will be exploited later in the book for computing the resultant stresses at each node for bar and beam finite elements by expressions similar to Eq.(1.9).

The assembled expression for vector $\mathbf{q}^{(e)}$ yields the reactions at the nodes with constrained displacements. The vector of nodal reactions can be computed from the global stiffness equations as

$$\mathbf{r} = \mathbf{q} = \mathbf{K}\mathbf{a} - \mathbf{f}^{ext} \tag{1.10a}$$

where $\mathbf{r}$ contains the reactions at the constrained nodes and $\mathbf{f}^{ext}$ contains global joint forces due to external loads only. Clearly the sum of the reactions and the external joint forces gives the global joint force vector $\mathbf{f}$, i.e.

$$\mathbf{f} = \mathbf{f}^{ext} + \mathbf{r} \tag{1.10b}$$

1.5.2 Analogy with the matrix analysis of other discrete systems

The steps between Eqs.(1.1) and (1.8) are very similar for many discrete systems. For instance, the study of a single resistance element 1-2 in an electric network (Figure 1.12a) yields the following relationship between the currents entering the resistance element and the voltages at the end points of the resistance (Ohm law)

$$I_1^{(e)} = -I_2^{(e)} = \frac{1}{R^{(e)}}(V_1^{(e)} - V_2^{(e)}) = k^{(e)}(V_1^{(e)} - V_2^{(e)}) \tag{1.11a}$$

This equation is identical to Eq.(1.4) for the bar element if the current intensities and the voltages are replaced by the joint forces and the joint displacements, respectively, and $1/R^{(e)}$ by $\left(\frac{EA}{l}\right)^{(e)}$. Indeed, if uniformly distributed external currents $t_x^{(e)}$ are supplied along the length of the element, the force term $\mathbf{f}^{(e)}$ of Eq.(1.6a) is found. The "assembly rule" is the well known Kirchhoff law stating that the sum of all the current intensities arriving at a joint must be equal to zero, i.e.

$$\sum_{e=1}^{n_e} I_i^{(e)} = I_j \quad , \quad j = 1, N \tag{1.11b}$$

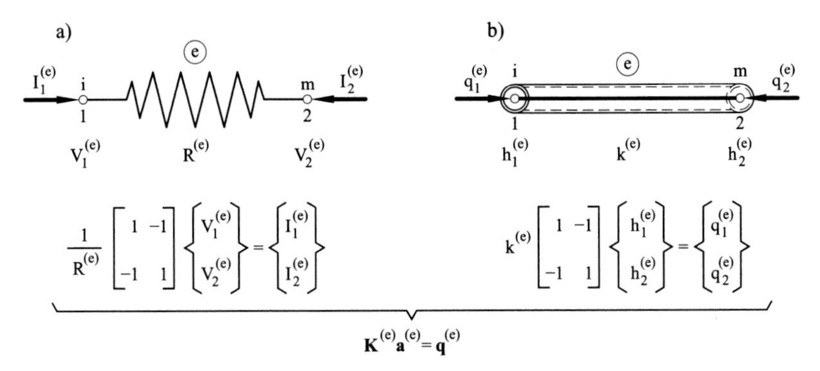

Fig. 1.12 a) Electrical resistance, b) Fluid carrying pipe. Equations of equilibrium

where I_i is the external current intensity entering joint i and N is the total number of joints. Note the analogy between Eqs.(1.11b) and (1.7).

The same analogy can be found for fluid carrying pipe networks. The equilibrium equation relating fluid flow q and hydraulic head h at the ends of a single pipe element can be written as (Figure 1.12b)

$$q_1^{(e)} = -q_2^{(e)} = k^{(e)}(h_1^{(e)} - h_2^{(e)}) \tag{1.12a}$$

where $k^{(e)}$ is a parameter which is a function of the pipe roughness and the hydraulic head. This implies that the terms of the stiffness matrix $\mathbf{K}^{(e)}$ for a pipe element are known functions of the joint heads $h_i^{(e)}$. The equilibrium equation for each pipe element is written as in Eq.(1.6) where $u_i^{(e)}$ and $F_{x_i}^{(e)}$ are replaced by $h_i^{(e)}$ and $q_i^{(e)}$, respectively and $t_x^{(e)}$ represents the input of a uniformly distributed flow source along the pipe length.

The assembly rule simply states that at each of the N pipe joints the sum of the flow contributed by the adjacent pipe elements should equal the external flow source, i.e.

$$\sum_{e=1}^{n_e} q_i^{(e)} = q_j \quad , \quad j = 1, N \tag{1.12b}$$

The global equilibrium equations are assembled similarly as for the bar element yielding the system of Eqs.(1.8a). In the general problem matrix $\mathbf{K}$ will be a function of the nodal hydraulic head via the $k^{(e)}$ parameter. Iterative techniques for solving the resulting non-linear system of equations are needed in this case.

1.5.3 Basic steps for matrix analysis of discrete systems

What we have seen this far leads us to conclude that the analysis of a discrete system (i.e. a bar structure) involves the following steps:

a) Definition of a network of discrete elements (bars) connected among themselves by joints adequately numbered. Each element e has known geometrical and mechanical properties. All these characteristics constitute the problem *data* and should be defined in the simplest possible way (preprocessing step).
b) Computation of the stiffness matrix $\mathbf{K}^{(e)}$ and the joint force vector $\mathbf{f}^{(e)}$ for each element of the system.
c) Assembly and solution of the resulting global matrix equilibrium equation ($\mathbf{Ka} = \mathbf{f}$) to compute the unknown parameters at each joint, i.e. the displacements for the bar system.
d) Computation of other relevant parameters for each element, i.e. the axial strain and the axial force, in terms of the joint parameters.

The results of the analysis should be presented in graphical form to facilitate the assessment of the system's performance (postprocessing step).

Example 1.1: Compute the displacements and axial forces in the three-bar structure of Figure 1.13 subjected to an horizontal force P acting at its right hand end.

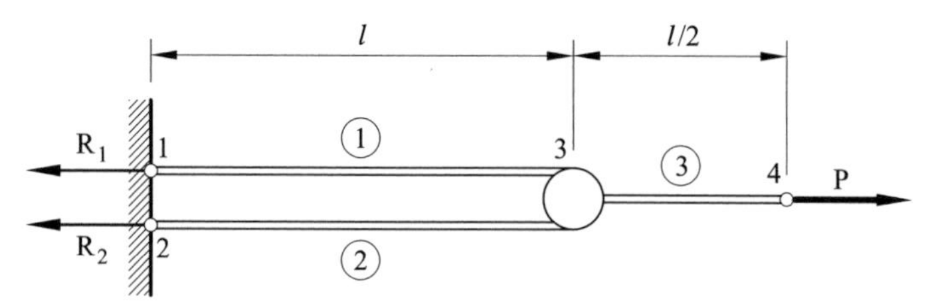

Fig. 1.13 Analysis of a simple three-bar structure under an axial load

- ***Solution***

The equilibrium equations for each joint are (see Eq.(1.5a))

$$\textit{Bar 1} \qquad \begin{Bmatrix} F_{x_1}^{(1)} \\ F_{x_2}^{(1)} \end{Bmatrix} = k^{(1)} \begin{bmatrix} 1 & -1 \\ -1 & 1 \end{bmatrix} \begin{Bmatrix} u_1^{(1)} \\ u_2^{(1)} \end{Bmatrix}$$

$$\textit{Bar 2} \qquad \begin{Bmatrix} F_{x_1}^{(2)} \\ F_{x_2}^{(2)} \end{Bmatrix} = k^{(2)} \begin{bmatrix} 1 & -1 \\ -1 & 1 \end{bmatrix} \begin{Bmatrix} u_1^{(2)} \\ u_2^{(2)} \end{Bmatrix}$$

$$\textit{Bar 3} \qquad \begin{Bmatrix} F_{x_1}^{(3)} \\ F_{x_2}^{(3)} \end{Bmatrix} = k^{(3)} \begin{bmatrix} 1 & -1 \\ -1 & 1 \end{bmatrix} \begin{Bmatrix} u_1^{(3)} \\ u_2^{(3)} \end{Bmatrix}$$

with $k^{(1)} = k^{(2)} = \frac{EA}{l}$ and $k^{(3)} = \frac{2EA}{l}$.
The compatibility equations between local and global displacements are

$$u_1^{(1)} = u_1 \quad ; \quad u_2^{(1)} = u_3 \quad ; \quad u_1^{(2)} = u_2$$

$$u_2^{(2)} = u_3 \quad ; \quad u_1^{(3)} = u_3 \quad ; \quad u_2^{(3)} = u_4$$

Applying the assembly equation (1.7) to each of the four joints we have

$$\text{joint 1:} \quad \sum_{e=1}^{3} F_{x_i}^{(1)} = -R_1 \quad , \quad \text{joint 2:} \quad \sum_{e=1}^{3} F_{x_i}^{(1)} = -R_2$$

$$\text{joint 3:} \quad \sum_{e=1}^{3} F_{x_i}^{(1)} = 0 \quad , \quad \text{joint 4:} \quad \sum_{e=1}^{3} F_{x_i}^{(1)} = P$$

Substituting the values of $F_{x_i}^{(e)}$ from the bar equilibrium equations gives

$$\text{joint 1}: \quad k^{(1)}(u_1^{(1)} - u_2^{(1)}) = -R_1 \quad , \quad \text{joint 2}: \quad k^{(2)}(u_1^{(2)} - u_2^{(2)}) = -R_2$$

$$\text{joint 3}: \quad k^{(1)}(-u_1^{(1)} + u_2^{(1)}) + k^{(2)}(-u_1^{(2)} + u_2^{(2)}) + k^{(3)}(u_1^{(3)} + u_2^{(3)}) = 0$$

$$\text{joint 4}: \quad k^{(3)}(-u_1^{(3)} + u_2^{(1)}) = P$$

Above equations can be written in matrix form using the displacement compatibility conditions as

$$\begin{array}{c} \\ 1 \\ 2 \\ 3 \\ 4 \end{array} \begin{array}{c} \begin{array}{cccc} 1 & 2 & 3 & 4 \end{array} \\ \begin{bmatrix} k^{(1)} & 0 & -k^{(1)} & 0 \\ 0 & k^{(2)} & -k^{(2)} & 0 \\ -k^{(1)} & -k^{(2)} & (k^{(1)} + k^{(2)} + k^{(3)}) & -k^{(3)} \\ 0 & 0 & -k^{(3)} & k^{(3)} \end{bmatrix} \end{array} \begin{Bmatrix} u_1 \\ u_2 \\ u_3 \\ u_4 \end{Bmatrix} = \begin{Bmatrix} -R_1 \\ -R_2 \\ 0 \\ P \end{Bmatrix}$$

Note that an external point load acting at node j can be placed directly in the jth position of the global joint force vector $\mathbf{f}$.
Substituting the values of $k^{(e)}$ for each bar and imposing the boundary conditions $u_1 = u_2 = 0$, the previous system can be solved to give

$$u_3 = \frac{Pl}{2EA} \quad ; \quad u_4 = \frac{Pl}{EA} \quad ; \quad R_1 = R_2 = \frac{P}{2}$$

The axial forces in each bar are finally obtained as

$$Bar\ 1:\ \mathcal{N}^{(1)} = \frac{EA}{l}(u_3-u_1) = \frac{P}{2} \quad , \quad Bar\ 2:\ \mathcal{N}^{(2)} = \frac{EA}{l}(u_3-u_2) = \frac{P}{2}$$

$$Bar\ 3:\ \mathcal{N}^{(3)} = \frac{2EA}{l}(u_4-u_3) = P$$

The joint axial force for each bar is computed from Eq.(1.8c) giving

$$\mathcal{N}_2^{(1)} = -\mathcal{N}_1^{(1)} = P/2 \quad ; \quad \mathcal{N}_2^{(2)} = -\mathcal{N}_1^{(2)} = P/2 \quad ; \quad \mathcal{N}_2^{(3)} = -\mathcal{N}_1^{(3)} = P$$

1.6 DIRECT ASSEMBLY OF THE GLOBAL STIFFNESS MATRIX

The stiffness and force contributions of each individual bar can be *directly* assembled in the global stiffness matrix by the following procedure. Consider a bar e connecting two joints with global numbers i and m (Figure 1.14). Each term in the position (i,m) of the bar stiffness matrix contributes to the same position (i,m) of the global stiffness matrix. Similarly, the nodal force components $f_{x_1}^{(e)}$ and $f_{x_2}^{(e)}$ corresponding to the

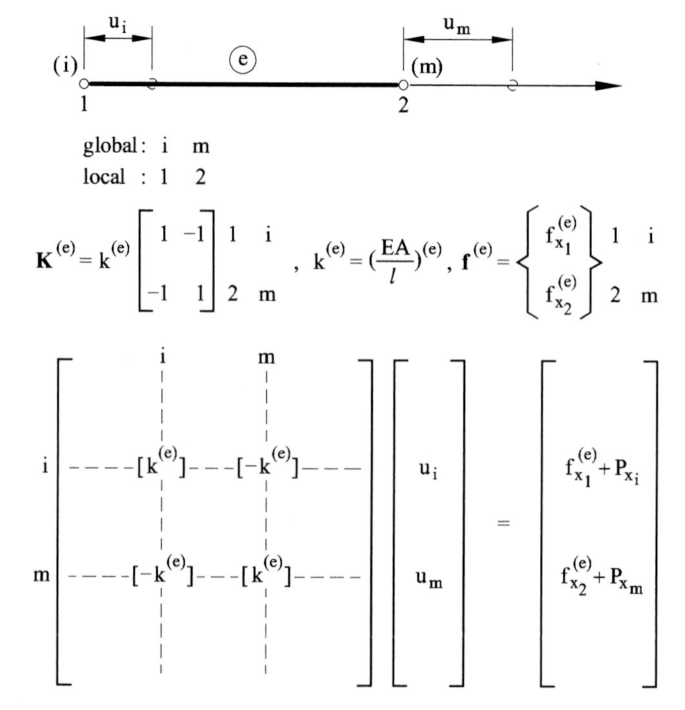

Fig. 1.14 Contributions to the global stiffness matrix and the global joint force vector from an individual bar

global joint numbers i and m are respectively placed in rows i and m of the global joint force vector $\mathbf{f}$. Also an external point load P_{x_i} acting at joint i is directly added to the component of the ith row of $\mathbf{f}$ (Figure 1.14). Thus, the global stiffness matrix and the global joint force vector can be computed by systematically adding the contributions from the different bars using information from the joint numbers. This assembly process can be programmed in a simple and general form [Hu,HO,HO2].

The cost of solving the global system of equations (1.8a) using a direct solver (Appendix B) is approximately equal to $N\frac{B^2}{4}$, where N is the order of $\mathbf{K}$ and B its *bandwidth* [CMPW]. For each row i of $\mathbf{K}$, the *semibandwidth* $\frac{B_i}{2}$ is equal to the number of columns from the diagonal to the right-most non zero term plus one. A root mean-square average of the B_i may be taken as representative B for the entire matrix (Example 1.2).

Example 1.2: Obtain the bandwidth of the stiffness matrix for the structure of the figure with the node numbering indicated below.

- ***Solution***

Numbering a)

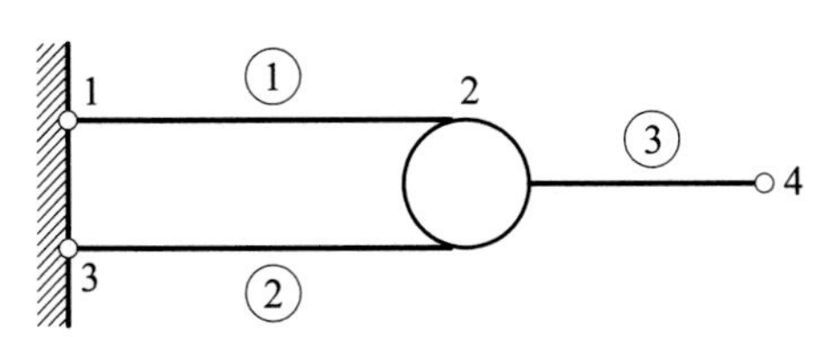

The local numbering for each bar element is always taken from left to right

$$\mathbf{K}^{(e)} = \begin{bmatrix} k_{11}^{(1)} & k_{12}^{(1)} & 0 & 0 \\ k_{21}^{(1)} & (k_{22}^{(1)} + k_{22}^{(2)} + k_{11}^{(3)}) & k_{21}^{(2)} & k_{12}^{(3)} \\ 0 & k_{12}^{(2)} & k_{11}^{(2)} & 0 \\ 0 & k_{21}^{(3)} & 0 & k_{22}^{(3)} \end{bmatrix}$$

Numbering b)

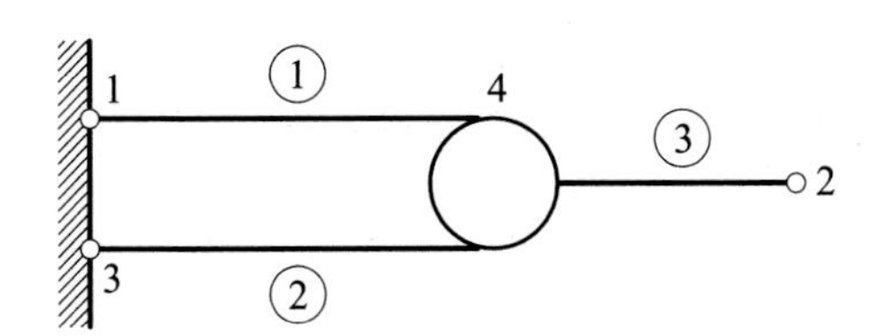

$$\mathbf{K}^{(e)} = \begin{bmatrix} k_{11}^{(1)} & 0 & 0 & k_{12}^{(1)} \\ 0 & k_{22}^{(3)} & 0 & k_{21}^{(3)} \\ 0 & 0 & k_{11}^{(2)} & k_{12}^{(2)} \\ k_{21}^{(1)} & 0 & k_{21}^{(2)} & (k_{11}^{(1)} + k_{22}^{(2)} + k_{11}^{(3)}) \end{bmatrix}$$

In numbering (a) the nodal bandwidths B_i are [4,6,4,2] and $B \simeq 4.1$. In numbering (b) the banded structure is lost and the bandwidths B_i are [8,6,4,2] and $B \simeq 5.5$. These differences can be very significant in practical problems where the order of $\mathbf{K}$ is much larger.

1.7 DERIVATION OF THE MATRIX EQUILIBRIUM EQUATIONS FOR THE BAR USING THE PRINCIPLE OF VIRTUAL WORK

A key step in the matrix analysis of bar structures is the derivation of the stiffness equations for the single bar element. These equations express the equilibrium between the loads acting at the bar joints and the displacements of the joint points (Eq.(1.5)). For the simple axially loaded bar these equations can be directly obtained using concepts from Strength of Materials [Ti,Ti2]. For complex structures more general procedures are needed. Among these, the Principle of Virtual Work (PVW) is the more powerful and widespread technique. This well known principle states that: "A structure is in equilibrium under a set of external loads if after imposing to the structure arbitrary (virtual) displacements compatible with the boundary conditions, the work performed by the external loads on the virtual displacements equals the work performed by the actual stresses on the strains induced by the virtual displacements".

The PVW is a necessary and sufficient condition for the equilibrium of the whole structure or any of its parts [Ti,Ti2,Was,ZT,ZTZ]. Next, we will apply this technique to the axially loaded bar of Figure 1.9. The PVW in this case is written as

$$\iiint_{V^{(e)}} \delta\varepsilon\sigma dV = \delta u_1^{(e)} F_{x_1}^{(e)} + \delta u_2^{(e)} F_{x_2}^{(e)} \tag{1.13}$$

where $\delta u_1^{(e)}$ and $\delta u_2^{(e)}$ are, respectively, the virtual displacements of ends 1 and 2 of a bar with volume $V^{(e)}$, and $\delta\varepsilon$ is the corresponding virtual strain which can be obtained in terms of $\delta u_1^{(e)}$ and $\delta u_2^{(e)}$ as

$$\delta\varepsilon = \frac{\delta u_2^{(e)} - \delta u_1^{(e)}}{l^{(e)}} \tag{1.14}$$

Substituting the values of σ and $\delta\varepsilon$ of Eqs.(1.2) and (1.14) into (1.13) and integrating the stresses over the cross sectional area of the bar gives

$$\int_{l^{(e)}} \frac{1}{l^{(e)}}\left[\delta u_2^{(e)} - \delta u_1^{(e)}\right](EA)^{(e)}\frac{1}{l^{(e)}}\left[u_2^{(e)} - u_1^{(e)}\right]dx = \delta u_1^{(e)} F_{x1}^{(e)} + \delta u_2^{(e)} F_{x2}^{(e)} \tag{1.15}$$

Integrating over the bar length, assuming the Young modulus $E^{(e)}$ and the area $A^{(e)}$ to be constant, yields

$$\left(\frac{EA}{l}\right)^{(e)}\left[u_1^{(e)} - u_2^{(e)}\right]\delta u_1^{(e)} + \left(\frac{EA}{l}\right)^{(e)}\left[u_2^{(e)} - u_1^{(e)}\right]\delta u_2^{(e)} =$$

$$= \delta u_1^{(e)} F_{x1}^{(e)} + \delta u_2^{(e)} F_{x2}^{(e)} \tag{1.16}$$

Since the virtual displacements are *arbitrary*, the satisfaction of Eq.(1.16) for any value of $\delta u_1^{(e)}$ and $\delta u_2^{(e)}$ requires that the terms multiplying each virtual displacement at each side of the equation should be identical. This leads to the following system of two equations

$$\text{For } \delta u_1^{(e)} \quad : \quad \left(\frac{EA}{l}\right)^{(e)}\left[u_1^{(e)} - u_2^{(e)}\right] = F_{x1}^{(e)} \tag{1.17a}$$

$$\text{For } \delta u_2^{(e)} \quad : \quad \left(\frac{EA}{l}\right)^{(e)}\left[u_2^{(e)} - u_1^{(e)}\right] = F_{x2}^{(e)} \tag{1.17b}$$

which are the equilibrium equations we are looking for.

These equations, written in matrix form, coincide with Eqs.(1.5a) directly obtained using more physical arguments. The effect of a uniformly distributed load (Figure 1.10) can easily be taken into account by adding to the right hand side (r.h.s.) of Eq.(1.13) the term $\int_{l^{(e)}} \delta u t_x^{(e)}\, dx$. Assuming a linear distribution of the virtual displacements in terms of the joint displacement values, the expression of Eq.(1.6a) is recovered is it can be verified by the reader.

The PVW will be used throughout this book to derive the matrix equilibrium equations for the different structures studied with the FEM.

1.8 DERIVATION OF THE BAR EQUILIBRIUM EQUATIONS VIA THE MINIMUM TOTAL POTENTIAL ENERGY PRINCIPLE

The equilibrium equations for a structure can also be derived via the principle of Minimum Total Potential Energy (MTPE). The resulting equations are identical to those obtained via the PVW. The applications of the MTPE principle are generally limited to elastic materials for which simple forms of the total potential energy can be derived [TG,Ti]. The PVW is more general as it is applicable to non linear problems (including both material and geometrical non linearities) and it is usually chosen as the starting variational form for deriving the finite element equations.

The total potential energy for a single bar e under joint forces $F_{x_i}^{(e)}$ is

$$\Pi^{(e)} = \frac{1}{2}\int_{l^{(e)}} \varepsilon \mathcal{N} dx - \sum_{i=1}^{2} u_i^{(e)} F_{x_i}^{(e)} \tag{1.18}$$

Substituting into Eq.(1.18) the expression for the elongation ε and the axial forces $\mathcal{N}$ in terms of the end displacements, i.e.

$$\varepsilon = \frac{u_2^{(e)} - u_1^{(e)}}{l^{(e)}} \quad , \quad \mathcal{N} = (EA)^{(e)} \frac{u_2^{(e)} - u_1^{(e)}}{l^{(e)}} \tag{1.19}$$

gives

$$\Pi^{(e)} = \frac{1}{2}\int_{l^{(e)}} \left(\frac{u_2^{(e)} - u_1^{(e)}}{l^{(e)}}\right) (EA)^{(e)} \left(\frac{u_2^{(e)} - u_1^{(e)}}{l^{(e)}}\right) dx - \left(u_1^{(e)} F_{x_1}^{(e)} + u_2^{(e)} F_{x_2}^{(e)}\right) \tag{1.20}$$

The MTPE principle states that a structure is in equilibrium for values of the displacement making Π stationary. The MTPE also holds for the equilibrium of any part of the structure. The equilibrium condition for the single bar is written as

$$\frac{\partial \Pi^{(e)}}{\partial u_i^{(e)}} = 0 \qquad i = 1,2 \tag{1.21}$$

i.e.

$$\begin{aligned} \frac{\partial \Pi^{(e)}}{\partial u_1^{(e)}} &= -\frac{1}{l^{(e)}}\int_{l^{(e)}} (EA)^{(e)} \left(\frac{u_2^{(e)} - u_1^{(e)}}{l^{(e)}}\right) dx - F_{x_1}^{(e)} = 0 \\ \frac{\partial \Pi^{(e)}}{\partial u_2^{(e)}} &= \frac{1}{l^{(e)}}\int_{l^{(e)}} (EA)^{(e)} \left(\frac{u_2^{(e)} - u_1^{(e)}}{l^{(e)}}\right) dx - F_{x_2}^{(e)} = 0 \end{aligned} \tag{1.22}$$

For a linear material, the above equations simplify to

$$\begin{aligned} \left(\frac{EA}{l}\right)^{(e)} \left[u_1^{(e)} - u_2^{(e)}\right] &= F_{x_1}^{(e)} \\ \left(\frac{EA}{l}\right)^{(e)} \left[u_2^{(e)} - u_1^{(e)}\right] &= F_{x_2}^{(e)} \end{aligned} \tag{1.23}$$

Note the coincidence between the above end force-displacement equilibrium equations and those obtained via the PVW (Eqs.(1.17)).

Eq.(1.20) can be rewritten as

$$\Pi^{(e)} = \frac{l}{2}[\mathbf{a}^{(e)}]^T \mathbf{K}^{(e)} \mathbf{a}^{(e)} - [\mathbf{a}^{(e)}]^T \mathbf{q}^{(e)} \tag{1.24}$$

where $\mathbf{K}^{(e)}$, $\mathbf{a}^{(e)}$ and $\mathbf{q}^{(e)}$ are respectively the stiffness matrix, the joint displacement vector and the joint equilibrium force vector for the bar.

The stationarity of $\Pi^{(e)}$ with respect to the joint displacements gives

$$\frac{\partial \Pi^{(e)}}{\partial \mathbf{a}^{(e)}} = \mathbf{0} \quad \rightarrow \quad \boxed{\mathbf{K}^{(e)} \mathbf{a}^{(e)} = \mathbf{q}^{(e)}} \tag{1.25}$$

Eq.(1.25) is the same matrix equilibrium equation between forces and displacements at the bar joints obtained in the previous section (Eq.(1.5a)).

The total potential energy for a bar structure can be written in a form analogous to Eq.(1.24) as

$$\Pi = \frac{1}{2}\mathbf{a}^T \mathbf{K} \mathbf{a} - \mathbf{a}^T \mathbf{f} \tag{1.26}$$

where $\mathbf{K}$, $\mathbf{a}$ and $\mathbf{f}$ are respectively the stiffness matrix, the joint displacement vector and the external joint force vector for the *whole structure.* The stationarity of Π with respect to $\mathbf{a}$ gives

$$\frac{\partial \Pi}{\partial \mathbf{a}} = \mathbf{0} \quad \rightarrow \quad \boxed{\mathbf{K}\mathbf{a} = \mathbf{f}} \tag{1.27}$$

Eq.(1.27) is the global matrix equilibrium equation relating the displacements and the external forces at all the joints of the structure. The global matrix equations can be obtained by assembly of the contributions from the individual bars, as previously explained.

1.9 PLANE FRAMEWORKS

1.9.1 Plane pin-jointed frameworks

We will briefly treat the case of plane pin-jointed frameworks as an extension of the concepts previously studied. Each joint has now two degrees of freedom (DOFs) corresponding to the displacements along the two cartesian axes. Eqs.(1.4) relating the joint displacements and the axial forces in the *local axis* of each bar still holds. However, the sum of the joint

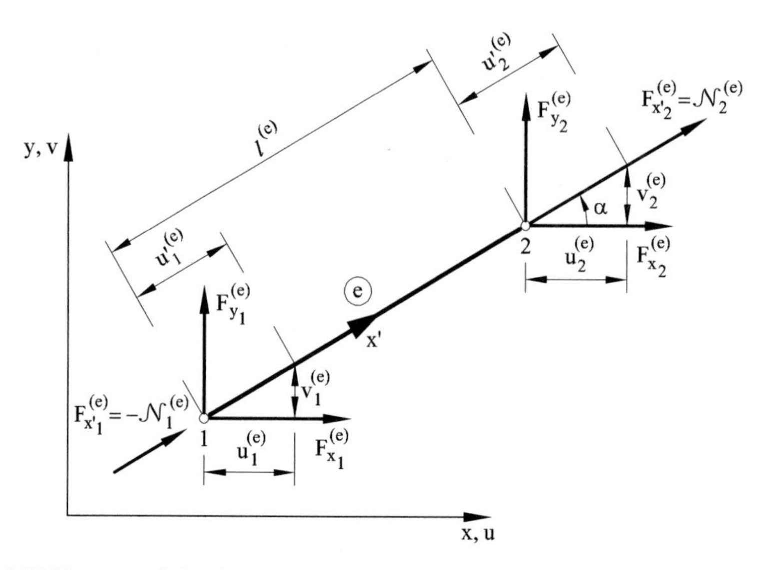

Fig. 1.15 Forces and displacements at the end points of a plane pin-jointed bar

forces for the different bars sharing a joint requires the force-displacement relationships to be expressed in a global cartesian system x, y.

Let us consider a bar 1-2 inclined an angle α with respect to the global axis x, as shown in Figure 1.15. For joint 1 we have

$$F_{x'_i}^{(e)} = F_{x_1}^{(e)} \cos\alpha + F_{y_1}^{(e)} \sin\ \alpha \quad , \quad u_1'^{(e)} = u_1^{(e)} \cos\alpha + v_1^{(e)} \sin\ \alpha \qquad (1.28)$$

where the primes denote the components in the direction of the local axis x'. In matrix form

$$\begin{aligned} F_{x'_1}^{(e)} &= [\cos\alpha, \sin\alpha] \begin{Bmatrix} F_{x_1} \\ F_{y_1} \end{Bmatrix}^{(e)} = \mathbf{L}^{(e)} \mathbf{q}_1^{(e)} \\ u_1'^{(e)} &= [\cos\alpha, \sin\alpha] \begin{Bmatrix} u_1 \\ v_1 \end{Bmatrix}^{(e)} = \mathbf{L}^{(e)} \mathbf{u}_1^{(e)} \end{aligned} \qquad (1.29)$$

where $\mathbf{u}_1^{(e)}$ and $\mathbf{q}_1^{(e)}$ contain the two displacements and the two forces of joint 1 expressed in the global cartesian system x, y, respectively and $\mathbf{L}^{(e)} = [\cos\alpha, \sin\alpha]$.

Analogous expressions can be found for node 2 as

$$F_{x'_2}^{(e)} = \mathbf{L}^{(e)} \mathbf{q}_2^{(e)} \quad \text{and} \quad u_2'^{(e)} = \mathbf{L}^{(e)} \mathbf{u}_2^{(e)} \qquad (1.30)$$

with

$$\mathbf{q}_2^{(e)} = \left[F_{x2}^{(e)}, F_{y2}^{(e)}\right]^T \quad \text{and} \quad \mathbf{u}_2^{(e)} = \left[u_2^{(e)}, v_2^{(e)}\right]^T$$

From Figure 1.15, we deduce

$$F_{x_1'}^{(e)} = -F_{x_2'}^{(e)} = k^{(e)}[u_1'^{(e)} - u_2'^{(e)}] \quad \text{with} \quad k^{(e)} = \left(\frac{EA}{l}\right)^{(e)} \tag{1.31}$$

Note that the local nodal forces $F_{x_1'}^{(e)}$ and $F_{x_2'}^{(e)}$ coincide, with the appropriate sign, with the nodal axial forces, i.e. $F_{x_1'}^{(e)} = -\mathcal{N}_1^{(e)}$ and $F_{x_2'}^{(e)} = \mathcal{N}_2^{(e)}$ (Figure 1.15).

Nothing that $\mathbf{q}_i^{(e)} = [\mathbf{L}^{(e)}]^T F_{x_i'}^{(e)}$, $i = 1, 2$ and using Eqs.(1.29)–(1.31) the following two equations are obtained

$$\begin{aligned} \mathbf{q}_1^{(e)} &= \left[\mathbf{L}^{(e)}\right]^T k^{(e)}\mathbf{L}^{(e)}\mathbf{u}_1^{(e)} - \left[\mathbf{L}^{(e)}\right]^T k^{(e)}\mathbf{L}^{(e)}\mathbf{u}_2^{(e)} \\ \mathbf{q}_2^{(e)} &= -\left[\mathbf{L}^{(e)}\right]^T k^{(e)}\mathbf{L}^{(e)}\mathbf{u}_1^{(e)} + \left[\mathbf{L}^{(e)}\right]^T k^{(e)}\mathbf{L}^{(e)}\mathbf{u}_2^{(e)} \end{aligned} \tag{1.32}$$

In matrix form

$$\begin{Bmatrix} \mathbf{q}_1^{(e)} \\ \\ \mathbf{q}_2^{(e)} \end{Bmatrix} = \begin{bmatrix} \mathbf{K}_{11}^{(e)} & \mathbf{K}_{12}^{(e)} \\ \\ \mathbf{K}_{21}^{(e)} & \mathbf{K}_{22}^{(e)} \end{bmatrix} \begin{Bmatrix} \mathbf{u}_1^{(e)} \\ \\ \mathbf{u}_2^{(e)} \end{Bmatrix} \tag{1.33a}$$

where

$$\begin{aligned} \mathbf{K}_{11}^{(e)} &= \mathbf{K}_{22}^{(e)} = -\mathbf{K}_{12}^{(e)} = -\mathbf{K}_{21}^{(e)} = \left[\mathbf{L}^{(e)}\right]^T k^{(e)}\mathbf{L}^{(e)} = \\ &= k^{(e)} \begin{bmatrix} \cos^2\alpha & \sin\alpha\cos\alpha \\ \sin\alpha\cos\alpha & \sin^2\alpha \end{bmatrix} \end{aligned} \tag{1.33b}$$

The assembly of the contributions of the individual bars into the global stiffness matrix follows precisely the steps explained in Section 1.6. Each joint contributes now a 2×2 matrix as shown in Figure 1.16. An example of the assembly process is presented in Figure 1.17.

1.9.2 Plane rigid jointed frames

The principles discussed above for pin-jointed frameworks can be readily extended to the case where the joints are all rigidly connected. A typical member (coinciding with a bar element) is shown in Figure 1.18 where the displacements and forces acting at the joints are illustrated. Now we

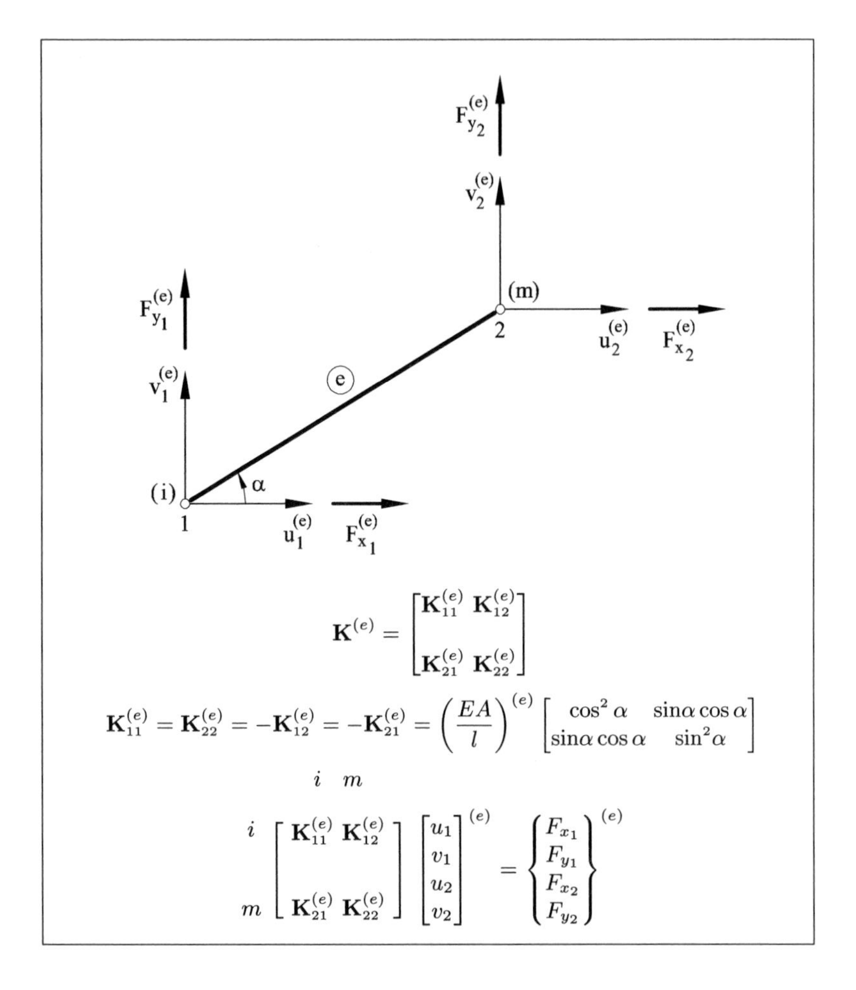

Fig. 1.16 Contributions to the global stiffness matrix from a general member of a pin-jointed framework

have three independent displacement and force components at each joint which can be collectively written as

$$\mathbf{q}'^{(e)} = \begin{Bmatrix} F_{x'_i} \\ F_{y'_i} \\ M_i \end{Bmatrix}^{(e)} \quad ; \quad \mathbf{u}_i'^{(e)} = \begin{Bmatrix} u'_i \\ v'_i \\ \theta_i \end{Bmatrix}^{(e)} \quad ; \quad i = 1, 2 \tag{1.34}$$

where $F_{x'_i}^{(e)}$, $F_{y'_i}^{(e)}$ and $u_i'^{(e)}$, $v_i'^{(e)}$ are, respectively, the force and displacement components of joint i in the local directions x', y' aligned as shown in

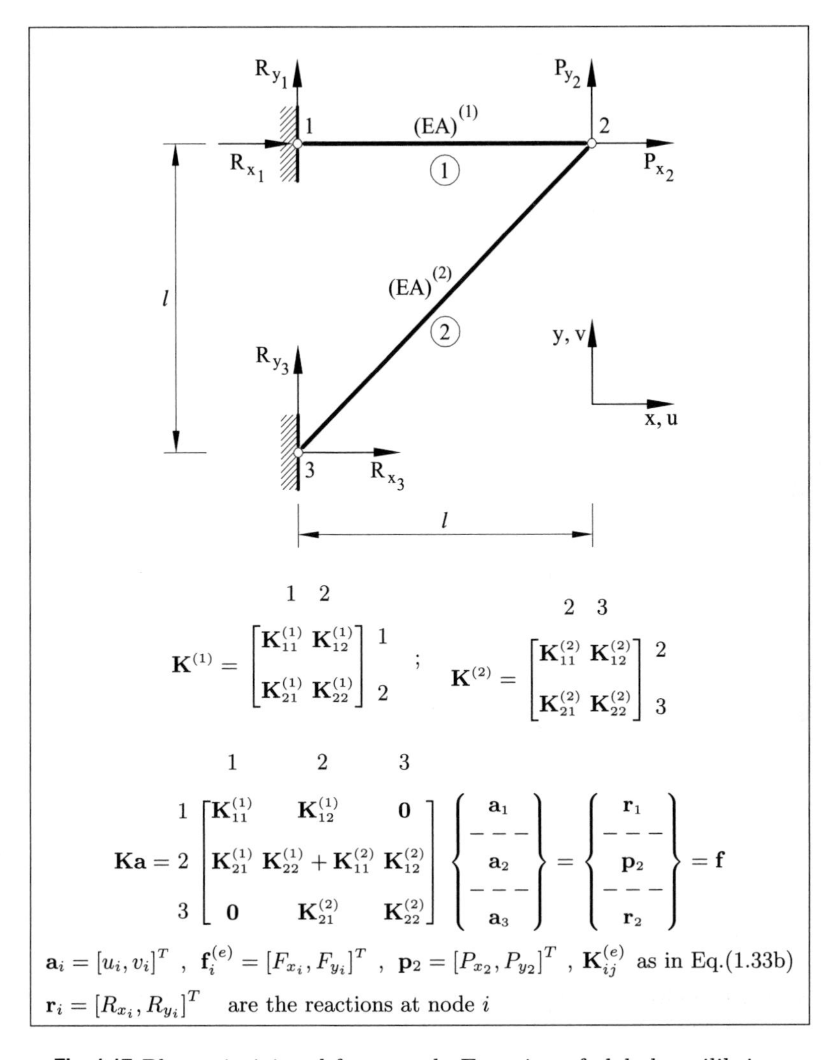

Fig. 1.17 Plane pin-jointed framework. Equation of global equilibrium

Figure 1.18a and $M_i^{(e)}$ and $\theta_i^{(e)}$ are, respectively, the bending moment and the rotation of joint i, where a positive sign corresponds to an anticlockwise direction. The relationship between the local joint forces $F_{x'_i}^{(e)}$, $F_{y'_i}^{(e)}$ and the joint bending moment $M_i^{(e)}$ with the resultant stresses $\mathcal{N}_i^{(e)}, Q_i^{(e)}$ and $M_i^{(e)}$ at the bar joint is shown in Figure 1.18b.

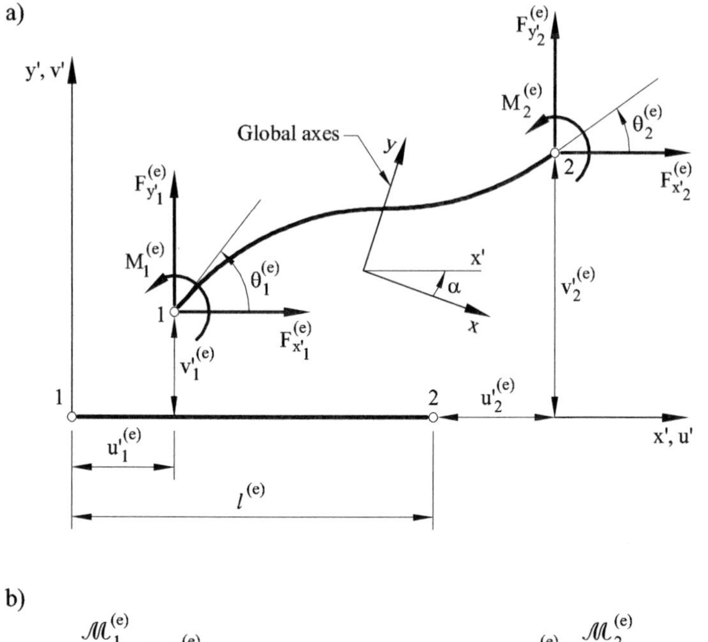

$\mathcal{N}_1^{(e)}$: axial force $\quad Q_1^{(e)}$: shear force $\quad \mathcal{M}_1^{(e)}$: bending moment

$$\underbrace{\left[\mathcal{N}_1^{(e)}, Q_1^{(e)}, \mathcal{M}_1^{(e)}, \mathcal{N}_2^{(e)}, Q_2^{(e)}, \mathcal{M}_2^{(e)}\right]^T}_{\text{resultant stresses at bar ends}} = \underbrace{\left[-F_{x_1'}^{(e)}, -F_{y_1'}^{(e)}, -M_1^{(e)}, F_{x_2'}^{(e)}, F_{y_2'}^{(e)}, M_2^{(e)}\right]^T}_{\text{end forces and moments}}$$

Fig. 1.18 (a) Components of end displacements, forces and moments for a general member in a rigid jointed plane framework; (b) Relationship between the resultant stresses and the forces and bending moments at the member ends

The axial behaviour of the member is identical to that of the axial bar and is defined by Eq.(1.3). Under the assumption of small displacements, the end moments are related to the end rotations and displacements by the well known slope-deflection equations of beam theory [Li,Pr,TY]

$$M_1^{(e)} = 2\left(\frac{EI}{l}\right)^{(e)} \left[2\theta_1^{(e)} + \theta_2^{(e)} + \frac{3(v_1'^{(e)} - v_2'^{(e)})}{l^{(e)}}\right]$$

$$M_2^{(e)} = 2\left(\frac{EI}{l}\right)^{(e)}\left[2\theta_2^{(e)} + \theta_1^{(e)} + \frac{3(v_1'^{(e)} - v_2'^{(e)})}{l^{(e)}}\right] \tag{1.35}$$

The equilibrium of moments at either end of the member requires that

$$\begin{aligned} F_{y_1'}^{(e)} = -F_{y_2'}^{(e)} &= \frac{(M_1^{(e)} + M_2^{(e)})}{l^{(e)}} = \\ &= \left(\frac{12EI}{l^3}\right)^{(e)}(v_1'^{(e)} - v_2'^{(e)}) + \left(\frac{6EI}{l^2}\right)^{(e)}(\theta_1^{(e)} + \theta_2^{(e)}) \end{aligned} \tag{1.36}$$

Equations (1.34), (1.35) and (1.36) can be written in matrix form as

$$\mathbf{q}'^{(e)} = \begin{Bmatrix} \mathbf{q}_1' \\ \mathbf{q}_2' \end{Bmatrix}^{(e)} = \begin{bmatrix} \mathbf{K}_{11}'^{(e)} & \mathbf{K}_{12}'^{(e)} \\ \mathbf{K}_{21}'^{(e)} & \mathbf{K}_{22}'^{(e)} \end{bmatrix} \begin{Bmatrix} \mathbf{u}_1' \\ \mathbf{u}_2' \end{Bmatrix}^{(e)} = \mathbf{K}'^{(e)}\mathbf{u}'^{(e)} \tag{1.37a}$$

where $\mathbf{K}'^{(e)}$ is the *local stiffness matrix* of the rigid jointed member, and

$$\mathbf{K}_{11}'^{(e)} = \begin{bmatrix} \frac{EA}{l} & 0 & 0 \\ 0 & \frac{12EI}{l^3} & \frac{6EI}{l^2} \\ 0 & \frac{6EI}{l^2} & \frac{4EI}{l} \end{bmatrix}^{(e)} ; \mathbf{K}_{12}'^{(e)} = \begin{bmatrix} \frac{-EA}{l} & 0 & 0 \\ 0 & \frac{-12EI}{l^3} & \frac{6EI}{l^2} \\ 0 & \frac{-6EI}{l^2} & \frac{2EI}{l} \end{bmatrix}^{(e)}$$

$$\mathbf{K}_{21}'^{(e)} = \begin{bmatrix} \frac{-EA}{l} & 0 & 0 \\ 0 & \frac{-12EI}{l^3} & \frac{-6EI}{l^2} \\ 0 & \frac{6EI}{l^2} & \frac{2EI}{l} \end{bmatrix}^{(e)} ; \mathbf{K}_{22}'^{(e)} = \begin{bmatrix} \frac{EA}{l} & 0 & 0 \\ 0 & \frac{12EI}{l^3} & \frac{-6EI}{l^2} \\ 0 & -\frac{6EI}{l^2} & \frac{4EI}{l} \end{bmatrix}^{(e)} \tag{1.37b}$$

Note that $\mathbf{K}'^{(e)}$ is symmetrical, as expected. The process by which these equations are transformed to a global coordinate system x, y for the assembly operations is identical to that described in the previous section. The local force and displacement components for each joint are expressed in terms of their global values as

$$\mathbf{q}_i'^{(e)} = \mathbf{L}_i^{(e)}\mathbf{q}_i^{(e)} \quad \text{and} \quad \mathbf{u}_i'^{(e)} = \mathbf{L}_i^{(e)}\mathbf{u}_i^{(e)} \tag{1.38}$$

where

$$\mathbf{q}^{(e)} = \left[F_{x_i}^{(e)}, F_{y_i}^{(e)}, M_i^{(e)}\right]^T \quad ; \quad \mathbf{u}_i^{(e)} = \left[u_i^{(e)}, v_i^{(e)}, \theta_i^{(e)}\right]^T \tag{1.39}$$

and $\mathbf{L}_i^{(e)}$ is the transformation matrix of joint i. Since the member is straight $\mathbf{L}_i^{(e)} = \mathbf{L}_j^{(e)} = \mathbf{L}^{(e)}$, with (Figure 1.18)

$$\mathbf{L}^{(e)} = \begin{bmatrix} \cos\alpha & \sin\alpha & 0 \\ -\sin\alpha & \cos\alpha & 0 \\ 0 & 0 & 1 \end{bmatrix} \tag{1.40}$$

From Eqs.(1.37) and (1.38) we deduce that

$$\begin{aligned}\mathbf{q}^{(e)} &= \begin{bmatrix} [\mathbf{L}^{(e)}]^T & \mathbf{0} \\ \mathbf{0} & [\mathbf{L}^{(e)}]^T \end{bmatrix} \mathbf{q}'^{(e)} = \left[\mathbf{T}^{(e)}\right]^T \mathbf{K}'^{(e)} \mathbf{u}'^{(e)} = \\ &= \left[\mathbf{T}^{(e)}\right]^T \mathbf{K}'^{(e)} \mathbf{T}^{(e)} \mathbf{u}^{(e)} = \mathbf{K}^{(e)} \mathbf{u}^{(e)}\end{aligned} \tag{1.41}$$

where

$$\mathbf{T}^{(e)} = \begin{bmatrix} \mathbf{L}^{(e)} & \mathbf{0} \\ \mathbf{0} & \mathbf{L}^{(e)} \end{bmatrix} \tag{1.42}$$

and

$$\mathbf{K}^{(e)} = \left[\mathbf{T}^{(e)}\right]^T \mathbf{K}'^{(e)} \mathbf{T}^{(e)} \tag{1.43}$$

is the global stiffness matrix of the member.

Eq.(1.41) can be written as

$$\begin{Bmatrix} \mathbf{q}_1 \\ \mathbf{q}_2 \end{Bmatrix}^{(e)} = \begin{bmatrix} \mathbf{K}_{11}^{(e)} & \mathbf{K}_{12}^{(e)} \\ \mathbf{K}_{21}^{(e)} & \mathbf{K}_{22}^{(e)} \end{bmatrix} \begin{Bmatrix} \mathbf{u}_1 \\ \mathbf{u}_2 \end{Bmatrix}^{(e)} \tag{1.44}$$

A typical submatrix $\mathbf{K}_{ij}^{(e)}$ in global axes is given by

$$\mathbf{K}_{ij}^{(e)} = \left[\mathbf{L}^{(e)}\right]^T \mathbf{K}'^{(e)}_{ij} \mathbf{L}^{(e)} \tag{1.45}$$

The assembly of the contributions from the individual members into the global stiffness matrix follows the steps described in the previous sections.

The analysis of rigid jointed bar structures will be dealt with again when we study beams, arches and rods in Volume 2 [On].

1.10 TREATMENT OF PRESCRIBED DISPLACEMENTS AND COMPUTATION OF REACTIONS

In this book we will not enter into the details of techniques for solving the system of algebraic equations $\mathbf{Ka} = \mathbf{f}$. This is a problem typical of matrix algebra and many well known direct and iterative solution procedures are available (i.e.: Gauss reduction, Choleski, modified Choleski, Frontal; Profile, etc.) [HO,PFTV,Ral]. A brief discussion of some of these methods is presented in Appendix B. We will just treat here briefly the problem of prescribed displacements and the computation of the corresponding reactions, as these are issues of general interest for the study of this book.

Let us consider the following system of algebraic equations

$$\begin{array}{ccccccccc} k_{11}u_1 & + & k_{12}u_2 & + & k_{13}u_3 & + \ldots + & k_{1n}u_n & = & f_1 \\ k_{21}u_1 & + & k_{22}u_2 & + & k_{23}u_3 & + \ldots + & k_{2n}u_n & = & f_2 \\ k_{31}u_1 & + & k_{32}u_2 & + & k_{33}u_3 & + \ldots + & k_{3n}u_n & = & f_3 \\ \vdots & & \vdots & & \vdots & & \vdots & & \vdots \\ k_{n1}u_1 & + & k_{n2}u_2 & + & k_{n3}u_3 & + \ldots + & k_{nn}u_n & = & f_n \end{array} \tag{1.46}$$

where f_i are external forces (which can be equal to zero) or reactions at points where the displacement is prescribed.

Let us assume that a displacement, for example u_2, is prescribed to the value $\overline{u}_2$, i.e.

$$u_2 = \overline{u}_2 \tag{1.47}$$

There are two basic procedures to introduce this condition in the above system of equations:

a) The second row and column of Eq.(1.46) are eliminated and the values of f_i in the r.h.s. are substituted by $f_i - k_{i2}\overline{u}_2$. That is, the system of n equations with n unknowns is reduced in one equation and one unknown as follows

$$\begin{array}{ccccccccc} k_{11}u_1 & + & k_{13}u_3 & + \ldots + & k_{1n}u_n & = & f_1 & - & k_{12}\overline{u}_2 \\ k_{31}u_1 & + & k_{33}u_3 & + \ldots + & k_{3n}u_n & = & f_3 & - & k_{32}\overline{u}_2 \\ \vdots & & \vdots & & \vdots & & \vdots & & \vdots \\ k_{n1}u_1 & + & k_{n3}u_3 & + \ldots + & k_{nn}u_n & = & f_n & - & k_{n2}\overline{u}_2 \end{array} \tag{1.48}$$

Once the values of $u_1, u_3, \ldots, u_n$ are obtained, the reaction f_2 is computed by the following equation (in the case that the external force acting at node 2 is equal to zero)

$$f_2 = k_{21}u_1 + k_{22}\overline{u}_2 + k_{23}u_3 + \ldots + k_{2n}u_n \tag{1.49}$$

If u_2 is zero, the procedure remains the same, although the values of f_i are not modified and f_2 is obtained by Eq.(1.49) with $\overline{u}_2 = 0$.

b) An alternative procedure which does not require the original system of equations to be modified substantially, is to add a very large number to the term of the main diagonal corresponding to the prescribed displacement. The force term in the modified row is substituted by the value of the prescribed displacement multiplied by the large number chosen. Thus, if we have $u_2 = \bar{u}_2$ we substitute k_{22} by $k_{22} + 10^{15}k_{22}$ (for instance), and f_2 by $10^{15}k_{22} \times \bar{u}_2$. The final system of equations is

$$\begin{array}{lllll}
k_{11}u_1 + & k_{12}u_2 & + k_{13}u_3 + \ldots + k_{1n}u_n = & f_1 \\
k_{21}u_2 + & (1+10^{15})k_{22}u_2 & + k_{23}u_3 + \ldots + k_{2n}u_n = & 10^{15}k_{22}\bar{u}_2 \\
k_{31}u_1 + & k_{32}u_2 & + k_{33}u_3 + \ldots + k_{3n}u_n = & f_3 \\
\vdots & \vdots & \vdots \qquad\qquad \vdots & \vdots \\
k_{n1}u_1 + & k_{n2}u_2 & + k_{n3}u_3 + \ldots + k_{nn}u_n = & f_n
\end{array} \tag{1.50}$$

In this way, the second equation is equivalent to

$$10^{15}k_{22}u_2 = 10^{15}k_{22}\bar{u}_2 \quad \text{or} \quad u_2 = \bar{u}_2 \tag{1.51}$$

which is the prescribed condition. The value of the reaction f_2 is computed "a posteriori" by Eq.(1.49).
The issue of prescribed displacements will be treated again in Chapter 9.

1.11 INTRODUCTION TO THE FINITE ELEMENT METHOD FOR STRUCTURAL ANALYSIS

Most structures in practice are of *continuous* nature and can not be accurately modelled by a collection of bars. Examples of "continuous" structures are standard in civil, mechanical, aeronautical and naval engineering. Amongst the more common we can list: plates, foundations, roofs, containers, bridges, dams, airplane fuselages, car bodies, ship hulls, mechanical components, etc. (Figure 1.19).

Although a continuous structure is inherently three-dimensional (3D), its behaviour can be accurately described in some cases by one- (1D) or two-dimensional (2D) structural models. This occurs, for instance, in the analysis of plates in bending, where only the deformation of the plate mid-plane is considered. Other examples are the structures modelled as 2D solids or as axisymmetric solids (i.e. dams, tunnels, water tanks, etc.)

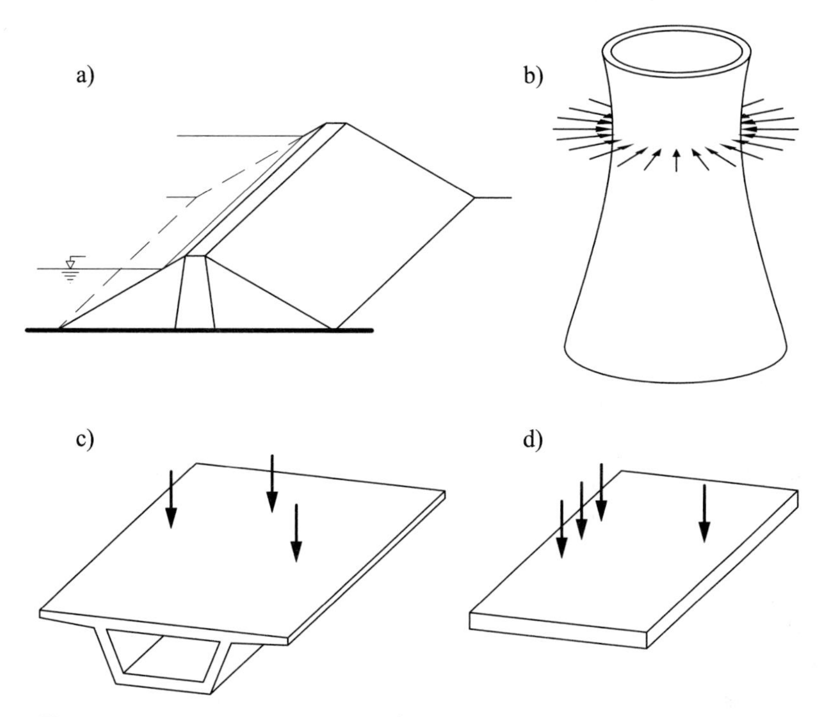

Fig. 1.19 Continuous structures: a) Dam, b) Shell, c) Bridge, d) Plate

The analytical solution of a continuous structure is very difficult (generally impossible) due to the complexities of the geometry, the boundary conditions, the material properties, the loading, etc. This explains the need for computational models to analyse continuous structures.

The FEM is the simpler and more powerful computational procedure for the analysis of structures with arbitrary geometry and general material properties subjected to any type of loading.

The FEM allows one the behaviour of a structure with an infinite number of DOFs to be modelled by that of another one with approximately the same geometrical and mechanical properties, but with a finite number of DOFs. The latter are related to the external forces by a system of algebraic equations expressing the equilibrium of the structure. We will find that the basic finite element methodology is analogous to the matrix analysis technique studied for bar structures. The analogies can be summarized by

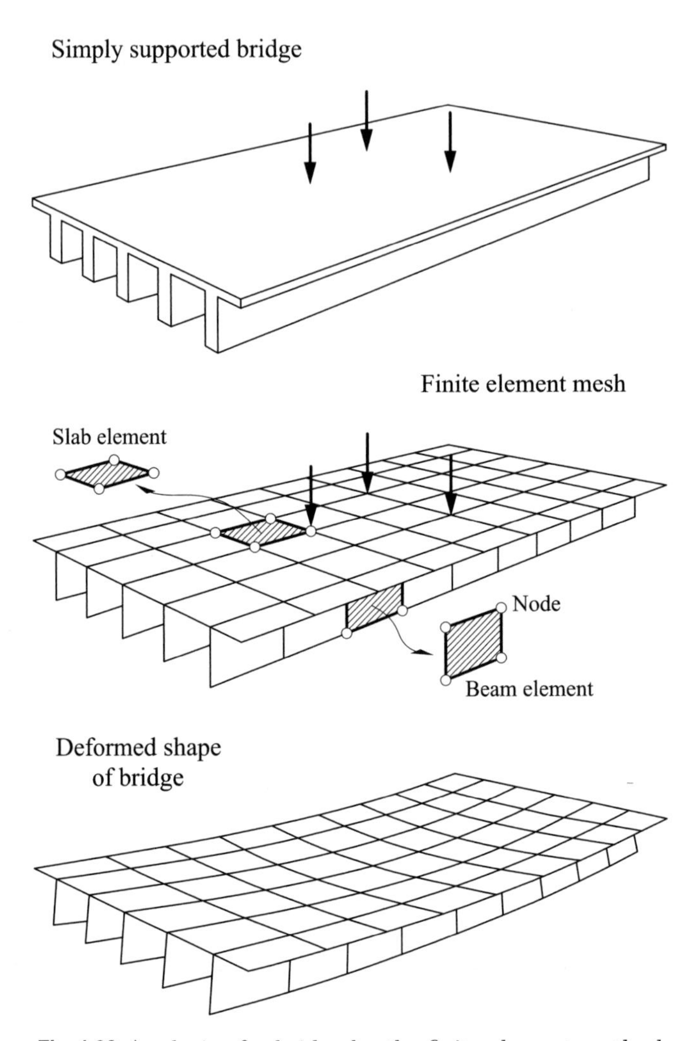

Fig. 1.20 Analysis of a bridge by the finite element method

considering the bridge shown in Figure 1.20. Without entering into the details, the basic steps in the finite element analysis are the following:

Step 1: Starting with the geometrical description of the bridge, its supports and the loading, the first step is to select a *structural model*. For example, we could use a 3D solid model (Chapter 8), a stiffened plate model (Chapter 10, Vol. 2 [On]) or a facet shell model (Chapter 7, Vol. 2 [On]). The material properties must also be defined, as well as the scope of the analysis (small or large displacements, static or dynamic analysis, etc.). As mentioned earlier, in this book we will focus on linear static analysis only.

Step 2: The structure is subdivided into a mesh of non-intersecting domains termed finite elements (*discretization process*). The problem variables (displacements) are interpolated within each element in terms of their values at a known set of points of the element called *nodes*. The number of nodes defines the approximation of the solution within each element. Some nodes are placed at the element boundaries and they can be interpreted as linking points between adjacent elements. However, nodes in the interior of the elements are needed for higher-order approximations and, hence, the nodes do not have a physical meaning as the connecting joints in bar structures. The mesh can include elements with different geometry, such as 2D plate elements coupled with 1D beam elements. The *discretization* process is an essential part of the *preprocessing* step which includes the definition of all the analysis data. The preprocessing step typically consumes a considerable amount of human effort. The use of efficient preprocessing tools is essential for the analysis of practical structures in competitive times. More details are given in Chapter 10.

Step 3: The stiffness matrices $\mathbf{K}^{(e)}$ and the load vectors $\mathbf{f}^{(e)}$ are obtained for each element. The computation of $\mathbf{K}^{(e)}$ and $\mathbf{f}^{(e)}$ is more complex than for bar structures and it usually requires the evaluation of integrals over the element domain.

Step 4: The element stiffness and the load terms are assembled into the overall stiffness matrix $\mathbf{K}$ and the load vector $\mathbf{f}$ for the structure.

Step 5: The global system of linear simultaneous equations $\mathbf{Ka} = \mathbf{f}$ is solved for the unknown displacement variables $\mathbf{a}$.

Step 6: Once the displacements $\mathbf{a}$ are computed, the strains and the stresses are evaluated within each element. Reactions at the nodes restrained against movement are also computed.

Step 7: Solving steps 3-6 requires a *computer implementation* of the FEM by means of a standard or specially developed program.

Step 8: After a successful computer run, the next step is *the interpretation and presentation of results*. Results are presented graphically to aid their interpretation and checking (*postprocessing step*). The use of specialized graphic software is essential in practice. More details are given in Chapter 10.

Step 9: Having assessed the finite element results, the analyst may consider several modifications which may be introduced at various stages of the analysis. For example, it may be found that the structural model selected is inappropriate and hence it should be adequately modified. Alternatively, the finite element mesh chosen may turn out to be too coarse to capture the expected stress distributions and must therefore be refined or a different, more accurate element used. Round-off problems arising from ill-conditioned equations, the equations solving algorithm or the computer word length employed in the analysis may cause difficulties and can require the use of double-precision arithmetic or some other techniques. Input data errors which occur quite frequently must be also corrected.

All these possible modifications are indicated by the feedback loop shown in Figure 1.21 taken from [HO2].

From the structural engineer's point of view, the FEM can be considered as an extension to continuous systems of the matrix analysis procedures for bar structures. The origins of the FEM go back to the early 1940's with the first attempts to solve problems of 2D elasticity using matrix analysis techniques by subdividing the continuum into bar elements [Hr,Mc]. In 1946 Courant [Co] introduced for the first time the concept of "continuum element" to solve 2D elasticity problems using a subdivision into triangular elements with an assumed displacement field. The arrival of digital computers in the 1960's contributed to the fast development of matrix analysis based techniques, free from the limitations imposed by the need to solve large systems of equations. It was during this period that the FEM rapidly established itself as a powerful approach to solve many problems in mathematics and physics. It is interesting that the first applications of the FEM were related to structural analysis and, in particular, to aeronautical engineering [AK,TCMT]. It is acknowledged that Clough first used the name "finite elements" in relation to the solution of 2D elasticity problems in 1960 [Cl]. Since then the FEM has had a tremendous expansion in its application to many different fields. Supported by the continuous upgrading of computers and by the increasing complexity of many areas in science and technology, today the FEM enjoys a unique position as a powerful technique for solving the most difficult problems in engineering and applied sciences.

It would be an impossible task to list here all the significant published work since the origins of the FEM. Only in 2008, the scientific publications in this field were estimated to number in excess of 25,000. The reader interested in bibliography on the FEM should consult the references listed

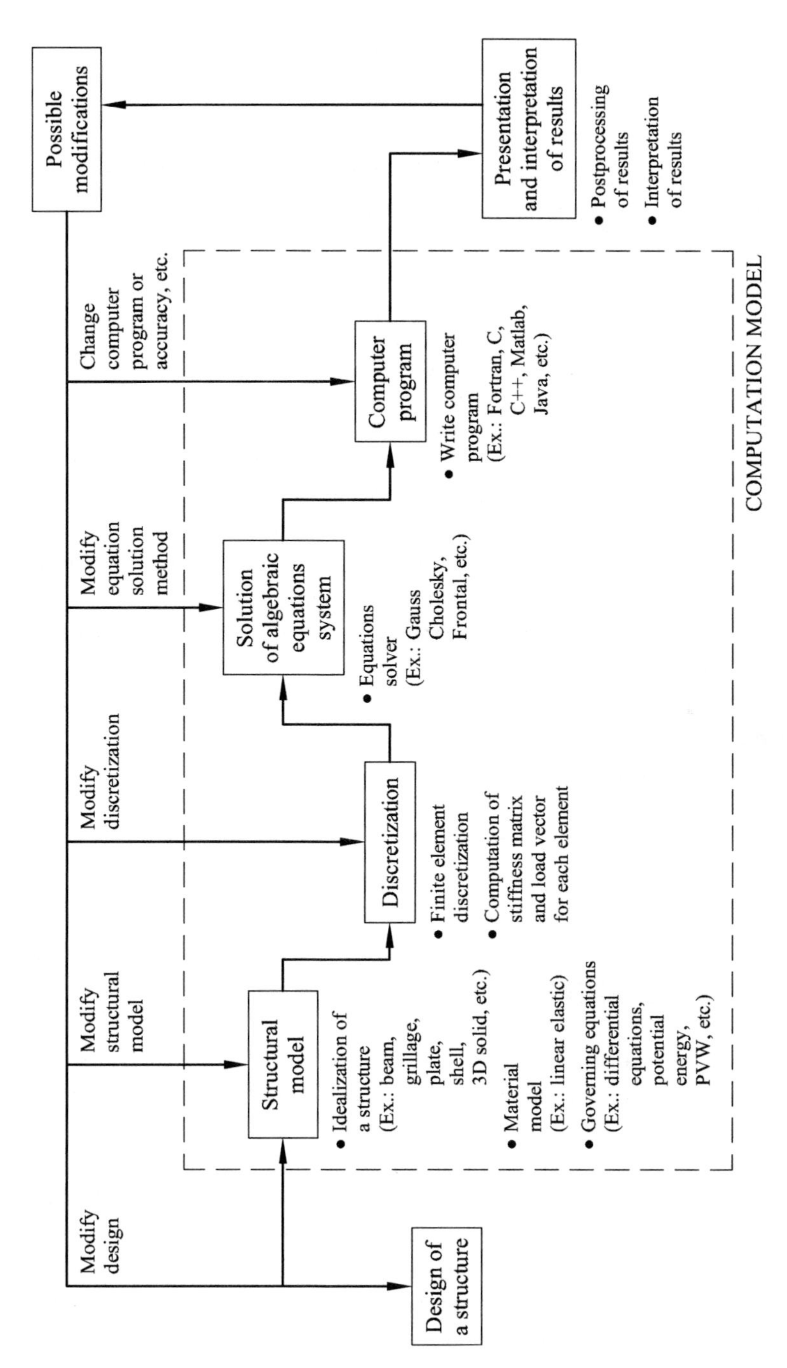

Fig. 1.21 Flow chart of the analysis of a structure by the FEM [HO2]

in [No,ZT,ZTZ] and in the Encyclopedia of Computational Mechanics [SDH,SDH2].

1.12 THE VALUE OF FINITE ELEMENT COMPUTATIONS FOR STRUCTURAL DESIGN AND VERIFICATION

The practical aim of finite element structural analysis is to verify the strength of existing constructions and the optimum design of new ones satisfying a number of specified criteria. A key objective is to prevent structural failure and guarantee the safety of structures under a set of loads.

The simpler failure criterium for bar structures states that failure will occur when the axial stress in any structural member exceeds a specified allowable value. In multidimensional stress fields typical of continuous structures, failure at a point is detected when a combination of the stresses (typically a stress invariant) reaches a critical value [ZT].

Other failure criteria for structures are based on setting up limits to the maximum displacement at any point in the structure. Alternatively the maximum strain at a point is used to control the onset of failure. For multidimensional stress states the failure bound is set on the value of an appropriate strain invariant (Section 8.2.5).

The optimum design of a safe structure typically involves a trial and error process in order to ensure that the shape, dimensions and materials chosen for the different structural members comply with the specified safety requirements.

1.13 CONCLUDING REMARKS

From the practical point of view of the structural engineer it should always be kept in mind that the FEM is a very powerful technique to obtain *approximate solutions* for structural problems. In the hands of a careful and expert user the FEM is an indispensable tool for the analysis, design and verification of complex structures which cannot be studied otherwise. However, being an approximate method it involves a certain *error* in the numerical values and users should always look upon FEM results with a critical eye. In this book we will try to facilitate the understanding of the theoretical and applied aspects of the FEM for the analysis of a wide range of structures.

2

1D FINITE ELEMENTS FOR AXIALLY LOADED RODS

2.1 INTRODUCTION

The objective of this chapter is to introduce the basic concepts of the FEM in its application to the analysis of simple one-dimensional (1D) axially loaded rods.

The organization of the chapter is as follows. In the first section the analysis of axially loaded rods using 2-noded rod elements is presented. Particular emphasis is put in the analogies with the solution of the same problem using the standard matrix analysis techniques studied in the previous chapter for bar structures. Here some examples of application are given. In the last part of the chapter the matrix finite element formulation adopted throughout this book is presented.

2.2 AXIALLY LOADED ROD

Let us consider a rod of length l subjected to a distributed axial load per unit length $t_x(x)$ and a set of axial point loads F_{x_i} acting at p different points x_i (Figure 2.1). The rod can also have prescribed displacements $\bar{u}_j$ at m points x_j. The displacement of the rod points produces the corresponding axial strain $\varepsilon(x) = du/dx$ (also called elongation) and the normal stress σ in the rod which are related by Hooke law, i.e.

$$\sigma = E\varepsilon = E\,\frac{du}{dx} \tag{2.1}$$

where E is the Young modulus of the material.

The axial force (or axial resultant stress) $\mathcal{N}$ is defined as the integral of the stress over the area of the transverse cross section (Figure 2.1). For

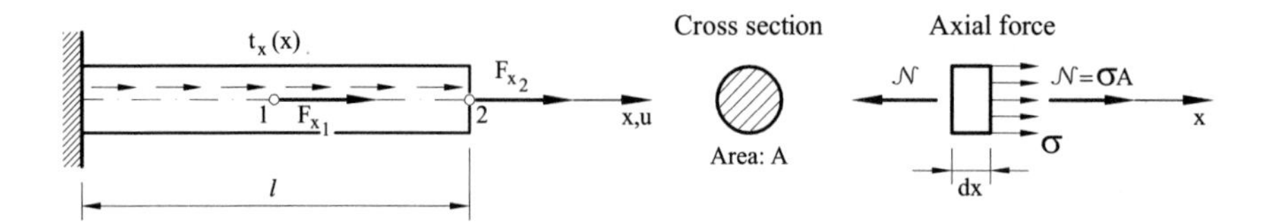

Fig. 2.1 Axially loaded rod

homogeneous material

$$\mathcal{N} = \iint_A \sigma \, dA = A\sigma = EA\frac{du}{dx} \tag{2.2}$$

In the equilibrium configuration the stresses and the external forces satisfy the *Principle of Virtual Work* (PVW) defined in Section 1.7. The PVW for the rod is written as [Was,ZT]

$$\iiint_V \delta\varepsilon\sigma \; dV = \int_0^l \delta u t_x dx + \sum_{i=1}^{p} \delta u_i F_{x_i} \tag{2.3}$$

where δu and $\delta\varepsilon$ are the virtual displacement and the virtual strain of an arbitrary point of the rod center line, δu_i is the virtual displacement of the point where the point load F_{x_i} acts, t_x is the distributed axial force and V is the rod volume. The left- and right-hand sides of Eq.(2.3) represent the internal and external virtual work carried out by the actual stresses and the external loads, respectively.

Eq.(2.3) can be rewritten after integration over the cross section area (note that $dV = dA \cdot dx$) and using Eq.(2.2) as

$$\int_0^l \delta\varepsilon\mathcal{N} \; dx = \int_0^l \delta u t_x \; dx + \sum_{i=1}^{p} \delta u_i F_{x_i} \tag{2.4}$$

where $\mathcal{N}$ is the axial force which is related to the displacement field via Eq.(2.2).

It can be proved [Ti2,Was,ZT] that the equilibrium solution of the rod problem is reduced to finding a displacement field $u(x)$ satisfying Eq.(2.4) and the displacement boundary conditions (kinematic conditions). The approximate solution using the FEM is set as follows: find an alternative displacement field $\hat{u}(x)$ which approximates $u(x)$ and which also satisfies Eq.(2.4) and the kinematic conditions.

Among the different options available to express the approximate displacement field $\hat{u}(x)$ *we will choose the simplest one using polinomials locally defined for each element.* Thus, after discretizing the rod in a mesh of finite elements we can write for each element

$$u(x) \simeq \hat{u}(x) = a_0 + a_1 x + a_2 x^2 + \cdots + a_n x^n = \sum_{i=1}^{n} a_i x^i \tag{2.5}$$

In Eq.(2.5) n is the number of points of the element where the displacement is assumed to be known. These points are called *nodes.* The parameters a_0, a_1, ..., a_n depend on the nodal displacements only. *In the following we will skip the "hat" over the approximate solution* and write Eq.(2.5) in the form

$$u(x) = N_1^{(e)}(x)u_1^{(e)} + N_2^{(e)}(x)u_2^{(e)} + \cdots + N_n^{(e)}(x)u_n^{(e)} = \sum_{i=1}^{n} N_i^{(e)}(x)u_i^{(e)} \tag{2.6}$$

where $N_1^{(e)}(x)$, ..., $N_n^{(e)}(x)$ are the polinomial interpolating functions defined over the domain of each element e and $u_i^{(e)}$ is the value of the (approximate) displacement of node i. The function $N_i^{(e)}(x)$ interpolates within each element the displacement of node i and it is called the *shape function of node i*. From Eq.(2.6) we deduce that $N_i^{(e)}(x)$ must take the value one at node i and zero at all other nodes so that $u(x_i) = u_i^{(e)}$. These concepts will be extended in the next section.

Substituting the displacement approximation for each element in the PVW allows us to express the equilibrium equations in terms of the nodal displacements of the finite element mesh. These algebraic equations can be written in the standard matrix form

$$\mathbf{K}\,\mathbf{a} = \mathbf{f} \tag{2.7}$$

where, by analogy with bar systems, $\mathbf{K}$ is termed the *stiffness matrix* of the finite element mesh, and $\mathbf{a}$ and $\mathbf{f}$ are the *vectors of nodal displacements* and of *equivalent nodal forces*, respectively. Both $\mathbf{K}$ and $\mathbf{f}$ are obtained by assembling the contributions from the individual elements, as in matrix analysis of bar structures. Solving Eq.(2.7) yields the values of the displacements at all the nodes in the mesh from which the axial strain, the axial force and the normal stress within each element can be found.

These concepts will be illustrated in the next section for the analysis of an axially loaded rod with constant cross sectional area using two meshes of one and two linear rod elements, respectively.

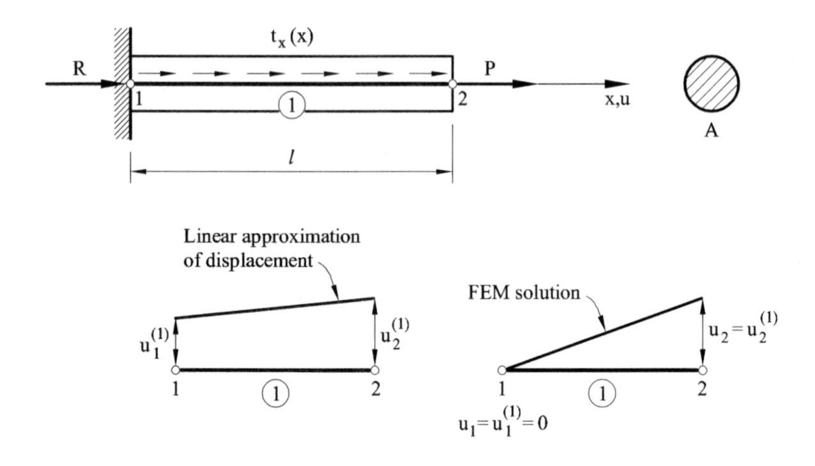

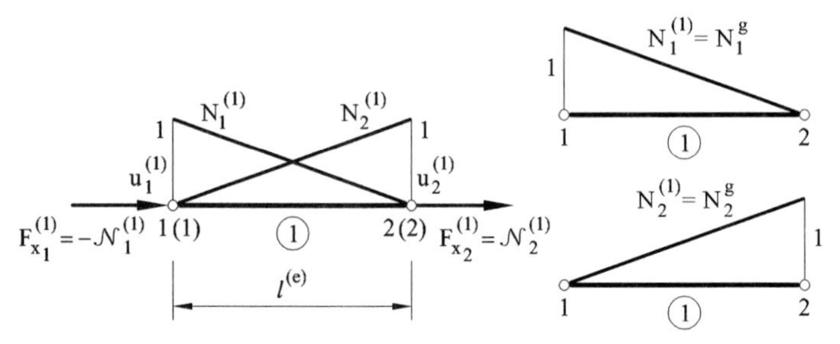

Fig. 2.2 Axially loaded rod. Discretization in a single 2-noded element. Number in brackets at node denotes global node number

2.3 AXIALLY LOADED ROD OF CONSTANT CROSS SECTION. DISCRETIZATION IN ONE LINEAR ROD ELEMENT

2.3.1 Approximation of the displacement field

Let us consider a rod with constant circular cross section under a distributed axial force $t_x(x)$ and an end axial point load P (Figure 2.2). The rod is discretized in a single element with two nodes which define a linear interpolation of the displacement field as

$$u(x) = \alpha_0 + \alpha_1 x \tag{2.8}$$

Is is clear that $u(x)$ must take the values $u_1^{(1)}$ and $u_2^{(1)}$ at nodes 1 and 2, i.e.

$$u(x_1^{(1)}) = u_1^{(1)} \qquad \text{and} \qquad u(x_2^{(1)}) = u_2^{(1)} \tag{2.9}$$

where $x_1^{(1)}$ and $x_2^{(1)}$ are the coordinates of nodes 1 and 2, respectively. Superindex 1 in Eq.(2.9) indicates that all the parameters refer to the element number one.

From Eqs.(2.8) and (2.7) the following system of equations is obtained

$$\begin{aligned} u_1^{(1)} &= \alpha_0 + \alpha_1 x_1^{(1)} \\ u_2^{(1)} &= \alpha_0 + \alpha_1 x_2^{(1)} \end{aligned} \tag{2.10a}$$

from which the parameters α_0 and α_1 are found as

$$\alpha_0 = \frac{u_1^{(1)} - u_2^{(1)}}{x_1^{(1)} - x_2^{(1)}} \quad \text{and} \quad \alpha_1 = \frac{x_2^{(1)} u_1^{(1)} - x_1^{(1)} u_2^{(1)}}{x_2^{(1)} x_1^{(1)}} \tag{2.10b}$$

Substituting Eq.(2.10b) into (2.8) allows us to rewrite the latter as

$$u = N_1^{(1)}(x) u_1^{(1)} + N_2^{(1)}(x) u_2^{(1)} \tag{2.11}$$

where $N_1^{(1)}$ and $N_2^{(1)}$ are the shape functions of nodes 1 and 2, respectively given by

$$N_1^{(1)}(x) = \frac{x_2^{(1)} - x}{l^{(1)}} \quad ; \quad N_2^{(1)}(x) = \frac{x - x_1^{(1)}}{l^{(1)}} \tag{2.12}$$

where $l^{(1)} = x_2^{(1)} - x_1^{(1)}$ is the element length. It is deduced from Eq.(2.12) that the shape functions $N_i^{(1)}$ $(i = 1, 2)$ vary linearly within the element and take the value one at node i and zero at the other node. This is a natural consequence of the interpolatory property of the shape functions that guarantees that the displacement $u(x)$ takes the values $u_1^{(1)}$ and $u_2^{(1)}$ at the element nodes. This let us anticipate in most cases the geometry of the shape functions, as it will be frequently seen throughout the book.

Before we proceed any further it is important to clarify the differences between local and global numbering. Table 2.1 shows an example of both numberings for the nodes, the nodal coordinate and the nodal displacement for the example in Figure 2.2.

Note that since we have taken in this case a single element, the local and global numbers coincide.

The derivatives of the shape functions are computed as

$$\frac{dN_1^{(1)}}{dx} = -\frac{1}{l^{(1)}} \quad \text{and} \quad \frac{dN_2^{(1)}}{dx} = \frac{1}{l^{(1)}} \tag{2.13}$$

Element	node		coordinate		displacement	
	local	global	local	global	local	global
1	1	1	$x_1^{(1)}$	x_1	$u_1^{(1)}$	u_1
	2	2	$x_2^{(1)}$	x_2	$u_2^{(1)}$	u_2

Table 2.1 Local and global parameters for the example in Figure 2.2

The axial strain and the axial force can be obtained at each point within the element as

$$\varepsilon^{(1)} = \left(\frac{du}{dx}\right)^{(1)} = \frac{dN_1^{(1)}}{dx}\,u_1^{(1)} + \frac{dN_2^{(1)}}{dx}\,u_2^{(1)} = -\frac{1}{l^{(1)}}u_1^{(1)} + \frac{1}{l^{(1)}}\,u_2^{(1)} \quad (2.14a)$$

$$\begin{aligned}\mathcal{N}^{(1)} &= (EA)^{(1)}\varepsilon^{(1)} = (EA)^{(1)}\left[\frac{dN_1^{(1)}}{dx}\,u_1^{(1)} + \frac{dN_2^{(1)}}{dx}\,u_2^{(1)}\right] = \\ &= (EA)^{(1)}\left[-\frac{1}{l^{(1)}}u_1^{(1)} + \frac{1}{l^{(1)}}u_2^{(1)}\right]\end{aligned} \quad (2.14b)$$

Obviously, the linear approximation for the displacement yields a constant field for the axial strain and the axial force over the element.

2.3.2 Derivation of equilibrium equations for the elements

The forces between elements are transmitted across the nodes. These forces denoted as $F_{x_i}^{(e)}$ are termed *equilibrating nodal forces* and can be obtained for each element using the PVW. The forces $F_{x_i}^{(e)}$ coincide with the appropriate sign with the axial forces at the element nodes, i.e. $F_{x_1}^{(e)} = -\mathcal{N}_1^{(e)}$ and $F_{x_2}^{(e)} = \mathcal{N}_2^{(e)}$ (Figure 2.2). For the single element of Figure 2.2 we have

$$\int_{x_1^{(1)}}^{x_2^{(1)}} \delta\varepsilon^{(1)}\mathcal{N}^{(1)}\,dx = \int_{x_1^{(1)}}^{x_2^{(1)}} \delta u^{(1)} t_x^{(1)}\,dx + \delta u_1^{(1)} F_{x_1}^{(1)} + \delta u_2^{(1)} F_{x_1}^{(1)} \quad (2.15)$$

where $\delta u_1^{(1)}$, $\delta u_2^{(1)}$, $F_{x_1}^{(1)}$ and $F_{x_2}^{(1)}$ are the virtual displacements and the equilibrating nodal forces for nodes 1 and 2 of the element, respectively. The virtual displacement can also be linearly interpolated in terms of the nodal values as

$$\delta u^{(1)} = N_1^{(1)}\,\delta u_1^{(1)} + N_2^{(1)}\,\delta u_2^{(1)} \quad (2.16)$$

The virtual axial strain is now expressed in terms of the virtual nodal displacements as

$$\delta\varepsilon^{(1)} = \frac{d}{dx}(\delta u) = \frac{dN_1^{(1)}}{dx}\,\delta u_1^{(1)} + \frac{dN_2^{(1)}}{dx}\,\delta u_2^{(1)} \tag{2.17}$$

Eq.(2.15) is rewritten, after substitution of (2.16) and (2.17), as

$$\int_{x_1^{(1)}}^{x_2^{(1)}} \left[\frac{dN_1^{(1)}}{dx}\delta u_1^{(1)} + \frac{dN_2^{(1)}}{dx}\delta u_2^{(1)}\right]\mathcal{N}^{(1)} - \int_{x_1^{(1)}}^{x_2^{(1)}} \left[N_1^{(1)}\delta u_1^{(1)} + N_2^{(1)}\delta u_2^{(1)}\right] t_x\, dx$$

$$= \delta u_1^{(1)} F_{x_1}^{(1)} + \delta u_2^{(1)} F_{x_2}^{(1)} \tag{2.18a}$$

Grouping terms gives

$$\delta u_1^{(1)}\left[\int_{x_1^{(1)}}^{x_2^{(1)}} \frac{dN_1^{(1)}}{dx}\mathcal{N}^{(1)}dx - \int_{x_1^{(1)}}^{x_2^{(1)}} N_1^{(1)} t_x\, dx - F_{x_1}^{(1)}\right] +$$

$$+\ \delta u_2^{(1)}\left[\int_{x_1^{(1)}}^{x_2^{(1)}} \frac{dN_2^{(1)}}{dx}\mathcal{N}^{(1)}dx - \int_{x_1^{(1)}}^{x_2^{(1)}} N_2^{(1)} t_x\, dx - F_{x_2}^{(1)}\right] = 0 \tag{2.18b}$$

Since the virtual displacements are arbitrary, the satisfaction of Eq. (2.18b) leads to the following system of two equations

$$\int_{x_1^{(1)}}^{x_2^{(1)}} \frac{dN_1^{(1)}}{dx}\mathcal{N}^{(1)}dx - \int_{x_1^{(1)}}^{x_2^{(1)}} N_1^{(1)} t_x\, dx - F_{x_1}^{(1)} = 0$$

$$\int_{x_1^{(1)}}^{x_2^{(1)}} \frac{dN_2^{(1)}}{dx}\mathcal{N}^{(1)} - \int_{x_1^{(1)}}^{x_2^{(1)}} N_2^{(1)} t_x\, dx - F_{x_2}^{(1)} = 0 \tag{2.18c}$$

Substituting the expression of $\mathcal{N}^{(1)}$ from Eq.(2.14b) into (2.18c) gives

$$\int_{x_1^{(1)}}^{x_2^{(1)}} \left(\frac{dN_1^{(1)}}{dx}(EA)^{(1)}\frac{dN_1^{(1)}}{dx}u_1^{(1)} + \frac{dN_1^{(1)}}{dx}(EA)^{(1)}\frac{dN_2^{(1)}}{dx}u_2^{(1)}\right)dx-$$

$$-\int_{x_1^{(1)}}^{x_2^{(1)}} N_1^{(1)} t_x\, dx - F_{x_1}^{(1)} = 0$$

$$\int_{x_1^{(1)}}^{x_2^{(1)}} \left(\frac{dN_2^{(1)}}{dx}(EA)^{(1)}\frac{dN_1^{(1)}}{dx}u_1^{(1)} + \frac{dN_2^{(1)}}{dx}(EA)^{(1)}\frac{dN_2^{(1)}}{dx}u_2^{(1)}\right)dx$$

$$-\int_{x_1^{(1)}}^{x_2^{(1)}} N_2^{(1)} t_x\, dx - F_{x_2}^{(1)} = 0 \tag{2.19}$$

From Eq.(2.19) the values of the equilibrating nodal forces $F_{x_1}^{(1)}$ are obtained. In matrix form

$$\left(\int_{x_1^{(1)}}^{x_2^{(1)}} (EA)^{(1)} \begin{bmatrix} \left(\frac{dN_1^{(1)}}{dx}\frac{dN_1^{(1)}}{dx}\right) & \left(\frac{dN_1^{(1)}}{dx}\frac{dN_2^{(1)}}{dx}\right) \\ \left(\frac{dN_2^{(1)}}{dx}\frac{dN_1^{(1)}}{dx}\right) & \left(\frac{dN_2^{(1)}}{dx}\frac{dN_2^{(1)}}{dx}\right) \end{bmatrix} dx\right) \begin{Bmatrix} u_1^{(1)} \\ u_2^{(1)} \end{Bmatrix} -$$

$$- \int_{x_1^{(1)}}^{x_2^{(1)}} \begin{Bmatrix} N_1^{(1)} \\ N_2^{(1)} \end{Bmatrix} t_x \, dx = \begin{Bmatrix} F_{x_1}^{(1)} \\ F_{x_2}^{(1)} \end{Bmatrix} \tag{2.20}$$

or

$$\mathbf{K}^{(1)} \, \mathbf{a}^{(1)} - \mathbf{f}^{(1)} = \mathbf{q}^{(1)} \tag{2.21a}$$

with

$$K_{ij}^{(1)} = \int_{x_1^{(1)}}^{x_2^{(1)}} \frac{dN_i{}^{(1)}}{dx} (EA)^{(1)} \frac{dN_j{}^{(1)}}{dx} \, dx$$

$$f_i^{(1)} = f_{x_i}^{(1)} = \int_{x_1^{(1)}}^{x_2^{(1)}} N_i^{(1)} \, t_x \, dx \qquad i,j = 1,2 \tag{2.21b}$$

$$\mathbf{a}^{(1)} = \left[u_1^{(1)}, u_2^{(1)}\right]^T \quad ; \quad \mathbf{q}^{(1)} = \left[F_{x_1}^{(1)}, F_{x_2}^{(1)}\right]^T = \left[-\mathcal{N}_1^{(1)}, \mathcal{N}_2^{(1)}\right]^T$$

In Eq.(2.21a) $\mathbf{K}^{(1)}$ is the *element stiffness matrix* and $\mathbf{f}^{(1)}$ is the *equivalent nodal force vector* for the element.

Eq.(2.21a) can be used to obtain the nodal axial forces $\mathcal{N}_1^{(1)}$ and $\mathcal{N}_2^{(1)}$ in terms of the nodal displacements and the external nodal forces by noting the relationship between the equilibrating nodal forces $F_{x_i}^{(e)}$ with the nodal axial forces $\mathcal{N}_i^{(e)}$ (Eq.(2.21b) and Figures 1.9 and 2.2).

If the Young modulus, the cross sectional area and the distributed loading are constant over the element, the following is obtained

$$\mathbf{K}^{(1)} = \left(\frac{EA}{l}\right)^{(1)} \begin{bmatrix} 1 & -1 \\ -1 & 1 \end{bmatrix} \quad ; \quad \mathbf{f}^{(1)} = \begin{Bmatrix} f_{x_1}^{(1)} \\ f_{x_2}^{(1)} \end{Bmatrix} = \frac{(lt_x)^{(1)}}{2} \begin{Bmatrix} 1 \\ 1 \end{Bmatrix} \tag{2.22}$$

Above expressions coincide with those obtained for the axially loaded bar in Chapter 1. This coincidence could have been anticipated if we had observed that in both cases the same linear displacement field is assumed. This obviously leads, via the PVW, to the same expressions for the element stiffness matrix and the nodal load vector.

2.3.3 Assembly of the global equilibrium equations

The global equilibrium equations $\mathbf{K}\ \mathbf{a} = \mathbf{f}$ are obtained by the same nodal load balancing procedure explained for bar structures in the previous chapter. Thus, for each of the N nodes in the mesh we have

$$\sum_e F_{x_i}^{(e)} = P_{x_j} \quad , \quad j = 1, N \tag{2.23}$$

where the sum is extended over all the elements sharing the node with global number j, $F_{x_i}^{(e)}$ is the equilibrating nodal force contributed by each element and P_{x_j} is the external point load acting at the node.

For the single element mesh considered, Eq.(2.23) is written as (see Figure 2.2)

$$\text{node } 1: \quad F_{x_1}^{(1)} = R$$

$$\text{node } 2: \quad F_{x_2}^{(1)} = P$$

Substituting the values of the equilibrating nodal forces from Eq.(2.20) and making use of Table 2.1 the global equilibrium equations are obtained as

$$\left(\frac{EA}{l}\right) \begin{bmatrix} 1 & -1 \\ -1 & 1 \end{bmatrix} \begin{Bmatrix} u_1 \\ u_2 \end{Bmatrix} = \begin{Bmatrix} R + \dfrac{lt_x}{2} \\ P + \dfrac{lt_x}{2} \end{Bmatrix}$$

or

$$\mathbf{K}\ \mathbf{a} = \mathbf{f} \tag{2.24}$$

where, as usual, $\mathbf{K}$, $\mathbf{a}$ and $\mathbf{f}$ are, respectively, the global stiffness matrix, the vector containing the displacements of all nodes in the mesh and the global equivalent nodal force vector. External point loads acting at a node j are assigned directly to the jth position of the global vector $\mathbf{f}$, as explained in Section 1.6 (see also Figure 1.13). Note that the reaction force R at node 1 has been assembled into vector $\mathbf{f}$.

Eq.(2.24) is solved after imposing the condition $u_1 = 0$, to give

$$u_2 = \frac{l}{EA}\left(P + \frac{lt_x}{2}\right) \quad ; \quad R = -(P + lt_x) \tag{2.25}$$

2.3.4 Computation of the reactions

The reaction R in Eq.(2.25) has been directly obtained from the first row of Eq.(2.24). In general, the reaction at the prescribed nodes can be computed "a posteriori" from Eq.(1.10a) as

$$\mathbf{r} = \mathbf{K}\ \mathbf{a} - \mathbf{f}^{\text{ext}} \tag{2.26a}$$

where $\mathbf{r}$ is the vector of nodal reactions and $\mathbf{f}^{ext}$ is obtained by assembling the equivalent nodal force vectors $\mathbf{f}^{(e)}$ due to external loads only (i.e. excluding the reactions).

Indeed the product $\mathbf{K}\,\mathbf{a}$ can be computed by assembly of the element contributions $\mathbf{K}^{(e)}\mathbf{a}^{(e)}$.

An alternative and useful expression for computing the nodal reaction vector $\mathbf{r}$ is

$$\mathbf{r} = \mathbf{f}_{\text{int}} - \mathbf{f}^{\text{ext}} \tag{2.26b}$$

where $\mathbf{f}_{\text{int}}$ is the *vector of internal nodal forces* which can be obtained by assembling the contributions of the individual elements given by

$$\mathbf{f}_{\text{int}}^{(e)} = \int_{l^{(e)}} \left[\frac{dN_1^{(e)}}{dx}, \frac{dN_2^{(e)}}{dx}\right]^T \mathcal{N}^{(e)} dx \tag{2.26c}$$

Eq.(2.26c) is deduced from the first integral in the l.h.s. of Eqs.(2.18c).

2.3.5 Computation of the axial strain and the axial force

The axial strain ε and the axial force $\mathcal{N}$ in the element are given by

$$\varepsilon^{(1)} = \frac{dN_1^{(1)}}{dx} u_1^{(1)} + \frac{dN_2^{(1)}}{dx} u_2^{(1)} = \frac{u_2^{(1)}}{l_1^{(1)}} = \frac{P + lt_x/2}{EA}$$
$$\mathcal{N}^{(1)} = (EA)^{(1)}\varepsilon^{(1)} = P + \frac{lt_x}{2} \tag{2.27a}$$

The nodal axial forces for the element can be obtained from the components of $\mathbf{q}^{(1)}$ in the element equilibrium equations (Eqs.(2.21a) and (2.21b))

$$\begin{aligned} \mathcal{N}_1^{(1)} &= -F_{x_1}^{(1)} = \left[\left(\frac{EA}{l}\right)^{(1)} u_2^{(1)} - f_{x_1}^{(1)}\right] = -(P + lt_x) \\ \mathcal{N}_2^{(1)} &= F_{x_2}^{(1)} = \left(\frac{EA}{l}\right)^{(1)} u_2^{(1)} - f_{x_2}^{(1)} = P \end{aligned} \tag{2.27b}$$

The exact solution for this simple problem is [Ti]

$$\begin{aligned} u &= \frac{1}{EA}\left[-\frac{x^2}{2}t_x + (P + lt_x)\,x\right] \\ \varepsilon &= \frac{1}{EA}\,[P + (l - x)t_x] \quad , \quad \mathcal{N} = P + (l - x)t_x \end{aligned} \tag{2.28}$$

The finite element and the exact solutions are compared in Figure 2.3 for $P = 0$ and $t_x = 1T/m$. Note that the value of the end displacement u_2 is the exact solution. This is an exceptional coincidence that only occurs

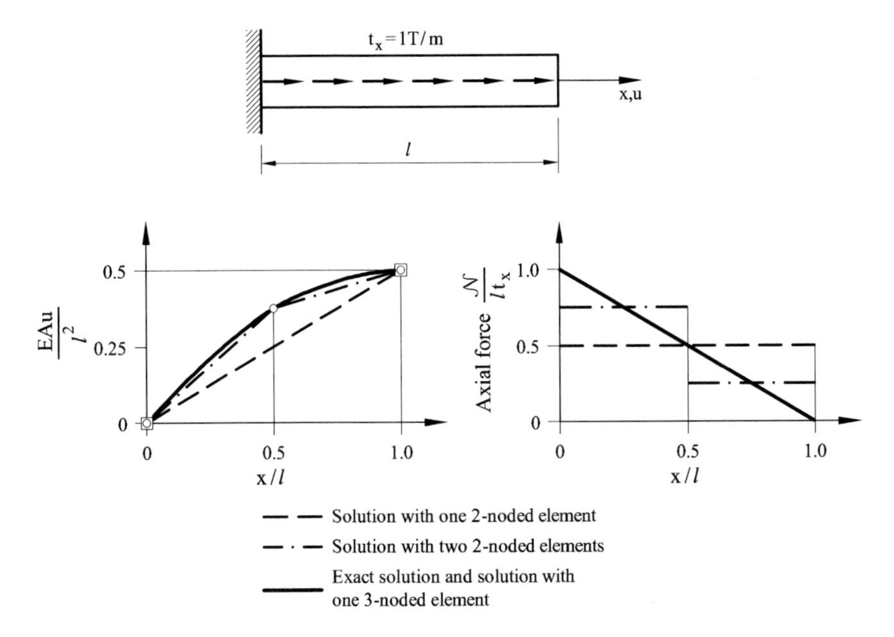

Fig. 2.3 Rod under uniformly distributed axial loading. Exact and approximate solutions using one and two linear rod elements and one 3-noded rod element (Example 3.4)

on very few occasions.* Within the rod the single element approximation yields a linear displacement field very different from the exact quadratic solution. Also note that the constant axial stress value obtained differs substantially from the linear exact solution. As expected, the numerical solution improves as the mesh is refined and this is shown in Section 2.3 for a mesh of two elements.

2.4 DERIVATION OF THE DISCRETIZED EQUATIONS FROM THE GLOBAL DISPLACEMENT INTERPOLATION FIELD

A general expression for the displacement interpolation field *for the whole mesh* can be obtained by simple superposition of the local approximations for each element. This let us define *global shape functions* which naturally

* It has been proved [ZTZ] that the finite element solution coincides with the exact one for 1D problems if the interpolation chosen satisfies exactly the homogeneous form of the differential equation of equilibrium. This is written for the rod problem as $d^2u/dx^2 = 0$, which is obviously satisfied by the linear approximation chosen.

coincide with the original local expressions within each element. The use of global shape functions leads to identical results as with the simpler local functions. It is important however to understand the conceptual differences between using local or global shape functions. For this purpose we will repeat the single rod element problem using a global interpolation for the displacement field.

The axial displacement can be written in the single element mesh as

$$u(x) = N_1^g(x)\, u_1 + N_2^g(x)\, u_2 \tag{2.29}$$

where $N_1^g(x)$ and $N_2^g(x)$ are the *global shape functions* of nodes 1 and 2, respectively, and u_1 and u_2 are the displacements of these nodes. Note that we skip the superindex e for the global displacements. We deduce from Eq.(2.29) that the global function of a node takes the value *one* at that node and *zero* at all other nodes. This provides the relationship between global and local shape functions as

$$\begin{aligned} N_i^g(x) &= N_i^{(e)}(x) && \text{if } x \text{ belongs to element } e \\ &= 0 && \text{if } x \text{ does not belong to element } e \end{aligned} \tag{2.30}$$

For the single element case considered, the global and local shape functions coincide (Figure 2.2). Thus,

$$N_1^g(x) = N_1^{(1)}(x) \quad \text{and} \quad N_2^g(x) = N_2^{(1)}(x) \tag{2.31}$$

The axial strain in the rod of Figure 2.2 can be obtained as

$$\varepsilon = \frac{du}{dx} = \frac{dN_1^g}{dx}\, u_1 + \frac{dN_2^g}{dx}\, u_2 \tag{2.32}$$

The virtual displacement and the virtual axial strain are expressed as

$$\begin{aligned} \delta u &= N_1^g \delta u_1 + N_2^g \delta u_2 \\ \delta\varepsilon &= \frac{dN_1^g}{dx}\, \delta u_1 + \frac{dN_2^g}{dx}\, \delta u_2 \end{aligned} \tag{2.33}$$

The PVW is written for the rod as

$$\begin{aligned} &\int_0^l \left[\frac{dN_1^g}{dx}\, \delta u_1 + \frac{dN_2^g}{dx}\, \delta u_2\right] (EA) \left[\frac{dN_1^g}{dx}\, u_1 + \frac{dN_2^g}{dx}\, u_2\right] dx \ - \\ &- \int_0^l [N_1^g \delta u_1 + N_2^g \delta u_2]\; t_x\, dx = \delta u_1 R + \delta u_2\, P \end{aligned} \tag{2.34}$$

After eliminating the virtual displacements, Eq.(2.34) leads to

$$\left(\int_0^l EA \begin{bmatrix} \left(\frac{dN_1^g}{dx}\frac{dN_1^g}{dx}\right) & \left(\frac{dN_1^g}{dx}\frac{dN_2^g}{dx}\right) \\ \left(\frac{dN_2^g}{dx}\frac{dN_1^g}{dx}\right) & \left(\frac{dN_2^g}{dx}\frac{dN_2^g}{dx}\right) \end{bmatrix} dx\right) \begin{Bmatrix} u_1 \\ u_2 \end{Bmatrix} -$$

$$- \int_0^l \begin{Bmatrix} N_1^g \\ N_2^g \end{Bmatrix} t_x \, dx = \begin{Bmatrix} R \\ P \end{Bmatrix} \tag{2.35}$$

The following relationships are important for the computation of the integrals in Eq.(2.35)

$$\left.\begin{aligned} N_1^g &= N_1^{(1)} \\ N_2^g &= N_2^{(1)} \\ \frac{dN_1^g}{dx} &= \frac{dN_1^{(1)}}{dx} \\ \frac{dN_2^g}{dx} &= \frac{dN_2^{(1)}}{dx} \end{aligned}\right\} \quad 0 \leq x \leq l \tag{2.36}$$

Using Eq.(2.36) we obtain

$$\begin{aligned} \int_0^l \frac{dN_1^g}{dx}\frac{dN_1^g}{dx}\, dx &= \int_0^{l^{(1)}} \left(\frac{dN_1^{(1)}}{dx}\right)^2 dx = \frac{1}{l} \\ \int_0^l \frac{dN_1^g}{dx}\frac{dN_2^g}{dx}\, dx &= \int_0^{l^{(1)}} \frac{dN_1^{(1)}}{dx}\frac{dN_2^{(1)}}{dx}\, dx = -\frac{1}{l} \\ \int_0^l \frac{dN_2^g}{dx}\frac{dN_2^g}{dx}\, dx &= \int_0^{l^{(1)}} \left(\frac{dN_2^{(1)}}{dx}\right)^2 dx = \frac{1}{l} \\ \int_0^l N_i^g \, dx &= \int_0^{l^{(1)}} N_1^{(1)}\, dx = \int_0^{l^{(1)}} N_2^{(1)}\, dx = \frac{l}{2} \end{aligned} \tag{2.37}$$

Substituting Eqs.(2.37) into the PVW expression (2.35) yields the *global equilibrium equation* (2.24) directly. Recall that in the previous section this equation was obtained from the assembly of the element contributions. From this point onwards the solution process is identical to that explained in Eqs.(2.24)-(2.27) and it will not be repeated here.

In the next section the same problem is solved using a mesh of two linear elements.

2.5 AXIALLY LOADED ROD OF CONSTANT CROSS SECTION. DISCRETIZATION IN TWO LINEAR ROD ELEMENTS

The same rod as for the previous example is discretized now in two linear rod elements as shown in Figure 2.4 where the local and global shape functions are also shown.

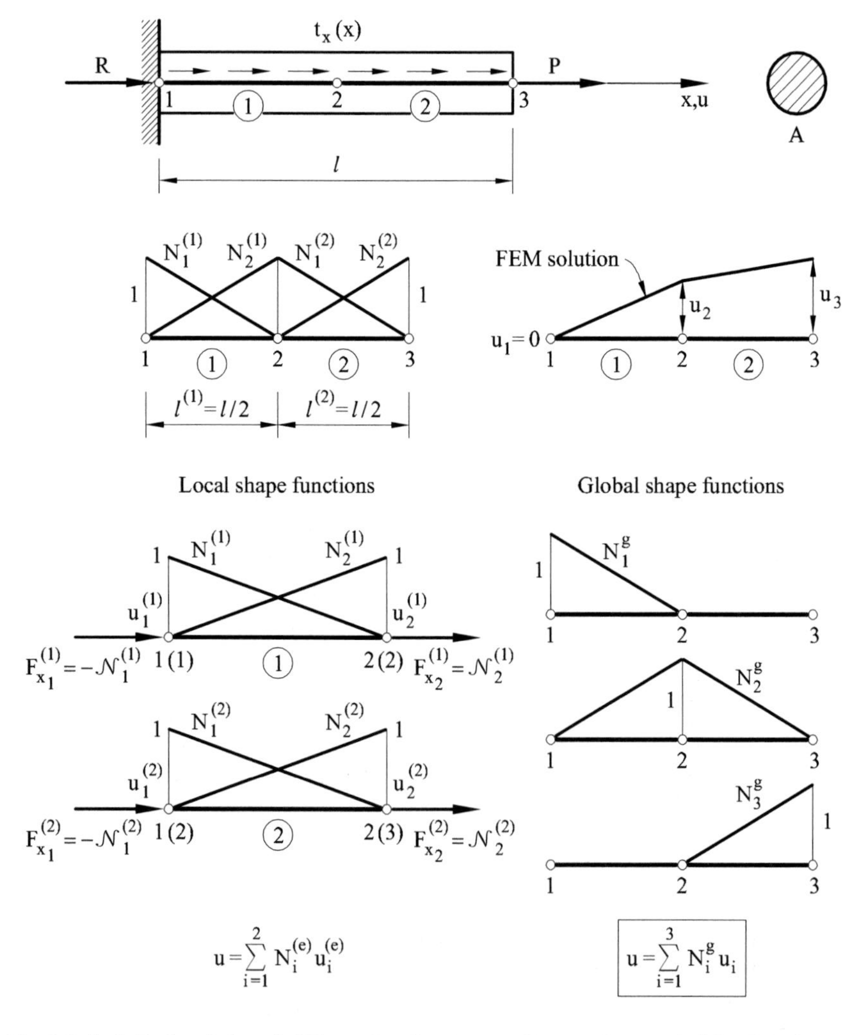

Fig. 2.4 Axially loaded rod. Discretization in two linear elements. Number in brackets at node denotes global node number

The discretized equilibrium equations will be obtained first using the local forms of the shape functions.

2.5.1 Solution using the element shape functions

The displacements within each element are interpolated as

Element 1 | **Element 2**

$$u(x) = N_1^{(1)}(x)u_1^{(1)} + N_2^{(1)}(x)u_2^{(1)} \quad \Big| \quad u(x) = N_1^{(2)}(x)u_1^{(2)} + N_2^{(2)}(x)u_2^{(2)} \tag{2.38}$$

The shape functions and their derivatives are

$$\begin{array}{l|l} N_1^{(1)} = \dfrac{x_2^{(1)} - x}{l^{(1)}} \; ; \; \dfrac{dN_1^{(1)}}{dx} = -\dfrac{1}{l^{(1)}} & N_1^{(2)} = \dfrac{x_2^{(2)} - x}{l^{(2)}} \; ; \; \dfrac{dN_1^{(2)}}{dx} = -\dfrac{1}{l^{(2)}} \\ N_2^{(1)} = \dfrac{x - x_2^{(1)}}{l^{(1)}} \; ; \; \dfrac{dN_2^{(1)}}{dx} = \dfrac{1}{l^{(1)}} & N_2^{(2)} = \dfrac{x - x_1^{(2)}}{l^{(2)}} \; ; \; \dfrac{dN_2^{(2)}}{dx} = \dfrac{1}{l^{(2)}} \end{array} \tag{2.39}$$

The axial strain in each element is

$$\varepsilon = \frac{du}{dx} = \frac{dN_1^{(1)}}{dx}u_1^{(1)} + \frac{dN_2^{(1)}}{dx}u_2^{(1)} \quad \Big| \quad \varepsilon = \frac{du}{dx} = \frac{dN_1^{(2)}}{dx}u_1^{(2)} + \frac{dN_2^{(2)}}{dx}u_2^{(2)} \tag{2.40}$$

The discretized equilibrium equations are obtained using the PVW as explained in the previous section for the single element case (see Eqs.(2.15)-(2.22)). We find that

$$\mathbf{q}^{(1)} = \mathbf{K}^{(1)}\mathbf{a}^{(1)} - \mathbf{f}^{(1)} \quad ; \quad \mathbf{q}^{(2)} = \mathbf{K}^{(2)}\mathbf{a}^{(2)} - \mathbf{f}^{(2)} \tag{2.41}$$

where

Element 1

$$\mathbf{K}^{(1)} = \int_{x_1^{(1)}}^{x_2^{(1)}} (EA)^{(1)} \begin{bmatrix} \left(\dfrac{dN_1^{(1)}}{dx}\dfrac{dN_1^{(1)}}{dx}\right) & \left(\dfrac{dN_1^{(1)}}{dx}\dfrac{dN_2^{(1)}}{dx}\right) \\ \left(\dfrac{dN_2^{(1)}}{dx}\dfrac{dN_1^{(1)}}{dx}\right) & \left(\dfrac{dN_2^{(1)}}{dx}\dfrac{dN_2^{(1)}}{dx}\right) \end{bmatrix} dx \tag{2.42}$$

$$\mathbf{f}^{(1)} = \left[f_{x_1}^{(1)}, f_{x_2}^{(1)}\right]^T = \int_{x_1^{(1)}}^{x_2^{(1)}} \left[N_1^{(1)}, N_2^{(1)}\right]^T t_x \, dx$$

$$\mathbf{q}^{(1)} = \left[F_{x_1}^{(1)}, F_{x_2}^{(1)}\right]^T = \left[-\mathcal{N}_1^{(1)}, \mathcal{N}_2^{(1)}\right]^T \quad , \quad \mathbf{a}^{(1)} = \left[u_1^{(1)}, u_2^{(1)}\right]^T$$

Element 2

$$\mathbf{K}^{(2)} = \int_{x_1^{(2)}}^{x_2^{(2)}} (EA)^{(2)} \begin{bmatrix} \left(\frac{dN_1^{(2)}}{dx}\frac{dN_1^{(2)}}{dx}\right) & \left(\frac{dN_1^{(2)}}{dx}\frac{dN_2^{(2)}}{dx}\right) \\ \left(\frac{dN_2^{(2)}}{dx}\frac{dN_1^{(2)}}{dx}\right) & \left(\frac{dN_2^{(2)}}{dx}\frac{dN_2^{(2)}}{dx}\right) \end{bmatrix} dx$$

$$\mathbf{f}^{(2)} = \left[f_{x_1}^{(2)}, f_{x_2}^{(2)}\right]^T = \int_{x_1^{(2)}}^{x_2^{(2)}} \left[N_1^{(2)}, N_2^{(2)}\right]^T t_x \, dx \tag{2.43}$$

$$\mathbf{q}^{(2)} = \left[F_{x_1}^{(2)}, F_{x_2}^{(2)}\right]^T = \left[-\mathcal{N}_1^{(2)}, \mathcal{N}_2^{(2)}\right]^T \quad , \quad \mathbf{a}^{(2)} = \left[u_1^{(2)}, u_2^{(2)}\right]^T$$

are respectively the stiffness matrices, the equivalent nodal force vectors, the equilibrating nodal force vectors and the nodal displacement vectors for elements 1 and 2.

The integrals in Eqs.(2.42) and (2.43) are computed keeping in mind the relationship between the local and global numbering of the element parameters summarized in Table 2.2.

Element	node		coordinate		displacement	
	local	global	local	global	local	global
1	1	1	$x_1^{(1)}$	x_1	$u_1^{(1)}$	u_1
	2	2	$x_2^{(1)}$	x_2	$u_2^{(1)}$	u_2
2	1	2	$x_1^{(2)}$	x_2	$u_1^{(2)}$	u_2
	2	3	$x_2^{(2)}$	x_3	$u_2^{(2)}$	u_3

Table 2.2 Local and global parameters for the example of Figure 2.4

Substituting Eqs.(2.39) into (2.42) and (2.43) and using Table 2.2 the following expressions are obtained for homogeneous material and uniformly distributed loading:

$$\mathbf{K}^{(1)} = \left(\frac{EA}{l}\right)^{(1)} \begin{bmatrix} 1 & -1 \\ -1 & 1 \end{bmatrix} \quad ; \quad \mathbf{K}^{(2)} = \left(\frac{EA}{l}\right)^{(2)} \begin{bmatrix} 1 & -1 \\ -1 & 1 \end{bmatrix}$$

$$\mathbf{f}^{(1)} = \frac{(lt_x)^{(1)}}{2} [1, 1]^T \quad ; \quad \mathbf{f}^{(2)} = \frac{(lt_x)^{(2)}}{2} [1, 1]^T \tag{2.44}$$

The equilibrium of nodal forces is written as (see Eq.(2.23) and Figure 2.4)

$$\begin{aligned}
\text{Node 1 :} \quad & F_{x_1}^{(1)} && = R \\
\text{Node 2 :} \quad & F_{x_2}^{(1)} + F_{x_1}^{(2)} && = 0 \\
\text{Node 3 :} \quad & F_{x_2}^{(2)} && = P
\end{aligned} \tag{2.45}$$

Substituting $F_{x_1}^{(e)}$ from Eq.(2.42) into (2.43) the following matrix equilibrium equation is obtained

$$\begin{bmatrix} \left(\frac{EA}{l}\right)^{(1)} & -\left(\frac{EA}{l}\right)^{(1)} & 0 \\ -\left(\frac{EA}{l}\right)^{(1)} & \left[\left(\frac{EA}{l}\right)^{(1)} + \left(\frac{EA}{l}\right)^{(2)}\right] & \left(\frac{EA}{l}\right)^{(2)} \\ 0 & -\left(\frac{EA}{l}\right)^{(2)} & \left(\frac{EA}{l}\right)^{(2)} \end{bmatrix} \begin{Bmatrix} u_1 \\ u_2 \\ u_3 \end{Bmatrix} = \begin{Bmatrix} \frac{lt_x}{4} + R \\ \frac{lt_x}{2} \\ \frac{lt_x}{4} + P \end{Bmatrix} \tag{2.46a}$$

$$\mathbf{Ka} = \mathbf{f} \tag{2.46b}$$

Note that the assembly process is identical to that explained in the previous chapter for bar structures.

Substituting $\left(\frac{EA}{l}\right)^{(1)} = \left(\frac{EA}{l}\right)^{(2)} = \frac{2EA}{l}$ into Eq.(2.46a) and solving the equation system we find

$$\begin{aligned}
& u_1 = 0 \quad ; \quad u_2 = \frac{l}{2EA}\left(P + \frac{3lt_x}{4}\right) \\
& u_3 = \frac{l}{2EA}\,(2P + lt_x) \quad ; \quad R_1 = -(P + lt_x)
\end{aligned} \tag{2.47}$$

The axial strain and the axial force are constant within each element and are obtained as

Element 1	**Element 2**
$\varepsilon^{(1)} = \left(\frac{du}{dx}\right)^{(1)} = \frac{u_2}{l^{(1)}} = \frac{P + \frac{3lt_x}{4}}{EA}$	$\varepsilon^{(2)} = \left(\frac{du}{dx}\right)^{(2)} = \frac{u_3 - u_2}{l^{(2)}} = \frac{1}{EA}\left(\frac{lt_x}{4} + P\right)$
$\mathcal{N}^{(1)} = (EA)^{(1)}\varepsilon^{(1)} = P + \frac{3lt_x}{4}$	$\mathcal{N}^{(2)} = (EA)^{(2)}\varepsilon^{(2)} = \frac{lt_x}{4} + P$

(2.48)

The nodal axial forces for each element can be computed from the components of $\mathbf{q}^{(e)}$ (see Eqs.(2.41)–(2.43)).

The distribution within each element of the displacement u and the constant axial force $\mathcal{N}^{(e)}$ is shown in Figure 2.3 for $P = 0$ and $t_x = 1\text{T/m}$. For the same reasons explained in Section 2.3 the nodal displacements coincide with the exact values. Some improvement in the approximation of the global displacement field is also observed. However, the error in the axial force is still considerable and its reduction requires a finer discretization and the nodal smoothing of the constant axial forces over each element. This can be done by simply averaging the nodal axial forces. Other stress smoothing techniques are described in Chapter 9.

A simple observation shows that the results obtained for the axial forces (and strains) are more inaccurate than those for the displacement field. This is a general rule which is a consequence from computing the strains and the stresses from the derivatives of the approximate displacement field. This, naturally, increases the solution error for those variables [ZTZ].

2.5.2 Solution using the global shape functions

The same problem is now solved using the global description of the shape functions.

The axial displacement can be expressed globally over the two elements mesh as (Figure 2.4)

$$u(x) = N_1^g(x)u_1 + N_2^g(x)u_2 + N_3^g(x)u_3 \tag{2.49}$$

and the axial strain is given by

$$\varepsilon = \frac{du}{dx} = \frac{dN_1^g}{dx}u_1 + \frac{dN_2^g}{dx}u_2 + \frac{dN_3^g}{dx}u_3 \tag{2.50}$$

The discretized form of the PVW is written using above equations as

$$\begin{aligned}\int_0^l \left(\frac{dN_1^g}{dx}\delta u_1 + \frac{dN_2^g}{dx}\delta u_2 + \frac{dN_3^g}{dx}\delta u_3\right)(EA)\\ \left(\frac{dN_1^g}{dx}u_1 + \frac{dN_2^g}{dx}u_2 + \frac{dN_3^g}{dx}u_3\right)dx-\\ -\int_0^l \left(N_1^g\delta u_1 + N_2^g\delta u_2 + N_3^g\delta u_3\right)t_x\,dx = \delta u_1 R + \delta u_3 P\end{aligned} \tag{2.51}$$

This leads, after eliminating the virtual displacements, to the following matrix system of equations

$$\left(\int_0^l \begin{bmatrix} \left(\frac{dN_1^g}{dx}\frac{dN_1^g}{dx}\right) & \left(\frac{dN_1^g}{dx}\frac{dN_2^g}{dx}\right) & \left(\frac{dN_1^g}{dx}\frac{dN_3^g}{dx}\right) \\ \left(\frac{dN_2^g}{dx}\frac{dN_1^g}{dx}\right) & \left(\frac{dN_2^g}{dx}\frac{dN_2^g}{dx}\right) & \left(\frac{dN_2^g}{dx}\frac{dN_3^g}{dx}\right) \\ \left(\frac{dN_3^g}{dx}\frac{dN_1^g}{dx}\right) & \left(\frac{dN_3^g}{dx}\frac{dN_2^g}{dx}\right) & \left(\frac{dN_3^g}{dx}\frac{dN_3^g}{dx}\right) \end{bmatrix} EA\,dx \right) \begin{Bmatrix} u_1 \\ u_2 \\ u_3 \end{Bmatrix} -$$

$$- \int_0^l \begin{Bmatrix} N_1^g \\ N_2^g \\ N_3^g \end{Bmatrix} t_x\,dx = \begin{Bmatrix} R \\ 0 \\ P \end{Bmatrix} \tag{2.52}$$

The computation of the integrals in Eq.(2.52) requires a correspondence between the global and local shape functions. The following relationships are deduced from Table 2.2 and Figure 2.4

$$\left. \begin{aligned} N_1^g &= N_1^{(1)} \\ \frac{dN_1^g}{dx} &= \frac{dN_1^{(1)}}{dx} \\ N_2^g &= N_2^{(1)} \\ \frac{dN_2^g}{dx} &= \frac{dN_2^{(1)}}{dx} \\ N_3^g &= 0 \\ \frac{dN_3^g}{dx} &= 0 \end{aligned} \right\} 0 \le x \le \frac{l}{2}\,; \quad \left. \begin{aligned} N_1^g &= 0 \\ \frac{dN_1^g}{dx} &= 0 \\ N_2^g &= N_1^{(2)} \\ \frac{dN_2^g}{dx} &= \frac{dN_1^{(2)}}{dx} \\ N_3^g &= N_2^{(2)} \\ \frac{dN_3^g}{dx} &= \frac{dN_2^{(2)}}{dx} \end{aligned} \right\} \frac{l}{2} < x \le l \tag{2.53}$$

Making use of the expressions (2.53) in (2.52) the *global* equilibrium equation is directly obtained. The reader can easily verify the coincidence of this equation with Eq.(2.46a) obtained by assembly of the element contributions.

This example clearly shows that the use of the global shape functions is less systematic and requires more detailed computations than the element by element approach. These differences are even more apparent for finer meshes. As a consequence, the assembly of the global equations from the elemental expressions derived via the local shape functions is the natural way to be followed.

2.6 GENERALIZATION OF THE SOLUTION WITH N LINEAR ROD ELEMENTS

The solution process explained in the previous sections can easily be generalized for a discretization using a mesh of N 2-noded (linear) rod elements. The stiffness equations for each element are

$$\mathbf{K}^{(e)}\mathbf{a}^{(e)} - \mathbf{f}^{(e)} = \mathbf{q}^{(e)} \tag{2.54a}$$

with

$$\mathbf{K}^{(e)} = \int_{x_1^{(e)}}^{x_2^{(e)}} (EA)^{(e)} \begin{bmatrix} \left(\frac{dN_1^{(e)}}{dx}\frac{dN_1^{(e)}}{dx}\right) & \left(\frac{dN_1^{(e)}}{dx}\frac{dN_2^{(e)}}{dx}\right) \\ \left(\frac{dN_2^{(e)}}{dx}\frac{dN_1^{(e)}}{dx}\right) & \left(\frac{dN_2^{(e)}}{dx}\frac{dN_2^{(e)}}{dx}\right) \end{bmatrix} dx$$

$$\mathbf{f}^{(e)} = \int_{x_1^{(e)}}^{x_2^{(e)}} \begin{Bmatrix} N_1^{(e)} \\ N_2^{(e)} \end{Bmatrix} t_x\, dx \quad , \quad \mathbf{q}^{(e)} = \begin{Bmatrix} F_{x_1}^{(e)} \\ F_{x_2}^{(e)} \end{Bmatrix} = \begin{Bmatrix} -N_1^{(e)} \\ N_2^{(e)} \end{Bmatrix} \tag{2.54b}$$

After substitution of the shape functions and their derivatives

$$\begin{aligned} N_1^{(e)} &= \frac{x_2^{(e)} - x}{l^{(e)}} \quad ; \quad \frac{dN_1^{(e)}}{dx} = -\frac{1}{l^{(e)}} \\ N_2^{(e)} &= \frac{x - x_1^{(e)}}{l^{(e)}} \quad ; \quad \frac{dN_2^{(e)}}{dx} = \frac{1}{l^{(e)}} \end{aligned} \tag{2.55}$$

gives (for homogeneous material and uniformly distributed loading)

$$\mathbf{K}^{(e)} = \left(\frac{EA}{l}\right)^{(e)} \begin{bmatrix} 1 & -1 \\ -1 & 1 \end{bmatrix} \quad ; \quad \mathbf{f}^{(e)} = \frac{(lt_x)^{(e)}}{2} \begin{Bmatrix} 1 \\ 1 \end{Bmatrix} \tag{2.56}$$

As usual, the assembly process is based on the global equilibrium of the element nodal forces $\mathbf{q}^{(e)}$ (Eq.(2.23)). This leads, after small algebra, to the following global matrix equation

$$\underbrace{\begin{bmatrix} k^{(1)} & -k^{(1)} & 0 & \cdots & 0 \\ -k^{(1)} & \left[k^{(1)}+k^{(2)}\right] & -k^{(2)} & \cdots & 0 \\ 0 & -k^{(2)} & \left[k^{(2)}+k^{(3)}\right] & \cdots & 0 \\ 0 & 0 & -k^{(3)} & \cdots & \vdots \\ \vdots & \vdots & \ddots & \cdots & \vdots \\ 0 & 0 & \cdots & \left[k^{(N-1)}+k^{(N)}\right] & -k^{(N)} \\ 0 & 0 & \cdots & -k^{(N)} & k^{(N)} \end{bmatrix}}_{\mathbf{K}} \underbrace{\begin{Bmatrix} u_1 \\ u_2 \\ u_3 \\ \vdots \\ \vdots \\ u_{N-1} \\ u_N \end{Bmatrix}}_{\mathbf{a}} =$$

$$= \underbrace{\begin{Bmatrix} \dfrac{(lt_x)^{(1)}}{2} + P_{x_1} \\ \dfrac{(lt_x)^{(1)}}{2} + \dfrac{(lt_x)^{(2)}}{2} + P_{x_2} \\ \dfrac{(lt_x)^{(2)}}{2} + \dfrac{(lt_x)^{(3)}}{2} + P_{x_3} \\ \vdots \\ \dfrac{(lt_x)^{(N-1)}}{2} + \dfrac{(lt_x)^{(N)}}{2} + P_{x_{N-1}} \\ \dfrac{(lt_x)^{(N)}}{2} + P_{x_N} \end{Bmatrix}}_{\mathbf{f}} \quad \text{with} \quad k^{(e)} = \left(\frac{EA}{l}\right)^{(e)} \tag{2.57}$$

Matrix $\mathbf{K}$ depends on the geometrical (l and A) and material (E) parameters for each element, while vector $\mathbf{f}$ depends on the intensity of the distributed load t_x, the element length and the external point forces P_{x_i} acting at the nodes. Recall that external nodal point forces are assigned directly to the rows of vector $\mathbf{f}$ corresponding to the number of the global node (Figure 1.14). The unknown reactions at the prescribed nodes are treated as point loads and they can be computed "a posteriori" as explained in Section 2.3.4.

Example 2.1: Analyse the axially loaded rod with exponentially varying cross sectional area of Figure 2.5 using three meshes of one, two and three linear rod elements.

- ***Solution***

The change in cross sectional area is defined by $A = A_0 e^{-\frac{x}{l}}$ where A_0 is the cross sectional area at the clamped end and l is the rod length. The rod is

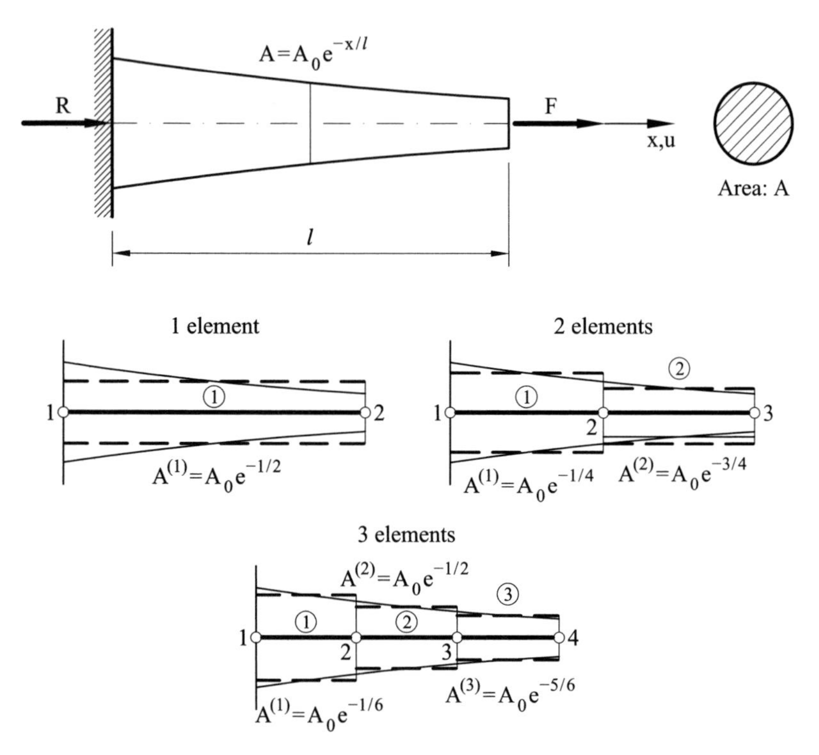

Fig. 2.5 Axially loaded road with exponentially varying cross section. Discretization in 3 meshes of two-noded rod elements

subjected to an axial force acting at the free end. The exact solution for this simple problem is

$$\sigma = \frac{F}{A} = \frac{F}{A_0}\, e^{\frac{x}{l}} \quad , \qquad \varepsilon = \frac{\sigma}{E} = \frac{F}{EA_0}\, e^{\frac{x}{l}}$$

$$u(x) = \int_0^x \varepsilon dx = \int_0^x \frac{F}{EA_0}\, e^{x/l}\, dx = \frac{Fl}{EA_0}(e^{x/l} - 1)$$

$$u(l) = \frac{Fl}{EA_0}(e-1) = 1.71828\frac{Fl}{EA_0} \quad ; \quad R = -F$$

Two options are possible for the finite element solution: a) to use the exact expression for the cross sectional area, and b) to assume a constant cross sectional area within each element. The second option has been chosen here for simplicity. The reader is encouraged to repeat this problem as an exercise using the first alternative.

One element solution

The cross sectional area is assumed to be constant and equal to $A = A_0 e^{-1/2}$. The element stiffness matrix is directly given by Eq.(2.56), i.e.

$$\mathbf{K}^{(1)} = \frac{EA_0}{l}\, e^{-1/2} \begin{bmatrix} 1 & -1 \\ -1 & 1 \end{bmatrix} = \frac{EA_0}{l}\, 0.60653 \begin{bmatrix} 1 & -1 \\ -1 & 1 \end{bmatrix}$$

The equilibrium equation is deduced from Eq.(2.57) (noting that the distributed loads $t_x^{(e)}$ are zero) as

$$\frac{EA_0}{l}\, 0.60653 \begin{bmatrix} 1 & -1 \\ -1 & 1 \end{bmatrix} \begin{Bmatrix} u_1 \\ u_2 \end{Bmatrix} = \begin{Bmatrix} R \\ F \end{Bmatrix} \; ; \; u_1 = 0$$

which when solved gives

$$u_2 = \frac{1}{0.60653}\frac{Fl}{EA_0} = 1.6487\,\frac{Fl}{EA_0} \quad ; \quad R = -F$$

The percentage of error with respect to the exact solution is 4.21%. This can be considered acceptable given the simplicity of the mesh.

Two elements solution

Now $A^{(1)} = A_0 e^{-1/4}$ and $A^{(2)} = A_0 e^{-3/4}$. The equilibrium equations for each element (Figure 2.5) are obtained as explained in Section 2.6.

Element 1

$$1.5576\left(\frac{EA_0}{l}\right) \begin{bmatrix} 1 & -1 \\ -1 & 1 \end{bmatrix} \begin{Bmatrix} u_1 \\ u_2 \end{Bmatrix} = \begin{Bmatrix} F_{x_1}^{(1)} \\ F_{x_2}^{(1)} \end{Bmatrix}$$

Element 2

$$0.9447\left(\frac{EA_0}{l}\right) \begin{bmatrix} 1 & -1 \\ -1 & 1 \end{bmatrix} \begin{Bmatrix} u_2 \\ u_3 \end{Bmatrix} = \begin{Bmatrix} F_{x_1}^{(2)} \\ F_{x_2}^{(2)} \end{Bmatrix}$$

After global assembly we have

$$\frac{EA_0}{l} \begin{bmatrix} 1.5576 & -1.5576 & 0 \\ -1.5576 & 2.5023 & -0.9447 \\ 0 & -0.9447 & 0.9447 \end{bmatrix} \begin{Bmatrix} u_1 \\ u_2 \\ u_3 \end{Bmatrix} = \begin{Bmatrix} R \\ 0 \\ F \end{Bmatrix} \quad u_1 = 0$$

which gives

$$\begin{aligned} u_3 &= 1.7005\,\frac{Fl}{EA_0} \quad (Error = 1.04\%) \\ u_2 &= 0.377541\, u_3 = 0.6419\frac{Fl}{EA_0} \\ R &= -F \end{aligned}$$

Three element mesh

For the three element mesh (Figure 2.5) $A^{(1)} = A_0 e^{-1/6}$, $A^{(2)} = A_0 e^{-1/2}$ and $A^{(3)} = A_0 e^{-5/6}$. The equilibrium equations for each element are

Element 1

$$2.5394 \frac{EA_0}{l} \begin{bmatrix} 1 & -1 \\ -1 & 1 \end{bmatrix} \begin{Bmatrix} u_1 \\ u_2 \end{Bmatrix} = \begin{Bmatrix} F_{x_1}^{(1)} \\ F_{x_2}^{(1)} \end{Bmatrix}$$

Element 2

$$1.8196 \frac{EA_0}{l} \begin{bmatrix} 1 & -1 \\ -1 & 1 \end{bmatrix} \begin{Bmatrix} u_2 \\ u_3 \end{Bmatrix} = \begin{Bmatrix} F_{x_1}^{(2)} \\ F_{x_2}^{(2)} \end{Bmatrix}$$

Element 3

$$1.3028 \frac{EA_0}{l} \begin{bmatrix} 1 & -1 \\ -1 & 1 \end{bmatrix} \begin{Bmatrix} u_3 \\ u_4 \end{Bmatrix} = \begin{Bmatrix} F_{x_1}^{(3)} \\ F_{x_2}^{(3)} \end{Bmatrix}$$

The global equilibrium equation after assembly is

$$\frac{EA_0}{l} \begin{bmatrix} 2.5394 & -2.5394 & 0 & 0 \\ -2.5394 & 4.3590 & -1.8196 & 0 \\ 0 & -1.8196 & 3.1234 & -1.3038 \\ 0 & 0 & -1.3038 & 1.3038 \end{bmatrix} \begin{Bmatrix} u_1 \\ u_2 \\ u_3 \\ u_4 \end{Bmatrix} = \begin{Bmatrix} R \\ 0 \\ 0 \\ F \end{Bmatrix} \quad u_1 = 0$$

and the solution is

$$u_4 = 1.71036 \frac{Fl}{EA_0} \; (Error = 0.46\%) \quad , \quad u_3 = 0.55156\, u_4 = 0.9432 \frac{Fl}{EA_0}$$

$$u_2 = 0.230241\, u_4 = 0.3938 \frac{Fl}{EA_0} \quad , \quad R = -F$$

Once the nodal displacements have been obtained, the axial strain and the axial stress can be computed for each element. For example, at the central point of element number 2 in the three element mesh we have

$$\begin{aligned} u(l/2) &= N_1^{(2)} \left(x^{(2)} = \frac{l^{(2)}}{2} \right) u_1^{(2)} + N_2^{(2)} \left(x^{(2)} = \frac{l^{(2)}}{2} \right) u_2^{(2)} = \\ &= \frac{1}{2} 0.23024 u_4 + \frac{1}{2} 0.55156 u_4 = 0.3909 u_4 = 0.6686 \frac{Fl}{EA_0} \end{aligned}$$

$$\left(\text{Exact value} = 0.6487 \frac{Fl}{EA_0}. \;\; \text{Error}: \; 3.07\%\right).$$

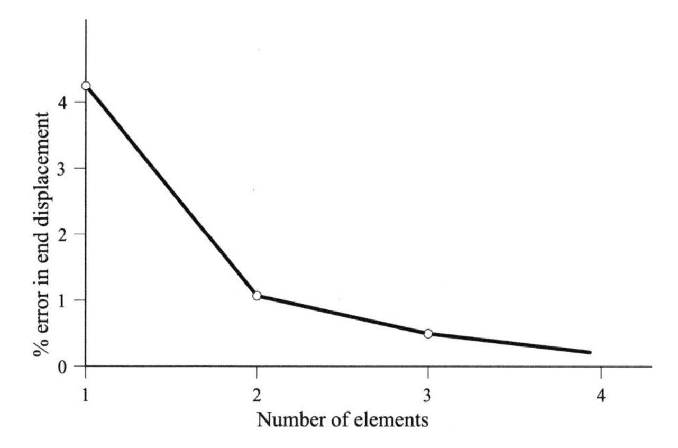

Fig. 2.6 Axially loaded rod with varying cross sectional area. Convergence of the end displacement value and the stress distribution with the number of elements

$$\varepsilon(l/2) = \left(\frac{dN_1^{(2)}}{dx}\right)_{x^{(2)}=\frac{l^{(2)}}{2}} u_1^{(2)} + \left(\frac{dN_2^{(2)}}{dx}\right)_{x^{(2)}=\frac{l^{(2)}}{2}} u_2^{(2)} = 1.6491\frac{F}{EA_0}$$
$$= \left(-\frac{3}{l}\ 0.2302 + \frac{3}{l}\ 0.5516\right) u_4 = 0.9642\ \frac{u_4}{l}$$
$$\sigma(l/2) = E\varepsilon_A = 1.649\ \frac{F}{A_0}\quad \text{(Exact value: } 1.6487\ \frac{F}{A_0}\text{ . Error: } 0.02\%\text{)}$$

The convergence of the end displacement value with the number of elements is shown in Figure 2.6. We see that the simple assumption of constant cross sectional area leads to percentage errors of less than 1% for meshes finer than two elements.

The displacement and stress distribution along the rod for the three meshes are plotted in Figure 2.7 together with the exact solution. The nodal displacements, and even the linear displacement field within each element, are quite accurate for the three meshes. However, the convergence of the constant axial stress field for each element to the exact exponential solution is slow.

2.7 EXTRAPOLATION OF THE SOLUTION FROM TWO DIFFERENT MESHES

Expanding in Taylor series the displacement in the vicinity of a node i gives

$$u = u_i + \left(\frac{\partial u}{\partial x}\right)_i (x - x_i) + \left(\frac{\partial^2 u}{\partial x^2}\right)_i (x - x_i)^2 + \cdots \tag{2.58}$$

If the shape functions $N_i(x)$ are polynomials of pth degree it is obvious that only the first p terms of the Taylor expansion can be approximated

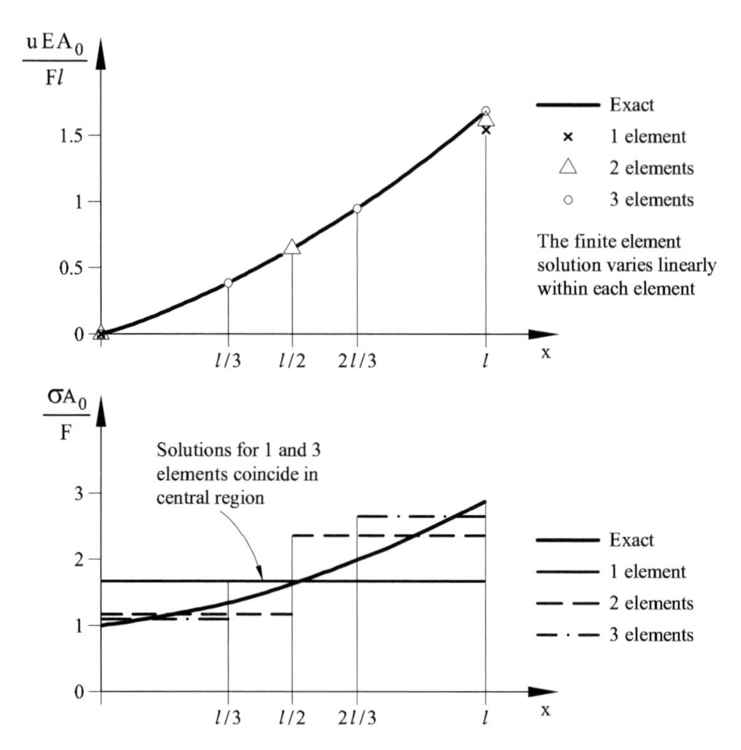

Fig. 2.7 Nodal displacements and axial stress distribution along the rod for the three meshes

exactly as the derivatives of order $p+1$, $p+2$, etc. are zero. The error of this approximation is then of the order of the first term disregarded in the above expansion, i.e.

$$\text{error} = u_{\text{exact}} - u_{\text{approx}} = O(x - x_i)^{p+1} \simeq O(l^{p+1}) \tag{2.59}$$

where $O(l^{p+1})$ is read as "of the order of l^{p+1}" and l is the element length.

Let us now consider two solutions u^1 and u^2 obtained with two meshes of uniform element sizes l and l/d, respectively. We can write

$$\begin{aligned} u_{\text{exact}} - u^1 &= O(l^{p+1}) \\ u_{\text{exact}} - u^2 &= O\Big[(\frac{l}{d})^{p+1}\Big] \end{aligned} \tag{2.60}$$

The approximate value of u_{exact} is obtained in terms of u^1 and u^2 from Eqs.(2.60) as

$$u_{\text{exact}} \cong \frac{(d^{p+1})u^2 - u^1}{(d^{p+1}) - 1} \tag{2.61}$$

This technique is known as Richardson extrapolation [Ral]. For the rod of Figure 2.5 we obtain for the end displacement value:

1. Extrapolated solution from meshes 1 and 2 ($d = 2$)

$$u(l) = \frac{4u^2 - u^1}{3} = 1.7178 \,\frac{Fl}{EA_0} \qquad \text{(Error: 0.03\%)}$$

2. Extrapolated solution using meshes 1 and 3 ($d = 3$)

$$u(l) = \frac{9u^2 - u^1}{8} = 1.71807 \,\frac{Fl}{EA_0} \qquad \text{(Error: 0.012\%)}$$

3. Extrapolated solution using meshes 2 and 3 ($d = 1.5$)

$$u(l) = \frac{(9.5)^2 u^2 - u^1}{(1.5)^2 - 1} = 1.71825 \,\frac{Fl}{EA_0} \qquad \text{(Error: 0.002\%)}$$

Richardson extrapolation is an effective technique to improve the displacement solution obtained from two meshes using elements of the same type. This simple procedure is also applicable for 2D and 3D problems. Obviously, the enhanced nodal displacement values can be used to obtain an improved solution for the stress field. Unfortunately the improvement is not so relevant as for the nodal displacements.

2.8 MATRIX FORMULATION OF THE ELEMENT EQUATIONS

The methodology explained in the previous sections is very useful for introducing the basic steps of the FEM. However, for problems with more than one displacement variable per node, a matrix formulation is much more convenient as it allows all variables and algebraic operations to be grouped together in a compact form. The matrix formulation also provides a systematic finite element methodology for *all* the structural problems treated in this book. The basic concepts of the matrix formulation will be presented next.

Most expressions used henceforth will be referred to an individual element only. Therefore, superindex "e" denoting element values will be omitted hereafter for simplicity, with the exception of a few significative element parameters such as the main geometrical dimensions ($l^{(e)}$, $A^{(e)}$ and $V^{(e)}$), the nodal displacement vector $\mathbf{a}^{(e)}$, the nodal coordinates vector $\mathbf{x}^{(e)}$, the nodal force vectors ($\mathbf{f}^{(e)}, \mathbf{q}^{(e)}$) and the stiffness matrix $\mathbf{K}^{(e)}$. All other parameters, vectors and matrices appearing in the text should be interpreted, unless otherwise mentioned, as referred also to an individual element.

For instance, the components of the displacement vector, the nodal force vectors and the stiffness matrix for the 2-noded rod element are denoted hereonwards as follows

$$\mathbf{a}^{(e)} = \begin{Bmatrix} u_1 \\ u_2 \end{Bmatrix} \ , \ \mathbf{f}^{(e)} = \begin{Bmatrix} f_{x_1} \\ f_{x_2} \end{Bmatrix} \ , \ \mathbf{q}^{(e)} = \begin{Bmatrix} F_{x_1} \\ F_{x_2} \end{Bmatrix} \ , \ \mathbf{K}^{(e)} = \begin{bmatrix} K_{11} & K_{12} \\ K_{21} & K_{22} \end{bmatrix}$$

In above expressions, indexes 1 and 2 refer to *local node numbers* for the element. The omission of the element superindex e in u_i, f_{x_i}, F_{x_i} and K_{ij} will simplify the notation when dealing with problems involving several DOFs per node.

2.8.1 Shape function matrix

Let us consider a general 2-noded rod element. The displacement field is expressed within the element as

$$u = N_1 u_1 + N_2 u_2 \tag{2.62}$$

Eq.(2.62) is written in matrix form as

$$\mathbf{u} = \{u\} = [N_1, N_2] \begin{Bmatrix} u_1 \\ u_2 \end{Bmatrix} = \mathbf{N}\, \mathbf{a}^{(e)} \tag{2.63}$$

where

$$\mathbf{N} = [N_1, N_2] \quad ; \quad \mathbf{a}^{(e)} = \begin{Bmatrix} u_1 \\ u_2 \end{Bmatrix} \tag{2.64}$$

are the *shape function matrix* and the *nodal displacement vector* for the element. Note that the superindex "e" denoting element values has been omitted for most terms in Eqs.(2.62)-(2.64).

2.8.2 Strain matrix

The strain vector contains the axial elongation and is written as

$$\boldsymbol{\varepsilon} = \{\varepsilon\} = \begin{Bmatrix} \frac{du}{dx} \end{Bmatrix} = \begin{Bmatrix} \frac{dN_1}{dx} u_1 + \frac{dN_2}{dx} u_2 \end{Bmatrix} = \begin{bmatrix} \frac{dN_1}{dx} \ , \ \frac{dN_2}{dx} \end{bmatrix} \begin{Bmatrix} u_1 \\ u_2 \end{Bmatrix} = \mathbf{B}\mathbf{a}^{(e)} \tag{2.65}$$

where

$$\mathbf{B} = \begin{bmatrix} \frac{dN_1}{dx} \ , \ \frac{dN_2}{dx} \end{bmatrix} \tag{2.66}$$

is the *strain matrix* for the element.

2.8.3 Constitutive matrix

The *stress vector* contains the axial force and is expressed as

$$\boldsymbol{\sigma} = \{\mathcal{N}\} = [EA]\,\varepsilon = \mathbf{D}\mathbf{B}\mathbf{a}^{(e)} \tag{2.67}$$

where

$$\mathbf{D} = [EA] \tag{2.68}$$

is the matrix of mechanical properties of the material, also called hereafter *constitutive matrix.*

For the axial rod problem vectors $\boldsymbol{\varepsilon}$ and $\boldsymbol{\sigma}$ and matrix $\mathbf{D}$ have a single component only. In general $\boldsymbol{\sigma}$ and $\boldsymbol{\varepsilon}$ will have t components. Thus, if n is the number of nodes of an individual finite element and d the number of DOFs for each node, the dimensions of the vectors and matrices in the constitutive equation are

$$\underset{t\times 1}{\boldsymbol{\sigma}} = \underset{t\times t}{\mathbf{D}} \cdot \underset{[t\times(n\times d)]}{\mathbf{B}} \cdot \underset{[(n\times d)\times 1]}{\mathbf{a}^{(e)}} \tag{2.69}$$

2.8.4 Principle of Virtual Work

The PVW for an individual element is written in matrix form as

$$\int_{l^{(e)}} \delta\boldsymbol{\varepsilon}^T\boldsymbol{\sigma}\,dx = \int_{l^{(e)}} \delta\mathbf{u}^T\mathbf{t}\,dx + [\delta\mathbf{a}^{(e)}]^T\,\mathbf{q}^{(e)} \tag{2.70a}$$

with

$$\begin{aligned} &\delta\boldsymbol{\varepsilon} = \{\delta\varepsilon\} \quad , \quad \delta\mathbf{u} = \{\delta u\} \quad , \quad \delta\mathbf{a}^{(e)} = [\delta u_1, \delta u_2]^T \\ &\mathbf{t} = \{t_x\} \quad \text{and} \quad \mathbf{q}^{(e)} = [F_{x_1}, F_{x_2}]^T = [-\mathcal{N}_1, \mathcal{N}_2]^T \end{aligned} \tag{2.70b}$$

In above $\delta\mathbf{u}$ and $\delta\boldsymbol{\varepsilon}$ are the virtual displacement vector and the virtual strain vector, respectively, $\delta\mathbf{a}^{(e)}$ is the virtual nodal displacement vector for the element, $\mathbf{t}$ is the distributed load vector and $\mathbf{q}^{(e)}$ is the equilibrating nodal force vector. Once again we recall that the components of $\mathbf{q}^{(e)}$ coincide with the appropriate sign with the axial forces at the element nodes (Figure 2.2).

The PVW is a *scalar equation*, i.e. both sides of Eq.(2.70a) are numbers representing the internal and external virtual work, respectively. This explains the organization of the terms in Eq.(2.70a), as a scalar number is obtained as product of a row vector times a column vector, i.e. if s is a scalar number we can write

$$s = a_1 b_1 + \cdots + a_n b_n = [a_1, a_2, \ldots, a_n] \begin{Bmatrix} b_1 \\ b_2 \\ \vdots \\ b_n \end{Bmatrix} = \mathbf{a}^T \mathbf{b} \tag{2.71}$$

Naturally, if vectors $\boldsymbol{\varepsilon}$ and $\boldsymbol{\sigma}$ have a single term, as in the axially loaded rod problem, the vector product (2.71) reduces to multiplying two numbers. Vectors $\boldsymbol{\varepsilon}$ and $\boldsymbol{\sigma}$ typically have several components and the matrix form of the PVW of Eq.(2.70a) will be used.

2.8.5 Stiffness matrix and equivalent nodal force vector

From Eqs.(2.63) and (2.65) we have

$$\begin{aligned} [\delta \mathbf{u}]^T &= [\delta \mathbf{a}^{(e)}]^T \mathbf{N}^T \\ [\delta \boldsymbol{\varepsilon}]^T &= [\delta \mathbf{a}^{(e)}]^T \mathbf{B}^T \end{aligned} \tag{2.72}$$

Substituting Eqs.(2.65), (2.67) and (2.72) into the PVW *written for a single element* gives

$$\int_{l^{(e)}} [\delta \mathbf{a}^{(e)}]^T \mathbf{B}^T \boldsymbol{\sigma} \, dx - \int_{l^{(e)}} [\delta \mathbf{a}^{(e)}]^T \mathbf{N}^T \mathbf{t} \, dx = \delta \mathbf{a}^{(e)} \mathbf{q}^{(e)} \tag{2.73}$$

where $\mathbf{t} = \{t_x\}$ is the vector of distributed axial loads acting on the element. Collecting the virtual displacements in Eq.(2.73) yields

$$[\delta \mathbf{a}^{(e)}]^T \left[\int_{l^{(e)}} \mathbf{B}^T \boldsymbol{\sigma} \, dx - \int_{l^{(e)}} \mathbf{N}^T \mathbf{t} \, dx - \mathbf{q}^{(e)} \right] = \mathbf{0} \tag{2.74}$$

As the virtual displacements are arbitrary, satisfaction of Eq.(2.74) implies

$$\int_{l^{(e)}} \mathbf{B}^T \boldsymbol{\sigma} \, dx - \int_{l^{(e)}} \mathbf{N}^T \mathbf{t} \, dx = \mathbf{q}^{(e)} \tag{2.75a}$$

Substituting now the constitutive equation for $\boldsymbol{\sigma}$ (Eq.(2.67)) into Eq.(2.75a) gives

$$\left(\int_{l^{(e)}} \mathbf{B}^T \mathbf{D} \mathbf{B} \, dx \right) \mathbf{a}^{(e)} - \int_{l^{(e)}} \mathbf{N}^T \mathbf{t} \, dx = \mathbf{q}^{(e)} \tag{2.75b}$$

Eq.(2.75) is a system of algebraic equations which can be written in the standard form

$$\mathbf{K}^{(e)} \mathbf{a}^{(e)} - \mathbf{f}^{(e)} = \mathbf{q}^{(e)} \tag{2.76a}$$

where

$$\boxed{\begin{aligned}\mathbf{K}^{(e)} &= \int_{l^{(e)}} \mathbf{B}^T\mathbf{D}\mathbf{B}\, dx\\ \mathbf{f}^{(e)} &= \int_{l^{(e)}} \mathbf{N}^T\mathbf{t}\, dx\end{aligned}} \tag{2.76b}$$

are respectively the stiffness matrix and the equivalent nodal force due to distributed loading for the element. Vector $\mathbf{q}^{(e)}$ in Eq.(2.76a) is the equilibrating nodal force vector for the element which is used for the global assembly process.

The above derivation of $\mathbf{K}^{(e)}$ and $\mathbf{f}^{(e)}$ is *completely general*. Expressions (2.76) will frequently appear throughout the book and will be particularized for each element.

The explicit form of $\mathbf{K}^{(e)}$ and $\mathbf{f}^{(e)}$ for the 2-noded axially loaded rod element is found by substituting into Eqs.(2.76b) the adequate expressions for $\mathbf{B}, \mathbf{D}, \mathbf{N}$ and $\mathbf{t}$. In this case we have

$$\begin{aligned}\mathbf{N} &= [\mathbf{N}_1, \mathbf{N}_2] = [N_1, N_2] = \left[\frac{x_2 - x}{l^{(e)}}, \frac{x - x_1}{l^{(e)}}\right]\\ \mathbf{B} &= [\mathbf{B}_1, \mathbf{B}_2] = \left[\frac{dN_1}{dx}, \frac{dN_2}{dx}\right] = \left[-\frac{1}{l^{(e)}}, \frac{1}{l^{(e)}}\right]\\ &\mathbf{D} = [EA] \qquad \text{and} \qquad \mathbf{t} = \{t_x\}\end{aligned} \tag{2.77}$$

Substituting Eqs.(2.77) into (2.76*b*) gives

$$\begin{aligned}\mathbf{K}^{(e)} &= \int_{l^{(e)}} \begin{Bmatrix} -\frac{1}{l^{(e)}} \\ \frac{1}{l^{(e)}} \end{Bmatrix} (EA)\left[-\frac{1}{l^{(e)}}, \frac{1}{l^{(e)}}\right] dx = \left(\frac{EA}{l}\right)^{(e)} \begin{bmatrix} 1 & -1 \\ -1 & 1 \end{bmatrix}\\ \mathbf{f}^{(e)} &= \begin{Bmatrix} f_{x1} \\ f_{x2} \end{Bmatrix} = \int_{l^{(e)}} \begin{Bmatrix} x_2 - x \\ x - x_1 \end{Bmatrix} \left(\frac{t_x}{l}\right)^{(e)} dx = \frac{(lt_x)^{(e)}}{2} \begin{Bmatrix} 1 \\ 1 \end{Bmatrix}\end{aligned} \tag{2.78}$$

Note the coincidence of these expressions with those obtained in Eq.(2.56).

Nodal computation of $K^{(e)}$ and $f^{(e)}$

It is interesting and useful that the element stiffness matrix and the equivalent nodal force vector can be obtained from the corresponding submatrices and subvectors.

Thus, from Eqs.(2.77) and (2.76b) we have

$$\mathbf{K}^{(e)} = \int_{l^{(e)}} \begin{Bmatrix} \mathbf{B}_1^T \\ \mathbf{B}_2^T \end{Bmatrix} \mathbf{D}[\mathbf{B}_1, \mathbf{B}_2]\, dx = \int_{l^{(e)}} \begin{bmatrix} \mathbf{B}_1^T\mathbf{D}\mathbf{B}_1 & \vdots & \mathbf{B}_1^T\mathbf{D}\mathbf{B}_2 \\ \cdots & \cdots & \cdots \\ \mathbf{B}_2^T\mathbf{D}\mathbf{B}_1 & \vdots & \mathbf{B}_2^T\mathbf{D}\mathbf{B}_2 \end{bmatrix} dx =$$

$$= \begin{bmatrix} \mathbf{K}_{11}^{(e)} & \mathbf{K}_{12}^{(e)} \\ \mathbf{K}_{21}^{(e)} & \mathbf{K}_{22}^{(e)} \end{bmatrix}$$

$$\mathbf{f}^{(e)} = \begin{Bmatrix} f_{x_1} \\ f_{x_2} \end{Bmatrix} = \int_{l^{(e)}} \begin{Bmatrix} \mathbf{N}_1^T \\ \mathbf{N}_2^T \end{Bmatrix} \mathbf{t}\, dx = \int_{l^{(e)}} \begin{Bmatrix} \mathbf{N}_1^T\mathbf{t} \\ \mathbf{N}_2^T\mathbf{t} \end{Bmatrix} dx \tag{2.79}$$

Matrix $\mathbf{K}_{ij}^{(e)}$ relating nodes i and j of element e is

$$\underset{d\times d}{\mathbf{K}_{ij}^{(e)}} = \int_{l^{(e)}} \underset{(d\times t)}{\mathbf{B}_i^T} \underset{(t\times t)}{\mathbf{D}} \underset{(t\times d)}{\mathbf{B}_j}\, dx \quad ; \quad i,j = 1,2 \tag{2.80}$$

and the equivalent nodal force vector for node i of element e is

$$\underset{(d\times 1)}{\mathbf{f}_i^{(e)}} = \int_{l^{(e)}} \underset{(d\times d)}{\mathbf{N}_i^T} \underset{(d\times 1)}{\mathbf{t}}\, dx \qquad i = 1,2 \tag{2.81}$$

Recall that d is the number of DOFs for each node (i.e. $d = 1$ for the axially loaded rod). For the 2-noded rod element

$$\begin{aligned} K_{ij}^{(e)} &= \int_{l^{(e)}} \frac{dN_i}{dx}\, EA\, \frac{dN_j}{dx}\, dx = (-1)^{i+j} \left(\frac{EA}{l}\right)^{(e)} \\ f_i^{(e)} &= f_{x_i} = \int_{l^{(e)}} N_i t_x\, dx = \frac{(lt_x)^{(e)}}{2} \end{aligned} \tag{2.82}$$

from which the expressions of $\mathbf{K}^{(e)}$ and $\mathbf{f}^{(e)}$ of Eq.(2.78) can be obtained.

The computation of the element stiffness matrix $\mathbf{K}^{(e)}$ and the equivalent nodal force vector $\mathbf{f}^{(e)}$ from the nodal contributions $\mathbf{K}_{ij}^{(e)}$ and $\mathbf{f}_i^{(e)}$ is simple and economical and it facilitates the organization of a computer program. We will verify this on many occasions throughout the book.

The global stiffness matrix $\mathbf{K}$ and the equivalent nodal force vector $\mathbf{f}$ are assembled from the element contributions in the standard manner into the global system $\mathbf{Ka} = \mathbf{f}$.

Once the nodal displacements $\mathbf{a}$ have been found the reactions at the prescribed nodes can be computed by Eq.(2.26a) or, what is usually more convenient, by Eq.(2.26b), with the following general expression for the internal nodal force vector for the element

$$\mathbf{f}_{\text{int}}^{(e)} = \int_{l^{(e)}} \mathbf{B}^T \boldsymbol{\sigma}\, dx \tag{2.83}$$

Eq.(2.83) is deduced from the first integral in the l.h.s. of Eq.(2.75a).

2.9 SUMMARY OF THE STEPS FOR THE ANALYSIS OF A STRUCTURE USING THE FEM

Let us summarize the main steps to be followed for the finite element analysis of a structure.

Step 1. Discretize the structure into a mesh of finite elements.

Step 2. Compute for each element the stiffness matrix and the equivalent nodal force vector due to external loads using expressions of the type

$$\mathbf{K}^{(e)} = \int_{l^{(e)}} \mathbf{B}^T \mathbf{D} \mathbf{B}\, dx \quad ; \quad \mathbf{K}_{ij}^{(e)} = \int_{l^{(e)}} \mathbf{B}_i^T \mathbf{D} \mathbf{B}_j\, dx$$

$$\mathbf{f}^{(e)} = \int_{l^{(e)}} \mathbf{N}^T \mathbf{t}\, dx \quad ; \quad \mathbf{f}_i^{(e)} = \int_{l^{(e)}} \mathbf{N}_i^T \mathbf{t}\, dx \tag{2.84}$$

For two and three dimensional structures the element integrals are computed over the element area and volume, respectively.

Step 3. Assemble the stiffness matrix and the equivalent nodal force vector for each element into the global system

$$\mathbf{K}\mathbf{a} = \mathbf{f} \tag{2.85a}$$

$$\mathbf{K} = \underset{e}{\mathrm{A}}\, \mathbf{K}^{(e)} \quad ; \quad \mathbf{f} = \underset{e}{\mathrm{A}}\, \mathbf{f}^{(e)} + \mathbf{p} + \mathbf{r} \tag{2.85b}$$

where $\underset{e}{\mathrm{A}}$ denotes the operator for the global assembly of all the individual matrices and vectors for each element in the mesh. In Eq.(2.85b) $\mathbf{p}$ is the vector of external point forces acting at the nodes and $\mathbf{r}$ is the vector of nodal reaction to be computed "a posteriori" once the nodal displacement are found.

The assembly of the reaction vector $\mathbf{r}$ into $\mathbf{f}$ is optional, as the reactions do not influence the solution for the nodal displacements [Li,Pr].

Step 4. The nodal displacements are computed by solving the equation system (2.85a) where the prescribed displacements must be adequately imposed, i.e.

$$\mathbf{a} = \mathbf{K}^{-1}\mathbf{f} \tag{2.86}$$

The nodal reactions are obtained at the prescribed nodes.

Step 5. The strains and stresses are computed within each element from the nodal displacements as

$$\boldsymbol{\varepsilon} = \mathbf{B}\,\mathbf{a} \quad ; \quad \boldsymbol{\sigma} = \mathbf{D}\,\mathbf{B}\,\mathbf{a} \tag{2.87}$$

The nodal axial forces for each element can be computed from

$$\mathbf{q}^{(e)} = [-\mathcal{N}_1^{(e)}, \mathcal{N}_2^{(e)}] = \mathbf{K}^{(e)}\mathbf{a}^{(e)} - \mathbf{f}^{(e)} \tag{2.88}$$

Details of above steps and of the precise form of the element vectors and matrices will be given for each of the structures studied in the book.

3

ADVANCED ROD ELEMENTS AND REQUIREMENTS FOR THE NUMERICAL SOLUTION

3.1 INTRODUCTION

The analysis of the simple axially loaded rod problem using the 2-noded rod element studied in the previous chapter is of big interest as it summarises the basic steps for the analysis of a structure by the FEM. However, a number of important questions still remain unanswered, such as: Can higher order rod elements be effectively used? What are their advantages versus the simpler 2-noded rod element? Can it be guaranteed that the numerical solution converges to the exact one as the mesh is refined? What are the conditions influencing the error in the numerical solution? The reader who faces the application of the FEM for the first time will certainly come across these and similar questions. In this chapter we will see that there are not definitive answers for many of the questions, and in some cases only some practical hints are possible. For simplicity we will mostly refer to the axially loaded rod problem as it allows a simple explanation of topics which are of general applicability to more complex problems.

The chapter is organized as follows. In the next section the derivation of the one-dimensional (1D) shape functions is presented. Such functions are very useful for obtaining the shape functions for two- (2D) and three- (3D) dimensional elements in the next chapters. An example of the derivation of the relevant matrix expressions for a quadratic 3-noded rod element is given. The concepts of isoparametric element and numerical integration are presented next. These concepts are essential for the derivation of high-order 2D and 3D elements. Finally, the requirements for the convergence of the numerical solution are discussed, together with a description of the more usual solution errors.

3.2 ONE DIMENSIONAL C^0 ELEMENTS. LAGRANGE ELEMENTS

In the previous chapter we introduced the basic concepts of the FEM using simple 2-noded 1D elements with linear shape functions. The polynomial interpolation guarantees that the axial displacement is continuous within the element and between adjacent elements. Elements satisfying this condition are termed "C^o continuous". Additionally, we could require continuity of the first derivative of the displacement and the approximation is then called "C^1 continuous". In general, an element is "C^k continuous" if the displacement field has continuous the $k-1$ first derivatives. In Section 3.8.3 we will come back to this subject. In this section we will derive the shape functions for C^o continuous 1D elements. These ideas will be very useful for deriving the shape functions of 2D elements in Chapter 5.

The approximation of a displacement unknown in 1D elements can be written as

$$u(x) = \alpha_0 + \alpha_1 x + \alpha_2 x^2 \ + \cdots \tag{3.1}$$

where α_0, α_1, etc., are constant parameters.

Let us choose a first degree polynomial (for example, the approximation introduced in Section 2.3)

$$u(x) = \alpha_0 + \alpha_1 x \tag{3.2}$$

The parameters α_0 and α_1 can be obtained from the value of $u(x)$ at two element points. This requires the element associated with the interpolation (3.2) to have two nodes. For a 2-noded element of length $\ell^{(e)}$ with node 1 at $x = x_1$ and node 2 at $x = x_2$ (Figure 3.1), we have

$$\begin{aligned} u(x_1) &= u_1 = \alpha_0 + \alpha_1 x_1 \\ u(x_2) &= u_2 = \alpha_0 + \alpha_1 x_2 \end{aligned} \tag{3.3}$$

where u_1 and u_2 are the values of the axial displacement at the two nodes. Substituting the values of α_o and α_1 obtained from Eq.(3.3) into Eq.(3.1) gives

$$u(x) = N_1(x)u_1 + N_2(x)u_2 \tag{3.4a}$$

where

$$N_1(x) = \frac{(x_2 - x)}{l^{(e)}} \quad ; \quad N_2(x) = \frac{(x - x_1)}{l^{(e)}} \tag{3.4b}$$

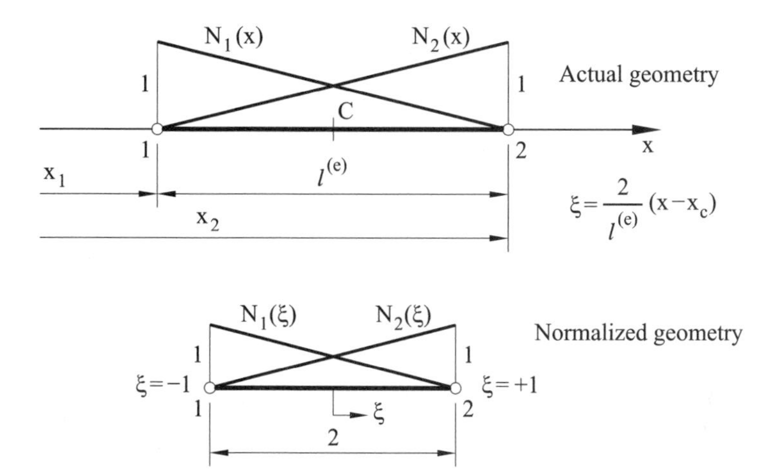

Fig. 3.1 Definition of the natural coordinate system ξ. Actual and normalized geometries for a 2-noded element

are the element shape functions. Note the coincidence with the expressions obtained in the previous chapter (see Eq.(2.12)).

The shape functions for C^o continuous 1D elements can be simply derived from the expressions of Lagrange polynomials. A $n-1th$ degree Lagrange polynomial $\ell_i^n(\mathrm{x})$ is defined in terms of n points with coordinates $x_1, x_2, \cdots x_n$ as follows

$$\ell_i^n(x) = (x-x_1)(x-x_2)\cdots(x-x_{i-1})(x-x_{i+1})\cdots(x-x_n) \tag{3.5a}$$

Note that $\ell_i^n(x_i) = y_i(\neq 0)$ and $\ell_i^n(x_j) = 0$ for $j = 1, 2, \cdots n (j \neq i)$. If the points coincide with the element nodes and the non-zero value y_i is normalized to the unity, the resulting normalized Lagrange polynomial is

$$l_i^n(x) = \frac{\ell_i^n(x)}{\ell_i^n(x_i)} = \prod_{j=1(j\neq i)}^{n} \left(\frac{x-x_j}{x_i-x_j}\right) \tag{3.5b}$$

The shape function N_i of a Lagrange element with n nodes coincides with the normalized Lagrange polynomial, i.e.

$$N_i(x) = l_i^n(x) \tag{3.6}$$

This explains why C° continuous 1D elements are also called Lagrange elements.

For a two-noded element we find again that

$$
\begin{aligned}
N_1 &= \frac{x - x_2}{x_1 - x_2} = \frac{x_2 - x}{l^{(e)}} \\
N_2 &= \frac{x - x_1}{x_2 - x_1} = \frac{x - x_1}{l^{(e)}}
\end{aligned}
\tag{3.7}
$$

A *natural coordinate* ξ is introduced for convenience as (Figure 3.1)

$$
\xi = 2\frac{x - x_c}{l^{(e)}} \tag{3.8}
$$

where x_c is the cartesian coordinate of the element midpoint. Eq. (3.8) gives

$\xi = -1$ at the left-hand end of the element
$\xi = \ \ 0$ at the element mid point
$\xi = \ \ 1$ at the right-hand end of the element

Eq. (3.8) transforms the actual element geometry into a *normalized geometry* of length equal to 2. The shape functions can now be written in terms of the natural coordinate ξ. By analogy with Eq.(3.6) we write

$$
N_i(\xi) = l_i^n(\xi) = \prod_{j=1(j\neq i)}^{n} (\frac{\xi - \xi_j}{\xi_i - \xi_j}) \tag{3.9}
$$

For a linear Lagrange rod element with two nodes at $\xi = -1$ and $\xi = +1$ we obtain

$$
\begin{aligned}
N_1 &= \frac{\xi - \xi_2}{\xi_1 - \xi_2} = \frac{1}{2}\,(1 - \xi) \\
N_2 &= \frac{\xi - \xi_1}{\xi_2 - \xi_1} = \frac{1}{2}\,(1 + \xi)
\end{aligned}
\tag{3.10}
$$

For a quadratic Lagrange rod element with three nodes at $\xi_1 = -1$, $\xi = 0$ and $\xi = +1$ (Figure 3.2) the shape functions are

$$
\begin{aligned}
N_1 &= \frac{(\xi - \xi_2)(\xi - \xi_3)}{(\xi_1 - \xi_2)} = \frac{1}{2}\xi(\xi - 1) \\
N_2 &= \frac{(\xi - \xi_1)(\xi - \xi_3)}{(\xi_2 - \xi_1)(\xi_2 - \xi_3)} = (1 - \xi^2) \\
N_3 &= \frac{(\xi - \xi_1)(\xi - \xi_2)}{(\xi_3 - \xi_1)(\xi_3 - \xi_2)} = \frac{1}{2}\xi(1 + \xi)
\end{aligned}
\tag{3.11}
$$

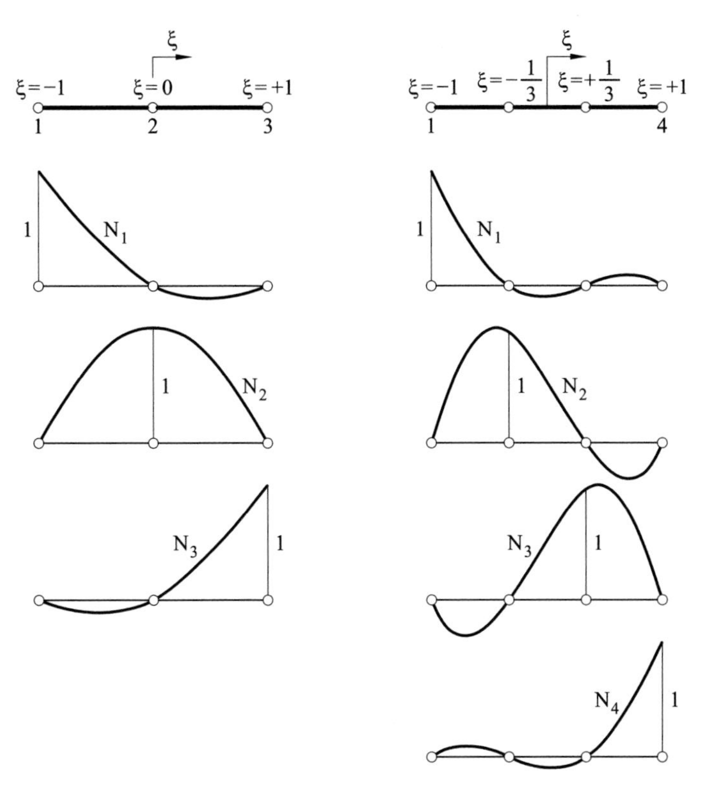

Fig. 3.2 Quadratic and cubic 1D elements with C^o continuity

For a cubic rod element with four nodes at $\xi_1 = -1$, $\xi_2 = -1/3$, $\xi_3 = 1/3$ and $\xi_4 = +1$ (Figure 3.2) the shape functions are

$$
\begin{aligned}
N_1 &= \frac{(\xi-\xi_2)(\xi-\xi_3)(\xi-\xi_4)}{(\xi_1-\xi_2)(\xi_1-\xi_3)(\xi_1-\xi_4)} = -\frac{9}{16}\left(\xi^2-\frac{1}{9}\right)(\xi-1)\\
N_2 &= \frac{(\xi-\xi_1)(\xi-\xi_3)(\xi-\xi_4)}{(\xi_2-\xi_1)(\xi_2-\xi_3)(\xi_2-\xi_4)} = \frac{27}{16}\left(\xi-\frac{1}{3}\right)(\xi^2-1)\\
N_3 &= \frac{(\xi-\xi_1)(\xi-\xi_2)(\xi-\xi_4)}{(\xi_3-\xi_1)(\xi_3-\xi_2)(\xi_3-\xi_4)} = -\frac{27}{16}\left(\xi+\frac{1}{3}\right)(\xi^2-1)\\
N_4 &= \frac{(\xi-\xi_1)(\xi-\xi_2)(\xi-\xi_3)}{(\xi_4-\xi_1)(\xi_4-\xi_2)(\xi_4-\xi_3)} = \frac{9}{16}(\xi+1)\left(\xi^2-\frac{1}{9}\right)
\end{aligned}
\tag{3.12}
$$

The cartesian expressions of the above shape functions can be obtained from the transformation (3.8). However, only the normalized forms are usually necessary in practice.

The reader is encouraged to derive by him/herself the expressions of the shape functions for higher order 1D Lagrange elements.

The shape functions for C^1 continuous 1D elements will be derived in Chapter 1 of Volume 2 when dealing with Euler-Bernouilli beams [On].

3.3 ISOPARAMETRIC FORMULATION AND NUMERICAL INTEGRATION

3.3.1 Introduction

We will now introduce two key concepts which have been essential for the development of the FEM. The first one is that of *isoparametric interpolation.* The basic idea is to interpolate the element geometry from the coordinates of the nodes. Such an interpolation yields a general relationship between the natural and cartesian coordinates.

The second concept is that of *numerical integration.* In most cases the exact analytical computation of the element integrals is not possible and numerical integration is the only option to evaluate them in a simple and precise way.

The application of these two techniques to C^o continuous 1D rod elements is presented in the next sections. The advantages of these procedures will become clearer when dealing with 2D and 3D elements.

3.3.2 The concept of parametric interpolation

Let us recall the displacement interpolation for a 2-noded axial rod element:

$$u(\xi) = N_1(\xi)u_1 + N_2(\xi)u_2 \tag{3.13}$$

In Eq.(3.13) we have used the expression of the shape functions in terms of the natural coordinate ξ. With a few exceptions this will be the usual procedure throughout the book.

The axial strain in the rod element is obtained from Eq.(3.13) as

$$\varepsilon = \frac{du}{dx} = \frac{dN_1(\xi)}{dx} u_1 + \frac{dN_2(\xi)}{dx} u_2 \tag{3.14}$$

The cartesian derivatives of the shape functions are therefore needed to compute the strain. This would be an easy task if the shape functions were expressed in terms of the cartesian coordinate x. However, as this will not generally be the case, some transformations are necessary. For the

1D problem we have

$$\begin{aligned}\frac{dN_1(\xi)}{dx} &= \frac{dN_1(\xi)}{d\xi}\frac{d\xi}{dx} = \frac{d}{d\xi}\left(\frac{1-\xi}{2}\right)\frac{d\xi}{dx} = -\frac{1}{2}\frac{d\xi}{dx}\\ \frac{dN_2(\xi)}{dx} &= \frac{dN_2(\xi)}{d\xi}\frac{d\xi}{dx} = \frac{d}{d\xi}\left(\frac{1+\xi}{2}\right)\frac{d\xi}{dx} = \frac{1}{2}\frac{d\xi}{dx}\end{aligned} \tag{3.15}$$

and the strain is obtained by

$$\varepsilon = -\frac{1}{2}\left(\frac{d\xi}{dx}\right)u_1 + \frac{1}{2}\left(\frac{d\xi}{dx}\right)u_2 \tag{3.16}$$

Eq.(3.16) involves the evaluation of $\frac{d\xi}{dx}$. This requires an explicit relationship between ξ and x which can be obtained using a parametric interpolation of the element geometry. This expresses the coordinate of any point within the element in terms of the coordinates of m element points $x_1, x_2, \cdots, x_m$ by the following interpolation

$$x = \hat{N}_1(\xi)x_1 + \hat{N}_2(\xi)x_1 \ + \cdots + \hat{N}_m(\xi)x_m \tag{3.17}$$

In Eq.(3.17) $\hat{N}_i(\xi)$ are *geometry interpolation functions* which satisfy the same requirements as the displacement shape functions; i.e. $\hat{N}_i(\xi)$ takes the value one at point i and zero at the other $m-1$ points for which the coordinates are known. Hence, the expression for $\hat{N}_i(\xi)$ can be obtained simply by changing n for m in Eq.(3.9), where ξ_i are the natural coordinates of the geometry interpolating points.

Eq. (3.17) yields precisely the relationship we are looking for between the coordinates ξ and x. This expression can also be interpreted as a transformation between the coordinates ξ and x, such that every point in the normalized space [-1,1] is mapped onto another point in the cartesian space $[x_1, x_2]$. It is essential that this mapping be unique and this generally depends on the element geometry. This issue will be discussed in some detail when studying 2D isoparametric elements in Chapter 6.

Example 3.1: Parametric interpolation of a cubic polinomial.

- ***Solution***

Let us consider the polynomial $y = x^3 - 2x^2 - x + 4$ plotted in Figure 3.3. Such a function can represent, for instance, the geometry of a curved beam or the boundary of a curved 2D element. We will assume that the coordinates

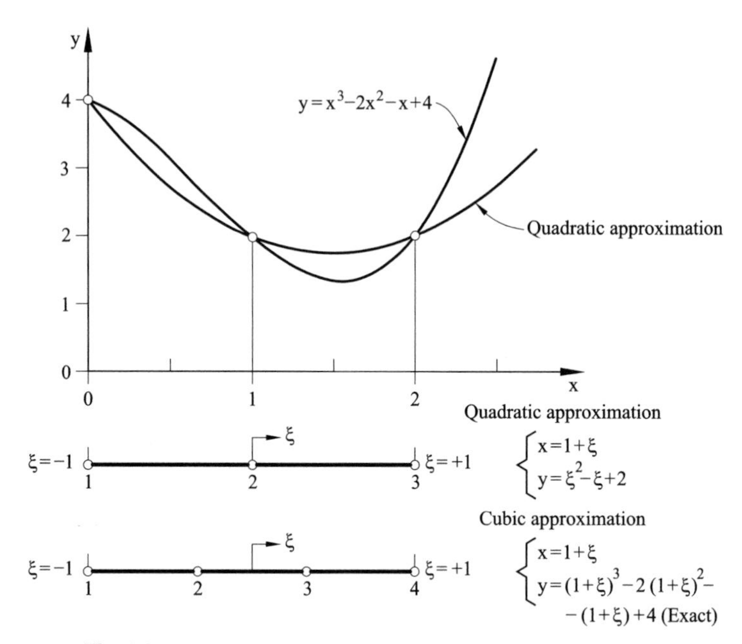

Fig. 3.3 Parametric interpolation of a cubic polynomial

of the three points at $x_1 = 0$, $x_2 = 1$, and $x_3 = 2$ are known.

The coordinates of the three points will be used to derive a quadratic approximation using a 3-noded 1D element. The relationship between the cartesian (x, y) coordinates and the natural coordinate ξ is obtained as a particular case of Eq.(3.17), i.e.

$$x = \sum_{i=1}^{3} N_i(\xi)x_i = \frac{1}{2}\xi(\xi-1)x_1 + (1-\xi^2)x_2 + \frac{1}{2}\xi(\xi+1)x_3 = 1-\xi$$

$$y = \sum_{i=1}^{3} N_i(\xi)y_i = \frac{1}{2}\xi(\xi-1)y_1 + (1-\xi^2)y_2 + \frac{1}{2}\xi(1+\xi)y_3 = \xi^2-\xi+2$$

Figure 3.3 shows the approximating quadratic function. Note the error with respect to the "exact" cubic function. Also note that this error is much larger outside the interval [0,2] which includes the three points selected.

The accuracy can be dramatically improved by using a cubic approximation in terms of the coordinates of four known points at $x_1 = 0$, $x_2 = 2/3$, $x_3 = 4/3$, and $x_4 = 2.0$, with $y(x_1) = 4.0$, $y(x_2) = 74/27$, $y(x_3) = 40/27$, and $y(x_4) = 2.0$, respectively. A cubic 1D element is now used giving

$$x = \sum_{i=1}^{4} N_i(\xi)x_i \quad ; \quad y = \sum_{i=1}^{4} N_i(\xi)y_i$$

where N_i are the cubic shape function of Eq.(3.12). After some easy algebra, the following is obtained

$$x = 1 + \xi \quad ; \quad y = (1+\xi)^3 - 2(1+\xi)^2 - (1+\xi) + 4$$

The reader can verify that the cubic field chosen exactly approximates the original cubic function, as expected.

Example 3.1 shows that important errors in the geometry approximation can occur unless a correct interpolation of the geometry is chosen. These errors are undesirable and should be avoided or, at least, minimized.

Two types of points must therefore be considered in an element: a) the n points used for interpolating the displacement field (nodes) by the shape functions $N_i(\xi)$; and b) the m points chosen for approximating the element geometry via the geometry interpolation functions $\hat{N}_i(\xi)$. These two sets of points can coincide depending on the problem. Complex structures might require a higher order interpolation of the geometry, whereas a simple geometry can be exactly approximated using a linear field for $\hat{N}_i$, independently of the interpolation used for the displacement field.

If the number of geometry points m is greater than that of element nodes, the geometry interpolation functions $\hat{N}_i$ will be polynomials of a higher degree than the displacement shape functions, and the element is termed *superparametric.* If m coincides with the number of nodes, then $N_i \equiv \hat{N}_i$ and the element is *isoparametric.* Finally, if the number of geometry points is less than that of nodes, the element is called *subparametric.*

In practice it is usual to choose an isoparametric formulation. However, it is important to have a clear picture of the two other options which are useful in some cases.

Isoparametric elements originate from the work of Taig [Ta,TK] who derived the first 4-noded isoparametric quadrilateral. Irons [IA,Ir] extended these ideas to formulate high order isoparametric elements. In Chapters 6 and 8 we will study 2D and 3D isoparametric elements.

3.3.3 Isoparametric formulation of the two-noded rod element

The geometry of the linear rod element is expressed in terms of the coordinates of the two nodes as

$$x(\xi) = N_1(\xi)x_1 + N_2(\xi)x_2 \tag{3.18}$$

where N_1 and N_2 are the same linear shape functions used for interpolating the displacement field (see Eq.(3.10)).

From Eq.(3.18) we obtain

$$\frac{dx}{d\xi} = \frac{dN_1}{d\xi}x_1 + \frac{dN_2}{d\xi}x_2 = -\frac{1}{2}x_1 + \frac{1}{2}x_2 = \frac{l^{(e)}}{2} \tag{3.19}$$

and

$$dx = \frac{l^{(e)}}{2}\, d\xi \quad \text{y} \quad \frac{d\xi}{dx} = \frac{2}{l^{(e)}} \tag{3.20}$$

Substituting Eq.(3.20) into (3.15) gives

$$\begin{aligned}\frac{dN_1}{dx} &= \frac{2}{l^{(e)}}\frac{dN_1}{d\xi} = -\frac{1}{l^{(e)}}\\ \frac{dN_2}{dx} &= \frac{2}{l^{(e)}}\frac{dN_2}{d\xi} = \frac{1}{l^{(e)}}\end{aligned} \tag{3.21}$$

and from Eqs.(3.21) and (3.14)

$$\varepsilon = \left[-1/l^{(e)},\ 1/l^{(e)}\right] \mathbf{a}^{(e)} = \mathbf{B}\ \mathbf{a}^{(e)} \tag{3.22a}$$

with

$$\mathbf{B} = \left[-1/l^{(e)}, 1/l^{(e)}\right] \tag{3.22b}$$

which naturally coincides with the expressions previously derived by a more direct procedure. The systematic approach chosen here is useful in order to understand the application of the isoparametric concept.

The stiffness matrix and the equivalent nodal force vector are expressed in the natural coordinate system combining Eqs.(3.20) and (2.76b) as

$$\mathbf{K}^{(e)} = \int_{-1}^{+1} \mathbf{B}^T (EA)\mathbf{B}\ \frac{l^{(e)}}{2}\ d\xi \quad , \quad \mathbf{f}^{(e)} = \int_{-1}^{+1} \mathbf{N}^T \mathbf{t}\ \frac{l^{(e)}}{2}\ d\xi \tag{3.23}$$

For homogeneous material and uniformly distributed loading the computation of the above integrals is simple, leading to the expressions (2.78) in the previous chapter.

3.3.4 Isoparametric formulation of the 3-noded quadratic rod element

We will now study the 3-noded rod element of Figure 3.2 with quadratic shape functions. The axial displacement is expressed by

$$u = N_1(\xi)u_1 + N_2(\xi)u_2 + N_3(\xi)u_3 \tag{3.24}$$

where the shape functions $N_1(\xi)$, $N_2(\xi)$ and $N_3(\xi)$are given by Eq.(3.11).

The x coordinate of a point within the element is written in the isoparametric formulation as

$$x = N_1(\xi)x_1 + N_2(\xi)x_2 + N_3(\xi)x_3 \tag{3.25}$$

The axial strain is obtained by

$$\varepsilon = \frac{du}{dx} = \sum_{i=1}^{3} \frac{dN_i}{d\xi} u_i = \left[\frac{dN_1}{d\xi}\frac{d\xi}{dx}, \frac{dN_2}{d\xi}\frac{d\xi}{dx}, \frac{dN_3}{d\xi}\frac{d\xi}{dx} \right] \begin{Bmatrix} u_1 \\ u_2 \\ u_3 \end{Bmatrix} = \mathbf{B}\,\mathbf{a}^{(e)} \tag{3.26}$$

From Eq.(3.11)

$$\frac{dN_1}{d\xi} = \xi - \frac{1}{2} \ ; \ \frac{dN_2}{d\xi} = -2\xi \ ; \frac{dN_3}{d\xi} = \xi + \frac{1}{2} \tag{3.27}$$

and the strain matrix $\mathbf{B}$ is

$$\mathbf{B} = \left(\frac{d\xi}{dx}\right) \left[(\xi - \frac{1}{2}), -2\xi, (\xi + \frac{1}{2}) \right] \tag{3.28}$$

The derivative $\frac{dx}{d\xi}$ is computed from Eq.(3.25) as

$$\begin{aligned} \frac{dx}{d\xi} &= \frac{dN_1}{d\xi}x_1 + \frac{dN_2}{d\xi}x_2 + \frac{dN_3}{d\xi}x_3 = (\xi - \frac{1}{2})\,x_1 - \\ &- 2\xi x_2 + (\xi + \frac{1}{2})\,x_3 = \frac{l^{(e)}}{2} + \xi\,(x_1 + x_3 - 2x_2) \end{aligned} \tag{3.29}$$

and

$$\frac{d\xi}{dx} = \frac{2}{l^{(e)} + 2\xi(x_1 + x_3 - 2x_2)} \tag{3.30}$$

Eq.(3.30) provides a relationship between dx and $d\xi$ in terms of the three nodal coordinates. In the (usual) case that the central node is located at the element midpoint, we have

$$\frac{d\xi}{dx} = \frac{2}{l^{(e)}} \tag{3.31}$$

and

$$\frac{dx}{d\xi} = \frac{2}{l^{(e)}} \quad \text{y} \quad dx = \frac{l^{(e)}}{2}\,d\xi \tag{3.32}$$

In this latter case the strain matrix of Eq.(3.28) is simply

$$\mathbf{B} = \frac{2}{l^{(e)}} \left[(\xi - \frac{1}{2}), -2\xi, (\xi + \frac{1}{2}) \right] \tag{3.33}$$

The expression of $\mathbf{B}$ for an arbitrary position of the central node is obtained by substituting Eq.(3.30) into (3.28).

The element stiffness matrix and the equivalent nodal force vector are obtained from the PVW as explained in Chapter 2 for the 2-noded element. It can easily be found that the element stiffness matrix has once again the general form

$$\mathbf{K}^{(e)} = \int_{l^{(e)}} \mathbf{B}^T \, (EA) \, \mathbf{B} \, dx \tag{3.34}$$

Substituting the above expressions for dx and $\mathbf{B}$ in terms of ξ into Eq.(3.34) leads to (for the case of the mid-node being central in the element)

$$\mathbf{K}^{(e)} = \int_{-1}^{+1} \frac{2}{l^{(e)}} \begin{Bmatrix} (\xi - \frac{1}{2}) \\ -2\xi \\ (\xi + \frac{1}{2}) \end{Bmatrix} (EA) \frac{2}{l^{(e)}} \left[(\xi - \frac{1}{2}), -2\xi, (\xi + \frac{1}{2}) \right] \frac{l^{(e)}}{2} \, d\xi \tag{3.35}$$

The computation of the above integral is straightforward if both E and A are constant over the element, giving

$$\mathbf{K}^{(e)} = \begin{bmatrix} K_{11} & K_{12} & K_{13} \\ K_{21} & K_{22} & K_{23} \\ K_{31} & K_{32} & K_{33} \end{bmatrix} = \left(\frac{EA}{6l} \right)^{(e)} \begin{bmatrix} 14 & -16 & 2 \\ -16 & 32 & -16 \\ 2 & -16 & 14 \end{bmatrix} \tag{3.36}$$

The equivalent nodal force vector for a distributed loading of intensity $\mathbf{t} = \{t_x\}$ is

$$\mathbf{f}^{(e)} = \begin{Bmatrix} f_{x_1} \\ f_{x_2} \\ f_{x_3} \end{Bmatrix} = \int_{l^{(e)}} \mathbf{N}^T \, \mathbf{t} \, dx = \int_{-1}^{+1} \begin{Bmatrix} \frac{1}{2}\xi(\xi - 1) \\ 1 - \xi^2 \\ \frac{1}{2}\xi(1 + \xi) \end{Bmatrix} t_x \, \frac{l^{(e)}}{2} \, d\xi \tag{3.37a}$$

For a uniformly distributed loading

$$\mathbf{f}^{(e)} = \frac{(l t_x)^{(e)}}{6} \begin{Bmatrix} 1 \\ 4 \\ 1 \end{Bmatrix} \tag{3.37b}$$

We note that the central node absorbes four times more loading than the end nodes. This result which is not the obvious one, is a natural consequence of the PVW and the quadratic approximation chosen.

The expressions of $\mathbf{K}^{(e)}$ and $\mathbf{f}^{(e)}$ for an arbitrary position of the central node are obtained using the relationship between of dx and $d\xi$ of Eq.(3.30). In this case rational algebraic functions in ξ are involved and, therefore, the analytical computation of the element integrals is not so simple.

The equilibrium equation for the 3-noded rod element is

$$\mathbf{q}^{(e)} = \mathbf{K}^{(e)}\mathbf{a}^{(e)} - \mathbf{f}^{(e)} \tag{3.38a}$$

where the equilibrating nodal force vector is

$$\mathbf{q}^{(e)} = [F_{x_1}, F_{x_2}, F_{x_3}]^T \tag{3.38b}$$

The axial forces at the element nodes can be obtained from the components of $\mathbf{q}^{(e)}$ as

$$[-\mathcal{N}_1, \mathcal{N}_2, \mathcal{N}_3]^T = [F_{x_1}, F_{x_2}, F_{x_3}]^T \tag{3.38c}$$

The global stiffness matrix $\mathbf{K}$ and the global equivalent nodal force vector $\mathbf{f}$ are assembled from the element contributions, as explained for bar structures and for the 2-noded rod element. The process is schematically shown in Figure 3.4. As usual P_{x_i} denotes the external point force acting at the node with global number i. The same assembly procedure applies for higher order rod elements. Example 3.4 presented in a next section shows an application of the 3-noded rod element.

The isoparametric formulation of higher order rod elements follows the rules explained for the quadratic element. The increasing complexity of the element integrals can be overcome by using numerical integration as explained in the next section.

3.4 NUMERICAL INTEGRATION

In some cases the exact analytical computation of the integrals appearing in $\mathbf{K}^{(e)}$ and $\mathbf{f}^{(e)}$ can be difficult and sometimes impossible. This typically occurs for 2D and 3D isoparametric elements, due to the complexity of the rational algebraic functions involved in the integrals. Numerical integration appears here as the only option to compute the element integrals in a simple and accurate way.

To enter into the mathematics of numerical integration falls outside the scope of this book. For simplicity we will only consider here the Gauss quadrature [PFTV,Ral] as this is the more popular numerical integration procedure used in the FEM. We will introduce the basic ideas for 1D problems which will be extended for 2D and 3D problems in subsequent chapters.

Let us assume that the integral of a function $f(x)$ in the interval [-1,1] is required, i.e.

$$I = \int_{-1}^{+1} f(\xi)\, d\xi \tag{3.39}$$

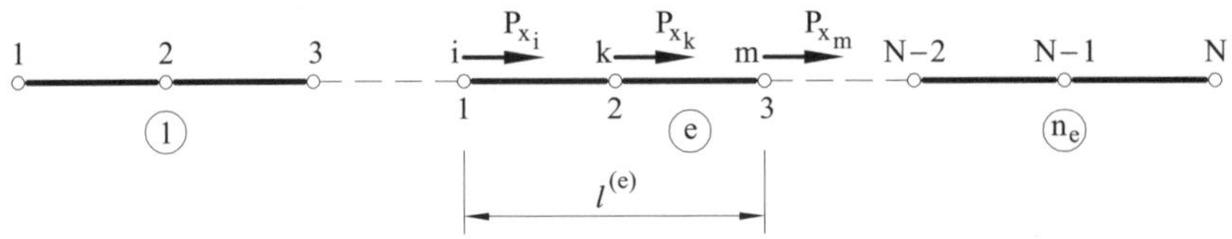

$$
\begin{array}{lccc}
\text{global} & i & k & m \\
\text{local} & 1 & 2 & 3
\end{array}
$$

$$
\mathbf{K}^{(e)} = \begin{bmatrix} K_{11} & K_{12} & K_{13} \\ & K_{22} & K_{23} \\ \text{Symm.} & & K_{33} \end{bmatrix} \begin{matrix} 1 \; i \\ 2 \; k \\ 3 \; m \end{matrix} \quad , \quad \mathbf{f}^{(e)} = \begin{Bmatrix} f_{x_1} \\ f_{x_2} \\ f_{x_3} \end{Bmatrix} \begin{matrix} 1 \; i \\ 2 \; k \\ 3 \; m \end{matrix}
$$

$$
\mathbf{K} = \begin{array}{c} \begin{matrix} 1\,2 \cdots i & \cdots & k & \cdots & m \cdots N \end{matrix} \\ \begin{bmatrix} \ddots & \vdots & \vdots & \vdots & \\ K_{11} & \cdots K_{12} & \cdots & K_{13} & \cdots \\ & \ddots & \vdots & \vdots & \\ & & K_{22} & \cdots K_{23} & \cdots \\ & & & \ddots \; \vdots & \\ \text{Symm.} & & & K_{33} & \cdots \\ & & & \vdots & \end{bmatrix} \end{array} \begin{matrix} 1 \\ 2 \\ \vdots \\ i \\ \vdots \\ k \\ \vdots \\ m \\ \vdots \\ N \end{matrix} \quad , \quad \mathbf{f} = \begin{Bmatrix} \vdots \\ f_{x_1} + P_{x_i} \\ \vdots \\ f_{x_2} + P_{x_k} \\ \vdots \\ f_{x_3} + P_{x_m} \\ \vdots \end{Bmatrix} \begin{matrix} 1 \\ 2 \\ \vdots \\ i \\ \vdots \\ k \\ \vdots \\ m \\ \vdots \\ N \end{matrix}
$$

Fig. 3.4 Three-noded rod element. Assembly of the global stiffness matrix **K** and the global equivalent nodal force vector **f** from the element contributions

The Gauss integration rule, or Gauss *quadrature*, expresses the value of the above integral as a sum the function values at a number of known points multiplied by prescribed weights. For a quadrature of order q

$$
I \simeq I_q = \sum_{i=1}^{q} f(\xi_i) W_i \tag{3.40}
$$

where W_i is the weight corresponding to the ith sampling point located at $\xi = \xi_i$ and q the number of sampling points. A *Gauss quadrature of qth order integrates exactly a polynomial function of degree* $2q - 1$ [Ral]. The error in the computation of the integral is of the order $0(\triangle^{2q})$, where $\triangle$ is the spacing between the sampling points. The coordinates of the sampling points and their weights for the first eight Gauss quadratures are shown in Table 3.1.

q	ξ_q	W_q
1	0.0	2.0
2	±0.5773502692	1.0
3	±0.774596697	0.5555555556
	0.0	0.8888888889
4	±0.8611363116	0.3478548451
	±0.3399810436	0.6521451549
5	±0.9061798459	0.2369268851
	±0.5384693101	0.4786286705
	0.0	0.5688888889
6	±0.9324695142	0.1713244924
	±0.6612093865	0.3607615730
	±0.2386191861	0.4679139346
7	±0.9491079123	0.1294849662
	±0.7415311856	0.2797053915
	±0.4058451514	0.3818300505
	0.0	0.4179591837
8	±0.9602898565	0.1012285363
	±0.7966664774	0.2223810345
	±0.5255324099	0.3137066459
	±0.1834346425	0.3626837834

Table 3.1 Coordinates and weights for Gauss quadratures

Note that the sampling points are all located within the normalized domain [-1,1]. This is useful for computing the element integrals expressed in terms of the natural coordinate ξ. The popularity of the Gauss quadrature derives from the fact that it requires the minimum number of sampling points to achieve a prescribed error in the computation of an integral. Thus, it minimizes the number of times the integrand function is computed. The reader can find further details in [PFTV,Rad,Ral].

Example 3.2: Applications of the Gauss quadrature.

- ***Solution***

Let us consider the fourth degree polynominal

$$f(x) = 1 + x + x^2 + x^3 + x^4$$

The exact integral of $f(x)$ over the interval $-1 \leq x \leq 1$ is

$$I = \int_{-1}^{+1} f(x)dx = 2 + \frac{2}{3} + \frac{2}{5} = 3.0666$$

- *First order Gauss quadrature*:

$$q=1 \;,\; x_1=0 \;,\; W_1=2 \quad ; \quad I=W_1 f(x_1)=2$$

- *Second order Gauss quadrature*:

$$q=2 \; \begin{cases} x_1=-0.57735 & , \quad W_1=1 \\ x_2=+0.57735 & , \quad W_2=2 \end{cases}$$

$$I=W_1 f(x_1)+W_2 f(x_2)=0.67464+2.21424=2.8888$$

- *Third order Gauss quadrature*:

$$q=3 \; \begin{Bmatrix} x_1=-0.77459 & , & W_1=0.5555 \\ x_2=0.57735 & , & W_2=0.8888 \\ x_3=+0.77459 & , & W_3=0.5555 \end{Bmatrix}$$

$$I=W_1 f(x_1)+W_2 f(x_2)+W_3 f(x_3)=0.7204\times 0.5555+$$
$$+1.0\times 0.8888+3.19931\times 0.5555=3.0666 \quad \textbf{Exact value!}$$

We see that the exact integration of a fourth order polynominal requires a third order Gauss quadrature as expected.

3.5 STEPS FOR THE COMPUTATION OF MATRICES AND VECTORS FOR AN ISOPARAMETRIC ROD ELEMENT

We will now present the basic steps for computing the stiffness matrix and the equivalent nodal vector for an isoparametric rod element with n nodes. The steps have been arranged so as to facilitate their implementation within a computer program.

3.5.1 Interpolation of the axial displacement

The axial displacement within the element is expressed as

$$u=N_1\,u_1+N_2\,u_2\,+\ldots+\,N_n\,u_n\;=$$

$$\sum_{i=1}^{n} N_i u_i=[N_1,N_2,\ldots,N_n]\begin{Bmatrix} u_1 \\ u_2 \\ \vdots \\ u_n \end{Bmatrix}=\mathbf{N}\,\mathbf{a}^{(e)} \tag{3.41}$$

3.5.2 Geometry interpolation

The coordinate x is interpolated using the isoparametric form as

$$x = N_1\, x_1 + N_2\, x_2\ + \ldots +\ N_n\, x_n = \sum_{i=1}^{n} N_i\, x_i\ =$$

$$= [N_1, N_2, \ldots, N_n] \begin{Bmatrix} x_1 \\ x_2 \\ \cdots \\ x_n \end{Bmatrix} = \mathbf{N}\, \mathbf{x}^{(e)} \tag{3.42}$$

3.5.3 Interpolation of the axial strain

The axial strain is expressed in terms of the nodal displacements as

$$\varepsilon = \frac{du}{dx} = \frac{dN_1}{dx}\, u_1 + \frac{dN_2}{dx}\, u_2 + \ldots +\ \frac{dN_n}{dx}\, u_n = \sum_{i=1}^{n} \frac{dN_i}{dx}\, u_i\ =$$

$$= \left[\frac{dN_1}{dx}, \frac{dN_2}{dx}, \ldots, \frac{dN_n}{dx}\right] \begin{Bmatrix} u_1 \\ u_2 \\ \vdots \\ u_n \end{Bmatrix} = \mathbf{B}\, \mathbf{a}^{(e)} \tag{3.43}$$

The cartesian derivative of the shape functions is obtained by

$$\frac{dN_i}{dx} = \frac{dN_i}{d\xi}\, \frac{d\xi}{dx} \tag{3.44}$$

From Eq.(3.42) we deduce

$$\frac{dx}{d\xi} = \sum_{i=1}^{n} \frac{dN_i}{d\xi}\, x_i = J^{(e)} \tag{3.45}$$

and, therefore

$$dx = J^{(e)}\, d\xi \qquad ; \qquad \frac{d\xi}{dx} = \frac{1}{J^{(e)}} \tag{3.46}$$

and

$$\frac{dN_i}{dx} = \frac{1}{J^{(e)}}\, \frac{dN_i}{d\xi} \tag{3.47}$$

Substituting Eq.(3.47) into the expression of **B** gives

$$\mathbf{B} = \left[\frac{dN_i}{dx}, \frac{dN_2}{dx}, \ldots, \frac{dN_n}{dx}\right] = \frac{1}{J^{(e)}}\left[\frac{dN_i}{d\xi}, \frac{dN_2}{d\xi}, \ldots, \frac{dN_n}{d\xi}\right] \tag{3.48}$$

In Eq.(3.48) $J^{(e)} = \frac{dx}{d\xi}$ is the Jacobian of the 1D transformation between dx and $d\xi$. For 2D and 3D problems $J^{(e)}$ is a matrix whose determinant relates the infinitesimal areas (for $2D$) and volumes (for $3D$) in the cartesian and natural coordinate systems.

3.5.4 Computation of the axial force

The axial force $\mathcal{N}$ for the element is obtained as

$$\mathcal{N} = (EA)\,\varepsilon = \mathbf{D}\,\mathbf{B}\,\mathbf{a}^{(e)} \quad \text{with} \quad \mathbf{D} = [EA] \tag{3.49}$$

3.5.5 Element stiffness matrix

The PVW leads to the following general expression for the element stiffness matrix (Section 2.8.5)

$$\mathbf{K}^{(e)} = \int_{l^{(e)}} \mathbf{B}^T\mathbf{D}\mathbf{B}\; dx = \int_{-1}^{+1} \mathbf{B}^T\mathbf{D}\mathbf{B}\; J^{(e)}\; d\xi \tag{3.50}$$

From Eqs.(3.48) and (3.49) we deduce

$$K_{ij}^{(e)} = \int_{-1}^{+1} \frac{1}{J^{(e)}}\frac{dN_i}{d\xi}\,(EA)\,\frac{dN_j}{d\xi}\; d\xi \tag{3.51}$$

The simplicity of the above integral depends on the expression of the shape functions and of $J^{(e)}$. In general, $\mathbf{K}^{(e)}$ is computed using the Gauss quadrature which evaluates (3.51) as exactly as possible. For a qth order Gauss quadrature we can write

$$\mathbf{K}^{(e)} = \sum_{r=1}^{q} [\mathbf{B}^T\mathbf{D}\mathbf{B}J^{(e)}]_r W_r \tag{3.52a}$$

or

$$K_{ij}^{(e)} = \sum_{r=1}^{q} \left[\frac{1}{J^{(e)}}\frac{dN_i}{d\xi}\,(EA)\,\frac{dN_j}{d\xi}\right]_r W_r \tag{3.52b}$$

where $[\cdot]_r$ denotes values computed at the sampling point $\xi = \xi_r$.

3.5.6 Equivalent nodal force vector

For a distributed loading of intensity $t_x(x)$ we have

$$\mathbf{f}^{(e)} = [f_{x_1}, f_{x_2}, \cdots, f_{x_n}]^T = \int_{l^{(e)}} \mathbf{N}^T t_x \, dx = \int_{-1}^{+1} \mathbf{N}^T t_x J^{(e)} \, d\xi \quad (3.53)$$

The computation of $\mathbf{f}^{(e)}$ can be performed using numerical integration as

$$\mathbf{f}^{(e)} = \sum_{r=1}^{q} [\mathbf{N}^T t_x J^{(e)}]_r W_r \quad (3.54a)$$

or

$$f_{x_i} = \sum_{r=1}^{q} [N_i \; t_x \; J^{(e)}]_r \; W_r \quad , \quad i = 1, 2, 3 \quad (3.54b)$$

The global stiffness matrix $\mathbf{K}$ and the equivalent nodal vector $\mathbf{f}$ are assembled from the element contributions $\mathbf{K}^{(e)}$ and $\mathbf{f}^{(e)}$, as usual (Figure 3.4).

Once the system of global equilibrium equations $\mathbf{Ka} = \mathbf{f}$ has been solved for the nodal displacements $\mathbf{a}$, the reactions at the prescribed nodes can be computed by Eq.(2.26a),or else by Eq.(2.26b) with the following expression for the internal force vector for each element

$$\mathbf{f}_{\text{int}}^{(e)} = \int_{l^{(e)}} \mathbf{B}^T \boldsymbol{\sigma} \, dx = \int_{l^{(e)}} \mathbf{B}^T \mathcal{N} \, dx \quad (3.55)$$

Example 3.3: Compute the term $K_{11}^{(e)}$ of $\mathbf{K}^{(e)}$ for the 3-noded rod element (Figure 3.2) using an isoparametric formulation and numerical integration.

- ***Solution***

The term $K_{11}^{(e)}$ is obtained from Eq.(3.51) as

$$K_{11}^{(e)} = \int_{-1}^{+1} \frac{1}{J^{(e)}} \frac{dN_1}{d\xi} (EA) \frac{dN_1}{d\xi} \, d\xi$$

The expression of N_1 for the 3-noded rod element is (Eq. (3.11))

$$N_1 = \frac{1}{2}(\xi - 1)\xi \quad \text{y} \quad \frac{dN_1}{d\xi} = \xi - \frac{1}{2}$$

Assuming that node 2 is centered in the element $J^{(e)} = \frac{l^{(e)}}{2}$.

Substituting $\frac{dN_1}{d\xi}$ and $J^{(e)}$ into $K_{11}^{(e)}$ we have

$$K_{11}^{(e)} = \int_{-1}^{+1} \frac{2EA}{l^{(e)}}(\xi - \frac{1}{2})^2 \, d\xi$$

The integrand is a quadratic function and hence, the exact integral requires a Gauss quadrature of 2nd order ($q = 2$). From Eq.(3.52) and Table 3.1 we obtain

$$K_{11}^{(e)} = \sum_{r=1}^{2}\Big[\frac{2EA}{l^{(e)}}(\xi - \frac{1}{2})^2\Big]_r W_r = \Big[\frac{2EA}{l^{(e)}}(\xi - \frac{1}{2})^2\Big]_{\xi=-\frac{\sqrt{3}}{3}} + \\ + \Big[\frac{2EA}{l^{(e)}}(\xi - \frac{1}{2})^2\Big]_{\xi=\frac{\sqrt{3}}{3}} = \frac{7}{3}(\frac{EA}{l})^{(e)}$$

The same procedure can be followed for computing the rest of terms of $\mathbf{K}^{(e)}$.

3.6 BASIC ORGANIZATION OF A FINITE ELEMENT PROGRAM

The steps presented in the previous section for computing the stiffness matrix and the equivalent nodal force vector for the simple rod element are general and almost identical to those required for more complex 2D and 3D elements. Also, these steps define naturally the basic subroutines of a computer program for structural analysis using the FEM. The programming aspects of the FEM will be studied in Chapter 10 and here we will just introduce the basic format of a finite element program for structural analysis.

Figure 3.5 shows the flow chart of a finite element program for the analysis of axially loaded rods. The first subroutine deals with the reading of the geometrical and material properties data required for the analysis (subroutine INPUT). Then, the stiffness matrix and the equivalent nodal force vector are computed for each element in subroutines STIFFNESS and LOAD, respectively. The next step is the assembly and solution of the global system of algebraic equilibrium equations to obtain the nodal displacement values in subroutine SOLVE. Finally, the strains and stresses are computed at selected points within each element in subroutine STRESS. Note the analogy of the program skeleton with that of a program for matrix analysis of bar structures [Hu,HO2,Li].

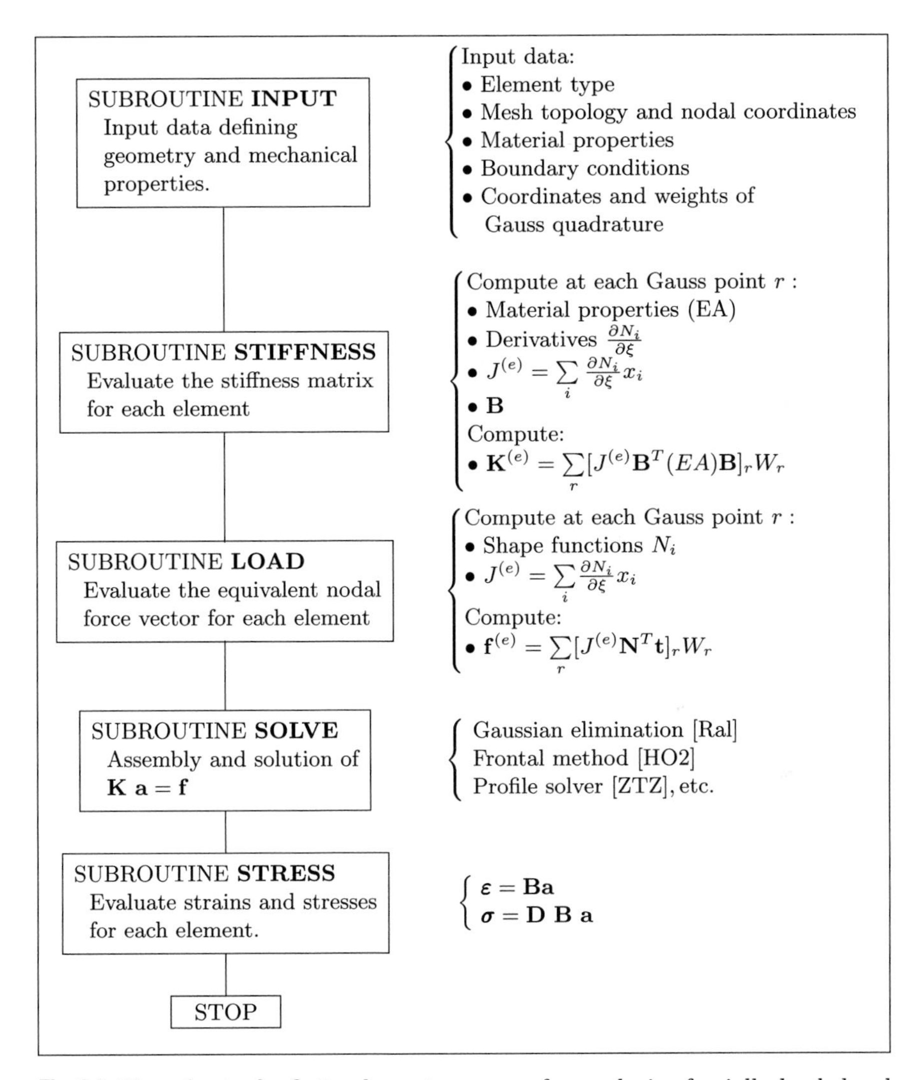

Fig. 3.5 Flow chart of a finite element program for analysis of axially loaded rods

3.7 SELECTION OF ELEMENT TYPE

The first task in the analysis of a structure by the FEM is to select the element to be used. This is an important decision and not a simple one, as there are many elements available for solving the same structural problem, and each one has different advantages and disadvantages with regard to simplicity, accuracy,cost, etc.

In most cases, the selection of an element for a particular problem is made by the analyst responsible for the computations. This decision should be based on: 1) the characteristics of the structure to be analysed; 2) the elements available in comercial on in-house computer programs and the type of computer to be used; and 3) the experience of the analyst in the solution of similar structures by the FEM.

Several rules for the selection of the best element for each particular structural problem will be given throughout this book. Nevertheless a few rules of "thumb" can be summarized at this stage. These are:

1. The element chosen must be *robust.* This simply means that there should be no danger of obtaining a spurious solution due to intrinsic bad behaviour of the element under general geometrical or mechanical conditions. A test for robustness of the element is provided by the *patch test* studied in a next section.
2. The mesh should account for the probable stress gradients in the solution, i.e. the mesh should be finer in zones where stress gradients are expected to be higher. Here the use of error estimators and adaptive mesh refinement procedures is recommended. These topics will be studied in Chapter 9.
3. The element should be as accurate as possible. The debate between using few elements of high order, or a finer mesh of simpler low order elements is still open in FEM practice. The growing popularity of adaptive mesh refinement strategies, and the continuing increase in computer power is favouring the use of low order elements.

The choice of low or high order order elements is schematically represented in Figure 3.6 showing the approximation of a third degree polynomial function representing the solution of an axial rod problem using different elements. Note that a large number of simple 2-noded elements is required, whereas a single 4-noded cubic element provides the exact solution.

An indicator to decide between two elements is the ratio between the accuracy and the number of nodal variables. This requires a definition of "accuracy", which is not obvious if the exact solution is not known "a priori" (see Chapter 9 for details). A guideline is that in case of doubt between two elements of different order, the analyst should always choose the simplest one (which is generally the low order one).

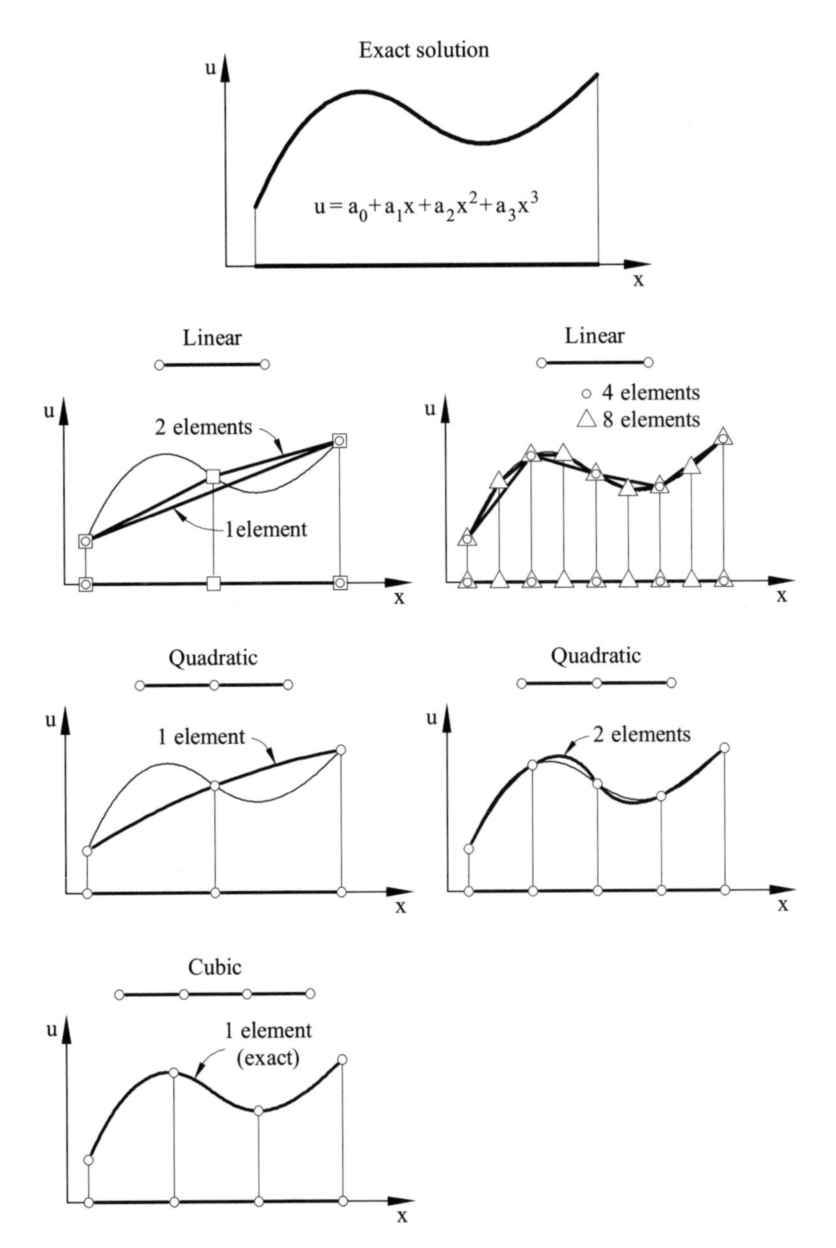

Fig. 3.6 Approximation of a cubic solution with different rod elements. For simplicity the finite element solution has been assumed to be exact at the nodes

A comparison between the quadratic and linear rod elements is presented next.

Example 3.4: Solve the problem of Figure 2.4 using a single 3-noded quadratic rod element.

- Solution

Since we have only one element the global equilibrium equation is written from Eqs.(3.36), (3.37b) and Figure 3.4 as

$$\frac{EA}{6l}\begin{bmatrix} 14 & -16 & 2 \\ -16 & 32 & -16 \\ 2 & -16 & 14 \end{bmatrix} \begin{Bmatrix} u_1 \\ u_2 \\ u_3 \end{Bmatrix} = \begin{Bmatrix} \frac{lt_x}{6} + R \\ \frac{2lt_x}{3} \\ \frac{lt_x}{6} + P \end{Bmatrix} \quad , \quad u_1 = 0$$

Solving the above system with the condition $u_1 = 0$, gives

$$u_2 = \frac{3t_x l^2}{8EA} + \frac{Pl}{2A}$$

$$u_3 = \frac{t_x l^2}{2EA} + \frac{Pl}{EA}$$

These values coincide with the exact solution (2.28) at the nodes.
The displacement field within the element is

$$u = (1 - \xi^2)u_2 + \frac{1}{2}(1 + \xi)\xi u_3$$

Substituting the values for u_2 and u_3 and making the change of variable ($\xi = \frac{2x-l}{l}$) gives

$$u = \frac{1}{EA}(-\frac{x^2}{2}t_x + (P + lt_x)x)$$

which coincides with the exact solution (2.28) everywhere. This could have been anticipated as the assumed displacement field contains the quadratic solution. The axial strain and axial force fields within the element are

$$\varepsilon^{(1)} = \frac{1}{EA}\Big(P + (l - x)t_x\Big) \quad , \quad \mathcal{N}^{(1)} = P + (l - x)t_x$$

which again coincide with the exact solution (see Eq.(2.28) and Figure 2.3 for $P = 0$ and $t_x = 1$).
The nodal axial forces are given by

$$\mathbf{q}^{(1)} = \mathbf{K}^{(1)}\,\mathbf{a}^{(1)} - \mathbf{f}^{(1)}$$

giving

$$\mathbf{q}^{(1)} = [-\mathcal{N}_1, \mathcal{N}_2, \mathcal{N}_3]^T = [-(lt_x + P), 0, P]^T$$

The axial forces at the element ends 1 and 3 are $\mathcal{N}_1 = (P + lt_x)$ and $\mathcal{N}_3 = P$, whereas $\mathcal{N}_2 = 0$, as no external point force is applied to node 2. The reaction value is $R = -(P + lt_x) = -\mathcal{N}_1$, as expected.

The previous example shows that the quadratic rod element has a better performance than the linear one (Figure 2.3). This can be taken as a general rule in favour of quadratic elements. However, in many cases (particularly for 3D problems) the increase in accuracy is counterbalanced by a greater complexity for mesh generation and a larger computing cost.

3.8 REQUIREMENTS FOR CONVERGENCE OF THE SOLUTION

The finite element approximation must satisfy certain conditions which guarantee that as the mesh is refined the numerical solution converges to the exact values. The satisfaction of these conditions is the basis for the success of mesh refinement strategies (Chapter 9).

3.8.1 Continuity condition

The displacement must be continuous *within* each element. This condition is automatically satisfed by using polynomial approximations for the displacement field. The issue of continuity of the displacements *along the element interfaces* is treated in Section 3.10.1.

3.8.2 Derivativity condition

The derivatives of the polynomial approximation should exist up to the order of the derivatives appearing in the element integrals.

For instance, for the axially loaded rod element the integrals derived from the PVW contain first order derivatives of the displacement only. Hence, the shape functions should be at least first order polynomials.

3.8.3 Integrability condition

Logically, the integrals appearing in the element expressions must have a primitive function. This condition can be explained by considering the simple example of Figure 3.7 where a continuous function $f(x)$ and its two first derivatives are represented. The integral of $f(x)$ in the interval considered exists and it is equal to the area shown in the figure. Also, the integral of $f'(x)$ exists, although it is not a continuous function. Finally, we observe that the second derivative $f''(x)$ has two singular points, due to the discontinuity of $f'(x)$, and it is not integrable. The general rule deduced from this simple example is the following. The derivative of a function is integrable if its $m-1$ first derivatives are continuous (C^{m-1} continuity).

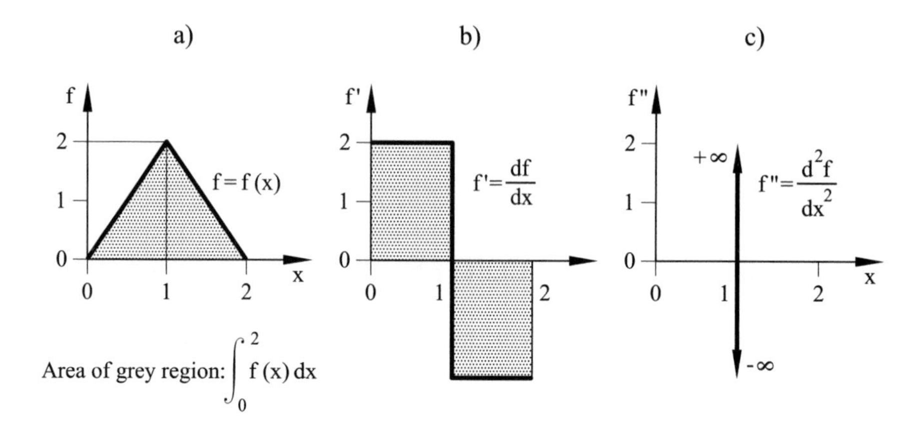

Fig. 3.7 Integral of a bilinear function and its two first derivatives

Thus, if *mth* order derivatives of the displacement field appear in the PVW, the displacement field (and also the shape functions) must be C^{m-1} continuous (Section 3.2). This condition ensures that the strains at the interfaces between elements are finite (even though they are discontinuous) [ZTZ].

As an example the PVW for the axial rod problem. (Eq.(2.3)) contains only first derivatives of u and hence just displacement continuity is required. The C^o continuity is guaranteed within each element by the polynomial approximation chosen, and between elements by the coincidence of the displacement at the common nodes.

All the elements derived in this volume for analysis of 2D solids, axisymmetric solids and 3D solids just require C^o continuity.

3.8.4 Rigid body condition

The displacement field closed should not allow straining of an element to occur when the nodal displacements are caused by a rigid body motion.

This physical condition is satisfied for a single element if the sum of the shape functions at any point is equal to one. To prove this, let us consider the simple 2-noded axially loaded rod element with equal prescribed nodal displacements $\bar{u}$. Within the element we have

$$u = N_1\bar{u} + N_2\bar{u} = (N_1 + N_2)\,\bar{u} \tag{3.56}$$

and for $u = \bar{u}$ then $N_1 + N_2 = 1$ must be satisfied.

3.8.5 Constant strain condition

The displacement function has to be such that if nodal displacements are compatible with a constant strain field, such constant strain will in fact be obtained. Clearly, as elements get smaller, nearly constant strain conditions will prevail in them. It is therefore desirable that a finite size element should be able to reproduce a constant strain condition [Sa,ZTZ].

The constant strain criterion incorporates the rigid body requirement, as a rigid body displacement is a particular case of a constant (zero) strain field. Strictly both criteria need only be satisfied in the limit as the size of the elements tends to zero. However, satisfaction of these criteria on elements of finite size leads to a convergent and more accurate solution.

3.9 ASSESSMENT OF CONVERGENCE REQUIREMENTS. THE PATCH TEST

The patch test was first introduced by Irons and Razzaque [IR] and has since then provided a necessary and sufficient condition for convergence [Dao,FdV,ZTZ]. The test is based on selecting an arbitrary patch of elements and imposing upon it nodal displacements corresponding to any state of constant strain. If nodal equilibrium is achieved without imposing external nodal forces, and if a state of constant stress is obtained, then clearly the constant strain criterion of the previous section is satisfied. Furthermore, displacement continuity is guaranteed, since no external work is lost through the interelement interfaces [Sa].

The patch test also includes the satisfaction of the rigid body condition by simply imposing a nodal displacement field corresponding to a zero strain value.

An alternative patch test is to prescribe a known linear displacement field at the boundary of the patch nodes only. It is then verified that the displacement solution at the interior nodes coincides with the exact values and that a constant strain field is obtained throughout the patch.

The patch test allows us to assess the convergence of elements with shape functions which are discontinuous along the element interfaces between adjacent elements. This issue will be discussed further in Section 3.10.1.

The application of the patch test to the simple 2-noded rod element is shown in the next example. The patch test for 2D elements is presented in Section 6.10.

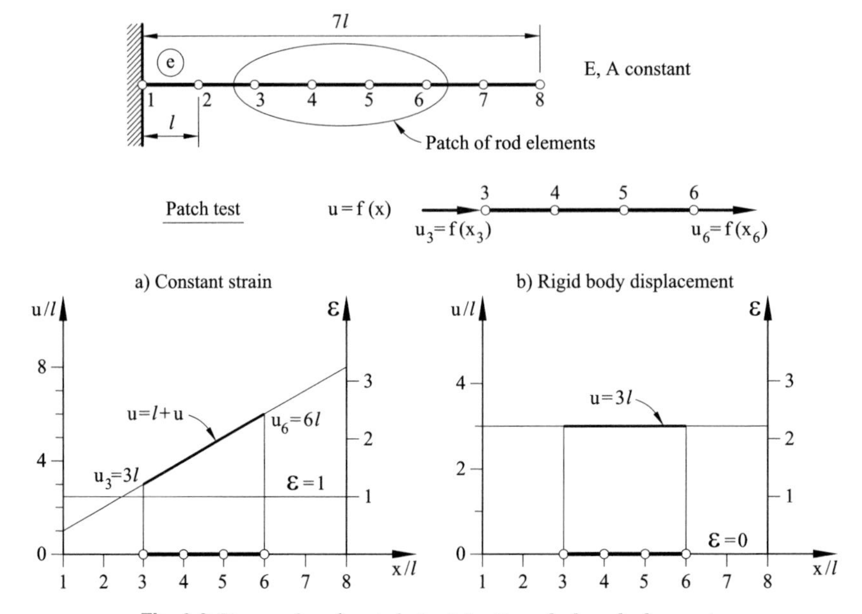

Fig. 3.8 Example of patch test in 2-noded rod elements

Example 3.5: Apply the patch test to the three element patch of 2-noded rod elements shown in Figure 3.8. All elements have equal length and the same material properties.

- ***Solution***

(a) Constant strain condition (Figure 3.8a)

We will assume a displacement field $u = l + x$ giving a constant strain field in the whole mesh, i.e. $\varepsilon = \frac{du}{dx} = 1$. The following displacements are prescribed at the end nodes of the patch:

$$u_3 = l + 3l = 4l$$
$$u_6 = l + 6l = 7l$$

We now look for the solution for the nodal displacements u_4 and u_5. The equation system to be solved is

$$\frac{EA}{l}\begin{bmatrix} 1 & -1 & 0 & 0 \\ -1 & 2 & -1 & 0 \\ 0 & -1 & 2 & -1 \\ 0 & 0 & -1 & 1 \end{bmatrix} \begin{Bmatrix} u_3 \\ u_4 \\ u_5 \\ u_6 \end{Bmatrix} = \begin{Bmatrix} R_3 \\ 0 \\ 0 \\ R_6 \end{Bmatrix} \quad \begin{matrix} u_3 = 4l \\ \\ \\ u_6 = 7l \end{matrix}$$

which gives $u_4 = 5l$ and $u_5 = 6l$. These values coincide with the exact ones given by the prescribed field. It can also be checked that

$$
\begin{aligned}
\varepsilon^{(3)} &= -\frac{1}{l}u_3 + \frac{1}{l}u_4 = 1 \\
\varepsilon^{(4)} &= -\frac{1}{l}u_4 + \frac{1}{l}u_5 = 1 \\
\varepsilon^{(5)} &= -\frac{1}{l}u_5 + \frac{1}{l}u_6 = 1
\end{aligned}
$$

which correspond to the exact constant field imposed. Therefore, the element satisfies the patch test.

(b) Rigid body condition (Figure 3.8b)

A particular case of the previous example is to study the patch subjected to a the constant displacement field $u = 3l$, corresponding to a rigid body movement of the patch. The FEM solution for $u_3 = u_6 = 3l$ yields $u_4 = u_5 = 3l$, which is the correct answer giving a zero strain field over the patch.

3.10 OTHER REQUIREMENTS FOR THE FINITE ELEMENT APPROXIMATION

Next, we will consider some requirements which, in fact, are not strictly necessary for the convergence of the finite element solution. However, their fulfilment is always desirable since, otherwise, the convergence and accuracy of the solution can deteriorate in some cases.

3.10.1 Compatibility condition

The elements must be compatible. This implies that the displacement field for C^0 elements, or its first derivative field for C^1 elements, must be continuous along interelemental boundaries. This is a consequence of the more general integrability condition of Section 3.8.3. Elements satisfying the compatibility condition are termed *compatible* or *conforming.* These elements, when integrated exactly, always converge to the exact solution from the stiffer side.

The compatibility condition is usually satisfied when the displacement field is defined by a polynomial taking a unique value at the nodes. This, however, is not sufficient in some particular cases, such as in some C^1 thin plate bending elements based in Kirchhoff theory where a discontinuity of the gradient of the deflection occurs at the element sides (see

Chapter 4 of Volume 2 [On]). These elements are termed *incompatible* or *non-conforming*. Incompatible elements can still converge to the exact solution if the patch test is satisfied. This guarantees that the compatibility condition is fulfilled in the limit as the mesh is refined.

Non-conforming elements can be still competitive in practice. The reason is that interelemental discontinuities introduce a greater flexibility in the element which counterbalances the intrinsic rigidity of the finite element approximation. This leads in some occasions to very good solutions with relatively coarse meshes.

In summary, the non-conformity is an undesirable deficiency which, however, does not automatically invalidate an element. The patch test is the critical proof for acceptance of the element for practical purposes. Although incompatible elements can sometimes be very attractive, they should be looked upon with caution since they can have unexpected features. For instance, some incompatible solid elements show a spurious dependency with the Poisson's ratio which varies with the mesh size [Na].

3.10.2 Condition of complete polynomial

This condition can be explained by recalling that the finite element approximation can reproduce only a finite number of the Taylor expansion terms of the exact solution, which can written in the vicinity of a point x_i as

$$u(x) = u(x_i) + \left(\frac{du}{dx}\right)_i (x - x_i) + \left(\frac{d^2u}{dx^2}\right)_i (x - x_i)^2 + \cdots + \left(\frac{d^nu}{dx^n}\right)_i (x - x_i)^n \tag{3.57a}$$

It is obvious that the finite element approximation written as

$$\bar{u}(x) = a_0 + a_1 x \ + a_2 x^2 \ + \cdots + \ a_m x^m \tag{3.57b}$$

can only reproduce exact results up to the mth term of the Taylor expansion (3.57a) when $\bar{u}(x)$ contains *all* the terms of the polynomial of mth degree. In such a case the approximation error is of the order Oh^{m+1} and this can be used to derive solution extrapolation rules (Section 2.7).

Therefore, the finite element approximation depends on the higher complete polynomial included in the shape functions. The approximation will be "optimal" if the shape functions are complete polynomials. Unfortunately this is not always possible, and in many cases the shape functions contain incomplete polynomial terms that do not contribute to a higher approximation of the element.

Example 3.6: Complete and incomplete polynomials and approximations.

- ***Solution***

a) Complete approximations of 2nd degree.

$$1D: \ \bar{u}(x) = a_0 + a_1 x + a_2 x^2$$
$$2D: \ \bar{u}(x,y) = a_0 + a_1 x + a_2 y + a_3 xy + a_4 x^2 + a_5 y^2$$

b) Incomplete approximations of 3rd degree.

$$1D: \ \bar{u}(x) = a_0 + a_1 x + a_2 x^3$$
$$2D: \ \bar{u}(x,y) = a_0 + a_1 x + a_2 y + a_3 x^2 + a_4 y^2 + a_5 x^3$$

The terms of a complete polynomial of high order can be deduced from the *Pascal triangle* and the *Pascal tetrahedron* [Ral]. This subject will be treated when studying the shape functions for 2D and 3D elements.

In conclusion, it is desirable for the shape functions be complete polynomials, or, if this is not possible, that they contain a small number of incomplete polynomials. An incomplete approximation does not preclude the convergence of the element.

3.10.3 Stability condition

The analysis of a structure requires prescribing enough displacements to prevent the appearance of unstable mechanisms. Lack of stability is usually detected by the existence of one or more mechanisms which correspond to the same number of zero eigenvalues in the stiffness matrix and the associated so-called *rigid body modes.* The same concept applies to the stability of an element. In consequence, the stiffness matrix of an individual element (and also that of a patch of elements) must have the correct rank [Ral]. This means that the number of zero eigenvalues of a single isolated element free of external constrains must be equal to the number of rigid body displacements of the element. The element is considered as stable if these zero eigenvalues disappear after prescribing the appropriate DOFs. Element stability is generally guaranteed if the stiffness matrix is integrated exactly. The inexact computation of some terms of the stiffness matrix (by using reduced integration, for instance) can introduce undesirable internal mechanisms in addition to those of rigid body motion.

These mechanisms should be avoided since they can lead to instability of the solution.

The existence of internal mechanisms is not always a reason to exclude an element as, in some cases, these mechanisms can not propagate themselves throughout the mesh. Eigenvalue tests must be performed to asses the existence of spurious mechanisms in an individual element, and also in patches of two or more element assemblies, in order to detect the capability of these mechanisms to propagate in a mesh.

3.10.4 Geometric –invariance condition

An element should not have preferent directions. This means that the elements must have what is usually called "geometric-invariance", also known as frame-invariance and geometric or spatial isotropy.

The lack of geometric-invariance is detected if different displacements or stresses are obtained when the element position is changed in space without changing the relative direction of the loading. In general, an element is geometric-invariant if all the displacement DOFs are interpolated with the same polynomial and this is not sensitive to the interchange of the coordinates. This can be achieved by using complete polynomial interpolations expressed in the natural coordinate system and an isoparametric formulation [CMPW].

Geometric-invariance can be lost in an element by underintegration of some of the terms in the stiffness matrix, such as in selective integration procedures (Section 4.4.2.1). The lack of geometric-invariance is a defect to be avoided. However, this does not necessarily destroies the convergence of the element.

3.11 SOME REMARKS ON THE COMPATIBILITY AND EQUILIBRIUM OF THE SOLUTION

We should keep in mind that the finite element solution is approximate and in general does not satisfy the equilibrium and compatibility requirements of the exact solution. In most cases we will find that:

1. *The solution is compatible within the elements.* This is always guaranteed by using continuous polynomial approximations.
2. *The solution can be incompatible along the interelemental boundaries.* As previously explained interelemental continuity can be violated if

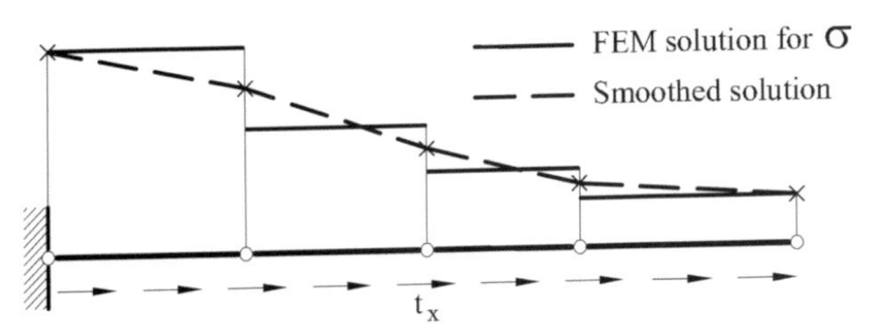

Fig. 3.9 Example of smoothing of nodal stresses in linear rod elements

the patch test is satisfied and this guarantees compatibility in the limit case of infinite refinement. Also, incompatible elements can sometimes produce excellent answers.

3. *Equilibrium of forces is satisfied at the nodes*, since these are the points where equilibrium is enforced during the assembly process and, therefore, at each node $\mathbf{Ka} - \mathbf{f} = 0$ is satisfied.

4. *There is not equilibrium of stresses along interelemental boundaries.* Nodal stresses can be directly obtained for each element in terms of the nodal displacements, or (what is more usual) by extrapolating the values computed at the Gauss points within the element (see Section 6.7). Stresses at interface nodes are different for each element and the global stress field is discontinuous between elements. *A continuous stress field can be obtained by smoothing the discontinuous nodal values* (for instance by simple nodal averaging) as shown in Figure 3.9. Also, the stresses computed at the free boundaries are usually not zero, although they are much smaller than the values inside the mesh. This incompatibility of the stress field is a consequence of the displacement formulation, where only displacement continuity is required and stresses can be discontinuous. Stress discontinuity does not violate the convergence requirements and it is usually corrected as the mesh is refined. The computation of nodal stresses is treated in Chapter 9.

5. *Stresses are not in equilibrium within elements.* The finite element values approximate the exact solution in an average integral form (by means of the PVW) [ZTZ]. Therefore, the differential equations of equilibrium in stresses are only approximately satisfied pointwise. An exception to this rule is for elements with linear shape functions, where the strain and stress fields are constant. The equilibrium differential equations involve the first derivatives of stresses and therefore are au-

tomatically satisfied for zero body forces [ZTZ]. The general lack of equilibrium of stresses is corrected as the mesh is refined and it does not preclude the convergence of the numerical solution.

3.12 CONVERGENCE REQUIREMENTS FOR ISOPARAMETRIC ELEMENTS

Isoparametric elements are based on the interpolation of the geometry field in terms of the nodal coordinate values using the same shape functions as for the displacement field. The coordinate transformation changes the derivatives of any function by a jacobian relation. For 1D problems we have

$$\frac{du}{dx} = \frac{du}{d\xi}\left(\frac{dx}{d\xi}\right)^{-1} = \frac{1}{J}\frac{du}{d\xi} \tag{3.58}$$

Therefore the PVW can be expressed in terms of the natural coordinate ξ with the maximum order of differentiation unchanged.

It follows immediately that if the displacement shape functions are so chosen in the natural coordinate system as to observe the usual rules of convergence (Section 3.8) then convergence of isoparametric elements will occur.

Furthermore, C° isoparametric elements always satisfy the rigid body conditions as defined in Section 3.8.4. The proof of this is simple; let us prescribe the following linear displacement field

$$u = a_1 + a_2 x \tag{3.59}$$

over a mesh of linear rod elements. The nodal displacements will take the values

$$u_i = a_1 + a_2 x_i \quad ; \quad i = 1, 2 \tag{3.60}$$

Inside the element $u = \sum_{i=1}^{2} N_i u_i$. Hence, making use of the prescribed field

$$u = \sum_{i=1}^{2} N_1 (a_1 + a_2 x_i) = a_1 \sum_{i=1}^{2} N_i + a_2 \sum_{i=1}^{2} N_i x_i \tag{3.61}$$

Since the element is isoparametric we have

$$x = \sum_{i=1}^{2} N_i x_i \tag{3.62}$$

From Eqs.(3.61) and (3.62) we deduce that the displacement field will coincide with the prescribed one (3.59) if

$$\sum_{i=1}^{2} N_i = 1 \tag{3.63}$$

is satisfied for any value of the natural coordinate ξ between -1 and $+1$. Eq.(3.63) is the usual rigid body requirement for the shape functions (Section 3.8.4). As a consequence, the constant derivative condition required in the patch test (Section 3.9) is satisfied for C° isoparametric elements.

Further details on 2D and 3D isoparametric elements will be given in Chapters 6 and 8.

3.13 ERROR TYPES IN THE FINITE ELEMENT SOLUTION

We recall once more that the finite element solution is approximate. This automatically implies that some kind of error in the numerical solution is unavoidable. Next, we will study the more usual sources of error.

3.13.1 Discretization error

This error is intrinsic to the polynomial form of the finite element approximation. We showed in Section 2.7 that the error involved in the approximation is of the order of the first term in the Taylor expansion of the solution not included in the complete shape function polynomial. Strang and Fix [SF] proposed the following general expression to estimate the error for 1D problems

$$e(\text{error}) = u_{\text{aprox}} - u_{\text{exact}} \leq Ch^{p+1} \,\text{Max} \left| \frac{\partial^{p+1} u_{\text{exact}}}{\partial x^{p+1}} \right| \tag{3.64}$$

where Max denotes the maximum value of the derivative over the element, C is a constant parameter depending on the element type, h is the maximum characteristic element dimension (i.e. the length in rod elements) and p the degree of the highest complete polynomial contained in the shape functions.

Eq.(3.64) shows that convergence is guaranteed if C and the $n+1th$ derivative of the solution are bounded. In this case the error will tend to zero as the element size diminishes.

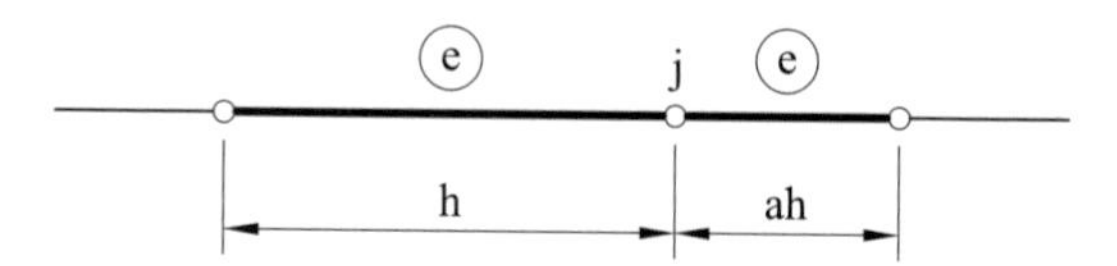

Fig. 3.10 Two rod elements of different sizes

The application of this concept to the 1D linear rod element gives for uniformly spaced meshes

$$e \propto h^2 \frac{\partial^2 u}{\partial x^2} \tag{3.65}$$

which implies that the error is proportional to the strain (or stress) gradient. Therefore, smaller elements should be used in zones where this gradient is expected to be higher. The reduction of the error by diminishing the element size is known in the mesh refinement literature as the *h method*.

The error can also be reduced by increasing the approximation order of the elements, while keeping their sizes constant. This results in a larger value of the exponent p in Eq.(3.64). This approach is known as the p *method*.

Eq.(3.64) assumed a mesh of equal element sizes. The effect of using elements of different sizes has been studied for analysis of axially loaded rods using linear elements of two different sizes (Figure 3.10). The error in the satisfaction of the differential equilibrium equation at the jth node is [SF]

$$e = -\frac{h}{3}(1-a)\left(\frac{\partial^3 u}{\partial x^3}\right)_j + \frac{h^2}{12}\left(\frac{1+a^3}{1+a}\right)\frac{\partial^4 u}{\partial x^4}(x_j) + \cdots \tag{3.66}$$

where h and a are the lengths of two adjacent elements (Figure 3.10). Eq.(3.66) shows that the error is of the order h^2 for a uniform mesh ($a \simeq 1$), whereas a higher error of order h is obtained when the element sizes are very different ($a \neq 1$). This indicates that drastic changes in the sizes of contiguous elements in a mesh should be avoided.

The same concepts apply for 2D and 3D problems. The estimation of the discretization error in two dimensions involves the Taylor expansion

$$\begin{aligned} u(x+h, y+k) = u(x,y) + \left[h\frac{\partial u}{\partial x} + k\frac{\partial u}{\partial y}\right] + \\ + \frac{1}{2!}\left[h^2\frac{\partial^2 u}{\partial x^2} + 2hk\frac{\partial^2 u}{\partial x \partial y} + k^2\frac{\partial^2 u}{\partial y^2}\right] + \cdots \end{aligned} \tag{3.67}$$

where u is the exact solution and h and k are a measure of the element sizes in the x and y directions respectively.

It can be shown that the discretization error for 2D linear elements, like the 3-noded triangle (Chapter 4), is proportional to the underlined term on the right-hand side of Eq.(3.67) [SF]. The second derivatives in Eq.(3.67) can be related to strain (or stress) gradients. Thus, for a constant strain field the error is very small.

The discretization error can also be expressed in terms of the ratio $\frac{k}{h}$.This is a measure of the relative dimensions of the element and it is known as the *element aspect ratio.* For an equilateral element its aspect ratio should be equal to one. However, it will take a large value for a long triangular element. It is recommended to keep the element aspect ratio as close to unity as possible through the mesh.

The estimation of the discretization error will be treated in more detail in Chapter 9, together with the techniques for reducing the error using adaptive mesh refinement.

3.13.2 Error in the geometry approximation

In many cases the interpolation of the geometry is unable to reproduce exactly the real shape of the structure. This can be due to a geometry approximation of a lower order than the exact one, or, what is more usual, to the ignorance of the exact analytical form for the geometry defined by the coordinates of a number of points. In both cases, there will be an error in the geometry approximation. This error can be reduced by refining the mesh, or by using higher order superparametric approximations. A compromise between these two options is to use isoparametric elements. This unavoidably introduces an error in the geometry approximation in some cases. An exception are structures with linear or planar boundaries where the geometry can always be exactly approximated.

3.13.3 Error in the computation of the element integrals

The exact numerical computation of the element integrals implies using an appropriate quadrature. Otherwise, an error occurs due to the underestimation of the integral value. In many cases, the exact numerical integration is not possible due to the rational algebraic functions appearing in the element integrals. Also, the approximation of the exact value may require a large number of integration points, which may be very expensive.

In such cases, it is usual to accept a certain error in the computation of the element integrals.

Paradoxically enough, this error can, on occasions, be beneficial. Usually by under-integrating the stiffness matrix terms the element becomes more flexible, and this balances the stiffening introduced by the approximation of the displacement field and the geometry. This explains why sometimes good results can be obtained with coarse meshes. In the following chapters we will see that the "reduced integration" quadrature is sometimes used to guarantee the correct solution. The inexact computation of the stiffness matrix can however modify its correct rank and introduce spurious mechanisms. Reduced integration is therefore a technique which should be used with extreme care.

3.13.4 Errors in the solution of the global equation system

Three type of errors are typical in the solution of the global system of FEM equations using a direct solution method (i.e. Gaussian elimination, Choleski, Frontal method, etc.): errors due to the *ill-conditioning* of the equations; *truncation* errors and *round-off* errors [Ral].

The equation system $\mathbf{Ka} = \mathbf{f}$ is ill-conditioned if small changes in the terms of $\mathbf{K}$ or $\mathbf{f}$ induce large changes in the solution $\mathbf{a}$. The main reason for ill-conditioning is the existence of an element, or a group of elements, of large stiffness connected to elements of much smaller stiffness. The behaviour of such a structure can be artificially altered and, unless the computer can store a sufficiently large number of digits, the stiffness matrix behaves as singular or quasi-singular. The error associated with ill-conditioning of the equation system therefore depends on the digit storage capacity of the computer, i.e. in the *truncation* and *round-off errors* which are the main contributors to the total error in the solution.

The *truncation error* is quite important. A computer using d digits to represent a number in simple precision can only store the first d digits of each term of $\mathbf{K}$ and $\mathbf{f}$. It is then possible that essential information for the correct solution is lost by truncating a number.

The *round-off error* is due to the adjustment automatically performed by the computer on the last digit of each number during computations. Experience shows that this error is less important than the truncation error. Nevertheless, unnecessary round-off errors, such as those in some parameters like the coordinates and weights of the numerical quadrature, should be avoided by defining these parameters with the maximum number of digits allowed by the computer.

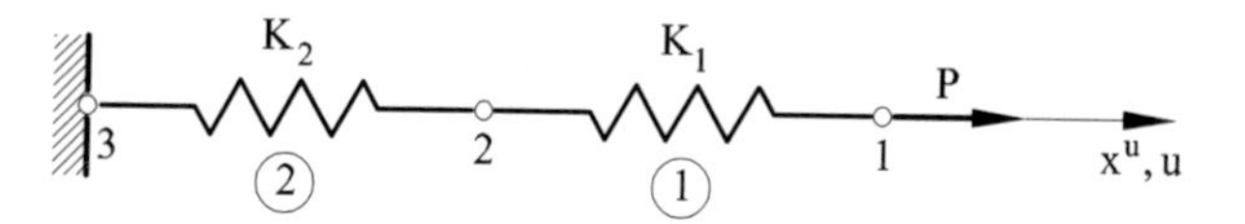

Fig. 3.11 Spring system with two degrees of freedom

Example 3.7: Study the influence of truncation error in the solution of the spring system shown in Figure 3.11 [CMPW].

- ***Solution***

The system of stiffness equations and its inverse after eliminating the prescribed DOF ($u_3 = 0$) are

$$\underbrace{\begin{bmatrix} K_1 & -K_1 \\ -K_1 & K_1+K_2 \end{bmatrix}}_{\mathbf{K}} \begin{Bmatrix} u_1 \\ u_2 \end{Bmatrix} = \begin{Bmatrix} P \\ 0 \end{Bmatrix} \quad ; \quad \underbrace{\begin{bmatrix} \frac{1}{K_1}+\frac{1}{K_2} & \frac{1}{K_2} \\ \frac{1}{K_2} & \frac{1}{K_2} \end{bmatrix}}_{\mathbf{K}^{-1}} \begin{Bmatrix} P \\ 0 \end{Bmatrix} = \begin{Bmatrix} u_1 \\ u_2 \end{Bmatrix}$$

If $K_1 >> K_2$, K_1 dominates in $\mathbf{K}$. However, K_2 dominates $\mathbf{K}^{-1}$ and therefore the value of the solution. The computation of $\mathbf{K}^{-1}$ is only correct if the terms in $\mathbf{K}$ are evaluated in a way such that K_2 is not lost during the solution. Thus, if $K_1 = 80$ and $K_2 = 0.0023$ the computer must retain at least six digits and K_1 must be represented as 80.0000 so that the last digit of K_2 is retained in the term $K_1 + K_2$. If only four digits are retained the sum $K_1 + K_2$ will give 80.00 and $\mathbf{K}$ will be singular. This problem is *ill conditioned* since the solution is sensitive to the changes (truncation) in the sixth digit of the term $K_1 + K_2$.
Also, if the system $\mathbf{Ka}=\mathbf{f}$ is solved using Gauss elimination (Appendix B), the elimination of the displacement u_1 changes the last diagonal term to $(K_1 + K_2) - K_1$. We see that information for a correct solution can again be lost if $K_1 >> K_2$.

A way to avoid truncation errors and to improve the solution is to use double precision throughout the solution process, i.e. for computing the terms of $\mathbf{K}$ and $\mathbf{f}$ during the solution of the equation system.

An indicator of how sensitive the system $\mathbf{Ka} = \mathbf{f}$ is to truncation and round-off errors is the condition number of $\mathbf{K}$. An estimation of the number of significative figures exactly computed in the solution process is [Bat]

$$s \approx t - \log_{10}[\text{cond}(\mathbf{K})] \tag{3.68}$$

where t is the maximum number of digits which can be stored by the computer and cond $(\mathbf{K})$ is the condition number of $\mathbf{K}$ defined as

$$\text{cond}(\mathbf{K}) = \frac{|\lambda\text{max}|}{|\lambda\text{min}|} \tag{3.69}$$

where λmax and λmin are respectively the larger and smaller eigenvalues of $\mathbf{K}$ (Appendix A). Although Eq.(3.68) is only approximate, it indicates that the accuracy of the solution decreases as the condition number increases.

A low condition number of $\mathbf{K}$ is also important in order to speed up the iterative solution of the system $\mathbf{Ka} = \mathbf{f}$ (Appendix B) [Ral].

It is therefore desirable that the condition number of $\mathbf{K}$ should be as low as possible. This can be achieved by an adequate scaling of the terms of $\mathbf{K}$ [Ral,RG,RGL].

3.13.5 Errors associated with the constitutive equation

The survey of the error sources in the finite element solution of a structure would be incomplete without referring to the errors arising from a wrong definition of the material properties. In this book only linear elastic materials are considered. The importance of the evaluation of the relevant parameters in the constitutive equation is obvious. For a structure with homogeneous and isotropic material the displacements are proportional to the Young modulus, although the stresses are not affected by this value and they depend only on the Poisson's ratio. For a structure with orthotropic or anisotropic materials both the displacements and the stresses depend on the Young modulus and the Poisson's ratio. We should be aware that an incorrect definition of the material parameters can lead to larger errors than those induced by all the error sources mentioned in the previous sections.

4

2D SOLIDS. LINEAR TRIANGULAR AND RECTANGULAR ELEMENTS

4.1 INTRODUCTION

This chapter initiates the application of the FEM to structures which satisfy the assumptions of two-dimensional (2D) elasticity (i.e. plane stress or plane strain). Many of the concepts here studied will be useful when dealing with other structural problems in the subsequent chapters. Therefore, this chapter is introductory to the application of the FEM to continuous 2D and 3D structures.

There are a wide number of structures of practical interest which can be analyzed following the assumptions of 2D elasticity. All these structures have a sort of prismatic geometry. Depending on the relative dimensions of the prism and the loading type, the following two categories can be distinguished:

Plane stress problems. A prismatic structure is under plane stress if one of its dimensions (thickness) is much smaller than the other two and all the loads are contained in the middle plane of the structure. The analysis domain is the middle section (Figure 4.1). Amongst the structural problems that can be included in the plane stress category we find the analysis of deep beams, plates and walls under in-plane loading, buttress dams, etc.

Plane strain problems. A prismatic structure is under plane strain if one of its dimensions (length) is larger than the other two and all the loads are uniformly distributed along its length and they act orthogonally to the longitudinal axis. The analysis domain is a cross section to this axis (Figure 4.2). Amongst the structures which follow the plane strain assumption we find containing walls, gravity dams, pressurised pipes and many problems of geotechnical engineering (tunnels, foundations, etc.).

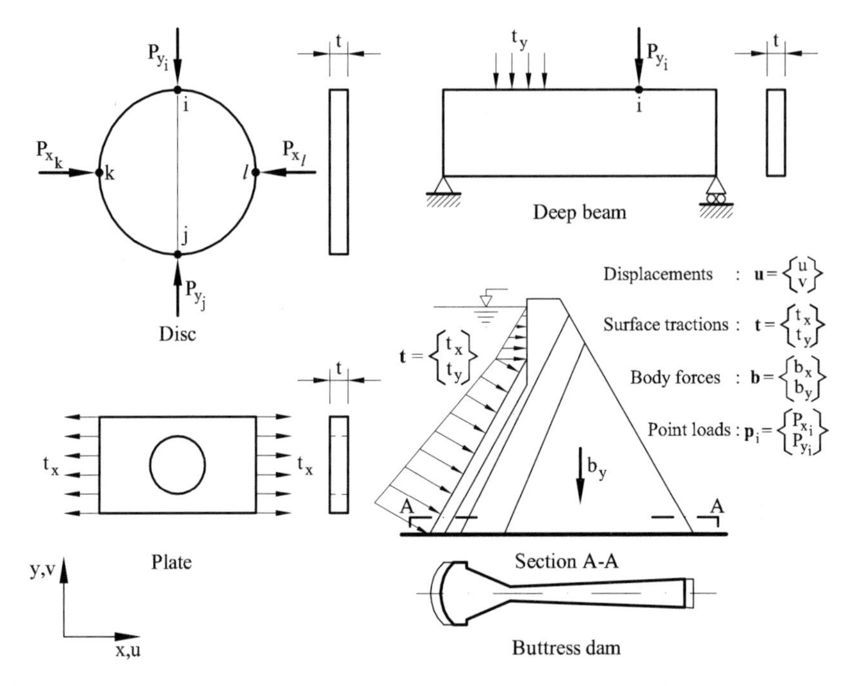

Fig. 4.1 Examples of plane stress problems. Displacement field and loads acting on the middle plane section

2D elasticity theory provides a mathematical model by which the behaviour of a real 3D structure is represented by that of a 2D solid. The FEM provides us with an approximation to the "exact" solution of the 2D elasticity equations using 2D solid elements. The accuracy of the numerical solution depends on the element type and the quality of the mesh chosen.

2D elasticity theory allows the FEM analysis of plane stress and plane strain problems in a unified manner. We should recall however that each of the two problems conceptually represents a class of very different structural types.

The chapter starts with a brief description of the basic concepts of 2D elasticity theory. Then the finite element solution using simple 3-noded triangles and 4-noded quadrilaterals is presented. Most of the finite element expressions are completely general and applicable to any other 2D solid element. The general derivation of the element shape functions and the formulation of higher order triangular and quadrilateral elements and of isoparametric elements are studied in the next chapter.

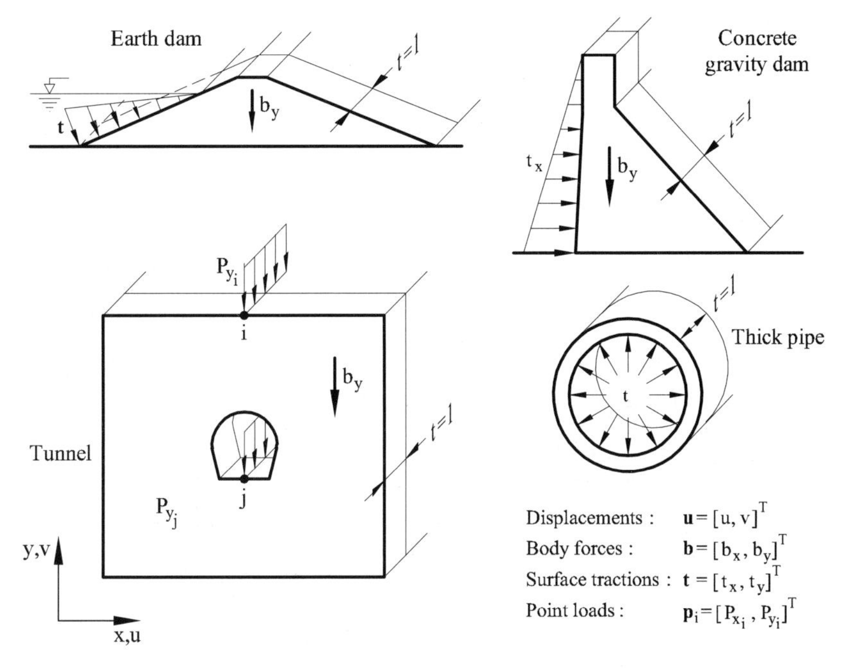

Fig. 4.2 Examples of plane strain problems. Displacement field and loads acting on a transverse section

4.2 TWO DIMENSIONAL ELASTICITY THEORY

Next, we present the concepts of 2D elasticity theory needed for the application of the FEM.

4.2.1 Displacement field

Both the plane stress and plane strain assumptions imply that the transversal sections to the prismatic axis z deform in the same manner and also that the displacement along the z axis is negligible. Therefore, only a generic 2D transverse section in the plane $x-y$ needs to considered for the analysis. The displacement field of the analysis section is defined by the displacements $u(x,y)$ and $v(x,y)$ in the x and y directions, respectively (Figures 4.1 and 4.2). The displacement vector of a point is

$$\mathbf{u}(x,y) = \begin{Bmatrix} u(x,y) \\ v(x,y) \end{Bmatrix} \tag{4.1}$$

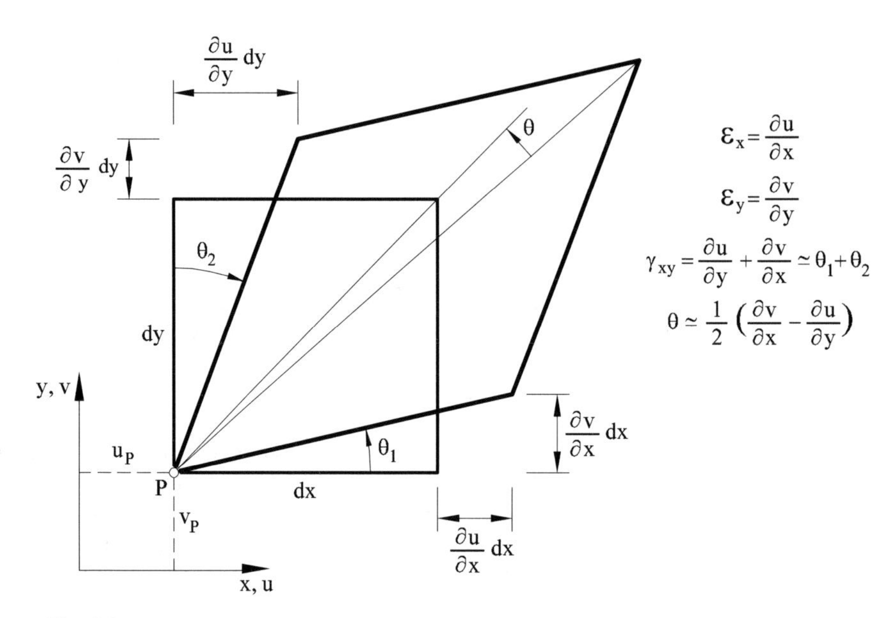

Fig. 4.3 Deformation of an infinitesimal 2D domain and definition of strains

4.2.2 Strain field

The displacement field (4.1) allows the corresponding strains to be derived from standard elasticity theory [TG]. This gives

$$\begin{aligned} \varepsilon_x &= \frac{\partial u}{\partial x} \quad , \quad \varepsilon_y = \frac{\partial v}{\partial y} \\ \gamma_{xy} &= \frac{\partial u}{\partial y} + \frac{\partial v}{\partial x} \quad , \quad \gamma_{xz} = \gamma_{yz} = 0 \end{aligned} \tag{4.2}$$

The longitudinal strain ε_z is assumed to be zero in the plane strain case. Conversely, ε_z is not zero in plane stress situations, although the conjugate stress σ_z is assumed to be zero. Therefore, ε_z needs not be considered for either plane stress or plane strain problems as the *work performed* by the longitudinal strain (i.e. $\sigma_z \varepsilon_z$) is always zero. Consequently, the *strain vector* is defined in both cases simply as

$$\boldsymbol{\varepsilon} = [\varepsilon_x, \varepsilon_y, \gamma_{xy}]^T \tag{4.3}$$

The graphical meaning of the strains for 2D problems is shown in Figure 4.3.

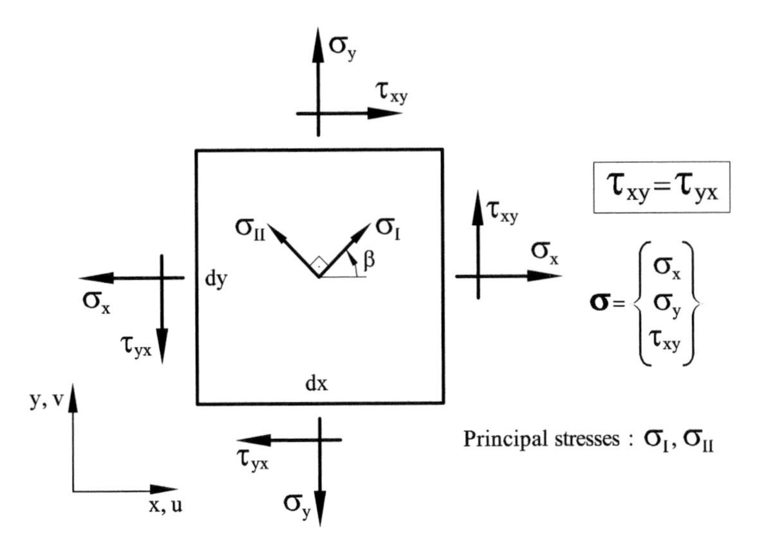

Fig. 4.4 Definition of stresses $\sigma_x, \sigma_y, \tau_{xy}$ and principal stresses σ_I, σ_{II} in 2D solids

4.2.3 Stress field

It is deduced from Eq.(4.2) that the shear stresses τ_{xz} and τ_{yz} are zero. Also, for the same reasons as explained above, the longitudinal stress σ_z does not contribute to the internal work and the stress vector is defined for both plane stress and plane strain cases as (Figure 4.4) [TG]

$$\boldsymbol{\sigma} = [\sigma_x, \sigma_y, \tau_{xy}]^T \tag{4.4}$$

4.2.4 Stress-strain relationship

The relationship between stresses and strains is derived from 3D elasticity theory [TG] using the assumptions stated above (i.e. $\sigma_z = 0$ for plane stress, $\varepsilon_z = 0$ for plane strain, and $\gamma_{xz} = \gamma_{yz} = 0$ in both cases). After same simple algebra (Example 4.1) the following matrix relationship can be obtained

$$\boldsymbol{\sigma} = \mathbf{D}\,\boldsymbol{\varepsilon} \tag{4.5}$$

where $\mathbf{D}$ is the elastic material matrix (or constitutive matrix)

$$\mathbf{D} = \begin{bmatrix} d_{11} & d_{12} & 0 \\ d_{21} & d_{22} & 0 \\ 0 & 0 & d_{33} \end{bmatrix} \tag{4.6}$$

It can be proved from the Maxwell-Betti theorem that $\mathbf{D}$ is always symmetrical [TG] and $d_{12} = d_{21}$. For isotropic elasticity we have

$$
\begin{array}{ll}
\textit{Plane stress} & \textit{Plane strain} \\
d_{11} = d_{22} = \dfrac{E}{1-\nu^2} & d_{11} = d_{22} = \dfrac{E(1-\nu)}{(1+\nu)(1-2\nu)} \\
d_{12} = d_{21} = \nu d_{11} & d_{12} = d_{21} = d_{11}\dfrac{\nu}{1-\nu} \\
d_{33} = \dfrac{E}{2(1+\nu)} = G & d_{33} = \dfrac{E}{2(1+\nu)} = G
\end{array}
\tag{4.7}
$$

where E is the Young modulus and ν the Poisson's ratio.

For an orthotropic material with principal orthotropy directions along the 1, 2, 3 axes (where 3 is the out-of-plane direction), matrix $\mathbf{D}$ has the following expression [BD,He,Le,TG]:

Plane-stress

$$
\mathbf{D} = \frac{1}{1-\nu_{12}\nu_{21}} \begin{bmatrix} E_1 & \nu_{21}E_1 & 0 \\ \nu_{12}E_2 & E_2 & 0 \\ 0 & 0 & (1-\nu_{12}\nu_{21})G_{12} \end{bmatrix} \tag{4.8a}
$$

Plane strain

$$
\mathbf{D} = \frac{1}{ad-bc} \begin{bmatrix} aE_1 & bE_1 & 0 \\ cE_2 & dE_2 & 0 \\ 0 & 0 & (ad-bc)G_{12} \end{bmatrix} \tag{4.8b}
$$

where

$$
\frac{1}{G_{12}} \simeq \frac{1+\nu_{21}}{E_1} + \frac{1+\nu_{12}}{E_2} \tag{4.9a}
$$

and

$$
\begin{array}{ll}
a = 1 - \nu_{23}\nu_{32} \;\; ; & b = \nu_{12} + \nu_{32}\nu_{13} \\
c = \nu_{21} + \nu_{23}\nu_{31} \; ; & d = 1 - \nu_{13}\nu_{31}
\end{array}
\tag{4.9b}
$$

The symmetry of $\mathbf{D}$ requires [BD]

$$
\frac{E_2}{E_1} = \frac{\nu_{12}}{\nu_{21}}(\text{plane stress}) \qquad \text{and} \qquad \frac{E_2}{E_1} = \frac{b}{c}(\text{plane strain}) \tag{4.10}
$$

If the in-plane orthotropy directions 1, 2 are inclined an angle β with respect to the global axes of the structure x, y (Figure 4.5) the constitutive

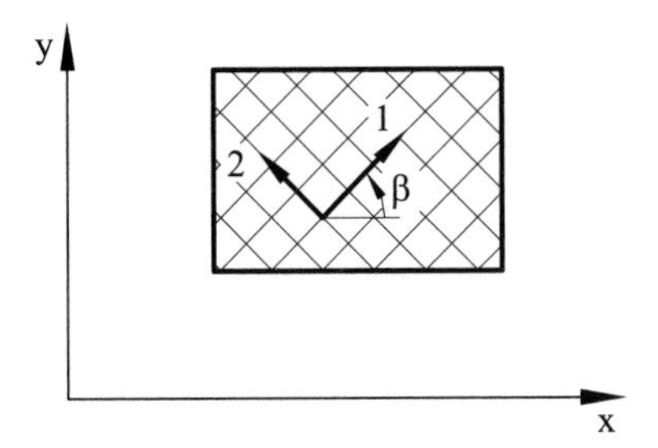

Fig. 4.5 Orthotropic material with principal orthotropy directions 1, 2

relationship is derived as follows. First, the strains in local axes 1, 2 are expressed in terms of the global strains by

$$\boldsymbol{\varepsilon}' = \mathbf{T}\boldsymbol{\varepsilon} \quad , \quad \boldsymbol{\varepsilon}' = [\varepsilon_1, \varepsilon_2, \gamma_{12}]^T \tag{4.11a}$$

with [CMPW]

$$\mathbf{T} = \begin{bmatrix} \cos^2\beta & \sin^2\beta & \sin\beta\,\cos\beta \\ \sin^2\beta & \cos^2\beta & -\sin\beta\,\cos\beta \\ -2\sin\beta\cos\beta & 2\sin\beta\,\cos\beta & \cos^2\beta - \sin^2\beta \end{bmatrix} \tag{4.11b}$$

We note that $|\mathbf{T}| = 1$.

The transformation for the stresses is obtained from the virtual work equivalent in global and local axes, i.e.

$$\delta\boldsymbol{\varepsilon}^T\boldsymbol{\sigma} = \delta\boldsymbol{\varepsilon}'^T\boldsymbol{\sigma}' = \delta\boldsymbol{\varepsilon}^T\mathbf{T}^T\boldsymbol{\sigma}' \tag{4.12a}$$

The later equation must be true for any virtual strain vector. Hence

$$\boldsymbol{\sigma} = \mathbf{T}^T\boldsymbol{\sigma}' \quad \text{and} \quad \boldsymbol{\sigma}' = [\mathbf{T}]^{-T}\boldsymbol{\sigma} \tag{4.12b}$$

where

$$\boldsymbol{\sigma}' = \begin{Bmatrix} \sigma_1 \\ \sigma_2 \\ \tau_{12} \end{Bmatrix} \quad , \quad [\mathbf{T}^{-1}]^T = \begin{bmatrix} \cos^2\beta & \sin^2\beta & 2\sin\beta\cos\beta \\ \sin^2\beta & \cos^2\beta & -2\sin\beta\cos\beta \\ -\sin\beta\cos\beta & \sin\beta\cos\beta & (\cos^2\beta - \sin^2\beta) \end{bmatrix} \tag{4.13}$$

The stress-strain relationship in the local axes is written as

$$\boldsymbol{\sigma}' = \mathbf{D}'\boldsymbol{\varepsilon}' \tag{4.14}$$

where $\mathbf{D}'$ is given by Eq.(4.8).

Finally, from Eqs.(4.1), (4.12b) and (4.14) we obtain

$$\boldsymbol{\sigma} = \mathbf{T}^T\mathbf{D}'\mathbf{T}\boldsymbol{\varepsilon} = \mathbf{D}\,\boldsymbol{\varepsilon} \tag{4.15}$$

with

$$\mathbf{D} = \mathbf{T}^T\mathbf{D}'\mathbf{T} \tag{4.16}$$

It is easy to check that matrix $\mathbf{D}$ resulting from Eq.(4.16) is symmetrical.

The d_{ij} coefficients for anisotropic elasticity can be found in [He,Le].

If the solid is subjected to initial strains such as thermal strains, the constitutive relationship (4.5) must be modified. The total strain $\boldsymbol{\varepsilon}$ is now equal to the sum of the elastic ($\boldsymbol{\varepsilon}^e$) and the initial ($\boldsymbol{\varepsilon}^0$) strains, whereas in Eq.(4.5) *all* the strains were considered to be elastic. Since the stresses are proportional to the elastic strains, the constitutive equation is now written as

$$\boldsymbol{\sigma} = \mathbf{D}\,\boldsymbol{\varepsilon}^e = \mathbf{D}(\boldsymbol{\varepsilon} - \boldsymbol{\varepsilon}^0) \tag{4.17}$$

For the case of initial strains due to thermal effects and isotropic material, vector $\boldsymbol{\varepsilon}^0$ has the following expressions

Plane stress *Plane strain*

$$\boldsymbol{\varepsilon}^0 = \begin{Bmatrix} \alpha\Delta T \\ \alpha\Delta T \\ 0 \end{Bmatrix} \qquad \boldsymbol{\varepsilon}^0 = (1+\nu)\begin{Bmatrix} \alpha\Delta T \\ \alpha\Delta T \\ 0 \end{Bmatrix} \tag{4.18}$$

where α is the thermal expansion coefficient and ΔT is the temperature increment at each point. Note that a temperature increment does not induce a shear strain.

The difference between the values of $\boldsymbol{\varepsilon}^0$ for plane stress and plane strain is due to the different assumptions for σ_z and ε_z in each case (see Examples 4.2 and 4.3).

Table 4.1 shows the basic constitutive properties for standard concrete, steel and aluminium materials. A more comprehensive list of material properties is given in [BD,Co2,PP] and in Annex 1 of Volume 2 [On].

For anisotropic materials, the initial strains due to thermal effects are considered first in the principal directions of the material and then are transformed to global axes to find the global components of $\boldsymbol{\varepsilon}^\circ$. In these cases the tangential strain τ_{xy}° is not longer zero [He,ZTZ].

The solid can also be initially subjected to stresses defined by a vector $\boldsymbol{\sigma}^0$. These *initial stresses* can have different sources. For instance, if a part

	$E\times10^{-3}$ Mpa	ν	Density ρ kg/m^3	$\alpha\times10^5$ $^\circ C^{-1}$	Limit tensile stress (MPa)
Concrete	20–40	0.15	2400	1.2	2–4
Steel	190–210	0.30	7800	1.3–1.6	400–1600
Aluminium	70	0.33	2710	2.2	140–600

Table 4.1 Basic material properties for standard concrete, steel and aluminium

of the material is removed from a deformed structure under a set of loads, then automatically a new deformation is originated due to the existence of initial stresses. The total stresses in the new equilibrium configuration are obtained by the sum of the initial ones and those originated in the deformation process. For the more general case

$$\boldsymbol{\sigma} = \mathbf{D}(\boldsymbol{\varepsilon} - \boldsymbol{\varepsilon}^0) + \boldsymbol{\sigma}^0 \tag{4.19a}$$

where

$$\boldsymbol{\sigma}^0 = [\sigma_x^0, \sigma_y^0, \tau_{xy}^0]^T \tag{4.19b}$$

is the initial stress vector. A practical example of initial stresses is the analysis of a tunnel in geotechnical engineering, where the equilibrium of the excavated zone depends on the initial stresses in the zone before the excavation. Initial stresses are also very common in welded mechanical parts and here they are usually termed "residual" stresses [ZTZ].

4.2.5 Principal stresses and failure criteria

The stress field in a 2D solid is better represented by the two principal stresses σ_I and σ_{II} (Figure 4.4). In general the principal stresses are the roots of the characteristic polynomial

$$\det\left([\sigma] - \lambda \mathbf{I}_2\right) = 0 \tag{4.20a}$$

where

$$[\sigma] = \begin{bmatrix} \sigma_x & \tau_{xy} \\ \tau_{xy} & \sigma_y \end{bmatrix} \quad \text{and} \quad \mathbf{I}_2 = \begin{bmatrix} 1 & 0 \\ 0 & 1 \end{bmatrix} \tag{4.20b}$$

From Eq.(4.20a) we deduce

$$\begin{aligned} \lambda_1 = \sigma_I &= \frac{\sigma_x + \sigma_y}{2} + \frac{1}{2}[(\sigma_x - \sigma_y)^2 + 4\tau_{xy}^2]^{1/2} \\ \lambda_2 = \sigma_{II} &= \frac{\sigma_x + \sigma_y}{2} - \frac{1}{2}[(\sigma_x - \sigma_y)^2 + 4\tau_{xy}^2]^{1/2} \end{aligned} \tag{4.21}$$

The angle that the direction of the principal stress σ_I forms with the x axis is defined by (Figure 4.4)

$$\tan 2\beta = \frac{2\tau_{xy}}{\sigma_x - \sigma_y} \tag{4.22}$$

Failure at a point in a 2D solid can be identified when the maximum principal stress reaches a prescribed limit value. This is typically used for detecting the onset of fracture at a point in fragile materials (plain concrete, glass, ceramics, etc.) [ZT]. Alternative failure criteria can be based on verifying the limit bound for an appropriate stress invariant. For more details see Section 8.2.5 and [ZT].

Example 4.1: Find the constitutive equation for an isotropic elastic material under plane stress and plane strain conditions.

- ***Solution***

The starting point is the constitutive equation for 3D isotropic elasticity [TG]

$$\varepsilon_x = \frac{(\sigma_x - \nu\sigma_y - \nu\sigma_z)}{E}; \ \varepsilon_y = \frac{(\sigma_y - \nu\sigma_x - \nu\sigma_z)}{E}; \ \varepsilon_z = \frac{(\sigma_z - \nu\sigma_x - \nu\sigma_y)}{E}$$

$$\gamma_{xy} = \frac{2(1+\nu)}{E}\tau_{xy} \quad ; \quad \gamma_{xz} = \frac{2(1+\nu)}{E}\tau_{xz} \quad ; \quad \gamma_{yz} = \frac{2(1+\nu)}{E}\tau_{yz}$$

These equations will be now simplified using the plane stress and plane strain assumptions.

Plane stress: $\sigma_z = 0; \gamma_{xz} = \gamma_{yz} = 0$

Substituting the plane stress conditions into the above equations we have

$$\varepsilon_x = \frac{1}{E}(\sigma_x - \nu\sigma_y), \ \varepsilon_y = \frac{1}{E}(\sigma_y - \nu\sigma_x), \ \gamma_{xy} = \frac{2(1+\nu)}{E}\tau_{xy}$$

$$\varepsilon_z = -\frac{\nu}{E}(\sigma_x + \sigma_y) \quad ; \quad \tau_{xz} = \tau_{yz} = 0$$

These equations yield the relationship between $\sigma_x, \sigma_y, \tau_{xy}$ and the corresponding strains as

$$\sigma_x = \frac{E}{1-\nu^2}(\varepsilon_x + \nu\varepsilon_y); \ \sigma_y = \frac{E}{1-\nu^2}(\varepsilon_y + \nu\varepsilon_x); \ \tau_{xy} = \frac{E}{(1+\nu)}\gamma_{xy}$$

from which the coefficients of $\mathbf{D}$ in Eqs.(4.6) and (4.7) can be deduced.

Substituting the expressions of σ_x and σ_y for ε_z we find that

$$\varepsilon_z = -\nu(\varepsilon_x + \varepsilon_y)$$

Therefore, the longitudinal strain ε_z can be obtained "a posteriori" in terms of ε_x and ε_y.

Plane strain: $\varepsilon_z = 0; \gamma_{xz} = \gamma_{yz} = 0$

From the general equations relating strains and stresses we have

$$\varepsilon_x = \frac{1}{E}(\sigma_x - \nu\sigma_y - \nu\sigma_z); \ \varepsilon_y = \frac{1}{E}(\sigma_y - \nu\sigma_x - \nu\sigma_z) = 0$$

$$\varepsilon_z = 0 = \frac{1}{E}(\sigma_z - \nu\sigma_x - \nu\sigma_y) \quad ; \quad \gamma_{xy} = \frac{2(1+\nu)}{E}\tau_{xy} \quad ; \quad \tau_{xz} = \tau_{yz} = 0$$

From the condition $\varepsilon_z = 0$ we find $\sigma_z = \nu(\sigma_x + \sigma_y)$. Substituting this value into the other equations we find

$$\begin{aligned}
\sigma_x &= \frac{E(1-\nu)}{(1+\nu)(1-2\nu)}\left(\varepsilon_x + \frac{\nu}{1-\nu}\varepsilon_y\right) \\
\sigma_y &= \frac{E(1-\nu)}{(1+\nu)(1-2\nu)}\left(\varepsilon_y + \frac{\nu}{1-\nu}\varepsilon_x\right) \\
\tau_{xy} &= \frac{E}{2(1+\nu)}\gamma_{xy}
\end{aligned}$$

from which the expression (4.7) for $\mathbf{D}$ can be obtained.
The same procedure can be used for orthotropic or anisotropic materials starting from the corresponding expressions of 3D elasticity [He,Le].

Example 4.2: Find the initial strain vectors due to thermal effects for 2D isotropic elasticity.

- ***Solution***

The main assumption is that the total strains are the sum of the elastic and the thermal ones. Also, it is assumed that a thermal expansion (or contraction) originates axial strains of value $\alpha \Delta T$, where α is the thermal expansion coefficient and ΔT the temperature increment. With these assumptions the total strains for 3D isotropic elasticity can be written as (see first equation of Example 4.1)

$$\varepsilon_x = \varepsilon_x^e + \varepsilon_x^0 = \frac{1}{E}(\sigma_x - \nu\sigma_y - \nu\sigma_z) + \alpha\Delta T$$

$$\varepsilon_y = \varepsilon_y^e + \varepsilon_y^0 = \frac{1}{E}(\sigma_y - \nu\sigma_z - \nu\sigma_x) + \alpha\Delta T$$
$$\varepsilon_z = \varepsilon_z^e + \varepsilon_z^0 = \frac{1}{E}(\sigma_z - \nu\sigma_x - \nu\sigma_y) + \alpha\Delta T$$
$$\gamma_{xy} = \frac{2(1+\nu)}{E}\tau_{xy} \quad ; \quad \gamma_{xz} = \frac{2(1+\nu)}{E}\tau_{xz} \quad ; \quad \gamma_{yz} = \frac{2(1+\nu)}{E}\tau_{yz}$$

Plane stress $\sigma_z = \gamma_{xz} = \gamma_{yz} = 0$

Substituting these conditions into above equations we have

$$\varepsilon_x = \frac{1}{E}(\sigma_x - \nu\sigma_y) + \alpha\Delta T \quad ; \quad \varepsilon_y = \frac{1}{E}(\sigma_y - \nu\sigma_x) + \alpha\Delta T$$

$$\varepsilon_z = -\frac{\nu}{E}(\sigma_x + \sigma_y) + \alpha\Delta T \quad ; \quad \gamma_{xy} = \frac{2(1+\nu)}{E}\tau_{xy} \quad ; \quad \tau_{xz} = \tau_{yz} = 0$$

Solving for σ_x, σ_y and τ_{xy} gives

$$\sigma_x = \frac{E}{1-\nu^2}\left[(\varepsilon_x - \varepsilon_x^0) + \nu(\varepsilon_y - \varepsilon_y^0)\right]$$
$$\sigma_y = \frac{E}{1-\nu^2}\left[(\varepsilon_y - \varepsilon_y^0) + \nu(\varepsilon_x - \varepsilon_x^0)\right]$$
$$\tau_{xy} = \frac{E}{2(1+\nu)}\gamma_{xy}$$

which can be written in the form $\boldsymbol{\sigma} = \mathbf{D}\ (\boldsymbol{\varepsilon} - \boldsymbol{\varepsilon}^0)$, with $\boldsymbol{\varepsilon}^0 = \alpha\Delta T[1,1,0]^T$ being the initial strain vector and $\mathbf{D}$ the matrix given in (4.6) and (4.7).

Plane strain $\varepsilon_z = \gamma_{xz} = \gamma_{yz} = 0$

From the general expressions we find

$$\varepsilon_x = \frac{1}{E}(\sigma_x - \nu\sigma_y - \nu\sigma_z) + \alpha\Delta T$$
$$\varepsilon_y = \frac{1}{E}(\sigma_y - \nu\sigma_x - \nu\sigma_z) + \alpha\Delta T$$
$$0 = \tfrac{1}{E}(\sigma_z - \nu\sigma_x - \nu\sigma_y) + \alpha\Delta T$$
$$\gamma_{xy} = \frac{2(1+\nu)}{E}\tau_{xy} \quad ; \quad \tau_{xz} = \tau_{yz} = 0$$

From the third equation we find

$$\sigma_z = \nu(\sigma_x + \sigma_y) - E\ \alpha\ \Delta T$$

Substituting this value into the first two equations yields

$$\begin{aligned}
\varepsilon_x &= \frac{1}{E}\Big[(1-\nu^2)\sigma_x - \nu(1+\nu)\sigma_y\Big] + (1+\nu)\alpha\Delta T\\
\varepsilon_y &= \frac{1}{E}\Big[(1-\nu^2)\sigma_y - \nu(1+\nu)\sigma_x\Big] + (1+\nu)\alpha\Delta T\\
\gamma_{xy} &= \frac{2(1+\nu)}{E}\tau_{xy}
\end{aligned}$$

Solving for σ_x, σ_y and τ_{xy} gives

$$\begin{aligned}
\sigma_x &= \frac{E(1-\nu)}{(1+\nu)(1-2\nu)}\Big[(\varepsilon_x - (1+\nu)\alpha\Delta T) + \frac{\nu}{1-\nu}(\varepsilon_y - (1+\nu)\alpha\Delta T)\Big]\\
\sigma_y &= \frac{E(1-\nu)}{(1+\nu)(1-1\nu)}\Big[(\varepsilon_y - (1+\nu)\alpha\Delta T) + \frac{\nu}{1-\nu}(\varepsilon_x - (1+\nu)\alpha\Delta T)\Big]\\
\tau_{xy} &= \frac{E}{2(1+\nu)}\gamma_{xy}
\end{aligned}$$

which can be written in matrix form as $\boldsymbol{\sigma} = \mathbf{D}(\boldsymbol{\varepsilon} - \boldsymbol{\varepsilon}^0)$, where

$$\boldsymbol{\varepsilon}^0 = (1+\nu)\alpha\Delta T[1,1,0]^T$$

is the initial strain vector and $\mathbf{D}$ the matrix given in Eqs.(4.6) and (4.7).

Example 4.3: Explain the meaning of the initial strains for the bar in Figure 4.6 subjected to a uniform temperature increase.

\- ***Solution***

Let us assume first that the bar is clamped at one end and free at the other end (Figure 4.6a). Under a uniform temperature increment the bar will increase in length by the amount

$$\Delta l = \alpha\ \Delta T\ l$$

and the corresponding "initial" strain is

$$\varepsilon_x^0 = \frac{\Delta l}{l} = \alpha\ \Delta T$$

Since the bar is free to move horizontally, the total elongation is equal to that produced by the thermal increment and, therefore, the elastic strain is equal to zero, i.e.

$$\varepsilon_x^e = \varepsilon_x - \varepsilon_x^0 = \alpha\ \Delta T - \alpha\ \Delta T = 0$$

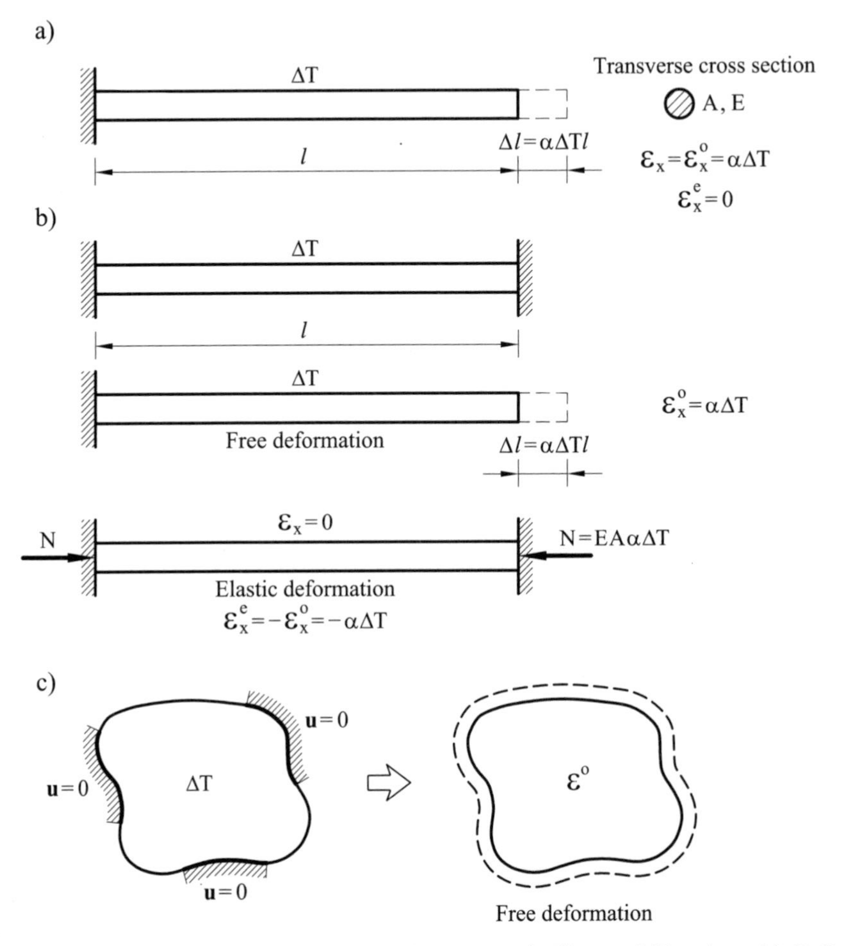

Fig. 4.6 Interpretation of initial thermal strains, a) Clamped/free bar, b) Fully clamped bar, c) 2D solid

Thus, from Eq.(4.17) it is deduced that the stresses in the deformed bar are zero.

Let us consider now the fully clamped bar of Figure 4.6b. To compute the initial strains let us assume that the bar points are free to move horizontally. Under these conditions the "initial" elongation of the bar will coincide with that of the clamped/free bar of Figure 4.6a, i.e. $\varepsilon_x^0 = \alpha \Delta T$. However, since the bar points have the horizontal displacement restrained (due to the two clamped ends), the "elastic strain" is now

$$\varepsilon_x^e = \varepsilon_x - \varepsilon_x^0 = 0 - \alpha \Delta T = -\alpha \Delta T$$

and by using Eq.(4.17) it is deduced that the bar is subjected to a uniform axial force of value $N = -\alpha EA\Delta T$.
Therefore, the initial thermal strains can be interpreted as the strains induced in the constraint-free body by a temperature increment occurring in some points (Figure 4.6c). Satisfaction of the kinematic (displacement) boundary conditions provides the values of the actual (total) strains. The difference between total and initial strains yields the "elastic" strains responsible for the actual stresses in the body.
It is also deduced from this example that a thermal increment produces no stresses in a body which can move freely in space.

4.2.6 Virtual work expression

The PVW is written for 2D elasticity problems as [Was,ZTZ]

$$\iint_A (\delta\varepsilon_x\sigma_x + \delta\varepsilon_y\sigma_y + \delta\gamma_{xy}\tau_{xy})t\, dA = \iint_A (\delta u b_x + \delta v b_y)t\, dA + \\ + \oint_l (\delta u t_x + \delta v t_y)t\, ds + \sum_i (\delta u_i\, P_{x_i} + \delta v_i\, P_{y_i}) \quad (4.23)$$

The terms in the r.h.s. of Eq.(4.23) represent the virtual work of the body forces (i.e. forces per unit area) b_x, b_y; the surface tractions t_x, t_y; and the external point loads P_{x_i}, P_{y_i}, respectively (Figures 4.1 and 4.2). The integral in the l.h.s. represents the work performed by the stresses $\sigma_x, \sigma_y, \tau_{xy}$ over the virtual strains $\delta\varepsilon_x, \delta\varepsilon_y$ y $\delta\gamma_{xy}$. A and l are respectively the area and the boundary of the transverse section of the solid and t its thickness. For plane stress problems t is the actual thickness of the solid. For plane strain situations the analysis domain is a unit slice and t is equal to one.

Eq.(4.23) can be written in matrix form as

$$\iint_A \delta\boldsymbol{\varepsilon}^T\boldsymbol{\sigma} t\, dA = \iint_A \delta\mathbf{u}^T\mathbf{b} t\, dA + \oint_l \delta\mathbf{u}^T\mathbf{t} t\, ds + \sum_i \delta\mathbf{u}_i^T\, \mathbf{p}_i \quad (4.24a)$$

where

$$\begin{aligned} \delta\boldsymbol{\varepsilon} &= \left[\delta\varepsilon_x, \delta\varepsilon_y, \delta\gamma_{xy}\right]^T ; \quad \delta\mathbf{u} = \left[\delta u, \delta v\right]^T \; ; \quad \mathbf{b} = \left[b_x, b_y\right]^T \\ \mathbf{t} &= \left[t_x, t_y\right]^T \; ; \quad \delta\mathbf{u}_i = \left[\delta u_i, \delta v_i\right]^T ; \quad \mathbf{p}_i = \left[P_{x_i}, P_{y_i}\right]^T \end{aligned} \quad (4.24b)$$

The above equations show that the PVW integrals involve up to first derivatives of the displacements only. Hence, C^o continuous elements can be used. This requirement holds for all elasticity elements studied in this book (i.e. 2D/3D solids and axisymmetric solids).

Eq.(4.23) is the starting point to derive the finite element equations as described in the next section.

4.3 FINITE ELEMENT FORMULATION. THREE-NODED TRIANGULAR ELEMENT

We will study first the simple 3-noded triangular element. This is the first element ever used for the analysis of structural problems. Prior to the finite element era, Courant successfully used linear polynomial approximations over triangular regions to solve differential equations in 2D domains [Co]. Some years later Turner *et al.* [TCMT] in their classic paper proposed the discretization of 2D solid domains into simple triangles as a way to analyze solids using matrix structural techniques. This explains why the 3-noded triangle is sometimes known as the Turner element. This element soon became very popular among engineers and it was widely used in the analysis of many structures in aeronautical and civil engineering [AFS,AK,ZTZ]. We note the impact of this element in the study of gravity dams and tunnels for practical civil engineering applications [ZT]. The key to the success of the 3-noded triangle is its simplicity which allows the assimilation of the FEM and the standard matrix method for bar structures known to most structural engineers. Conversely, it has limited accuracy due to the linear displacement approximation yielding constant strain and stress fields. Hence, fine meshes are required to capture accurate solutions in zones of high displacement gradients. This is however not a serious problem due to its versatile geometry, which is also very adequate for adaptive mesh refinement, as shown in Chapter 9. In summary, the 3-noded triangular element has the ideal features to introduce the application of the FEM to the analysis of 2D solids.

4.3.1 Discretization of the displacement field

Figure 4.6 shows the transverse section of a solid analized under the assumptions of plane elasticity. As usual the first step is the discretization of the analysis domain as a mesh of finite elements. Figure 4.7 shows the mesh

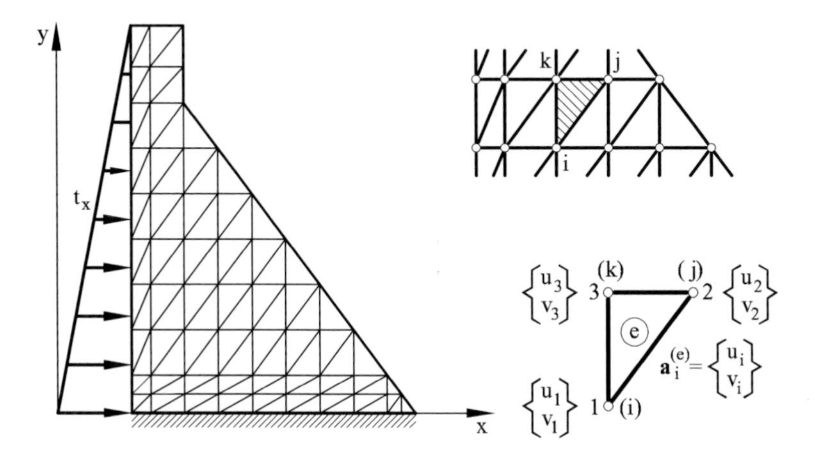

Element	Node		Nodal variables		Nodal coordinates	
	Local	Global	Local	Global	Local	Global
(e)	1	i	u_1 v_1	u_i v_i	x_1 y_1	x_i y_i
	2	j	u_2 v_2	u_j v_j	x_2 y_2	x_j y_j
Area: $A^{(e)}$	3	k	u_3 v_3	u_k v_k	x_3 y_3	x_k y_k

Fig. 4.7 Discretization of a structure in 3-noded triangular elements

of 3-noded triangles chosen. The accuracy of the finite element solution can obviously be improved by using a finer mesh.

A typical 3-noded triangular element is characterized by the numbering of its nodes and their coordinates x, y. The three nodes have a global numbering i, j, k which corresponds to the local numbers 1, 2, 3 respectively (Figure 4.7). It is convenient to use the local numbering to compute the element matrices and vectors and the correspondence between local and global numbering for the assembly process, as in matrix analysis of bar structures [HO2,Li].

Let us consider an individual triangle like that shown in Figure 4.7. The two cartesian displacements of an arbitrary point within the element can be expressed in terms of the nodal displacements as

$$\begin{aligned} u &= N_1u_1 + N_2u_2 + N_3u_3 \\ v &= N_1v_1 + N_2v_2 + N_3v_3 \end{aligned} \tag{4.25}$$

where (u_i, v_i) and N_i are the horizontal and vertical displacements and the shape function of node i, respectively. There is not a fundamental reason to choose the same approximation for the vertical and horizontal displacements. However, the same interpolation for both displacements is typically used in practice.

Eq.(4.25) is written in matrix form as

$$\mathbf{u} = \begin{Bmatrix} u \\ v \end{Bmatrix} = \begin{bmatrix} N_1 & 0 & N_2 & 0 & N_3 & 0 \\ 0 & N_1 & 0 & N_2 & 0 & N_3 \end{bmatrix} \begin{Bmatrix} u_1 \\ v_1 \\ u_2 \\ v_2 \\ u_3 \\ v_3 \end{Bmatrix} \tag{4.26}$$

or

$$\mathbf{u} = \mathbf{N}\ \mathbf{a}^{(e)} \tag{4.27}$$

where

$$\mathbf{u} = \begin{Bmatrix} u \\ v \end{Bmatrix} \tag{4.28a}$$

is the displacement vector of a point,

$$\mathbf{N} = [\mathbf{N}_1, \mathbf{N}_2, \mathbf{N}_3] \quad ; \quad \mathbf{N}_i = \begin{bmatrix} N_i & 0 \\ 0 & N_i \end{bmatrix} \tag{4.28b}$$

are the shape function matrices of the element and the ith node, respectively, and

$$\mathbf{a}^{(e)} = \begin{Bmatrix} \mathbf{a}_1^{(e)} \\ \mathbf{a}_2^{(e)} \\ \mathbf{a}_3^{(e)} \end{Bmatrix} \quad \text{with} \quad \mathbf{a}_i^{(e)} = \begin{Bmatrix} u_i \\ v_i \end{Bmatrix} \tag{4.29}$$

are the nodal displacement vectors of the element and of the ith node, respectively.

Note that $\mathbf{N}$ and $\mathbf{a}^{(e)}$ contain as many matrices $\mathbf{N}_i$ and vectors $\mathbf{a}_i^{(e)}$ as element nodes. This is a general rule, as we will see throughout the book.

The shape functions for the 3-noded triangular element is found as follows.

The three nodes define a linear displacement field which can be written as

$$\begin{aligned} u &= \alpha_1 + \alpha_2 x + \alpha_3 y \\ v &= \alpha_4 + \alpha_5 x + \alpha_6 y \end{aligned} \tag{4.30}$$

Since we have assumed the same interpolation for u and v, it suffices to derive the shape functions for one of the two displacements. For instance, the horizontal nodal displacements are deduced from Eq.(4.30) as

$$\begin{aligned} u_1 &= \alpha_1 + \alpha_2 x_1 + \alpha_3 y_1 \\ u_2 &= \alpha_1 + \alpha_2 x_2 + \alpha_3 y_2 \\ u_3 &= \alpha_1 + \alpha_2 x_3 + \alpha_3 y_3 \end{aligned} \tag{4.31}$$

Solving for α_1, α_2 and α_3 and substituting into Eq.(4.30) yields

$$u = \frac{1}{2A^{(e)}}\Big[(a_1 + b_1 x + c_1 y)u_1 + (a_2 + b_2 x + c_2 y)u_2 + (a_3 + b_3 x + c_3 y)u_3\Big] \tag{4.32a}$$

where $A^{(e)}$ is the element area and

$$a_i = x_j y_k - x_k y_j \quad , \quad b_i = y_j - y_k \quad , \quad c_i = x_k - x_j \quad ; \quad i,j,k = 1,2,3 \tag{4.32b}$$

The parameters a_i, b_i and c_i are obtained by cyclic permutation of the indexes i, j, k.

Comparing Eqs.(4.32) and (4.25) the expression for the shape functions is found as

$$\boxed{N_i = \frac{1}{2A^{(e)}}(a_i + b_i x + c_i y)} \quad , \quad i = 1,2,3 \tag{4.33}$$

The form of the linear shape functions is shown in Figure 4.8. It can be checked that the shape function N_i takes the value one at node i and zero at the other two nodes.

4.3.2 Discretization of the strain field

Substituting Eq.(4.25) into (4.2) gives the three characteristic strains as

$$\begin{aligned} \varepsilon_x &= \frac{\partial u}{\partial x} = \frac{\partial N_1}{\partial x}u_1 + \frac{\partial N_2}{\partial x}u_2 + \frac{\partial N_3}{\partial x}u_3 \\ \varepsilon_y &= \frac{\partial v}{\partial y} = \frac{\partial N_1}{\partial y}v_1 + \frac{\partial N_2}{\partial y}v_2 + \frac{\partial N_3}{\partial y}v_3 \\ \gamma_{xy} &= \frac{\partial u}{\partial y} + \frac{\partial v}{\partial x} = \frac{\partial N_1}{\partial y}u_1 + \frac{\partial N_1}{\partial x}v_1 + \frac{\partial N_2}{\partial y}u_2 + \frac{\partial N_2}{\partial x}v_2 + \frac{\partial N_3}{\partial y}u_3 + \frac{\partial N_3}{\partial x}v_3 \end{aligned} \tag{4.34}$$

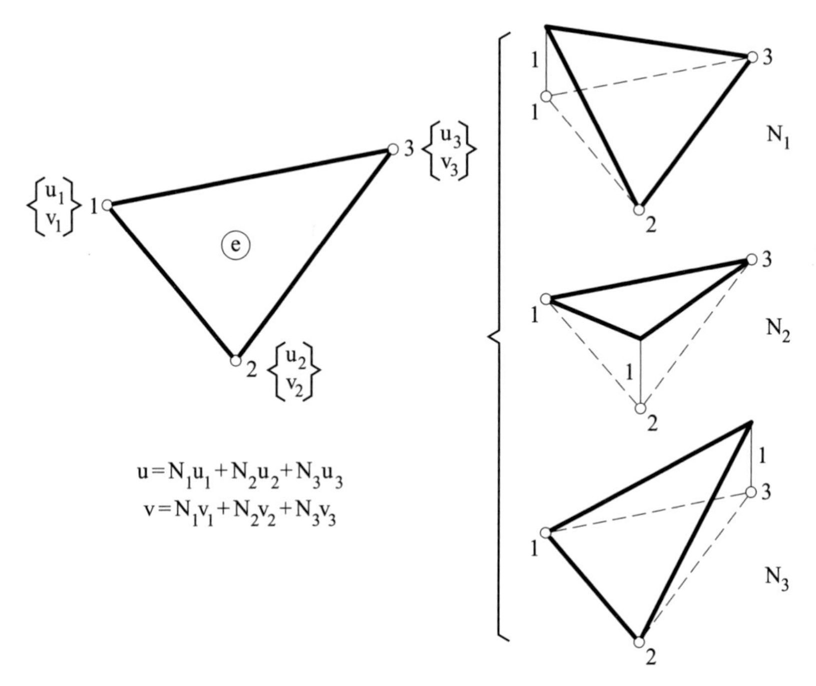

Fig. 4.8 Shape functions for the 3-noded triangular element

In matrix form

$$\boldsymbol{\varepsilon} = \left\{ \begin{matrix} \dfrac{\partial u}{\partial x} \\ \dfrac{\partial v}{\partial y} \\ \dfrac{\partial u}{\partial y} + \dfrac{\partial v}{\partial x} \end{matrix} \right\} = \begin{bmatrix} \dfrac{\partial N_1}{\partial x} & 0 & \vdots & \dfrac{\partial N_2}{\partial x} & 0 & \vdots & \dfrac{\partial N_3}{\partial x} & 0 \\ 0 & \dfrac{\partial N_1}{\partial y} & \vdots & 0 & \dfrac{\partial N_2}{\partial y} & \vdots & 0 & \dfrac{\partial N_3}{\partial y} \\ \dfrac{\partial N_1}{\partial y} & \dfrac{\partial N_1}{\partial x} & \vdots & \dfrac{\partial N_2}{\partial y} & \dfrac{\partial N_2}{\partial x} & \vdots & \dfrac{\partial N_3}{\partial y} & \dfrac{\partial N_3}{\partial x} \end{bmatrix} \left\{ \begin{matrix} u_1 \\ v_1 \\ u_2 \\ v_2 \\ u_3 \\ v_3 \end{matrix} \right\} \tag{4.35}$$

or

$$\boldsymbol{\varepsilon} = \mathbf{B}\mathbf{a}^{(e)} \tag{4.36}$$

where

$$\mathbf{B} = [\mathbf{B}_1, \mathbf{B}_2, \mathbf{B}_3] \tag{4.37}$$

is the element strain matrix, and

$$\mathbf{B}_i = \begin{bmatrix} \dfrac{\partial N_i}{\partial x} & 0 \\ 0 & \dfrac{\partial N_i}{\partial y} \\ \dfrac{\partial N_i}{\partial y} & \dfrac{\partial N_i}{\partial x} \end{bmatrix} \tag{4.38}$$

is the strain matrix of node i.

The expression for $\mathbf{B}_i$ in Eq.(4.38) is completely general and applicable to any 2D solid element.

Matrix $\mathbf{B}$ contains as many $\mathbf{B}_i$ matrices as element nodes. This is also a general property. Particularizing Eqs.(4.37) and (4.38) for the 3-noded triangle we obtain (using Eq.(4.33))

$$\mathbf{B} = \frac{1}{2A^{(e)}} \begin{bmatrix} b_1 & 0 & \vdots & b_2 & 0 & \vdots & b_3 & 0 \\ 0 & c_1 & \vdots & 0 & c_2 & \vdots & 0 & c_3 \\ c_1 & b_1 & \vdots & c_2 & b_2 & \vdots & c_3 & b_3 \end{bmatrix} \tag{4.39}$$

and, therefore

$$\mathbf{B}_i = \frac{1}{2A^{(e)}} \begin{bmatrix} b_i & 0 \\ 0 & c_i \\ c_i & b_i \end{bmatrix} \tag{4.40}$$

4.3.3 Discretization of the stress field

The discretized expression for the stress field within the element is obtained by substituting Eq.(4.36) into (4.5) as

$$\boldsymbol{\sigma} = \mathbf{D}\boldsymbol{\varepsilon} = \mathbf{D}\mathbf{B}\mathbf{a}^{(e)} \tag{4.41}$$

If initial strains and stresses are considered we deduce from Eq.(4.17)

$$\boldsymbol{\sigma} = \mathbf{D}(\boldsymbol{\varepsilon} - \boldsymbol{\varepsilon}^0) + \boldsymbol{\sigma}^0 = \mathbf{D}\mathbf{B}\ \mathbf{a}^{(e)} - \mathbf{D}\boldsymbol{\varepsilon}^0 + \boldsymbol{\sigma}^0 \tag{4.42}$$

The strain matrix for the 3-noded triangle is constant (Eq.(4.39)). *This implies that both the strain and stress fields are constant within the element.* This is a consequence of the linear displacement interpolation chosen which, naturally, has constant first derivatives. Therefore, a finer mesh will be needed in zones where stress gradients are higher, so that the strain and stress fields are accurately approximated.

4.3.4 Discretized equilibrium equations

The discretized equilibrium equations for the 3-noded triangle will be derived by applying the PVW to an individual element. It is interesting that the expressions obtained hereafter are completely general and aplicable to any 2D solid element.

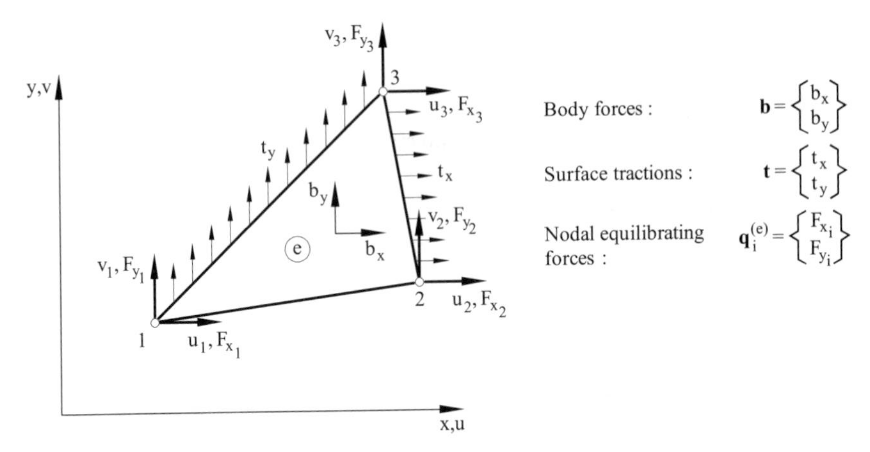

Fig. 4.9 Forces acting on a 3-noded triangle. The sides 13 and 23 belong to the external boundary

Let us assume that the following *external* forces act on the element (Figure 4.9): a) distributed forces $\mathbf{b}$ acting per unit area (body forces), and b) distributed forces $\mathbf{t}$ acting along the element sides belonging to a boundary line (surface tractions).

The surface tractions due to the interaction of adjacent elements are excluded "a priori", as they cancel themselves out during the assembly process.

As usual in the FEM, the equilibrium of the forces acting on the element is enforced point-wise at the nodes only. We therefore define nodal point loads F_{x_i} and F_{y_i} which balance the external forces and the internal forces due to the element deformation (Figure 4.9). These "equilibrating nodal forces" are obtained by applying the PVW to an individual element as

$$\begin{aligned}\iint_{A^{(e)}} \delta\boldsymbol{\varepsilon}^T \boldsymbol{\sigma} t\, dA = \iint_{A^{(e)}} \delta\mathbf{u}^T \mathbf{b} t\, dA + \oint_{l^{(e)}} \delta\mathbf{u}^T \mathbf{t} t\, ds + \\ + \sum_{i=1}^{3} \delta u_i F_{x_i} + \sum_{i=1}^{3} \delta v_i F_{y_i}\end{aligned} \tag{4.43}$$

where δu_i and δv_i are the nodal virtual displacements and F_{x_i} and F_{y_i} the equilibrating nodal forces along the horizontal and vertical directions, respectively. The virtual work performed by these forces is obtained from Eq.(4.43) as

$$\iint_{A^{(e)}} \delta\boldsymbol{\varepsilon}^T \boldsymbol{\sigma} t\, dA - \iint_{A^{(e)}} \delta\mathbf{u}^T \mathbf{b} t\, dA - \oint_{l^{(e)}} \delta\mathbf{u}^T\ \mathbf{t} t\, ds = [\delta\mathbf{a}^{(e)}]^T \mathbf{q}^{(e)} \tag{4.44}$$

For the 3-noded triangular element

$$
\begin{aligned}
[\delta \mathbf{a}^{(e)}]^T &= [\delta \mathbf{a}_1^T, \delta \mathbf{a}_2^T, \delta \mathbf{a}_3^T]^{(e)} = [\delta u_1, \delta v_1, \delta u_2, \delta v_2, \delta u_3, \delta v_3] \\
[\mathbf{q}^{(e)}]^T &= [\mathbf{q}_1^T, \mathbf{q}_2^T, \mathbf{q}_3^T]^{(e)} = [F_{x_1}, F_{y_1}, F_{x_2}, F_{y_2}, F_{x_3}, F_{y_3}]
\end{aligned} \tag{4.45}
$$

Next we interpolate the virtual displacements in terms of the nodal values. Following the same procedure as for deriving Eqs.(4.26) and (4.36) we obtain

$$\delta \mathbf{u} = \mathbf{N} \delta \mathbf{a}^{(e)} \quad ; \quad \delta \boldsymbol{\varepsilon} = \mathbf{B} \delta \mathbf{a}^{(e)} \tag{4.46a}$$

and

$$\delta \mathbf{u}^T = [\delta \mathbf{a}^{(e)}]^T \mathbf{N}^T \quad ; \quad \delta \boldsymbol{\varepsilon}^T = [\delta \mathbf{a}^{(e)}]^T \mathbf{B}^T \tag{4.46b}$$

Substituting the last equations into Eq.(4.44) gives

$$[\delta \mathbf{a}^{(e)}]^T \left[\iint_{A^{(e)}} \mathbf{B}^T \boldsymbol{\sigma} t \, dA - \iint_{A^{(e)}} \mathbf{N}^T \mathbf{b} t \, dA - \oint_{l^{(e)}} \mathbf{N}^T \mathbf{t} t \, ds \right] = [\delta \mathbf{a}^{(e)}]^T \mathbf{q}^{(e)} \tag{4.47}$$

Since the virtual displacements are arbitrary it is deduced that

$$\iint_{A^{(e)}} \mathbf{B}^T \boldsymbol{\sigma} t \, dA - \iint_{A^{(e)}} \mathbf{N}^T \mathbf{b} t \, dA - \oint_{l^{(e)}} \mathbf{N}^T \mathbf{t} t \, ds = \mathbf{q}^{(e)} \tag{4.48}$$

Eq.(4.48) yields the equilibrating nodal forces $\mathbf{q}^{(e)}$ in terms of the nodal forces due to the element deformation (first integral), the body forces (second integral) and the surface tractions (third integral). Substituting the stresses in terms of the nodal displacements from Eq.(4.42) gives

$$\iint_{A^{(e)}} \mathbf{B}^T (\mathbf{D}\mathbf{B}\mathbf{a}^{(e)} - \mathbf{D}\boldsymbol{\varepsilon}^0 + \boldsymbol{\sigma}^0) t \, dA - \iint_{A^{(e)}} \mathbf{N}^T \mathbf{b} t \, dA - \oint_{l^{(e)}} \mathbf{N}^T \mathbf{t} t \, ds = \mathbf{q}^{(e)} \tag{4.49}$$

and

$$
\begin{aligned}
&\left[\iint_{A^{(e)}} \mathbf{B}^T \mathbf{D} \mathbf{B} t \, dA \right] \mathbf{a}^{(e)} - \iint_{A^{(e)}} \mathbf{B}^T \mathbf{D} \boldsymbol{\varepsilon}^0 t \, dA + \\
&+ \iint_{A^{(e)}} \mathbf{B}^T \boldsymbol{\sigma}^0 t \, dA - \iint_{A^{(e)}} \mathbf{N}^T \mathbf{b} t \, dA - \oint_{l^{(e)}} \mathbf{N}^T \mathbf{t} t \, ds = \mathbf{q}^{(e)}
\end{aligned} \tag{4.50}
$$

or

$$\mathbf{K}^{(e)} \mathbf{a}^{(e)} - \mathbf{f}^{(e)} = \mathbf{q}^{(e)} \tag{4.51}$$

where

$$\boxed{\mathbf{K}^{(e)} = \iint_{A^{(e)}} \mathbf{B}^T \mathbf{D} \, \mathbf{B} t \, dA} \tag{4.52}$$

is the element stiffness matrix, and

$$\boxed{\mathbf{f}^{(e)} = \mathbf{f}_\varepsilon^{(e)} + \mathbf{f}_\sigma^{(e)} + \mathbf{f}_b^{(e)} + \mathbf{f}_t^{(e)}} \tag{4.53}$$

is the equivalent nodal force vector for the element where

$$\mathbf{f}_\varepsilon^{(e)} = \iint_{A^{(e)}} \mathbf{B}^T \; \mathbf{D}\boldsymbol{\varepsilon}^0 t \, dA \tag{4.54}$$

$$\mathbf{f}_\sigma^{(e)} = -\iint_{A^{(e)}} \mathbf{B}^T \; \boldsymbol{\sigma}^0 t \, dA \tag{4.55}$$

$$\mathbf{f}_b^{(e)} = \iint_{A^{(e)}} \mathbf{N}^T \mathbf{b} t \, dA \tag{4.56}$$

$$\mathbf{f}_t^{(e)} = \oint_{l^{(e)}} \mathbf{N}^T \mathbf{t} t \, ds \tag{4.57}$$

are the equivalent nodal force vectors due to initial strains, initial stresses, body forces and surface tractions, respectively.

The expressions for the element stiffness matrix and the equivalent nodal force vectors given by Eqs.(4.52) - (4.57) are completely general and are applicable to any 2D solid element. The particularization for the 3-noded triangular element is given in the next section.

The global equilibrium equations for the whole mesh are obtained by establishing that the nodes are in equilibrium, similarly as for 1D problems; i.e. the sum of all the equilibrating nodal forces at each node j balance the point loads $\mathbf{p}_j = [P_{x_j}, P_{y_j}]^T$ acting at the node and

$$\sum_e \mathbf{q}_i^{(e)} = \mathbf{p}_j \quad , \quad j = 1, N \tag{4.58}$$

where the sum refers to all elements sharing the node with global number j and N is the total number of nodes in the mesh. Vector $\mathbf{p}_j$ typically includes the reactions at the prescribed nodes. Eq.(4.58) is identical to the equation of equilibrium of joint forces in bar structures (Chapter 1). The matrix equilibrium equations for the whole mesh can thus be obtained following identical procedures as for bar structures as

$$\mathbf{K}\mathbf{a} = \mathbf{f} \tag{4.59}$$

where $\mathbf{K}$ and $\mathbf{f}$ are the stiffness matrix and the equivalent nodal force vector for the whole mesh. Both $\mathbf{K}$ and $\mathbf{f}$ are assembled from the element contributions in the standard manner (Eq.(2.85)). The assembly process

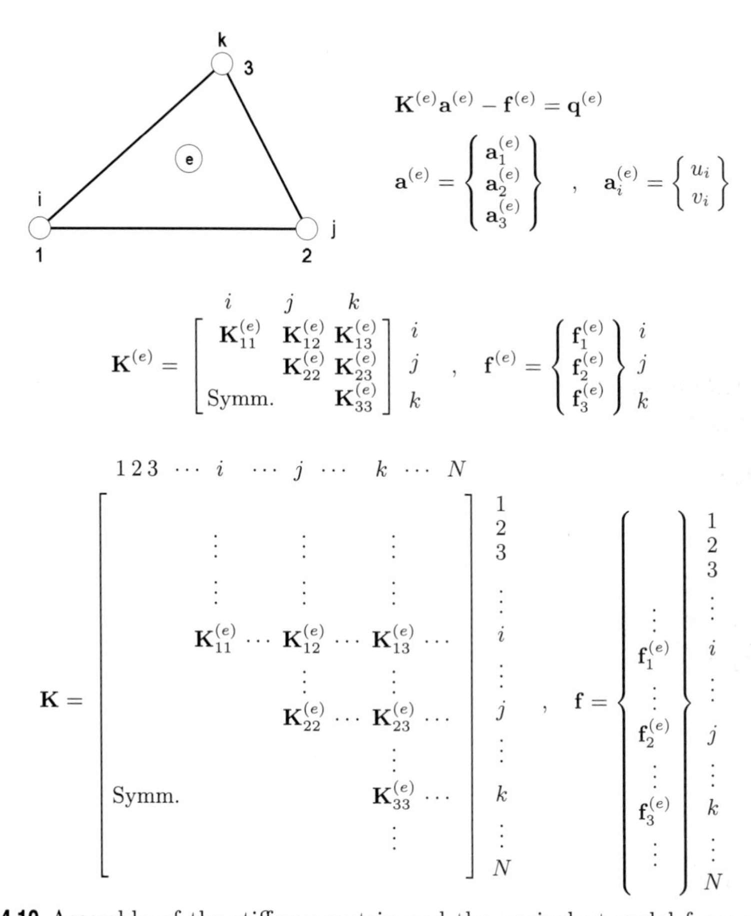

Fig. 4.10 Assembly of the stiffness matrix and the equivalent nodal force vector into the global equation system for the 3-noded triangle with global node numbers i, j, k

is schematically shown in Figure 4.10.

We note once more that the equilibrating nodal forces due to the surface tractions along the element interface cancel themselves out during the assembly process. Therefore, *only the surface tractions acting on element sides belonging to the external boundaries of the structure must be considered in the analysis.*

The reactions at the prescribed nodes are computed "a posteriori" using Eq.(2.26a). As already mentioned an alternative procedure is to compute the nodal reaction vector $\mathbf{r}$ from

$$\mathbf{r} = \mathbf{f}_{\text{int}} - \mathbf{f}_{\text{ext}} \tag{4.60a}$$

where $\mathbf{f}_{\text{ext}}$ contains contributions from the external forces only (i.e. the body forces, the surface tractions and the point loads) and the internal nodal force vector $\mathbf{f}_{\text{int}}$ is assembled from the element contributions

$$\mathbf{f}_{\text{int}}^{(e)} = \iint_{A^{(e)}} \mathbf{B}^T \boldsymbol{\sigma} t \, dA \tag{4.60b}$$

The above expression for $\mathbf{f}_{\text{int}}^{(e)}$ is deduced from the first integral in the l.h.s. of Eq.(4.48).

4.3.5 Stiffness matrix and equivalent nodal force vectors for the 3-noded triangular element

Stiffness matrix

Eq.(4.52) can be written for the 3-noded triangle using (4.37) as

$$\mathbf{K}^{(e)} = \iint_{A^{(e)}} \begin{Bmatrix} \mathbf{B}_1^T \\ \mathbf{B}_2^T \\ \mathbf{B}_3^T \end{Bmatrix} \mathbf{D} \, [\mathbf{B}_1, \mathbf{B}_2, \mathbf{B}_3] t \, dA =$$

$$= \iint_{A^{(e)}} \begin{bmatrix} \mathbf{B}_1^T\mathbf{D}\mathbf{B}_1 & \mathbf{B}_1^T\mathbf{D}\mathbf{B}_2 & \mathbf{B}_1^T\mathbf{D}\mathbf{B}_3 \\ \ddots & \mathbf{B}_2^T\mathbf{D}\mathbf{B}_2 & \mathbf{B}_2^T\mathbf{D}\mathbf{B}_3 \\ \text{Symm.} & \ddots & \mathbf{B}_3^T\mathbf{D}\mathbf{B}_3 \end{bmatrix} t \, dA \tag{4.61}$$

A typical element stiffness submatrix, $\mathbf{K}_{ij}^{(e)}$, linking nodes i and j of the element can be obtained as

$$\mathbf{K}_{ij}^{(e)} = \iint_{A^{(e)}} \mathbf{B}_i^T \mathbf{D} \mathbf{B}_j t \, dA \tag{4.62a}$$

Substituting Eqs.(4.6) and (4.40) into Eq.(4.61) gives

$$\mathbf{K}_{ij}^{(e)} = \iint_{A^{(e)}} \frac{1}{2A^{(e)}} \begin{bmatrix} b_i & 0 & c_i \\ 0 & c_i & b_i \end{bmatrix} \begin{bmatrix} d_{11} & d_{12} & 0 \\ d_{21} & d_{22} & 0 \\ 0 & 0 & d_{33} \end{bmatrix} \frac{1}{2A^{(e)}} \begin{bmatrix} b_j & 0 \\ 0 & c_j \\ c_j & b_j \end{bmatrix} t \, dA \tag{4.63a}$$

For an homogeneous material the integrand of Eq.(4.63a) is constant and this gives

$$\mathbf{K}_{ij}^{(e)} = \left(\frac{t}{4A}\right)^{(e)} \begin{bmatrix} b_i b_j d_{11} + c_i c_j d_{33} & b_i c_j d_{12} + b_j c_i d_{33} \\ c_i b_j d_{21} + b_i c_j d_{33} & b_i b_j d_{33} + c_i c_j d_{22} \end{bmatrix} \tag{4.63b}$$

The form of $\mathbf{K}_{ij}^{(e)}$ for plane stress and plane strain situations is simply obtained by introducing the adequate values of the coefficients d_{ij} from Eq.(4.7). Note that $\mathbf{K}_{ij}^{(e)}$ is always symmetrical as $d_{12} = d_{21}$.

Equivalent nodal force vectors

a) Body forces

$$\mathbf{f}_b^{(e)} = \begin{Bmatrix} \mathbf{f}_{b_1}^{(e)} \\ \mathbf{f}_{b_2}^{(e)} \\ \mathbf{f}_{b_3}^{(e)} \end{Bmatrix} = \iint_{A^{(e)}} \mathbf{N}^T \mathbf{b} t \, dA = \iint_{A^{(e)}} \begin{Bmatrix} \mathbf{N}_1^T \mathbf{b} \\ \mathbf{N}_2^T \mathbf{b} \\ \mathbf{N}_3^T \mathbf{b} \end{Bmatrix} t \, dA \tag{4.64}$$

The nodal contribution of vector $\mathbf{f}_b^{(e)}$ is

$$\mathbf{f}_{b_i}^{(e)} = \iint_{A^{(e)}} \mathbf{N}_i^T \mathbf{b} \; t \, dA = \iint_{A^{(e)}} \begin{Bmatrix} N_i \; b_x \\ N_i \; b_y \end{Bmatrix} t \, dA \tag{4.65}$$

If the body forces $\mathbf{b}$ are uniformly distributed over the element we obtain using Eq.(4.33)

$$\mathbf{f}_{b_i}^{(e)} = \frac{(At)^{(e)}}{3} \begin{Bmatrix} b_x \\ b_y \end{Bmatrix} \tag{4.66}$$

i.e. the total force acting over the element is split into equal parts between the three nodes of the triangle, as expected.

A particular case of body force is *self-weight* with gravity acting in the direction of the y-axis. In this case $b_x = 0$ and $b_y = -\rho g$ where ρ and g are the material density and the value of the gravity constant, respectively.

b) Surface tractions

$$\mathbf{f}_t^{(e)} = \oint_{l^{(e)}} \mathbf{N}^T \mathbf{t} t \, ds \tag{4.67}$$

For a node i belonging to a loaded external boundary we have

$$\mathbf{f}_{t_i}^{(e)} = \oint_{l^{(e)}} \mathbf{N}_i^T \mathbf{t} t \, ds = \oint_{l^{(e)}} \begin{Bmatrix} N_i \; t_x \\ N_i \; t_y \end{Bmatrix} t \, ds \tag{4.68}$$

We note that the shape function of a node not belonging to the loaded boundary takes a zero value. Thus, if the element side 1-2 is loaded with uniformly distributed tractions t_x and t_y, vector $\mathbf{f}_t^{(e)}$ is simply

$$\mathbf{f}_t^{(e)} = \frac{(l_{12} \; t)^{(e)}}{2} \left[t_x, t_y, t_x, t_y, 0, 0 \right]^T \tag{4.69}$$

where $l_{12}^{(e)}$ is the side length. Eq.(4.69) shows that the traction force acting along the element side is split in equal parts between the two side nodes. The expressions of $\mathbf{f}_t^{(e)}$ for loaded sides 1-3 and 2-3 are

$$\mathbf{f}_t^{(e)} = \frac{(l_{13}t)^{(e)}}{2}\left[t_x, t_y, 0, 0, t_x, t_y\right]^T$$

$$\mathbf{f}_t^{(e)} = \frac{(l_{23}t)^{(e)}}{2}\left[0, 0, t_x, t_y, t_x, t_y\right]^T \tag{4.70}$$

c) Forces due to initial strains

Substituting Eq.(4.37) into (4.54) gives

$$\mathbf{f}_\varepsilon^{(e)} = \begin{Bmatrix} \mathbf{f}_{\varepsilon 1}^{(e)} \\ \mathbf{f}_{\varepsilon 2}^{(e)} \\ \mathbf{f}_{\varepsilon 3}^{(e)} \end{Bmatrix} = \iint_{A^{(e)}} \mathbf{B}^T \mathbf{D}\boldsymbol{\varepsilon}^0 t\, dA = \iint_{A^{(e)}} \begin{Bmatrix} \mathbf{B}_1^T \mathbf{D}\boldsymbol{\varepsilon}^0 \\ \mathbf{B}_2^T \mathbf{D}\boldsymbol{\varepsilon}^0 \\ \mathbf{B}_3^T \mathbf{D}\boldsymbol{\varepsilon}^0 \end{Bmatrix} t\, dA \tag{4.71}$$

and the equivalent nodal force of node i due to the initial strains is

$$\mathbf{f}_{\varepsilon_i}^{(e)} = \iint_{A^{(e)}} \mathbf{B}_i^T\, \mathbf{D}\, \boldsymbol{\varepsilon}^0\, t\, dA \tag{4.72}$$

If $\boldsymbol{\varepsilon}^0$ is constant over the element and the material is homogeneous we obtain using Eqs.(4.6) and (4.40)

$$\mathbf{f}_{\varepsilon_i}^{(e)} = \iint_{A^{(e)}} \frac{1}{2A^{(e)}} \begin{bmatrix} b_i & 0 & c_i \\ 0 & c_i & b_i \end{bmatrix} \begin{bmatrix} d_{11} & d_{12} & 0 \\ d_{21} & d_{22} & 0 \\ 0 & 0 & d_{33} \end{bmatrix} \begin{Bmatrix} \varepsilon_x^0 \\ \varepsilon_y^0 \\ \gamma_{xy}^0 \end{Bmatrix} t\, dA =$$

$$= \frac{t^{(e)}}{2} \begin{Bmatrix} b_i(d_{11}\varepsilon_x^0 + d_{12}\varepsilon_y^0) + c_i d_{33}\gamma_{xy}^0 \\ c_i(d_{21}\varepsilon_x^0 + d_{22}\varepsilon_y^0) + b_i d_{33}\gamma_{xy}^0 \end{Bmatrix} \tag{4.73}$$

For initial thermal strains, the expressions (4.18) for $\boldsymbol{\varepsilon}^0$ should be used.

d) Forces due to initial stresses

Substituting Eq.(4.37) into (4.55) gives

$$\mathbf{f}_\sigma^{(e)} = \begin{Bmatrix} \mathbf{f}_{\sigma 1}^{(e)} \\ \mathbf{f}_{\sigma 2}^{(e)} \\ \mathbf{f}_{\sigma 3}^{(e)} \end{Bmatrix} = -\iint_{A^{(e)}} \mathbf{B}^T \boldsymbol{\sigma}^0 t\, dA = -\iint_{A^{(e)}} \begin{Bmatrix} \mathbf{B}_1^T\, \boldsymbol{\sigma}^0 \\ \mathbf{B}_2^T\, \boldsymbol{\sigma}^0 \\ \mathbf{B}_3^T\, \boldsymbol{\sigma}^0 \end{Bmatrix} t\, dA \tag{4.74}$$

and the equivalent nodal force of node i due to the initial stresses is

$$\mathbf{f}_{\sigma_i}^{(e)} = -\iint_{A^{(e)}} \mathbf{B}_i^T \boldsymbol{\sigma}^0 t\, dA \tag{4.75}$$

For $\boldsymbol{\sigma}^0$ being constant over the element, we obtain using Eqs.(4.19b) and (4.40)

$$\mathbf{f}_{\sigma_i}^{(e)} = -\iint_{A^{(e)}} \frac{1}{2A^{(e)}} \begin{bmatrix} b_i & 0 & c_i \\ 0 & c_i & b_i \end{bmatrix} \begin{Bmatrix} \sigma_x^0 \\ \sigma_y^0 \\ \tau_{xy}^0 \end{Bmatrix} t\, dA = -\frac{t^{(e)}}{2} \begin{Bmatrix} b_i\sigma_x^0 + c_i\tau_{xy}^0 \\ c_i\sigma_y^0 + b_i\tau_{xy}^0 \end{Bmatrix} \tag{4.76}$$

The above expressions allow us to compute explicitly the matrices and vectors for the 3-noded triangle for 2D elasticity applications. Examples showing the behaviour of the element are given in Section 4.7. An example illustrating the assembly and solution process is presented next.

Example 4.4: Analyze the plane structure of the figure below under self-weight.

- ***Solution***

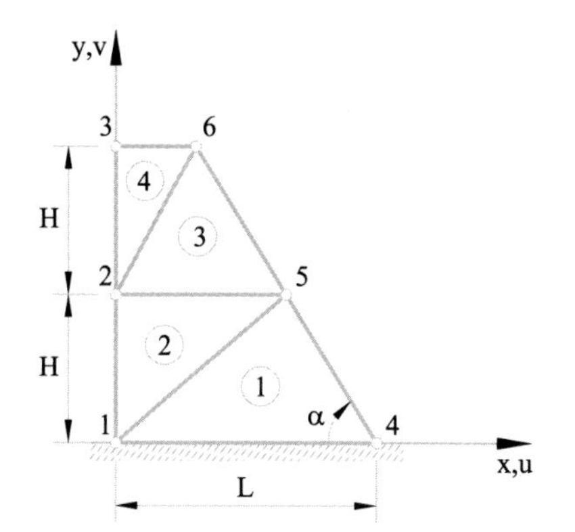

Mesh topology

Element	Nodal connections		
1	1	4	5
2	1	5	2
3	2	5	6
4	2	6	3

Plane strain situation
$\mathbf{u}_1 = \mathbf{u}_4 = \mathbf{0}$

The assembly process is similar to that for matrix analysis of bar structures (Figure 4.10). The global system of equations has the following form

$$\begin{array}{c} \\ 1 \\ 2 \\ 3 \\ 4 \\ 5 \\ 6 \end{array}
\begin{bmatrix}
(\mathbf{K}_{11}^{(1)}+\mathbf{K}_{11}^{(2)}) & \mathbf{K}_{13}^{(2)} & \mathbf{0} & \mathbf{K}_{12}^{(1)} & (\mathbf{K}_{13}^{(1)}+\mathbf{K}_{12}^{(2)}) & \mathbf{0} \\
 & (\mathbf{K}_{33}^{(2)}+\mathbf{K}_{11}^{(3)}+\mathbf{K}_{11}^{(4)}) & \mathbf{K}_{13}^{(4)} & \mathbf{0} & (\mathbf{K}_{32}^{(2)}+\mathbf{K}_{12}^{(3)}) & (\mathbf{K}_{13}^{(3)}+\mathbf{K}_{12}^{(4)}) \\
 & & \mathbf{K}_{33}^{(4)} & \mathbf{0} & \mathbf{0} & \mathbf{K}_{32}^{(4)} \\
 & & & \mathbf{K}_{22}^{(1)} & \mathbf{K}_{23}^{(1)} & \mathbf{0} \\
 & \text{Symm.} & & & (\mathbf{K}_{33}^{(1)}+\mathbf{K}_{22}^{(2)}+\mathbf{K}_{22}^{(3)}) & \mathbf{K}_{23}^{(3)} \\
 & & & & & (\mathbf{K}_{22}^{(4)}+\mathbf{K}_{33}^{(3)})
\end{bmatrix}
\begin{Bmatrix} \mathbf{a}_1 \\ \mathbf{a}_2 \\ \mathbf{a}_3 \\ \mathbf{a}_4 \\ \mathbf{a}_5 \\ \mathbf{a}_6 \end{Bmatrix}
=
\begin{Bmatrix}
(\mathbf{r}_1+\mathbf{f}_1^{(1)}+\mathbf{f}_1^{(2)}) \\
(\mathbf{f}_3^{(2)}+\mathbf{f}_1^{(3)}+\mathbf{f}_1^{(4)}) \\
\mathbf{f}_3^{(4)} \\
\mathbf{r}_4+\mathbf{f}_2^{(1)} \\
(\mathbf{f}_3^{(1)}+\mathbf{f}_2^{(2)}+\mathbf{f}_2^{(3)}) \\
\mathbf{f}_3^{(3)}+\mathbf{f}_2^{(4)}
\end{Bmatrix}$$

where $\mathbf{K}_{ij}^{(e)}$ is obtained from Eq.(4.63) and $\mathbf{f}_i^{(e)}$ from Eq.(4.66) with $b_x = 0$ and $b_y = -\rho g$. In both cases $t = 1$ should be taken.
The above system can be solved in the usual way by eliminating the rows and columns corresponding to the prescribed displacements $\mathbf{a}_1$ and $\mathbf{a}_4$. Once the nodal displacements have been obtained the corresponding reaction vectors $\mathbf{r}_1$ and $\mathbf{r}_4$ can be computed.
The constant strains and stresses within each element can be found "a posteriori" from the known nodal displacements by Eqs.(4.36) and (4.41).
The reader is encouraged to repeat this exercise by him/herself.

4.4 THE FOUR NODED RECTANGULAR ELEMENT

4.4.1 Basic formulation

The 4-noded rectangle is the simplest quadrilateral element. This element was developed by Argyris and Kelsey [AK] almost simultaneously to the 3-noded triangle. The general quadrilateral form is attributed to Taig [Ta]. However, the irregular behaviour of the standard 4-noded rectangle has motivated much research which we will summarize here.

Figure 4.11 shows a deep beam discretized in a mesh of 4-noded rectangles. Let us consider an isolated element with the local coordinate system r and s shown in Figure 4.11. The four nodal displacements define a four-term polynomial interpolation for the displacement field. The simplest interpolation satisfying the condition of interelement compatibility and geometric-invariance is

$$
\begin{aligned}
u(x,y) &= \alpha_1 + \alpha_2\, r + \alpha_3\, s + \alpha_4 rs \\
v(x,y) &= \alpha_5 + \alpha_6 r + \alpha_7 s + \alpha_8 rs
\end{aligned}
\tag{4.77}
$$

Eq.(4.77) implies a linear distribution of u and v along each element side, thus guaranteeing the continuity of the displacement field between adjacent elements. Note that the displacements vary as an incomplete quadratic polynomial within the element. The four constants α_i for each displacement component are obtained from the following conditions expressed in the local system r, s.

$$
\begin{array}{lllllll}
u = u_1 & \text{and} & v = v_1 & \text{for} & r = -a & , & s = -b \\
u = u_2 & \text{and} & v = v_2 & \text{for} & r = -a & , & s = -b \\
u = u_3 & \text{and} & v = v_3 & \text{for} & r = a & , & s = b \\
u = u_4 & \text{and} & v = v_4 & \text{for} & v = a & , & s = b
\end{array}
\tag{4.78}
$$

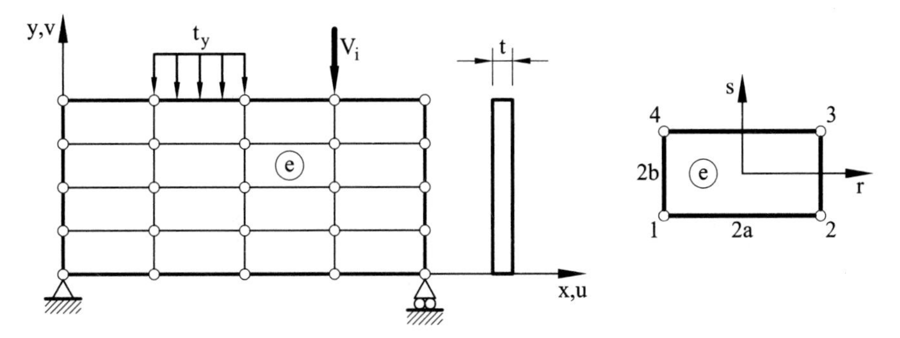

Fig. 4.11 Discretization of a deep beam with 4-noded rectangular elements. Definition of the local axes r and s for an element

Substituting these conditions into Eq.(4.77) and solving for the α_i parameters, Eq.(4.77) can be rewritten as follows (note that only the α_i parameters for one of the two displacements are needed, as the same interpolation is used for u and v)

$$u = \sum_{i=1}^{4} N_i \, u_i \quad ; \quad v = \sum_{i=1}^{4} N_i \, v_i \tag{4.79}$$

The shape functions N_i are

$$\begin{aligned} N_1 &= \frac{1}{4}\left(1 - \frac{r}{a}\right)\left(1 - \frac{s}{b}\right) \quad ; \quad N_2 = \frac{1}{4}\left(1 + \frac{r}{a}\right)\left(1 - \frac{s}{b}\right) \\ N_3 &= \frac{1}{4}\left(1 + \frac{r}{a}\right)\left(1 + \frac{s}{b}\right) \quad ; \quad N_4 = \frac{1}{4}\left(1 - \frac{r}{a}\right)\left(1 + \frac{s}{b}\right) \end{aligned} \tag{4.80}$$

Eqs.(4.79) can be rewritten in matrix form as

$$\mathbf{u} = \begin{Bmatrix} u \\ v \end{Bmatrix} = \begin{bmatrix} N_1 & 0 & \vdots & N_2 & 0 & \vdots & N_3 & 0 & \vdots & N_4 & 0 \\ 0 & N_1 & \vdots & 0 & N_2 & \vdots & 0 & N_3 & \vdots & 0 & N_4 \end{bmatrix} \begin{Bmatrix} u_1 \\ v_1 \\ u_2 \\ v_2 \\ u_3 \\ v_3 \\ u_4 \\ v_4 \end{Bmatrix} = \mathbf{N}\,\mathbf{a}^{(e)} \tag{4.81}$$

where

$$\mathbf{N} = [\mathbf{N}_1, \mathbf{N}_2, \mathbf{N}_3, \mathbf{N}_4] \quad ; \quad \mathbf{N}_i = \begin{bmatrix} N_i & 0 \\ 0 & N_i \end{bmatrix}$$

$$\mathbf{a}^{(e)} = \begin{Bmatrix} \mathbf{a}_1^{(e)} \\ \mathbf{a}_2^{(e)} \\ \mathbf{a}_3^{(e)} \\ \mathbf{a}_4^{(e)} \end{Bmatrix} \quad ; \quad \mathbf{a}_i^{(e)} = \begin{Bmatrix} u_i \\ v_i \end{Bmatrix} \tag{4.82}$$

are the shape function matrix and the displacement vector for the element and the node i, respectively.

The element strain matrix is obtained from Eqs.(4.2) and (4.79) as

$$\boldsymbol{\varepsilon} = \sum_{i=1}^{4} \mathbf{B}_i \mathbf{a}_i^{(e)} = [\mathbf{B}_1, \mathbf{B}_2, \mathbf{B}_3, \mathbf{B}_4] \begin{Bmatrix} \mathbf{a}_1^{(e)} \\ \mathbf{a}_2^{(e)} \\ \mathbf{a}_3^{(e)} \\ \mathbf{a}_4^{(e)} \end{Bmatrix} = \mathbf{B}\mathbf{a}^{(e)} \tag{4.83}$$

where $\mathbf{B}_i$ is given by precisely the same expression (4.38) derived for the 3-noded triangle. For the computation of $\mathbf{B}_i$ note that

$$\frac{\partial N_i}{\partial x} = \frac{\partial N_i}{\partial r} \quad \text{and} \quad \frac{\partial N_i}{\partial y} = \frac{\partial N_i}{\partial s} \tag{4.84}$$

The expression of $\mathbf{B}$ is shown in Box 4.1.

The stiffness matrix and the equivalent nodal force vectors for the element are obtained via the PVW as explained for the linear triangle. The element stiffness matrix is

$$\mathbf{K}^{(e)} = \iint_{A^{(e)}} \mathbf{B}^T\ \mathbf{D}\ \mathbf{B} t\, dr\, ds =$$

$$= \iint_{A^{(e)}} \begin{bmatrix} \mathbf{B}_1^T\mathbf{D}\mathbf{B}_1 & \mathbf{B}_1^T\mathbf{D}\mathbf{B}_2 & \mathbf{B}_1^T\mathbf{D}\mathbf{B}_3 & \mathbf{B}_1^T\mathbf{D}\mathbf{B}_4 \\ & \ddots & \mathbf{B}_2^T\mathbf{D}\mathbf{B}_2 & \mathbf{B}_2^T\mathbf{D}\mathbf{B}_3 & \mathbf{B}_2^T\mathbf{D}\mathbf{B}_4 \\ & & \ddots & \mathbf{B}_3^T\mathbf{D}\mathbf{B}_3 & \mathbf{B}_3^T\mathbf{D}\mathbf{B}_4 \\ \text{Symm.} & & & & \mathbf{B}_4^T\mathbf{D}\mathbf{B}_4 \end{bmatrix} t\, dr\, ds \tag{4.85}$$

Box 4.1 shows that the strain matrix contains linear terms in r and s. Therefore, the integrand of Eq.(4.85) contains quadratic terms. However, the simplicity of the element geometry allows an explicit integration of all terms. The resulting expression for $\mathbf{K}^{(e)}$ is also shown in Box 4.1.

In the same way, the equivalent nodal force vectors for the element are obtained by Eqs.(4.54)-(4.57) using the above expressions for $\mathbf{N}_i$ and $\mathbf{B}_i$. The nodal contributions of a uniformly distributed load over the element (Eq.(4.65)) are

$$\mathbf{f}_{b_i}^{(e)} = \frac{(tA)^{(e)}}{4} \begin{Bmatrix} b_x \\ b_y \end{Bmatrix} \tag{4.86}$$

$$\mathbf{B} = \begin{bmatrix} -b_2 & 0 & | & b_2 & 0 & | & b_1 & 0 & | & -b_1 & 0 \\ 0 & -a_2 & | & 0 & -a_1 & | & 0 & a_1 & | & 0 & a_2 \\ -a_2 & b_2 & | & -a_1 & b_2 & | & a_1 & b_1 & | & a_2 & b_1 \end{bmatrix}$$

$$a_1 = \frac{1}{4b}(1+\frac{r}{a}) \quad , \quad a_2 = \frac{1}{4b}(1-\frac{r}{a})$$
$$b_1 = \frac{1}{4a}(1+\frac{1}{b}) \quad , \quad b_2 = \frac{1}{4a}(1-\frac{1}{b})$$

$$\mathbf{K}^{(e)} = \begin{bmatrix} 2a_{11} & a_{36} & c_{41} & b_{36} & -a_{14} & -a_{36} & c_{14} & b_{63} \\ & 2a_{35} & b_{63} & c_{25} & -a_{36} & -a_{25} & b_{36} & c_{52} \\ & & 2a_{14} & -a_{36} & c_{14} & b_{36} & -a_{14} & a_{63} \\ & & & 2a_{25} & b_{63} & c_{52} & a_{36} & -a_{52} \\ & & & & 2a_{14} & a_{36} & c_{41} & b_{63} \\ \text{Symmetric} & & & & & 2a_{25} & b_{63} & c_{25} \\ & & & & & & 2a_{11} & -b_{36} \\ & & & & & & & 2a_{25} \end{bmatrix}$$

$$a_{ij} = a_i + a_j \quad , \quad b_{ij} = a_i - a_j \quad , \quad c_{ij} = a_i - 2a_j$$
$$a_1 = \frac{tbd_{11}}{6a} \; , \; a_2 = \frac{tad_{22}}{6b} \; , \; a_3 = \frac{td_{12}}{4} \; , a_4 = \frac{tad_{33}}{6b} \; , \; a_5 = \frac{tbd_{33}}{6a} \; , \; a_6 = \frac{td_{33}}{4}$$

Box 4.1 Strain and stiffness matrices for a 4-noded rectangular element of dimensions $2a \times 2b$

i.e. the total force is distributed in equal parts between the four nodes, like for the 3-noded triangle.

Similarly, a uniformly distributed traction acting over a side is distributed in equal parts between the two side nodes.

4.4.2 Some remarks on the behaviour of the 4-noded rectangle

Both the 3-noded triangle and the 4-noded rectangle perform excellently in problems where traction (or compression) is important. Conversely, the accuracy of both elements deteriorates in situations where bending movements are involved, and very fine meshes are needed to obtain accurate solutions in these cases (Section 4.7).

The fact that the 4-noded rectangle cannot be used to model bending dominated fields has a very instructive explanation. Let us consider the behaviour of an isolated element subjected to pure bending (Figure 4.12).

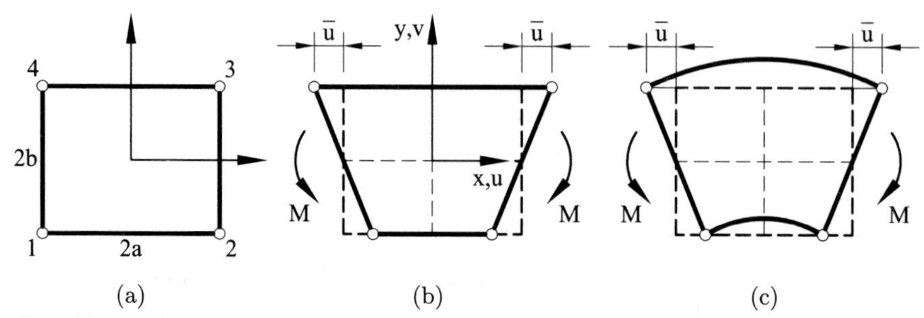

Fig. 4.12 4-noded rectangle subjected to pure bending, a) Initial geometry, b) Element distortion, c) Correct deformation of a beam segment in pure bending

The exact solution from beam theory is [TG]

$$
\begin{aligned}
u(r,s) &= \frac{M}{EI}\,rs \\
v(r,s) &= \frac{Ma^2}{2EI}\left(1-\frac{r^2}{a^2}\right)+\frac{Mb^2}{2EI}\left(1-\frac{s^2}{b^2}\right)
\end{aligned}
\tag{4.87}
$$

Since the element sides are straight, the 4-noded rectangle can only represent the following bending mode (Figure 4.12b)

$$
u = \bar{u}rs \qquad ; \qquad v = 0 \tag{4.88}
$$

It is obvious from the above that the element cannot correctly reproduce the quadratic distribution of vertical displacements for the pure bending case. This leads to excessive stiffness, which is a natural consequence of the inability of the element sides to be curved.

Additionally it is deduced from Eq.(4.87) that

$$
\gamma_{xy} = \frac{\partial u}{\partial y} + \frac{\partial v}{\partial x} = 0
$$

i.e. the "exact" shear strain vanishes and only normal strains (and stresses) exist.

The shear strain field for the element is obtained from Eq.(4.88) as

$$
\gamma_{xy} = \bar{u}r \tag{4.89}
$$

i.e. the element has an "excess" of shear strain. This introduces an undesirable stiffness which contributes to the poor ability of the element to reproduce bending modes. Similar results are obtained for moments acting on the horizontal sides simply by changing the coordinate r for s in Eq.(4.89).

The deficiencies of the 4-noded rectangle also appear for more irregular quadrilateral shapes. These drawbacks are usually overcome in practice by using very fine meshes. Other alternatives are possible, however, and some are presented in the following sections.

4.4.2.1 Reduced integration of the shear stiffness terms

Eq.(4.89) clearly shows that the shear strain is zero at the element center only. Therefore, the excess of shear strain can be eliminated by sampling the shear strain at the element center ($r = s = 0$). This is simply achieved by using a reduced one point Gauss quadrature for the shear terms in the stiffness matrix. For this purpose the element stiffness matrix is split as

$$\mathbf{K}^{(e)} = \mathbf{K}_a^{(e)} + \mathbf{K}_s^{(e)} \tag{4.90}$$

where $\mathbf{K}_a^{(e)}$ and $\mathbf{K}_s^{(e)}$ include the "axial" and "shear" contributions, respectively given by

$$\mathbf{K}_{a_{ij}}^{(e)} = \iint_{A^{(e)}} \mathbf{B}_{a_i}^T \mathbf{D}_a \mathbf{B}_{a_j} t\, dA \;\; ; \;\; \mathbf{K}_{s_{ij}}^{(e)} = \iint_{A^{(e)}} \mathbf{B}_{s_i}^T \mathbf{D}_t \mathbf{B}_{s_j} t\, dA \tag{4.91}$$

with

$$\begin{aligned} \mathbf{B}_{a_i} &= \begin{bmatrix} \dfrac{\partial N_i}{\partial x} & 0 \\ 0 & \dfrac{\partial N_i}{\partial y} \end{bmatrix} \;\; ; \;\; \mathbf{B}_{s_i} = \left[\dfrac{\partial N_i}{\partial y}, \;\; \dfrac{\partial N_i}{\partial x} \right] \\ \mathbf{D}_a &= \begin{bmatrix} d_{11} \; d_{12} \\ d_{12} \; d_{22} \end{bmatrix} \;\; ; \;\; \mathbf{D}_s = [d_{33}] \end{aligned} \tag{4.92}$$

Matrix $\mathbf{K}_a^{(e)}$ is integrated exactly, either analytically or via a 2×2 Gauss quadrature, whereas a single integration point is used for $\mathbf{K}_s^{(e)}$. This "selective integration" technique also improves the behaviour of 4-noded quadrilaterals of arbitrary shape.

The reduced integration of $\mathbf{K}_s^{(e)}$ can also be interpreted as a simple procedure to mitigate the excessive influence of the shear terms in the element stiffness matrix. A disadvantage of reduced integration is that it produces a quadrilateral element that is not geometric-invariant (Section 3.10.4), although it passes the patch test and, therefore, it converges to the exact solution [CMPW]. In Chapter 2 of Volume 2 [On] we will apply reduced integration to alleviate the influence of the transverse shear stiffness in Timoshenko beam elements. However, the reduced integration of the stiffness matrix terms should always be looked upon with extreme

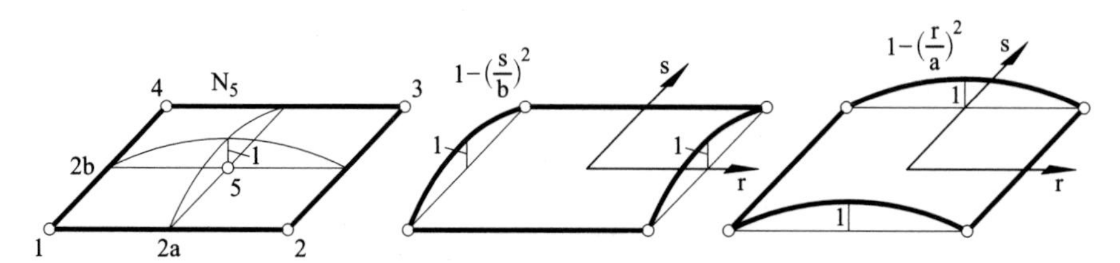

Fig. 4.13 Four-noded rectangle, a) Addition of a central node, b) Shape functions for the incompatible modes

caution, as it can lead to internal mechanisms and to the violation of the patch test in some cases. Reduced integration techniques will be further studied when dealing with plates and shells in Volume 2 [On].

4.4.2.2 Addition of internal modes

The flexibility of the 4-noded rectangle can be enhanced by adding to the original interpolation internal displacement modes which vanish at the element boundaries. The simplest mode is a "bubble" function associated with an extra central node (Figure 4.13a). The displacement field is expressed as

$$u = \sum_{i=1}^{5} N_i u_i \quad ; \quad v = \sum_{i=1}^{5} N_i v_i \tag{4.93}$$

where N_1, N_2, N_3, N_4 are the linear functions of (4.80) and

$$N_5 = \left[1 - \left(\frac{r}{a}\right)^2\right]\left[1 - \left(\frac{s}{b}\right)^2\right] \tag{4.94}$$

The internal DOFs u_5 and v_5 (also called hierarchical DOFs [Cr]) can be eliminated after the element stiffness matrix is obtained. Note that u_5 and v_5 are not absolute displacements and they represent the difference between the total displacements of the central node and the bilinear field defined by the four corner displacements. For instance, the horizontal displacement of the central node is given by

$$u(0,0) = \left(\sum_{i=1}^{4} N_i u_i\right)_{0,0} + u_5 \tag{4.95}$$

The behaviour of the modified 4-noded element can be improved by using a reduced single point quadrature for the shear terms as described in the previous section.

4.4.2.3 Addition of incompatible modes

The 4-noded rectangle can also be improved by adding to the original displacement field the displacement modes $1-\left(\frac{r}{a}\right)^2$ and $1-\left(\frac{s}{b}\right)^2$ (also called incompatible modes) which are needed to reproduce the exact solution (4.87) (Figure 4.12b). The new displacement field is

$$
\begin{aligned}
u &= \sum_{i=1}^{4} N_i u_i + \left[1-\left(\frac{r}{a}\right)^2\right]u_5 + \left[1-\left(\frac{s}{b}\right)^2\right]u_6 \\
v &= \sum_{i=1}^{4} N_i v_i + \left[1-\left(\frac{r}{a}\right)^2\right]v_5 + \left[1-\left(\frac{s}{b}\right)^2\right]v_6
\end{aligned}
\tag{4.96}
$$

The additional variables u_5, v_5, u_6, v_6 (also called "nodeless" DOFs) are internal to each element and can be eliminated by static condensation. However, the displacements along the interelemental boundaries are discontinuous and the element is incompatible. Incompatible 4-noded quadrilaterals formulated in this way fail to pass the patch test under constant stress (or constant strain) states unless they are rectangular.

Fortunately, the element satisfies the patch test for arbitrary quadrilateral shapes if the shear stiffness terms are evaluated using a reduced single point Gauss quadrature, whereas the rest of the stiffness terms can be exactly integrated. The resulting element is geometric-invariant and does not have spurious mechanisms. Box 4.2 shows the stiffness matrix for an homogeneous and isotropic element of this kind with reduced integration after eliminating the internal incompatible DOFs by static condensation [CMPW,FNS,TBW].

The incompatible modes technique can also be successfully applied to 4-noded quadrilaterals of arbitrary shape [CMPW,ZHZ].

4.4.2.4 Use of an assumed strain field

Another procedure to enhance the performance of the 4-noded quadrilateral is to impose over the element an assumed strain field compatible with the condition $\gamma_{xy} = 0$ for the pure bending case.

Dvorkin and Vassolo [DV] proposed the following assumed strain field

$$
\varepsilon_x = \alpha_1 + \alpha_2 x + \alpha_3 y \ ; \qquad \varepsilon_y = \alpha_4 + \alpha_5 x + \alpha_6 y \ ; \qquad \gamma_{xy} = \alpha_7 \tag{4.97}
$$

The α_i parameters are expressed in terms of the nodal displacements by sampling the assumed strains at a number of element points and equaling their values to those given by the strains deduced from the original

$$\begin{Bmatrix} u \\ v \end{Bmatrix} = \sum_{i=1}^{4} N_i \begin{Bmatrix} u_i \\ v_i \end{Bmatrix} + \left[1 - \left(\frac{\mathrm{r}}{a}\right)^2\right] \begin{Bmatrix} u_5 \\ v_5 \end{Bmatrix} + \left[1 - \left(\frac{s}{a}\right)^2\right] \begin{Bmatrix} u_6 \\ v_6 \end{Bmatrix}$$

$$\mathbf{K}^{(e)} = \frac{D\,t}{12(1-\alpha^2)} \begin{bmatrix} C_1 & C_5 & C_2 & -C_6 & C_4 & -C_5 & C_3 & C_6 \\ & C_7 & C_6 & C_9 & -C_5 & C_{10} & -C_6 & C_8 \\ & & C_1 & -C_5 & C_3 & -C_6 & C_4 & C_5 \\ & & & C_7 & C_6 & C_8 & C_5 & C_{10} \\ & & & & C_1 & C_5 & C_2 & -C_6 \\ \text{Symmetric} & & & & & C_7 & C_6 & C_9 \\ & & & & & & C_1 & -C_5 \\ & & & & & & & C_7 \end{bmatrix}$$

Plane stress : $D = E$; $\alpha = \nu$

Plane strain : $D = \dfrac{E}{1-\nu^2}$; $\alpha = \dfrac{\nu}{1-\nu}$

$C_1 = \frac{a}{b}\,(-m^2 - 1.5m + 5.5)$, $C_4 = \frac{a}{b}(-m^2 + 1.5m - 3.5)$

$C_2 = \frac{a}{b}\,(m^2 - 1.5m - 2.5)$, $C_5 = 1.5\,(1 + m)$

$C_3 = \frac{a}{b}(m^2 + 1.5m - 0.5m)$, $C_6 = 1.5\,(1 + 3m)$

$C_7 - C_{10}$ are obtained from $C_1 - C_4$ interchanging a for b

Box 4.2 Stiffness matrix for an homogeneous and isotropic 4-noded rectangular element of dimensions 2a×2b with incompatible modes

displacement field. This leads to a substitute strain matrix from which the element stiffness matrix can be directly obtained on [DV].

4.5 PERFORMANCE OF THE 3-NODED TRIANGLE AND THE 4-NODED RECTANGLE

The 3-noded triangle and the 4-noded rectangle perform well under pure tension or compression dominated situations. In general the 4-noded element is more accurate than the 3-noded triangle for the same number of DOFs in these cases. However, as mentioned earlier the behaviour of both elements is relatively poor in bending situations. Still, the 4-noded element has a superior performance for such problems. This is clearly seen in the examples shown in Figures 4.14 and 4.15 of a thick cantilever beam under an end point load and a simply supported beam under self weight analyzed with different meshes of 3-noded triangles and 4-noded rectangles. All units are in the International System. The values compared are: 1) the

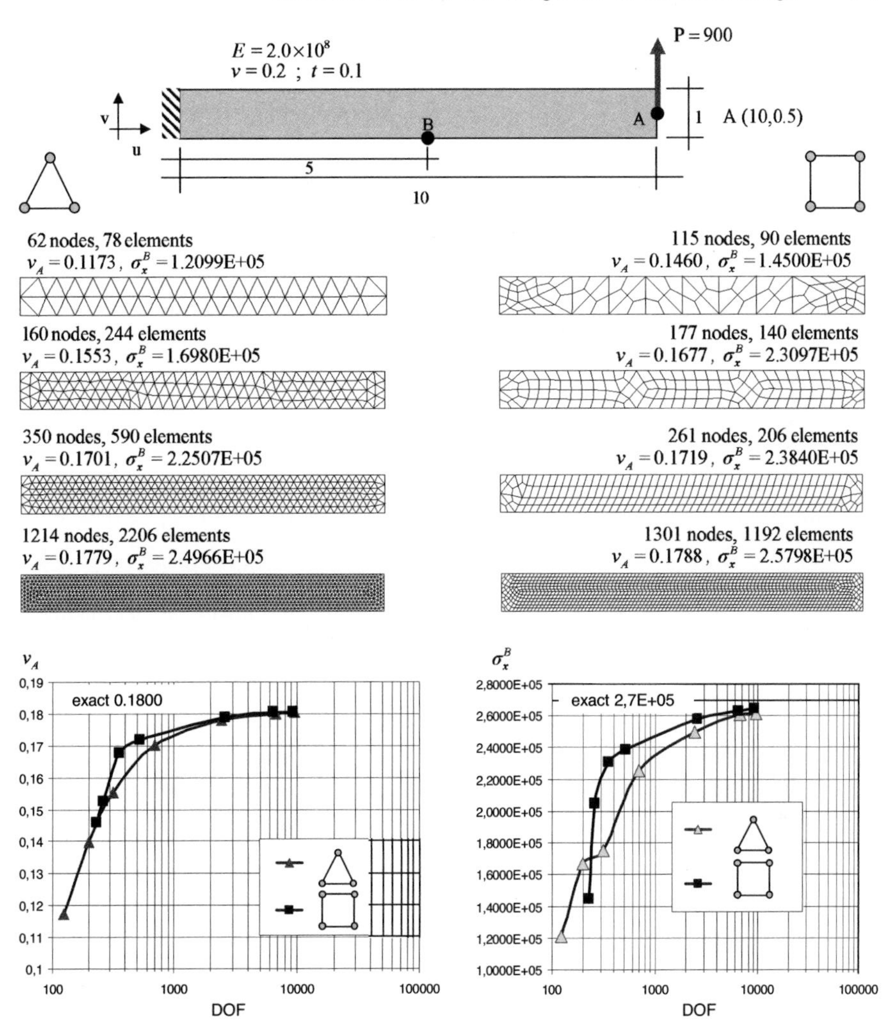

Fig. 4.14 Cantilever beam under end point load. Convergence of the vertical deflection at the free end and the horizontal stress σ_x at the lower fiber of the middle section for unstructured meshes of 3-noded triangles and 4-noded rectangles

vertical deflection at the center of the free end for the clamped beam and at the lower point of the middle section for the simply supported case, and 2) the σ_x stress at the lower fiber of the middle section for both problems. Note the higher accuracy of the 4-noded rectangle in accordance with that previously explained. The accuracy of the 4-noded rectangle increases by adding the two incompatible modes as described in Section 4.4.2.3.

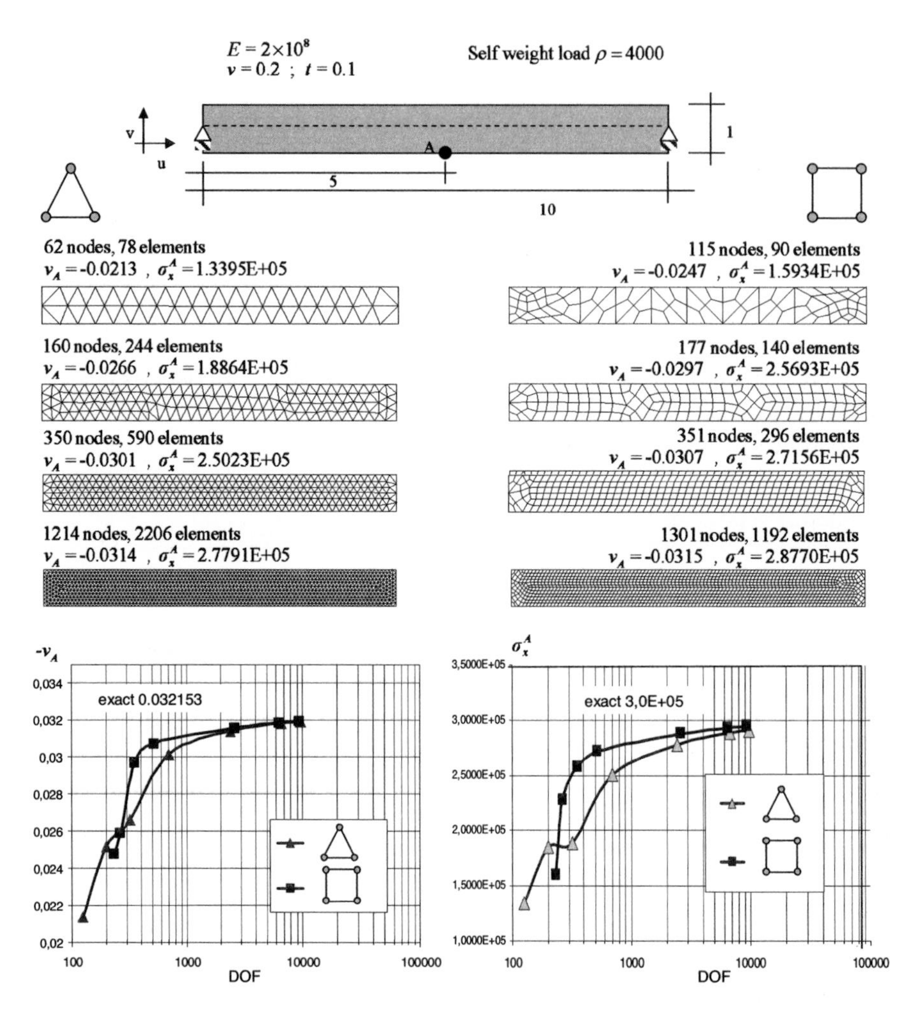

Fig. 4.15 Simply supported beam under self-weight. Convergence of the vertical deflection and the horizontal stress σ_x at the lower fiber of the middle section for different unstructured meshes of 3-noded triangles and 4-noded rectangles

The accuracy of both elements also increases by using a higher order approximation for the displacement field. This also allows curve sided elements to be derived using an isoparametric formulation as described in the next chapter.

The poorer performance of the 3-noded triangle is compensated by its versatility to discretize complex geometries using unstructured meshes and so it remains very popular.

More examples of the performance of the 3-noded triangle and the 4-noded quadrilateral, as well as of other higher order triangular and quadrilateral elements are presented in Section 5.7.

4.6 CONCLUDING REMARKS

This chapter has presented the basic concepts for the analysis of 2D solids with the FEM. The steps followed in the formulation of the kinematic variables, the strain and stress fields, the equilibrium expressions via the PVW and the discretization process are completely general and will repeatedly appear when considering the finite element analysis of any other structure. The study of this chapter is therefore essential as a general introduction to the analysis of continuous structures with the FEM.

The procedure for deriving the element stiffness matrix and the equivalent nodal force vector from the PVW has been detailed. The basic expressions for the different matrices and vectors have general applicability to any element type. The particular form of these matrices for the 3-noded linear triangle and the 4-noded rectangle has been given. The linear triangle has limited accuracy for coarse meshes although its simplicity and versatility for discretizing any geometrical shape make it probably the most popular element for practical analysis of 2D solids. The standard 4-noded rectangle has some limitations when it comes to capturing pure bending modes. These deficiencies can be overcome by "ad hoc" procedures such as reduced integration, the addition of internal nodes and the use of an assumed strain field.

The derivation of higher order triangular and quadrilateral elements of arbitrary shape requires a systematic procedure to obtain the shape functions, the use of an isoparametric formulation and numerical integration. These topics will be studied in the next chapter.

5

HIGHER ORDER 2D SOLID ELEMENTS. SHAPE FUNCTIONS AND ANALYTICAL COMPUTATION OF INTEGRALS

5.1 INTRODUCTION

This chapter extends the concepts studied in the previous one for the analysis of solids under the assumptions of 2D elasticity using higher order triangular and quadrilateral elements.

The chapter is organized as follows. In the first sections we detail the systematic derivation of the shape functions for rectangular and triangular elements of different order of approximation. Next, some rules for the analytical computation of the element integrals over rectangles and straight side triangles are given. Finally the performance of linear and quadratic triangular and rectangular elements is compared in some academic examples.

5.2 DERIVATION OF THE SHAPE FUNCTIONS FOR C^o TWO DIMENSIONAL ELEMENTS

Next, we will derive the shape functions for different triangular and rectangular elements with C^0 continuity. The possibilities of distorting these elements into arbitrary shapes including curve sides will be treated in Chapter 6 using the concept of isoparametric interpolation.

5.2.1 Complete polynomials in two dimensions. Pascal triangle

The chosen displacement field can only reproduce exactly a polynomial solution of an order equal to or less than that of the complete polynomial contained in the shape functions (Section 3.10.2). Consequently, the solution will improve as the degree of such a complete polynomial increases.

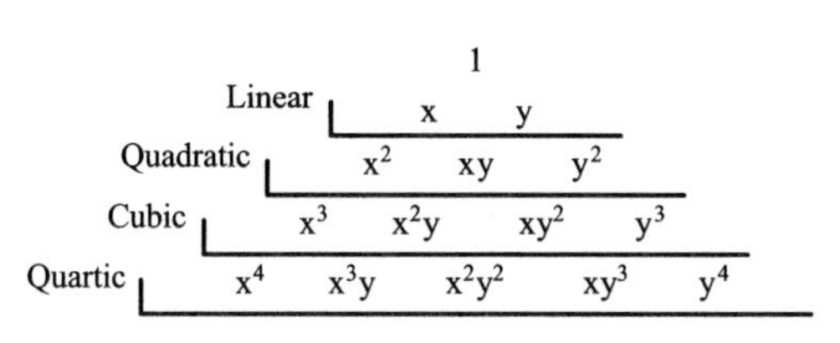

Polynomial degree	Nº of terms
Constant	1
Linear	2
Quadratic	6
Cubic	10
Quartic	15

Fig. 5.1 Pascal triangle in two dimensions

A 2D complete polynomial of nth degree can be written as

$$f(x,y) = \sum_{i=1}^{p} \alpha_i x^j y^k \qquad ; \qquad j+k \leq n \tag{5.1}$$

where the number of terms is

$$p = (n+1)(n+2)/2 \tag{5.2}$$

For a linear polynomial ($p = 3$)

$$f(x,y) = \alpha_1 + \alpha_2 x + \alpha_3 y \tag{5.3a}$$

and for a quadratic polynomial ($p = 6$)

$$f(x,y) = \alpha_1 + \alpha_2 x + \alpha_3 y + \alpha_4 xy + \alpha_5 x^2 + \alpha_6 y^2 \tag{5.3b}$$

The terms of a 2D complete polynomial can be readily identified by means of the *Pascal triangle* (Figure 5.1).

The shape functions of triangles and tetrahedra are complete polynomials, whereas those of quadrilateral and hexahedral elements contain incomplete polynomial terms. For instance, the shape functions of the 4-noded rectangle include the term $\alpha_4 xy$ from the quadratic polynomial (Eq.(4.77)). These terms in general do not contribute to increasing the order of the approximation.

5.2.2 Shape functions of C^o rectangular elements. Natural coordinates in two dimensions

A local coordinate system ξ, η is defined for each element in order to facilitate the derivation of the shape functions. Such a *natural* or *intrinsic* coordinate system is normalized so that the element sides are located at

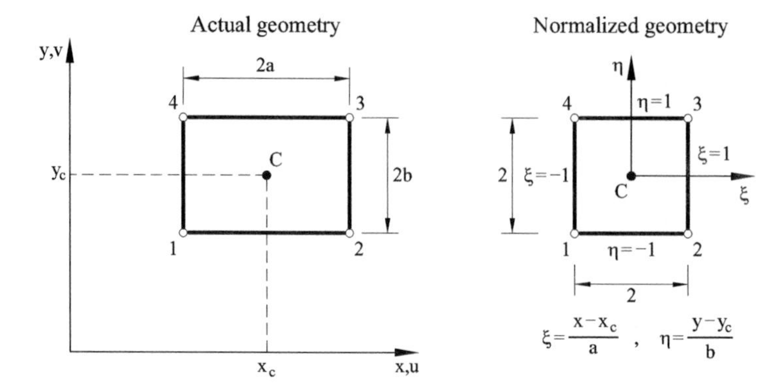

Fig. 5.2 Rectangular element. Cartesian and natural coordinate systems

$\xi = \pm 1$ and $\eta = \pm 1$ as shown in Figure 5.2. The natural coordinate ξ was introduced for 1D rod elements in Section 3.2. From Figure 5.2 we deduce

$$\xi = \frac{x - x_c}{a} \quad ; \quad \eta = \frac{y - y_c}{b} \tag{5.4}$$

where x_c and y_c are the coordinates of the element centroid. Note that in the natural coordinate system the rectangular element becomes a square of side equal to two. From Eq.(5.4)

$$\frac{d\xi}{dx} = \frac{1}{a} \quad ; \quad \frac{d\eta}{dy} = \frac{1}{b} \tag{5.5}$$

The differentials of area in the cartesian and natural systems are related by

$$dx\, dy = ab\, d\xi\, d\eta \tag{5.6}$$

The integration of a function $f(x, y)$ over a rectangular element can be expressed in the natural coordinate system by

$$\iint_{A^{(e)}} f(x, y)\, dx\, dy = \int_{-1}^{+1} \int_{-1}^{+1} g(\xi, \eta) ab\, d\xi\, d\eta \tag{5.7}$$

The shape functions when expressed in the natural coordinates must satisfy the same requirements as in cartesian coordinates. Therefore, the shape functions for C^0 continuous elements must satisfy:

a) Condition of nodal compatibility

$$N_i(\xi_j, \eta_j) = \begin{matrix} 1 & \quad i = j \\ 0 & \quad i \neq j \end{matrix} \tag{5.8}$$

b) Rigid body condition (Section (3.8.4))

$$\sum_{i=1}^{n} N_i(\xi,\eta) = 1 \tag{5.9}$$

Two element families can be clearly identified within C^0 continuous rectangular elements, i.e. the Lagrange family and the Serendipity family. The derivation of the shape functions for each of these two element families is presented next.

5.3 LAGRANGE RECTANGULAR ELEMENTS

The shape functions for 2D Lagrange elements can be obtained by the simple product of the two normalized 1D Lagrange polynomials corresponding to the natural coordinates ξ and η of the node. Thus, if $l_I^i(\xi)$ is the 1D Lagrange polynomial of order I in the ξ direction for node i and $l_J^i(\eta)$ is the normalized 1D Lagrange polynomial of order J in the η direction for node i, the shape function of node i is

$$N_i(\xi,\eta) = l_I^i(\xi) \;\; l_J^i(\eta) \tag{5.10}$$

The normalized 1D Lagrange polynomials for each node can be obtained by Eq.(3.9) which can be indistinctly used for the coordinates ξ and η. Figure 5.3 shows some of the more popular Lagrange rectangular elements. Note that the number of nodes in each of the two directions ξ and η are the same along any nodal line. This is a particular feature of Lagrange elements.

The polynomial terms contained in the shape functions can be directly obtained from the Pascal triangle as shown in Figure 5.3. The shape functions are not complete polynomials and all contain some incomplete polynomial terms.

The derivation of the shape functions for the more popular Lagrange rectangular elements is presented next.

5.3.1 Four-noded Lagrange rectangle

This is the simplest element of the Lagrange family and it coincides precisely with that studied in Section 4.4. For consistency we will derive its shape functions again using natural coordinates (Figure 5.4).

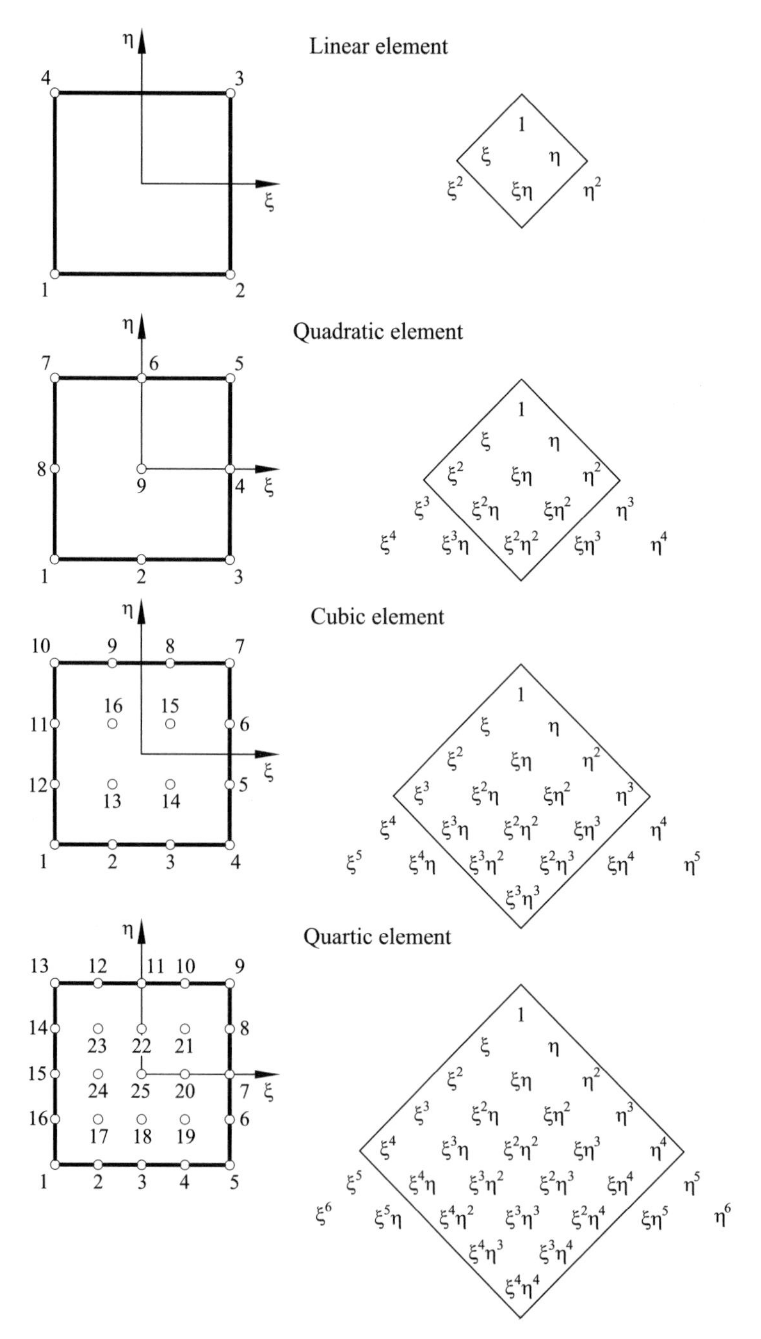

Fig. 5.3 Some Lagrange rectangular elements. Polynomial terms contained in the shape functions

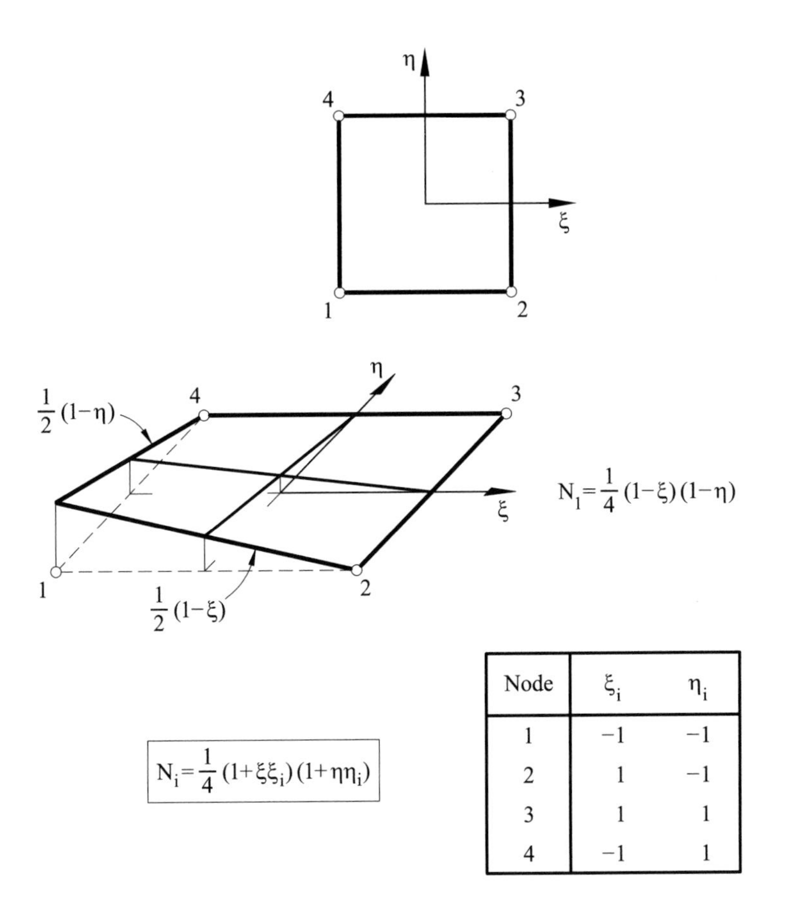

Node	ξ_i	η_i
1	−1	−1
2	1	−1
3	1	1
4	−1	1

Fig. 5.4 Four-noded Lagrange rectangular element

Let us consider a generic node i. The 1D shape functions corresponding to the local directions ξ and η coincide with the shape functions for the 2-noded bar element. Thus, it is easy to find

$$l_1^i(\xi) = \frac{1}{2}(1 + \xi\xi_i) \ \ ; \ \ l_1^i(\eta) = \frac{1}{2}(1 + \eta\eta_i) \tag{5.11}$$

where ξ_i and η_i take the values given in the table of Figure 5.4. The shape function of any node can be written in compact form as

$$N_i(\xi, \eta) = l_1^i(\xi) l_1^i(\eta) = \frac{1}{4}(1 + \xi\xi_i)(1 + \eta\eta_i) \tag{5.12}$$

A simple change of coordinates shows that the above shape functions coincide with those directly obtained in the local system r, s in Section 4.4.

Figure 5.4 shows in graphic form the shape function of node 1. It is easy to verify that the shape functions (5.12) satisfy the conditions (5.8) and (5.9).

5.3.2 Nine-noded quadratic Lagrange rectangle

The shape functions for the 9-noded Lagrange rectangle (Figure 5.5) are obtained by the product of two normalized 1D quadratic polynomials in ξ and η. These polynomials are obtained from the shape functions of the quadratic rod element (Eq.(3.11)). Thus, for node 1

$$l_2^1(\xi) = \frac{1}{2}(\xi - 1)\xi \quad ; \quad l_2^1(\eta) = \frac{1}{2}(\eta - 1)\eta \tag{5.13}$$

and the shape function is

$$N_2(\xi, \eta) = l_2^1(\xi) l_2^1(\eta) = \frac{1}{4}(\xi - 1)(\eta - 1)\xi\ \eta \tag{5.14}$$

Following a similar procedure for the rest of the nodes, the shape functions can be written in compact form as

a) Corner nodes

$$N_i = \frac{1}{4}(\xi^2 + \xi\xi_i)(\eta^2 + \eta\eta_i) \quad ; \quad i = 1, 3, 5, 7 \tag{5.15}$$

b) Mid-side nodes

$$N_i = \frac{1}{2}\eta_i^2(\eta^2 - \eta\eta_i)(1 - \xi^2) + \frac{1}{2}\,\xi_i^2(\xi^2 - \xi\xi_i)(1 - \eta^2) \quad ; \quad i = 2, 4, 6, 8 \tag{5.16}$$

c) Central node

$$N_9(\xi, \eta) = (1 - \xi^2)(1 - \eta^2) \tag{5.17}$$

We can verify that these shape functions satisfy Eqs.(5.8) and (5.9).

Figure 5.5 shows the shape function of three characteristic nodes. These functions contain the polynomial terms shown in Figure 5.3. The 9-noded Lagrange rectangle contains all the terms of a complete quadratic polynomial plus three additional terms of the cubic and quartic polynomials ($\xi^2\eta$, $\xi\eta^2$, $\xi^2\eta^2$). Therefore, the approximation is simply of quadratic order.

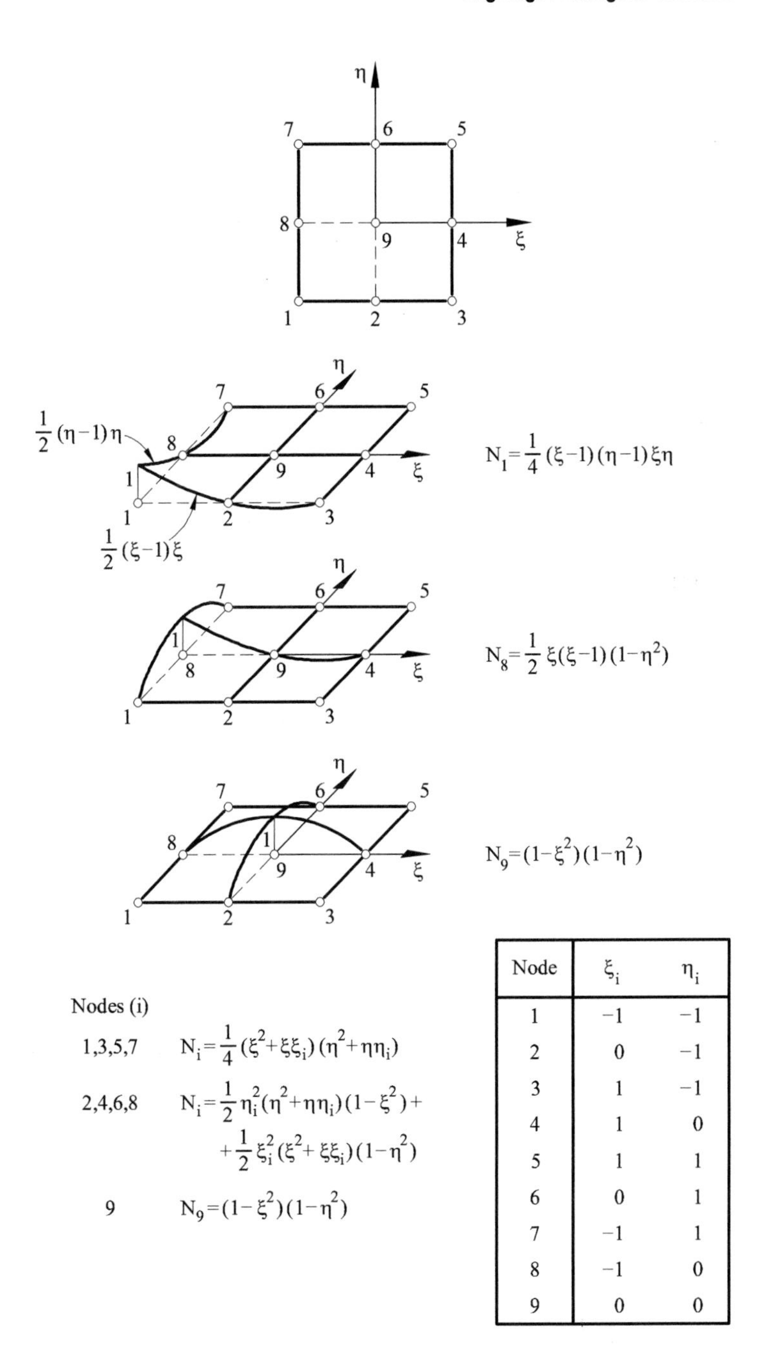

Node	ξ_i	η_i
1	−1	−1
2	0	−1
3	1	−1
4	1	0
5	1	1
6	0	1
7	−1	1
8	−1	0
9	0	0

Fig. 5.5 Nine-noded quadratic Lagrange rectangle

5.3.3 Sixteen-noded cubic Lagrange rectangle

This element has four nodes in each of the two directions ξ and η. The shape functions are obtained by the product of two normalized 1D cubic Lagrange polynomials in the ξ and η directions deduced from Eq.(3.12). Figure 5.6 shows the expressions for the shape functions that are *complete cubic* polynomials and contain the following additional terms: $\xi^3\eta$, $\xi^2\eta^2$, $\xi\eta^3$, $\xi^3\eta^2$, $\xi^2\eta^3$ and $\xi^3\eta^3$ from the quartic, quintic and sextic polynomials (Figure 5.3). The shape functions satisfy Eqs.(5.8) and (5.9).

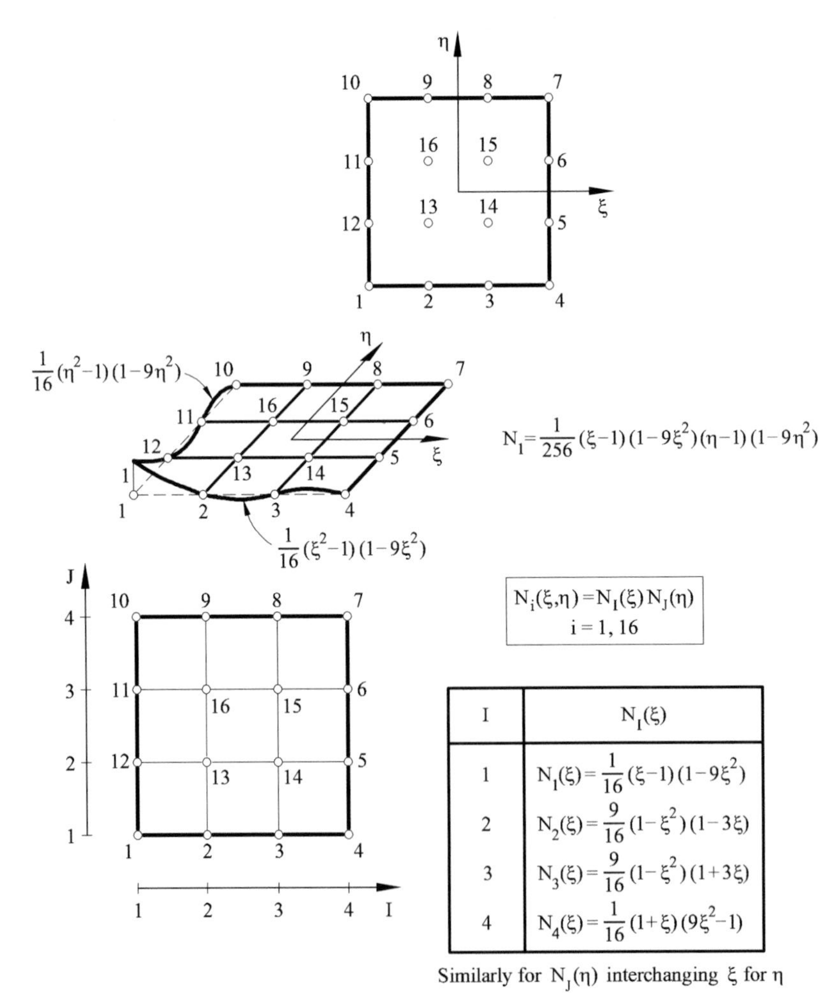

I	$N_I(\xi)$
1	$N_1(\xi)=\frac{1}{16}(\xi-1)(1-9\xi^2)$
2	$N_2(\xi)=\frac{9}{16}(1-\xi^2)(1-3\xi)$
3	$N_3(\xi)=\frac{9}{16}(1-\xi^2)(1+3\xi)$
4	$N_4(\xi)=\frac{1}{16}(1+\xi)(9\xi^2-1)$

Similarly for $N_J(\eta)$ interchanging ξ for η

Fig. 5.6 Shape functions for the sixteen-noded cubic Lagrange rectangle

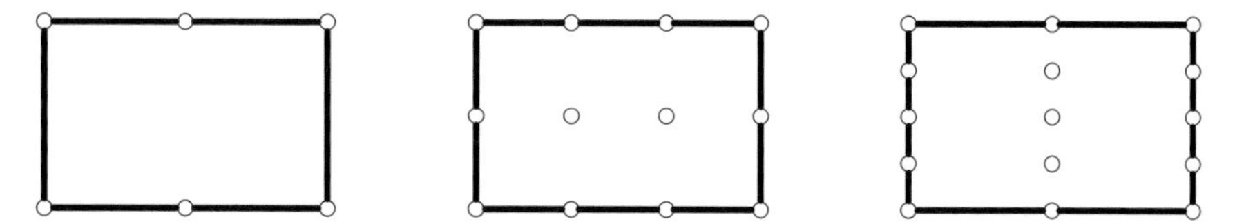

Fig. 5.7 Lagrange rectangles with different number of nodes in the local directions

5.3.4 Other Lagrange rectangular elements

The shape functions of higher order Lagrange rectangular elements with 5, 6, 7, etc. nodes in each of the ξ and η directions are obtained by the product of fourth, fifth, sixth, etc. degree normalized 1D Lagrange polynomials in ξ and η, similar to the linear, quadratic and cubic elements previously studied. It is easy to verify that the shape functions of a Lagrange element with n nodes in each of the two local directions ξ and η contain a complete nth degree polynomial and $n(n+1)/2$ terms of incomplete polynomials up to a $\xi^n\eta^n$ degree which can be deduced from the Pascal triangle.

Lagrange elements can have different number of nodes in each local direction ξ or η (Figure 5.7). The shape functions in this case are obtained by the product of the adequate 1D polynomials in ξ and η corresponding to the number of nodes in each direction. The shape functions now contain a complete 2D polynomial of a degree equal to the smallest degree of the two 1D polynomials in each local direction. Therefore, the degree of approximation of the element does not change by simply increasing the number of nodes in one of the two local directions only. This explains why these elements are not very popular and they are only occasionally used as a transition between elements of two different orders.

5.4 SERENDIPITY RECTANGULAR ELEMENTS

Serendipity elements are obtained as follows. First the number of nodes defining a 1D polynomial of a given degree along each side is chosen. Then, the minimum number of nodes *within* the element is added so that a complete and symmetrical 2D polynomial *of the same degree* as the 1D polynomial chosen along the sides is obtained. Figure 5.8 shows some of the more popular Serendipity elements and the polynomial terms contained in the shape functions. The simplest element of the Serendipity family, i.e. the 4-noded rectangle, coincides with the same element of the Lagrange family. Also note that the quadratic and cubic elements of 8 and 12 nodes,

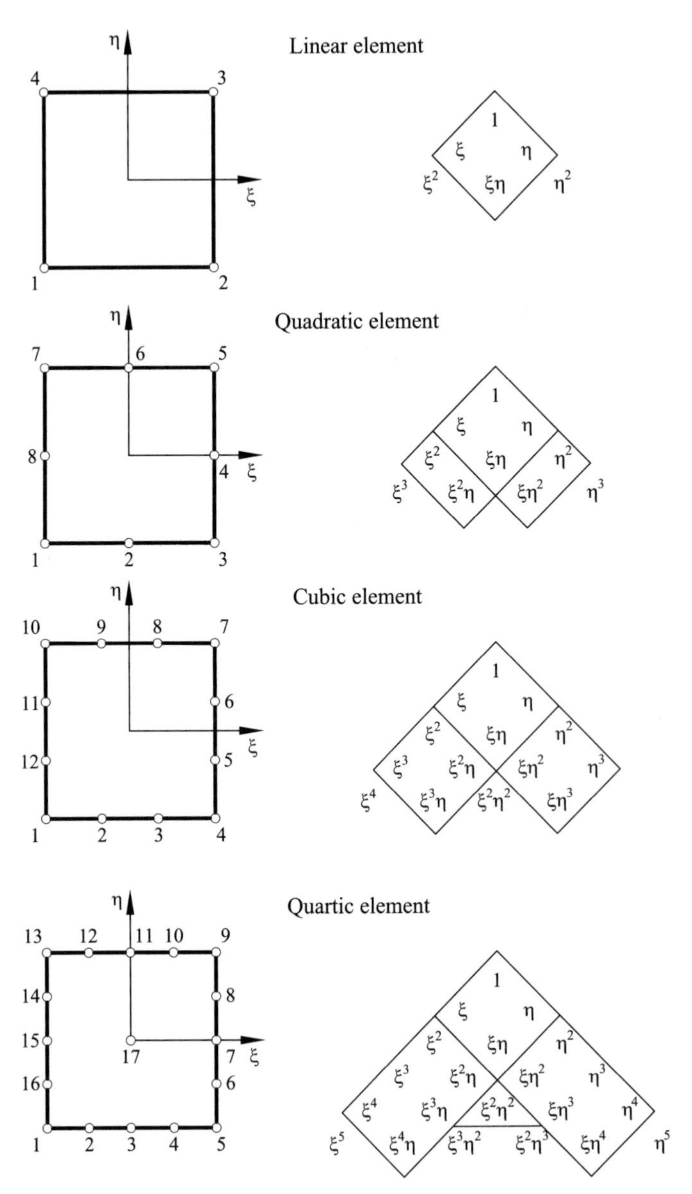

Fig. 5.8 Some Serendipity elements and terms contained in their shape functions

respectively, have not interior nodes, whereas the 17 node element requires a central node to guarantee the complete quartic approximation, as is explained next.

The derivation of the shape functions for Serendipity elements is not as straightforward as for Lagrange elements. In fact, some ingenuity is needed and this explains the name Serendipity, after the ingenuous discoveries of the Prince of Serendip quoted in the romances of Horace Walpole in the eighteenth century [EIZ,ZTZ].

5.4.1 Eigth-noded quadratic Serendipity rectangle

The shape functions for the side nodes are readily obtained as the product of a second degree polynomial in ξ (or η) and another one in η (or ξ). It can be checked that this product contains the required complete quadratic terms (Figure 5.9). For these nodes we obtain

$$\begin{aligned} N_i(\xi,\eta) &= \frac{1}{2}(1+\xi\xi_i)(1-\eta^2) \quad ; \quad i=4,8 \\ N_i(\xi,\eta) &= \frac{1}{2}(1+\eta\eta_i)(1-\xi^2) \quad ; \quad i=2,6 \end{aligned} \tag{5.18}$$

Unfortunately this strategy can not be applied for the corner nodes, since in this case the product of two quadratic polynomials will yield a zero value at the center and thus the criterion of Eq.(5.9) would be violated. Consequently, a different procedure is followed as detailed below.

Step 1. The shape function for the corner node is initially assumed to be bi-linear, i.e. for node 1 (Figure 5.9) we have

$$N_1^L = \frac{1}{4}(1-\xi)(1-\eta) \tag{5.19}$$

This shape function takes the value one at the corner node and zero at all the other nodes, except for the two nodes 2 and 8 adjacent to node 1 where it takes the value 1/2.

Step 2. The shape function is made zero at node 2 by subtracting from N_1^L one half of the quadratic shape function of node 2:

$$\overline{N}_1(\xi,\eta) = N_1^L - \frac{1}{2}N_2 \tag{5.20}$$

Step 3. Function $\overline{N}_1$ still takes the value 1/2 at node 8. The final step is to substract from $\overline{N}_1$ one half of the quadratic shape function of node 8

$$N_1(\xi,\eta) = N_1^L - \frac{1}{2}\,N_2 - \frac{1}{2}N_8 \tag{5.21}$$

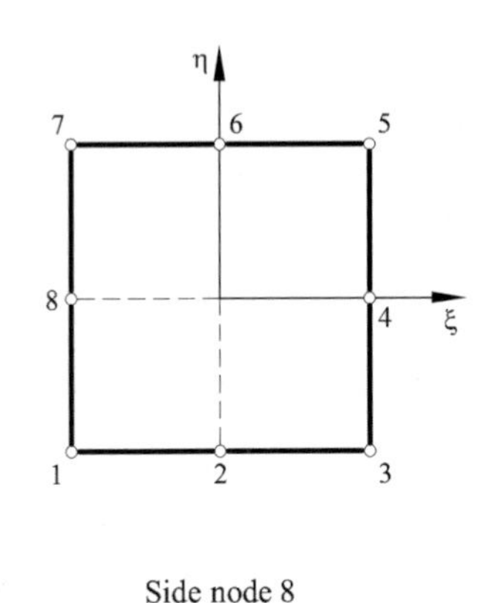

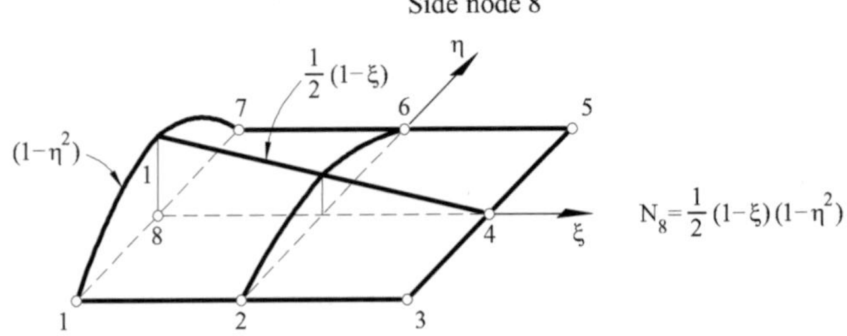

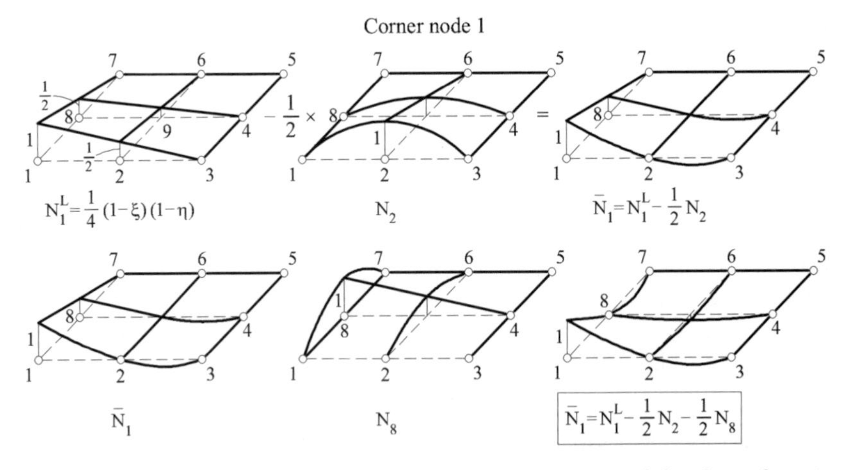

Fig. 5.9 8-noded quadratic Serendipity rectangle. Derivation of the shape functions for a mid-side node and a corner node

The resulting shape function N_1 satisfies the conditions (5.8) and (5.9) and contains the desired (quadratic) polynomial terms. Therefore, it is the shape function of node 1 we are looking for.

Following the same procedure for the rest of the corner nodes yields the following general expression

$$N_i(\xi,\eta) = \frac{1}{4}(1+\xi\xi_i)(1+\eta\eta_i)(\xi\xi_i+\eta\eta_i-1) \quad ; \quad i=1,3,5,7 \tag{5.22}$$

Figure 5.9 shows that the shape functions for the 8-noded Serendipity element contain a complete quadratic polynomial and two terms $\xi^2\eta$ and $\xi\eta^2$ of the cubic polynomial. Therefore, this element has the same approximation as the 9-noded Lagrange element *and it has one node less.* This makes the 8-noded quadrilateral in principle more competitive for practical purposes (see Section 5.9.2 for further details).

5.4.2 Twelve-noded cubic Serendipity rectangle

This element has four nodes along each side and a total of twelve side nodes which define the twelve terms polynomial approximation shown in Figure 5.8. The shape functions are derived following the same procedure explained for the 8-noded element. Thus, the shape functions for the side nodes are obtained by the simple product of two Lagrange cubic and linear polynomials. For the corner nodes the starting point is again the bilinear approximation. This initial shape function is forced to take a zero value at the two side nodes adjacent to the corner node by subtracting the shape functions of those nodes weighted by the factors 2/3 and 1/3. Figure 5.10 shows the expression of the shape functions which can be derived by the reader as an exercise.

It is simple to check that the element satisfies conditions (5.8) and (5.9). Figure 5.8 shows that the shape functions contain a complete cubic approximation plus two terms $(\xi^3\eta, \xi\eta^3)$ of the quartic polynomial. This element compares very favourably with the 16-noded Lagrange element, since both have a cubic approximation but the Serendipity element has fewer nodes (12 nodes versus 16 nodes for the cubic Lagrange rectangle).

5.4.3 Seventeen-noded quartic Serendipity rectangle

The quartic Serendipity rectangle has five nodes along each side and a total of seventeen nodes (sixteen side nodes plus a central node, Figure 5.10). The central node is necessary to introduce the "bubble" function $(1-\xi^2)(1-\eta^2)$ as shown in Figure 5.5. This function contributes the term $\xi^2\eta^2$ to complete a quartic approximation.

The derivation of the shape functions follows a procedure similar to that for the 8 and 12 node Serendipity elements. The shape functions

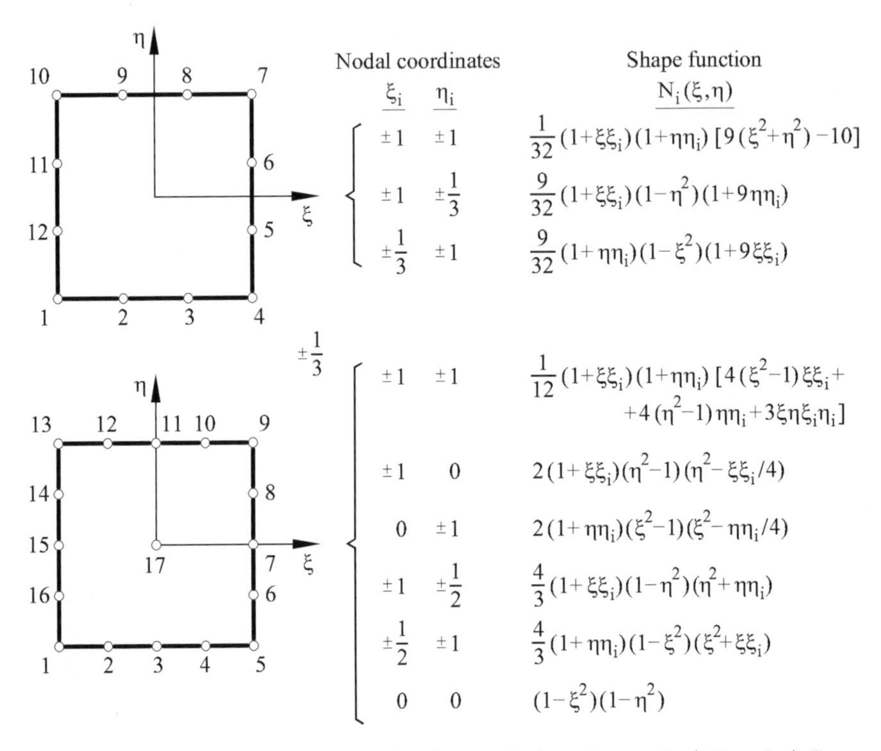

Fig. 5.10 Shape functions for the cubic (12 nodes) and quartic (17 nodes) Serendipity rectangles

for the side nodes are obtained by the product of a quartic and a linear polynomial. An exception are nodes 3, 7, 11 and 15 for which the function $1/2\ (1 - \xi^2)(1 - \eta^2)$ is subtracted from that product so that the resulting shape function takes a zero value at the central node. The starting point for the corner nodes is the bilinear function to which a proportion of the shape functions of the side nodes is subtracted so that the final shape function takes a zero value at these nodes. Finally, the shape function for the central node is the bubble function. Figure 5.10 shows the shape functions for this element.

Figure 5.8 shows that the shape functions contain a complete quartic approximation plus two additional terms ($\xi^4\eta$ and $\xi\eta^4$) from the quintic polynomial. The corresponding quartic Lagrange element has 25 nodes (Figure 5.3) and hence the 17-noded Serendipity rectangle is more economical for practical purposes.

5.5 SHAPE FUNCTIONS FOR C^0 CONTINUOUS TRIANGULAR ELEMENTS

The shape functions for the more popular C^0 continuous triangular elements are *complete polynomials* whose terms can be readily identified by the Pascal triangle. This also defines the position of the nodes within the element. We recall, for instance, that the shape function for the 3-noded triangle is linear. Similarly, the six and ten-noded triangles define the following complete quadratic and cubic approximations

6-noded triangle

$$\phi = \alpha_0 + \alpha_1 x + \alpha_2 y + \alpha_3 xy + \alpha_4 x^2 + \alpha_5 y^2 \tag{5.23}$$

10-noded triangle

$$\phi = \alpha_0 + \alpha_1 x + \alpha_2 y + \alpha_3 xy + \alpha_4 x^2 + \alpha_5 y^2 + \alpha_6 x^3 + \alpha_7 x^2 y + \alpha_8 xy^2 + \alpha_9 y^3 \tag{5.24}$$

The α_i parameters are obtained by the same procedure as described in Section 4.3.1 for the 3-noded triangle. This method has obvious difficulties for higher order elements and it is simpler to apply the technique based on area coordinates that is explained below.

5.5.1 Area coordinates

Let us join a point P within a triangle of area A with the three vertices 1, 2, 3 (Figure 5.11). This defines three sub-areas A_1, A_2 and A_3 corresponding to the triangles $P13$, $P12$ and $P23$, respectively (note that $A_1+A_2+A_3 = A$). The area coordinates are defined as

$$L_1 = \frac{A_1}{A} \quad ; \quad L_2 = \frac{A_2}{A} \quad ; \quad L_3 = \frac{A_3}{A} \tag{5.25}$$

Obviously

$$L_1 + L_2 + L_3 = \frac{A_1 + A_2 + A_3}{A} = \frac{A}{A} = 1 \tag{5.26}$$

The position of point P can be defined by any two of these three coordinates. The area coordinates of a node can be interpreted as the ratio between the distance from point P to the opposite side divided by the distance from the node to that side (Figure 5.11). Thus, area coordinates of the centroid are $L_1 = L_2 = L_3 = 1/3$. Area coordinates are also known as barycentric, triangular or trilinear coordinates and they are typical of

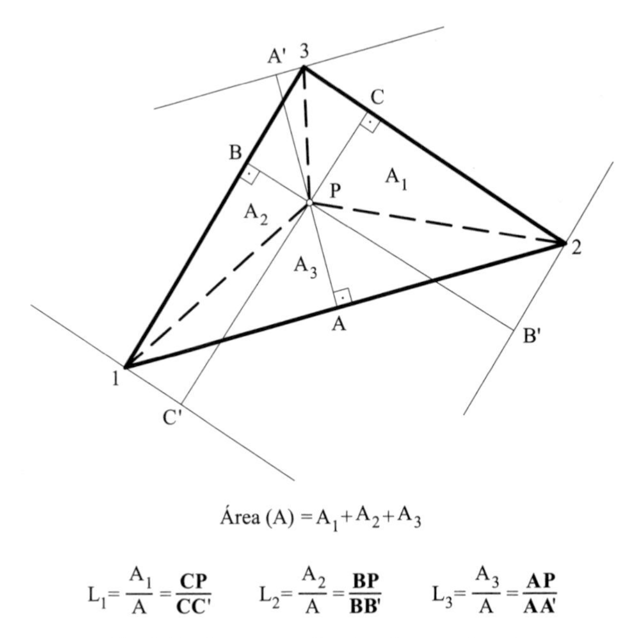

Área (A) = $A_1 + A_2 + A_3$

$L_1 = \frac{A_1}{A} = \frac{\mathbf{CP}}{\mathbf{CC'}} \qquad L_2 = \frac{A_2}{A} = \frac{\mathbf{BP}}{\mathbf{BB'}} \qquad L_3 = \frac{A_3}{A} = \frac{\mathbf{AP}}{\mathbf{AA'}}$

Fig. 5.11 Area coordinates for a triangle

geometry treatises [Fe]. In the FEM context area coordinates have proved to be very useful for deriving the shape functions of triangular finite elements.

Area coordinates are also of interest to define a parametric interpolation of the element geometry. For a straight side triangle the following relationship between the area and cartesian coordinates can be written

$$\begin{aligned} x &= L_1 x_1 + L_2 x_2 + L_3 x_3 \\ y &= L_1 y_1 + L_2 y_2 + L_3 y_3 \end{aligned} \tag{5.27}$$

This equation system is completed with Eq.(5.26) so that L_1, L_2 and L_3 can be obtained in terms of the cartesian coordinates by

$$L_i = \frac{1}{2A^{(e)}}(a_i + b_i x + c_i y) \tag{5.28}$$

where $A^{(e)}$ is the area of the triangle and a_i, b_i, c_i coincide with the values given in Eq.(4.32b). It is therefore concluded that *the area coordinates coincide precisely with the shape functions for the 3-noded triangular element.*

5.5.2 Derivation of the shape functions for C^0 continuous triangles

The shape functions for triangles containing complete Mth degree polynomials can be obtained in terms of the area coordinates as follows. Let us consider a node i characterized by the position (I, J, K) where I, J and K are the powers of the area coordinates L_1, L_2 and L_3, respectively in the expression of the shape function. Thus, $I + J + K = M$ and the shape function of node i is given by

$$N_i = l_I^i(L_1)\, l_J^i(L_2)\, l_K^i(L_3) \tag{5.29}$$

where $l_I^i(L_1)$ is the *normalized* 1D Lagrange Ith degree polynomial in L_1 which takes the value one at node i (Eq.(3.5b)), i.e.

$$l_I^i(L_1) = \prod_{\substack{j=1,I \\ j\neq i}} \frac{\left(L_1 - L_1^j\right)}{\left(L_1^i - L_1^j\right)} \tag{5.30}$$

with identical expressions for $l_J^i(L_2)$ and $l_K^i(L_3)$. In Eq.(5.30) L_1^i is the value of L_1 at node i.

The values of I, J, K for each node can be deduced by noting that: a) the shape function of a corner node depends on a single area coordinate only and thus the corresponding I, J or K power for that node is equal to M; b) all nodes located on the lines L_1 = constant have the same value for I and the same occurs with L_2 and J and L_3 and K; and c) the values of I, J and K associated with L_1, L_2 and L_3, respectively, decrease progressively from the maximum value equal to M for the lines $L_i = 1$ at the corner nodes, to a value equal to zero at the lines $L_i = 0$ which coincide with the opposite side to each corner node i (Figure 5.12a).

This application of Eqs.(5.29) and (5.30) will be clarified next with some examples.

5.5.3 Shape functions for the 3-noded linear triangle

The shape functions for the 3-noded triangle are linear polynomials ($M = 1$). The area coordinates and the values of I, J, K for each node can be seen in Figure 5.12b.

Node 1

Position $(I, J, K) : (1, 0, 0)$

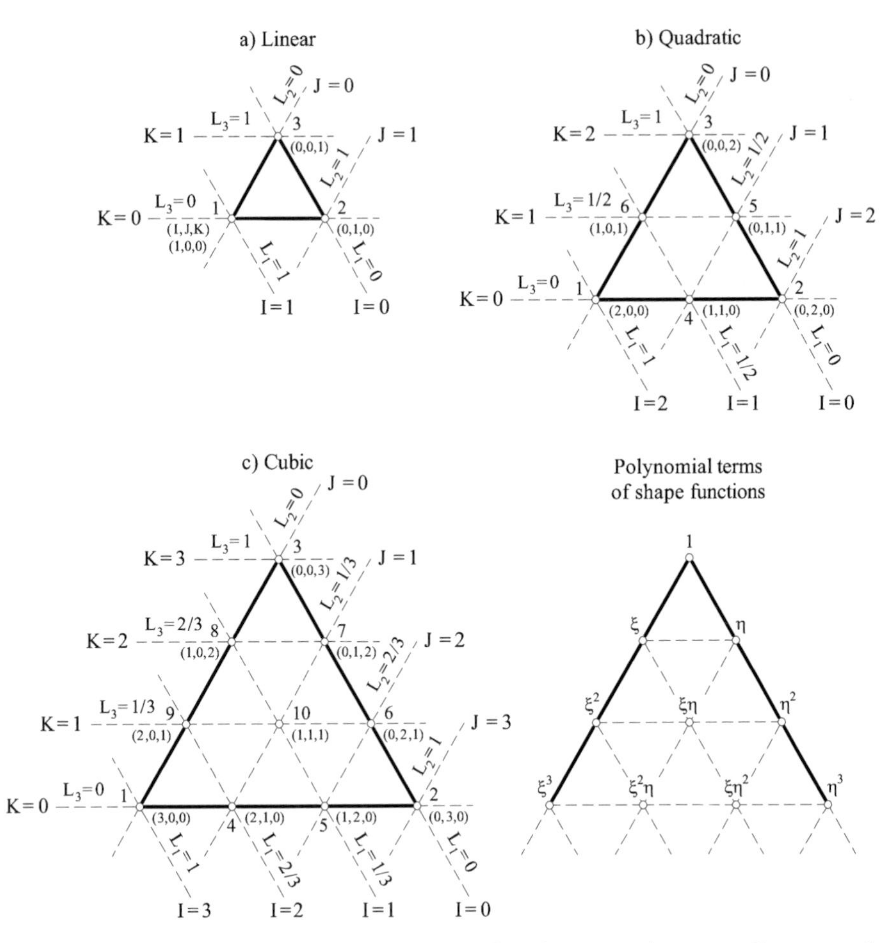

Fig. 5.12 Linear, quadratic and cubic triangular elements. Area coordinates and values of (I, J, K) for each node

Area coordinates: (1, 0, 0)

$$N_1 = l_1^1(L_1) = L_1 \tag{5.31}$$

It is straight-forward to find $N_2 = L_2$ and $N_1 = L_3$ as expected.

5.5.4 Shape functions for the six-noded quadratic triangle

The shape functions for this element are complete quadratic polynomials ($M = 2$). The area coordinates and the values of I, J and K for each node are shown in Figure 5.12b.

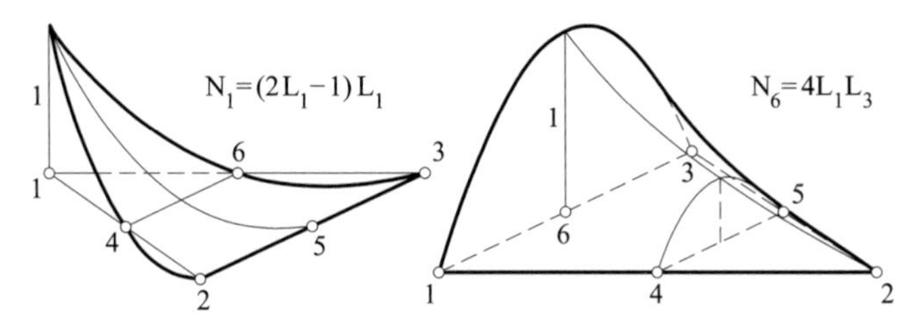

Fig. 5.13 Shape functions for a corner node and a side node for a quadratic triangle

Node 1

Position (I, J, K) : (2, 0, 0)
Area coordinates: (1, 0, 0)

$$N_1 = l_2^1(L_1) = \frac{(L_1 - 1/2)L_1}{(1 - 1/2)\ 1} = (2L_1 - 1)L_1 \tag{5.32}$$

Node 4

Position $(I, J, K) : (1, 1, 0)$
Area coordinates: $(1/2, 1/2, 0)$

$$N_4 = l_1^2(L_1)\ l_1^2(L_2) = \frac{L_1}{1/2}\ \frac{L_2}{1/2} = 4L_1L_2 \tag{5.33}$$

Following the same procedure for the rest of nodes we find

$$\begin{aligned} N_1 &= (2L_1 - 1)L_1\ ;\ N_2 = (2\ L_2 - 1)L_2\ ;\ N_3 = (2L_3 - 1)L_3 \\ N_4 &= 4L_1L_2 \qquad ;\ N_5 = 4\ L_2L_3 \qquad ;\ N_6 = 4L_1L_3 \end{aligned} \tag{5.34}$$

Figure 5.13 shows two characteristic shape functions for this element.

5.5.5 Shape functions for the ten-noded cubic triangle

The shape functions for this element are complete cubic polynomials ($M = 3$). Figure 5.12c shows the area coordinates and the values of I, J and K at each node:

Node 1

Position $(I, J, K) : (3, 0, 0)$
Area coordinates: $(1, 0, 0)$

$$N_1 = l_3^1(L_1) = \frac{(L_1 - 2/3)\ (L_1 - 1/3)\ L_1}{(1 - 2/3)\ (1 - 1/3)\ 1} = \frac{1}{2}\ L_1\ (3L_1 - 1)\ (3L_1 - 2) \tag{5.35}$$

Node 4

Position (I,J,K): (2, 1, 0)
Area coordinates: $(2/3, 1/3, 0)$

$$N_4 = l_2^2(L_1)\ l_1^2(L_2) = \frac{(L_1 - 1/3)\ L_1}{(2/3 - 1/3)\ 2/3} \cdot \frac{L_2}{1/3} = \frac{9}{2}(3L_1 - 1)\ L_1 L_2 \quad (5.36)$$

The same procedure applied to the rest of the nodes gives

$$\begin{aligned}
N_1 &= \frac{1}{2}L_1(3L_1-1)(3L_1-2) \quad ; \quad N_2 = \frac{1}{2}L_2(3L_2-1)(3L_2-2)\\
N_3 &= \frac{1}{2}L_3(3L_3-1)(3L_3-2) \quad ; \quad N_4 = \frac{9}{2}(3L_1-1)L_1L_2\\
N_5 &= \frac{9}{2}(3L_2-1)L_1\ L_2 \quad ; \quad N_6 = \frac{9}{2}(3L_2-1)L_2\ L_3\\
N_7 &= \frac{9}{2}(3L_3-1)\ L_2L_3 \quad ; \quad N_8 = \frac{9}{2}(3L_3-1)L_3L_1\\
N_9 &= \frac{9}{2}(3L_2-1)L_3\ L_1 \quad ; \quad N_{10} = 27L_1L_2L_3
\end{aligned} \quad (5.37)$$

A similar technique can be employed to derive the shape functions for higher order triangular elements.

5.5.6 Natural coordinates for triangles

It is convenient to define a normalized coordinate system α, β (also called natural coordinate system), such that the triangle has the sides over the lines $\alpha = 0, \beta = 0$ and $1 - \alpha - \beta = 0$ as shown in Figure 5.14. The shape functions for the 3-noded triangle can then be written as

$$N_1 = 1 - \alpha - \beta \quad ; \quad N_2 = \alpha \quad , \quad N_3 = \beta \quad (5.38)$$

Clearly as $L_i = N_i$ (Eq.(5.28)) the area coordinates L_2 and L_3 coincide with the natural coordinates α and β, respectively and $L_1 = 1 - \alpha - \beta$.

These coincidences allow us to express the shape functions of triangular elements in terms of the natural coordinates. This is particularly useful for deriving isoparametric triangular elements (Section 5.10).

5.6 ANALYTIC COMPUTATION OF INTEGRALS OVER RECTANGLES AND STRAIGHT-SIDED TRIANGLES

For irregular and straight-sided element, the analytical computation of the element integrals is possible as simple polynomial forms are involved

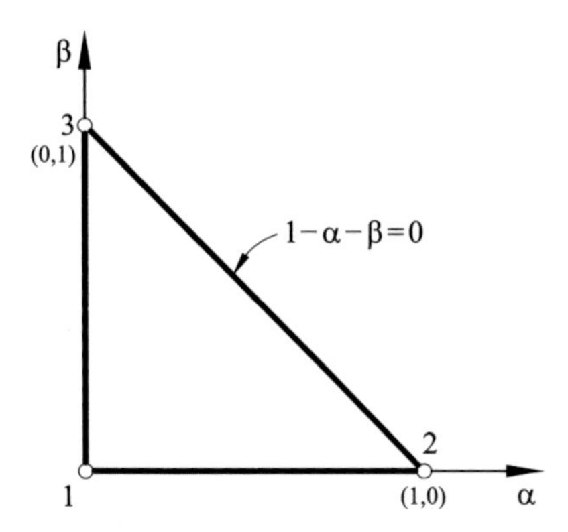

Fig. 5.14 Natural coordinates for a triangular element

in the integrand. Some interesting analytical expressions for the element integrals in terms of the local cartesian coordinates $\overline{x}, \overline{y}$ shown in Figure 5.15 exist for rectangles and straight-sided triangles. A typical integral term such as

$$\overline{C}_{ij} = D \iint_{A^{(e)}} \overline{x}^m \overline{y}^n \, dA \tag{5.39}$$

can be directly integrated by the following expressions:

Straight-sided triangular element

$$\overline{C}_{ij} = D \; c^{n+1} \left[a^{m+1} - (-b)^{m+1} \right] \frac{m!n!}{(m+n+2)!} \tag{5.40}$$

Rectangular element

$$\overline{K}_{ij} = D \frac{(2a)^{m+1} \; (2b)^{n+1}}{(m+1) \; (n+1)} \tag{5.41}$$

In the above m and n are integers and a, b and c are typical element dimensions (Figure 5.15). Once $\overline{\mathbf{K}}$ and $\overline{\mathbf{f}}$ have been found in the local coordinate system $\overline{x}, \overline{y}$ using the above expressions, they can be transformed into the global axes using the standard relationships (see Chapter 1)

$$\mathbf{K}_{ij} = \mathbf{T}^T \; \overline{\mathbf{K}}_{ij} \; \mathbf{T} \quad , \quad \mathbf{f}_i = \mathbf{T}^T \; \overline{\mathbf{f}}_i \tag{5.42}$$

where $\mathbf{T}$ is the 2×2 coordinate transformation matrix

$$\mathbf{T} = \begin{bmatrix} \cos\,(\overline{x}x) & \cos\,(\overline{x}y) \\ \cos\,(\overline{y}x) & \cos\,(\overline{y}y) \end{bmatrix} \tag{5.43}$$

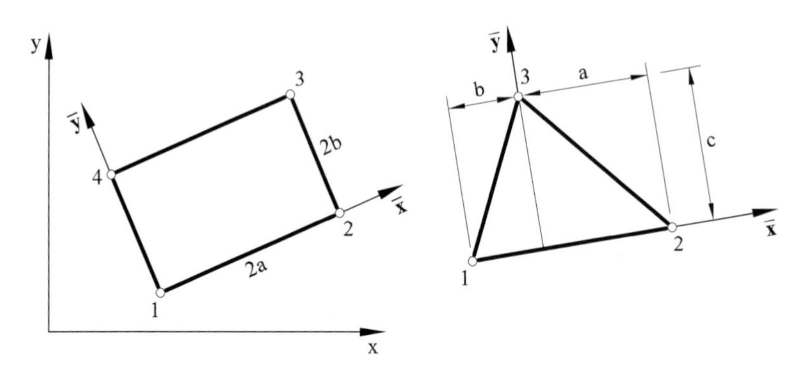

Fig. 5.15 Local coordinate system $\bar{x}$, $\bar{y}$ for the analytical computation of integrals over triangular and rectangular elements

with $(\overline{x}x)$ being the angle between the local $\overline{x}$ axis and the global x axis, etc.

Other simple analytical forms in terms of area coordinates can be found for area integrals over straight-sided triangles. A typical stiffness matrix term involves the cartesian derivatives of the shape functions. This is expressed in terms of area coordinates by the standard chain rule

$$\frac{\partial N_1(L_1,L_2,L_3)}{\partial x} = \frac{\partial N_1}{\partial L_1}\frac{\partial L_1}{\partial x} + \frac{\partial N_1}{\partial L_2}\frac{\partial L_2}{\partial x} + \frac{\partial N_1}{\partial L_3}\frac{\partial L_3}{\partial x} \tag{5.44}$$

As the element sides are assumed to be straight, Eq.(5.28) leads to

$$\frac{\partial L_i}{\partial x} = \frac{b_i}{2A^{(e)}} \quad \text{and} \quad \frac{\partial L_i}{\partial y} = \frac{c_i}{2A^{(e)}} \tag{5.45}$$

where b_i and c_i are given by Eq.(4.32b). Combining Eqs.(5.44) and (5.45) gives

$$\frac{\partial N_i}{\partial x} = \frac{1}{2A^{(e)}}\sum_{i=1}^{3} b_i \frac{\partial N_i}{\partial L_i} \quad ; \quad \frac{\partial N_i}{\partial y} = \frac{1}{2A^{(e)}}\sum_{i=1}^{3} c_i \frac{\partial N_i}{\partial L_i} \tag{5.46}$$

Thus, the element integrals can be easily expressed in terms of area coordinates and they can be directly computed by the following expressions

$$\iint_{A^{(e)}} L_1^k\, L_2^l\, L_3^m\, dA = 2A^{(e)}\, \frac{k!\; l!\; m!}{(2+k+l+m)!} \tag{5.47a}$$

$$\oint_{l^{(e)}} L_i^k \, L_j^l \, ds = l^{(e)} \, \frac{k! \; l!}{(1+k+l)!} \tag{5.47b}$$

If one of the area coordinates is missing in the integrand, the corresponding power is omitted in the denominator of Eqs.(5.47) and it is made equal to a unit value in the numerator.

The use of the natural coordinates α and β does not introduce any additional difficulty. Exact expressions for the integrals over straight-sided triangles can be found as

$$I = \iint_{A^{(e)}} \alpha^m \beta^n \, dA = \frac{2A^{(e)} \, \Gamma(m+1) \, \Gamma(n+1)}{\Gamma\,(3+m+n)} \tag{5.48}$$

where Γ is the Gamma function [Ral]. If m and n are positive integers

$$I = 2A^{(e)} \frac{m! \; n!}{(2+m+n)!} \tag{5.49}$$

This is just a particular case of Eq.(5.47a) when one of the area coordinates is missing. Similarly, it is deduced from (5.47b) that

$$\oint_{l^{(e)}} \alpha^m ds = l^{(e)} \frac{m!}{(2+m)!} \tag{5.50}$$

For curve-sided elements the computation of the cartesian derivatives of the shape functions requires a parametric formulation. This generally introduces rational algebraic functions in the integrals and makes numerical integration unavoidable. This topic is explained in Chapter 6.

Example 5.1: Compute the stiffness matrix $\mathbf{K}_{11}^{(e)}$ for a quadratic triangle with straight sides and unit thickness.

- ***Solution***

The first step is to obtain the cartesian derivatives for the shape function N_1 expressed in terms of the area coordinates as

$$\frac{\partial N_1}{\partial x} = \frac{\partial N_1}{\partial L_1} \frac{\partial L_1}{\partial x} + \frac{\partial N_1}{\partial L_2} \frac{\partial L_2}{\partial x} + \frac{\partial N_1}{\partial L_3} \frac{\partial L_3}{\partial x}$$

$$\frac{\partial N_1}{\partial y} = \frac{\partial N_1}{\partial L_1} \frac{\partial L_1}{\partial y} + \frac{\partial N_1}{\partial L_2} \frac{\partial L_2}{\partial y} + \frac{\partial N_1}{\partial L_3} \frac{\partial L_3}{\partial y}$$

From Eq.(5.32) we deduce

$$\frac{\partial N_1}{\partial L_1} = 4L_1 - 1 \quad ; \quad \frac{\partial N_1}{\partial L_2} = \frac{\partial N_1}{\partial L_3} = 0$$

and from Eq.(5.45)

$$\frac{\partial L_i}{\partial x} = \frac{b_i}{2A^{(e)}} \quad ; \quad \frac{\partial L_i}{\partial y} = \frac{c_i}{2A^{(e)}}$$

Therefore

$$\frac{\partial N_1}{\partial x} = \frac{b_1}{2A^{(e)}}(4L_1 - 1) \quad ; \quad \frac{\partial N_1}{\partial y} = \frac{c_1}{2A^{(e)}}(4L_1 - 1)$$

and

$$\mathbf{B}_1 = \frac{(4L_1 - 1)}{2A^{(e)}} \begin{bmatrix} b_1 & 0 \\ 0 & c_1 \\ c_1 & b_1 \end{bmatrix}$$

Matrix $\mathbf{K}_{11}^{(e)}$ is thus obtained by

$$\mathbf{K}_{11}^{(e)} = \iint_{A^{(e)}} \mathbf{B}_1^T \, \mathbf{D} \, \mathbf{B} \, t dA = \frac{t}{(2A^{(e)})^2} \begin{bmatrix} b_1 & 0 & b_1 \\ 0 & c_1 & c_1 \end{bmatrix} \begin{bmatrix} d_{11} & d_{12} & 0 \\ d_{12} & d_{22} & 0 \\ 0 & 0 & d_{33} \end{bmatrix} \times$$

$$\times \begin{bmatrix} b_1 & 0 \\ 0 & c_1 \\ c_1 & b_1 \end{bmatrix} \iint_{A^{(e)}} (4L_1 - 1)^2 \, dA$$

We deduce from Eq.(5.47a)

$$\iint_{A^{(e)}} (4L_1 - 1)^2 \, dA = 2A^{(e)} \left[\frac{16 \cdot 2!}{4!} - \frac{8 \cdot 1!}{3!} + \frac{1}{2} \right] = A^{(e)}$$

which leads to

$$\mathbf{K}_{11}^{(e)} = \frac{1}{4A^{(e)}} \begin{bmatrix} b_1^2 \, d_{11} + c_1^2 \, d_{33} & b_1 c_1 \, (d_{33} + d_{12}) \\ b_1 c_1 (d_{33} + d_{12}) & b_1^2 \, d_{33} + c_1^2 d_{22} \end{bmatrix}$$

The rest of the $\mathbf{K}_{ij}^{(e)}$ matrices are obtained following an identical procedure. The complete expression of the stiffness matrix for the quadratic triangle can be found in [CMPW] and [WJ].

5.7 GENERAL PERFORMANCE OF TRIANGULAR AND RECTANGULAR ELEMENTS

We present next two examples which lead us to draw some conclusions on the behaviour of rectangular and triangular elements. The first example shown in Figure 5.16 is the analysis of a square plate under a parabolic traction acting symmetrically on two opposite sides. Different meshes of 3 and 6-noded triangles and 4 and 9-noded rectangles are used for the analysis. Numerical results for the horizontal displacement of the central point on the loaded side show that the 3-noded triangle is the less accurate of all elements studied. Nevertheless 1% error with respect to the "exact" analytical solution is obtained with a fine mesh [Ga,Ya].

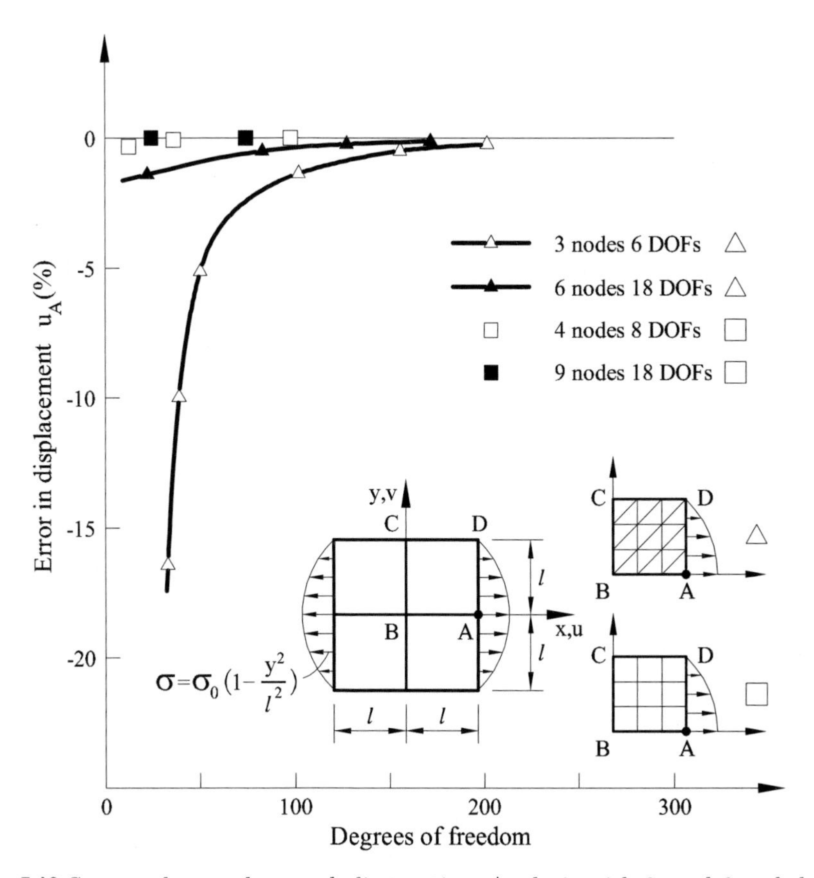

Fig. 5.16 Square plate under parabolic traction. Analysis with 3- and 6-noded triangles and 4 and 9-noded rectangles

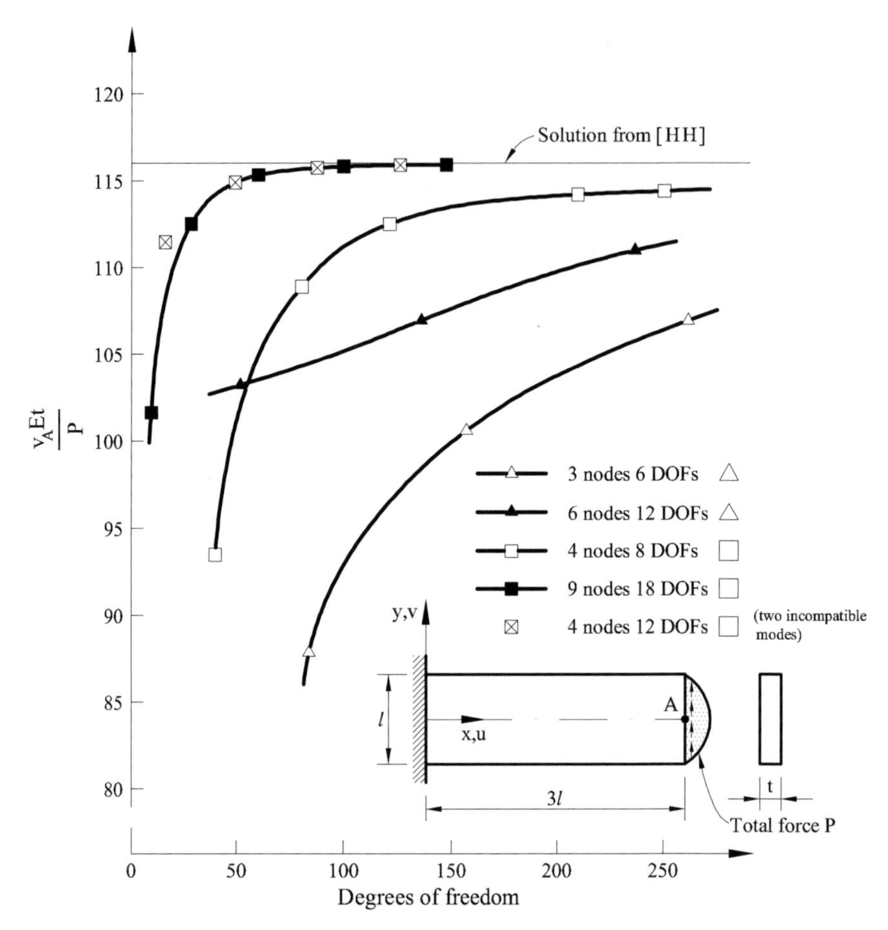

Fig. 5.17 Cantilever deep beam under parabolic edge load ($\nu = 0.2$). Analysis with 3- and 6-noded triangles, 4- and 9-noded rectangles and the 4-noded rectangle with two incompatible modes

The accuracy increases notably for the same number of DOFs when 6-noded triangles are used and, even more, when either the 4- or the 9-noded rectangles are used, as shown in Figure 5.16. Similarly good results are obtained with the 8-noded rectangle.

The second example is a cantilever deep beam under a parabolic edge load (Figure 5.17) [HH]. The analysis is performed using the same elements as in the previous example and, in addition, the 4-noded rectangle enhanced with two incompatible modes studied in Section 4.4.2.3. Results plotted in Figure 5.17 show clearly the poor accuracy of the 3-noded triangle for bending dominated problems. The accuracy improves slightly for the 6-noded quadratic triangular element. The 4-noded rectangle has an overstiff behaviour, as expected from its inability to reproduce pure bending situations. Its accuracy improves however when finer meshes are used. Note the excellent performance of the 4-noded rectangle with incompatible modes and the 9-noded Lagrange rectangle. Similar good results are obtained using the 8-noded quadratic Serendipity rectangle.

These results can be generalized to other situations (see for instance Section 4.5). Typically, rectangles are more accurate than triangles for the same number of DOFs. However, triangular elements are more versatile due to their better ability to model complex geometries with unstructured meshes.

As a rule, low order elements are simpler to use, although finer meshes are needed in zones where high stress gradients exist. Higher order elements are more competitive in these regions.

5.8 ENHANCEMENT OF 2D ELASTICITY ELEMENTS USING DRILLING ROTATIONS

The flexural behaviour of 2D elasticity elements can be substantially improved by using the so-called "drilling rotations". This technique was originally developed to enhance the performance of plane stress triangles and quadrilaterals for shell analysis (Chapter 7 of Volume 2 [On]). However, it can be also applied to derive improved plane stress and plane strain elements.

The basic idea is to introduce a mechanical in-plane rotation defined as (Figure 4.3)

$$\theta_z = \frac{1}{2}\left(\frac{\partial v}{\partial x} - \frac{\partial u}{\partial y}\right) \tag{5.51}$$

The corresponding rotational stiffness for each element is introduced by adding to the PVW the term

$$\iint_{A^{(e)}} \alpha_r E t(\delta\theta_z - \delta\bar{\theta}_z)(\theta_z - \bar{\theta}_z)\, dA \tag{5.52}$$

where α_r is a user-defined parameter (typically $\alpha_r \simeq 10^{-2} - 10^{-3}$) and $\bar{\theta}_z$ is a mean element in-plane rotation. Substituting Eq.(5.51) into Eq.(5.52) allows the resulting stiffness equation to be expressed in terms of the nodal displacement DOFs only.

Details on the derivation of triangles and quadrilaterals using this approach can be found in [ZT] and in Chapter 7 of Volume 2 [On].

5.9 CONCLUDING REMARKS

We have presented in this chapter the derivation of the shape functions for rectangular and triangular solid elements of any order of approximation. The element integrals appearing in the expressions of the stiffness matrix and the equivalent nodal force vector can be computed analytically for rectangular elements and straight-side triangular elements and some useful integration rules have been given.

The examples presented show the superiority of quadratic elements versus linear ones. Also, rectangles show a better performance than triangles. The simple 3-noded triangle is however the more versatile element for modelling complex structures with unstructured meshes.

6

ISOPARAMETRIC 2D SOLID ELEMENTS. NUMERICAL INTEGRATION AND APPLICATIONS

6.1 INTRODUCTION

In the previous chapter we have described how to obtain the shape functions for 2D solid elements of triangular and rectangular shape and how to compute analytically the stiffness matrix and the equivalent nodal force vector for straight-sided triangular elements and rectangular elements.

This chapter explains how to derive 2D solid elements of arbitrary shape (i.e. irregular quadrilaterals and curve-sided triangles) using an isoparametric formulation and numerical integration. The basis of the isoparametric formulation for 2D solid elements is described in the next section. Then, the quadrature rules for the numerical integration of the stiffness matrix and the equivalent nodal force vector for triangular and quadrilateral elements are explained. The patch test for 2D solid elements is presented. Some hints on the organization of a computer program for FEM analysis of 2D solids are given. The chapter concludes with examples of the application of some of the 2D solid elements studied to the analysis of real structures.

6.2 ISOPARAMETRIC QUADRILATERAL ELEMENTS

We recall that the term "isoparametric" means that the displacement shape functions are used to interpolate the element geometry in terms of the nodal coordinates. Thus, the geometry of a 2D isoparametric quadrilateral with n nodes is expressed as

$$x = \sum_{i=1}^{n} N_i(\xi, \eta)\, x_i \quad ; \quad y = \sum_{i=1}^{n} N_i(\xi, \eta)\, y_i \tag{6.1}$$

where $N_i(\xi, \eta)$ are the standard displacement shape functions. Eqs.(6.1) relate the cartesian and the natural coordinates at each point. Such a relationship must be unique and this is satisfied if the Jacobian of the transformation of the partial derivatives of a function in the natural and cartesian coordinate systems has a constant positive sign over the element [SF].

It can be shown that this condition is satisfied for linear quadrilateral elements if no internal angle between two element sides is equal or greater than 180° [SF]. For quadratic elements it is additionally required that the side nodes are located within the "middle third" of the distance between adjacent corners [Jor]. There are no practical rules for higher order quadrilateral elements and the constant sign of the determinant of the Jacobian matrix is the only possible verification in this case. Figure 6.1 shows some examples of 2D isoparametric elements.

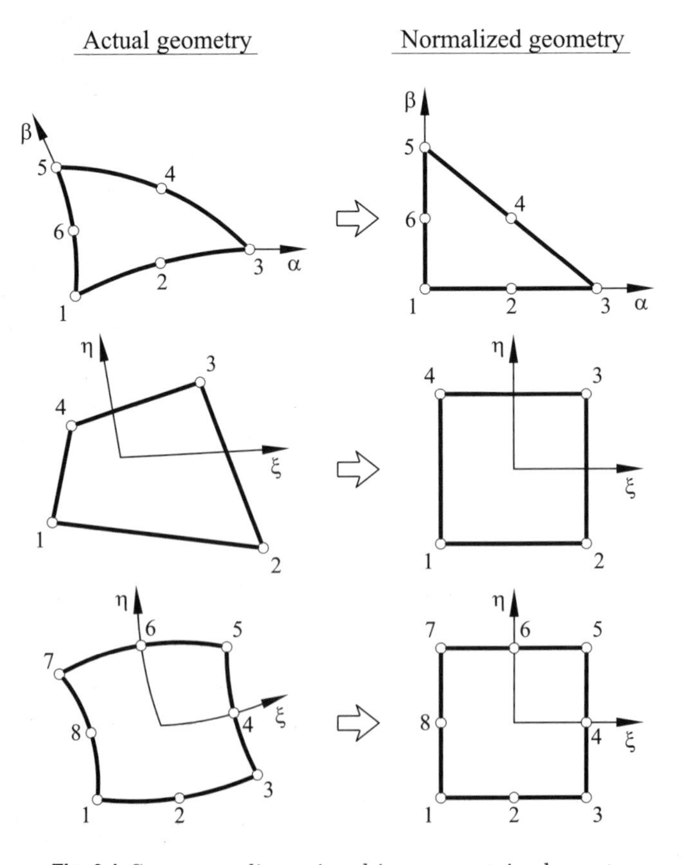

Fig. 6.1 Some two-dimensional isoparametric elements

Most of the isoparametric formulation ideas originated from the work of Taig [Ta,TK] who derived the 4-noded isoparametric quadrilateral. These concepts were generalized to more complex elements by Irons [IA,Ir].

Eq.(6.1) allows us to obtain a relationship between the derivatives of the shape functions with respect to the cartesian and the natural coordinates. In general, N_i is expressed in terms of the natural coordinates ξ and η and the chain rule of derivation yields

$$\frac{\partial N_i}{\partial \xi} = \frac{\partial N_i}{\partial x}\frac{\partial x}{\partial \xi} + \frac{\partial N_i}{\partial y}\frac{\partial y}{\partial \xi} \quad ; \quad \frac{\partial N_i}{\partial \eta} = \frac{\partial N_i}{\partial x}\frac{\partial x}{\partial \eta} + \frac{\partial N_i}{\partial y}\frac{\partial y}{\partial \eta} \tag{6.2}$$

In matrix form

$$\begin{Bmatrix} \dfrac{\partial N_i}{\partial \xi} \\ \dfrac{\partial N_i}{\partial \eta} \end{Bmatrix} = \underbrace{\begin{bmatrix} \dfrac{\partial x}{\partial \xi} & \dfrac{\partial y}{\partial \xi} \\ \dfrac{\partial x}{\partial \eta} & \dfrac{\partial y}{\partial \eta} \end{bmatrix}}_{\mathbf{J}^{(e)}} \begin{Bmatrix} \dfrac{\partial N_i}{\partial x} \\ \dfrac{\partial N_i}{\partial y} \end{Bmatrix} = \mathbf{J}^{(e)} \begin{Bmatrix} \dfrac{\partial N_i}{\partial x} \\ \dfrac{\partial N_i}{\partial y} \end{Bmatrix} \tag{6.3}$$

where $\mathbf{J}^{(e)}$ is the *Jacobian matrix* of the transformation of the derivatives of N_i in the natural and global axes. The superindex e in $\mathbf{J}$ denotes that this matrix is always computed at element level. We deduce from Eq.(6.3)

$$\begin{Bmatrix} \dfrac{\partial N_i}{\partial x} \\ \dfrac{\partial N_i}{\partial y} \end{Bmatrix} = \begin{bmatrix} \mathbf{J}^{(e)} \end{bmatrix}^{-1} \begin{Bmatrix} \dfrac{\partial N_i}{\partial \xi} \\ \dfrac{\partial N_i}{\partial \eta} \end{Bmatrix} = \frac{1}{\left|\mathbf{J}^{(e)}\right|} \begin{bmatrix} \dfrac{\partial y}{\partial \eta} & -\dfrac{\partial y}{\partial \xi} \\ -\dfrac{\partial x}{\partial \eta} & \dfrac{\partial x}{\partial \xi} \end{bmatrix} \begin{Bmatrix} \dfrac{\partial N_i}{\partial \xi} \\ \dfrac{\partial N_i}{\partial \eta} \end{Bmatrix} \tag{6.4}$$

where $\left|\mathbf{J}^{(e)}\right|$ is the determinant of the Jacobian matrix (also simply called "the Jacobian"). This determinant relates the differential of area in the two coordinate systems, i.e.

$$dx\, dy = \left|\mathbf{J}^{(e)}\right| d\xi\, d\eta \tag{6.5}$$

The terms of $\mathbf{J}^{(e)}$ are computed using the isoparametric approximation (6.1), i.e.

$$\frac{\partial x}{\partial \xi} = \sum_{i=1}^{n} \frac{\partial N_i}{\partial \xi} x_i \quad ; \quad \frac{\partial x}{\partial \eta} = \sum_{1=1}^{n} \frac{\partial N_i}{\partial \eta} x_i; \text{ etc.} \tag{6.6}$$

and

$$\mathbf{J}^{(e)} = \sum_{i=1}^{n} \begin{bmatrix} \dfrac{\partial N_i}{\partial \xi} x_i & \dfrac{\partial N_i}{\partial \xi} y_i \\ \dfrac{\partial N_i}{\partial \eta} x_i & \dfrac{\partial N_i}{\partial \eta} y_i \end{bmatrix} \tag{6.7}$$

For a rectangular element

$$\mathbf{J}^{(e)} = \begin{bmatrix} a & 0 \\ 0 & b \end{bmatrix} \quad \text{and} \quad |\mathbf{J}^{(e)}| = ab \tag{6.8}$$

6.2.1 Stiffness matrix and load vector for the isoparametric quadrilateral

Substituting the cartesian derivatives of the shape functions from Eq.(6.4) into (4.38) yields the general form of the strain matrix for an isoparametric quadrilateral element in terms of the natural coordinates by

$$\mathbf{B}_i(\xi,\eta) = \begin{bmatrix} \frac{\partial N_i}{\partial x} & 0 \\ 0 & \frac{\partial N_i}{\partial y} \\ \frac{\partial N_i}{\partial y} & \frac{\partial N_i}{\partial x} \end{bmatrix} = \frac{1}{\left|\mathbf{J}^{(e)}\right|} \begin{bmatrix} \bar{b}_i & 0 \\ 0 & \bar{c}_i \\ \bar{c}_i & \bar{b}_i \end{bmatrix} \tag{6.9}$$

where

$$\bar{b}_i = \frac{\partial y}{\partial \eta}\frac{\partial N_i}{\partial \xi} - \frac{\partial y}{\partial \xi}\frac{\partial N_i}{\partial \eta} \quad ; \quad \bar{c}_i = \frac{\partial x}{\partial \xi}\frac{\partial N_i}{\partial \eta} - \frac{\partial x}{\partial \eta}\frac{\partial N_i}{\partial \xi} \tag{6.10}$$

The integrals in the element stiffness matrix are transformed to the normalized natural coordinate space as

$$\begin{aligned}
\mathbf{K}_{ij}^{(e)} &= \iint_{A^{(e)}} \mathbf{B}_i^T \mathbf{D} \mathbf{B}_j t \, dx\, dy = \int_{-1}^{+1}\int_{-1}^{+1} \mathbf{B}_i^T(\xi,\eta)\mathbf{D}\mathbf{B}_j(\xi,\eta)\left|\mathbf{J}^{(e)}\right| t\, d\xi\, d\eta \\
&= \int_{-1}^{+1}\int_{-1}^{+1} \begin{bmatrix} d_{11}\bar{b}_i\bar{b}_j + d_{33}\bar{c}_i\bar{c}_j & d_{12}\bar{b}_i\bar{c}_j + d_{33}\bar{c}_j\bar{b}_i \\ d_{21}\bar{c}_i\bar{b}_j + d_3\bar{b}_i\bar{c}_j & d_{33}\bar{b}_i\bar{b}_j + d_{22}\bar{c}_i\bar{c}_j \end{bmatrix} \frac{t}{\left|\mathbf{J}^{(e)}\right|} d\xi\, d\eta \\
&= \int_{-1}^{+1}\int_{-1}^{+1} \mathbf{G}_{ij}(\xi,\eta) \frac{t}{\left|\mathbf{J}^{(e)}\right|} d\xi\, d\eta
\end{aligned} \tag{6.11}$$

Eq.(6.11) shows that the integrand of $\mathbf{K}_{ij}^{(e)}$ contains rational algebraic functions in ξ and η. An exception to this rule is when the determinant of the Jacobian matrix is constant. This only occurs for rectangular elements and for straight side triangles. In these cases the element integrals contain simple polynomials and the analytical expressions of Section 5.6 can be applied. For general quadrilateral shapes the analytical integration of $\mathbf{K}_{ij}^{(e)}$ in the natural coordinate system ξ, η is difficult (and in some cases impossible!) and the best option is to use numerical integration.

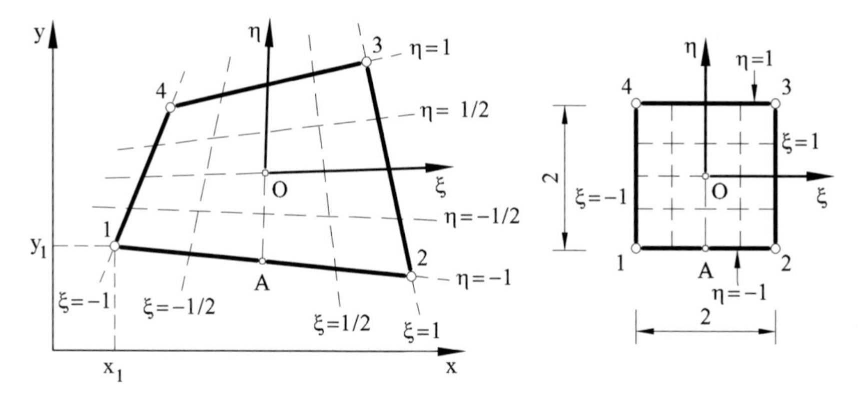

Fig. 6.2 Four-noded isoparametric quadrilateral. Actual and normalized geometry

A similar procedure is followed to compute the equivalent nodal force vectors for isoparametric quadrilateral elements. For example, for the case of body forces (see Eq.(4.65))

$$\mathbf{f}_{b_i}^{(e)} = \iint_{A^{(e)}} \mathbf{N}_i^T \; \mathbf{b}t \, dx \, dy = \int_{-1}^{+1} \int_{-1}^{+1} \mathbf{N}_i^T \mathbf{b} \left| \mathbf{J}^{(e)} \right| t \, d\xi \, d\eta \tag{6.12}$$

Numerical integration is also used to compute integrals such as that of Eq.(6.12).

Example 6.1: Formulate an isoparametric quadrilateral of 4 nodes.

- ***Solution***

The actual and normalized geometries of the element are shown in Figure 6.2. The isoparametric description of the geometry is written as

$$\mathbf{x} = \begin{Bmatrix} x \\ y \end{Bmatrix} = \sum_{i=1}^{4} N_i(\xi,\eta) \begin{Bmatrix} x_i \\ y_i \end{Bmatrix} \tag{6.13}$$

where $N_i(\xi,\eta) = \frac{1}{4}(1+\xi\xi_i)(1+\eta_i\eta)$.

The above expression maps the natural coordinates of each element point onto the cartesian space. For example, the cartesian position of the central point O with natural coordinates (0,0) is

$$\mathbf{x}_0 = \sum_{i=1}^{4} N_i(0,0) \begin{Bmatrix} x_i \\ y_i \end{Bmatrix} = \frac{1}{4} \begin{Bmatrix} x_1 + x_2 + x_3 + x_4 \\ y_1 + y_2 + y_3 + y_4 \end{Bmatrix} \tag{6.14a}$$

Point A at the center of side 1–2 with natural coordinates (0,-1) is located in the cartesian space at

$$\mathbf{x}_A = \sum_{i=1}^{4} N_i(0,-1)\begin{Bmatrix} x_i \\ y_i \end{Bmatrix} = \frac{1}{2}\begin{Bmatrix} x_1 + x_2 \\ y_1 + y_2 \end{Bmatrix} \tag{6.14b}$$

The isoparametric description is useful to express all the element expressions in the natural coordinate system.
For instance, the Jacobian matrix of Eq.(6.7) is given by

$$\mathbf{J}^{(e)} = \sum_{i=1}^{4} \begin{bmatrix} \frac{\xi_i}{4}(1+\eta\eta_i)x_i & \frac{\xi_i}{4}(1+\eta\eta_i)y_i \\ \frac{\eta_i}{4}(1+\xi\xi_i)x_i & \frac{\eta_i}{4}(1+\xi\xi_i)y_i \end{bmatrix} \tag{6.15}$$

The cartesian derivatives of the shape functions are obtained from Eq.(6.4) as

$$\begin{Bmatrix} \frac{\partial N_i}{\partial x} \\ \frac{\partial N_i}{\partial y} \end{Bmatrix} = \frac{1}{|\mathbf{J}^{(e)}|}\sum_{j=1}^{4} \begin{bmatrix} \frac{\eta_j}{4}(1+\xi\xi_j)y_j & -\frac{\xi_j}{4}(1+\eta\eta_j)y_j \\ -\frac{\eta_j}{4}(1+\xi\xi_j)x_j & \frac{\xi_j}{4}(1+\eta\eta_j)x_j \end{bmatrix} \begin{Bmatrix} \frac{\xi_i}{4}(1+\eta\eta_i) \\ \frac{\eta_i}{4}(1+\xi\xi_i) \end{Bmatrix} \tag{6.16}$$

From above equations we deduce

$$\begin{Bmatrix} \frac{\partial N_i}{\partial x} \\ \frac{\partial N_i}{\partial y} \end{Bmatrix} = \begin{Bmatrix} \hat{b}_i \\ \hat{c}_i \end{Bmatrix} \tag{6.17a}$$

with

$$\hat{b}_i = \frac{\alpha_1^i + \alpha_2^i\xi + \alpha_3^i\eta + \alpha_4^i\xi\eta}{\beta_1^i + \beta_2^i\xi + \beta_3^i\eta + \beta_4^i\xi\eta} \quad , \quad \hat{c}_i = \frac{\alpha_5^i + \alpha_6^i\xi + \alpha_7^i\eta + \alpha_8^i\xi\eta}{\beta_5^i + \beta_6^i\xi + \beta_7^i\eta + \beta_8^i\xi\eta} \tag{6.17b}$$

where $\alpha_1^i \cdots \alpha_8^i, \beta_1^i \cdots \beta_8^i$ are nodal parameters depending on the nodal coordinates.
The strain matrix and the element stiffness matrix now contain rational algebraic functions making exact integration over an arbitrary quadrilateral domain difficult. This problem is overcome by using numerical integration as only the numerical values of all expressions at the integration points are needed (Section 6.4).
The above equations simplify considerably for rectangular shapes. For a rectangular element of size $2a \times 2b$ we have

$$\mathbf{J}^{(e)} = \begin{bmatrix} a & 0 \\ 0 & b \end{bmatrix} \quad , \quad |\mathbf{J}^{(e)}| = ab \quad \text{and} \quad dx\,dy = ab\,d\xi\,d\eta \tag{6.18}$$

The cartesian derivatives of the shape functions are now simply given by

$$\begin{Bmatrix} \dfrac{\partial N_i}{\partial x} \\ \dfrac{\partial N_i}{\partial y} \end{Bmatrix} = \begin{Bmatrix} \dfrac{\xi_i}{4b}(1+\eta\eta_i) \\ \dfrac{\eta_i}{4a}(1+\xi\xi_i) \end{Bmatrix} \tag{6.19}$$

The element stiffness matrix is expressed in the natural system by

$$\mathbf{K}_{ij}^{(e)} = \int_{-1}^{+1}\int_{-1}^{+1} \mathbf{B}_i^T \mathbf{D} \mathbf{B}_i \, ab \, d\xi \, d\eta \tag{6.20}$$

The integral expression of $\mathbf{K}_{ij}^{(e)}$ now contains polynomial expressions in ξ, η, ξ^2, η^2 and $\xi\eta$ which can be directly computed, noting that

$$\int_{-1}^{+1}\int_{-1}^{+1} C[1, \xi, \eta, \xi^2, \eta^2, \xi\eta] \, d\xi \, d\eta = C\left[4, 0, 0, \frac{4}{3}, \frac{4}{3}, 0\right] \tag{6.21}$$

where C is a constant parameter.
The reader is encouraged to find matrix $\mathbf{K}^{(e)}$ of Box 4.1 as an exercise.

6.2.2 A comparison between the 8- and 9-noded isoparametric quadrilaterals

It is interesting to assess the circumstances under which the linearly distorted 8 and 9-noded quadrilaterals can fully represent any quadratic cartesian expansion. The straight-sided element geometry is exactly approximated by the bilinear (subparametric) interpolation

$$\mathbf{x} = \sum_{i=1}^{4} N_i \mathbf{x}_i \tag{6.22}$$

where $N_i = \frac{1}{4}(1+\xi\xi_i)(1+\eta\eta_i)$ are the shape functions for the 4-noded rectangle.

We wish to be able to reproduce

$$u = \alpha_1 + \alpha_2 x + \alpha_3 y + \alpha_4 x^2 + \alpha_5 xy + \alpha_6 y^2 \tag{6.23}$$

Noting that the bilinear form contains terms such as $1, \xi, \eta$ and $\xi\eta$ and by substituting Eq.(6.22) into (6.23) the above can be written as

$$u = \beta_1 + \beta_2\xi + \beta_3\eta + \beta_4\xi^2 + \beta_5\xi\eta + \beta_6\eta^2 + \beta_7\eta^2 + \beta_8\xi^2 + \beta_9\xi^2\eta^2 \tag{6.24}$$

where β_1 to β_9 depend on the values of α_1 to α_6.

We shall now try to match the terms arising from the quadratic expansion of the 8-noded Serendipity shown in Figure 5.9. Noting the terms occuring in the Pascal triangle of Figure 5.8, the interpolation can be written as

$$u = b_1 + b_2\xi + b_3\eta + b_4\xi^2 + b_5\xi\eta + b_6\eta^2 + b_7\xi\eta^2 + b_8\xi^2\eta \tag{6.25}$$

It is evident that for arbitrary values of η_1 to η_9 it is impossible to match the coefficients b_1 to b_8 due to the absence of the term $\xi^2\eta^2$ in Eq.(6.25).

For the 9-noded Lagrange element (Figure 5.5) the expansion similar to Eq.(6.25) gives

$$u = b_1 + b_2\xi + b_3\eta + b_4\xi^2 + b_5\xi\eta + b_6\eta^2 + b_7\xi\eta^2 + b_8\xi^2\eta + b_9\xi^2\eta^2 \tag{6.26}$$

and the matching of the coefficients in Eqs.(6.24) and (6.26) can be made directly.

We conclude that the 9-noded element can better represent quadratic cartesian polynomials on linearly distorted shapes and therefore is generally preferable for modelling smooth solutions. Figure 6.3 taken from [ZTZ] shows an example of this for the analysis, with 8- and 9-node elements respectively, of a simple beam solution where exact answers are quadratic. With no distorsion both elements with a full (3×3) integration rule give exact results but when distorted only the 9-node element does so, the 8-noded element giving a significant stress fluctuation. More examples of this kind are presented in Section 6.10.

A similar argument leads to the conclusion that in 3D again only 27-noded Lagrange elements are capable of reproducing fully a quadratic function in cartesian coordinates when trilinearity distorted.

Lee and Bathe [LB] have investigated the problem for cubic and quartic Serendipity and Lagrange quadrilaterals and showed that under bilinear distorsions the full order cartesian polynomial terms remain in Lagrange elements but not in Serendipity ones.

6.3 ISOPARAMETRIC TRIANGULAR ELEMENTS

The isoparametric interpolation for triangular elements is written in similar form to Eq.(6.1) by

$$x = \sum_{i=1}^{n} N_i(L_1, L_2, L_3)\, x_i \quad ; \quad \mathrm{y} = \sum_{i=1}^{n} N_i(L_1, L_2, L_3)\, y_i \tag{6.27}$$

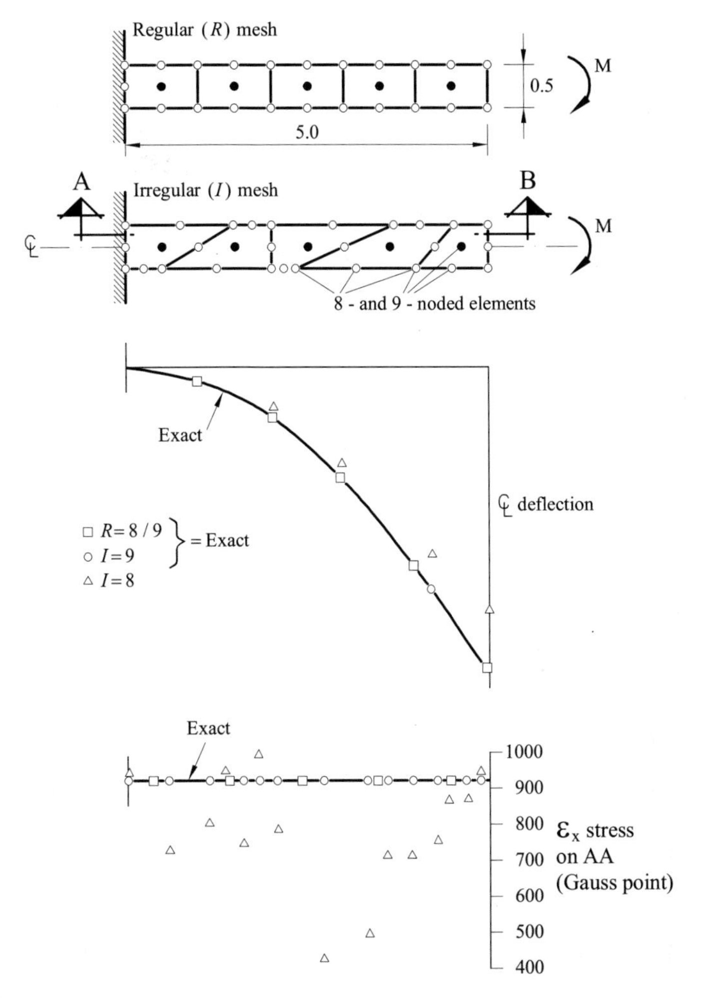

Fig. 6.3 Quadratic Serendipity and Lagrange 8- and 9-noded elements in regular and distorted form. Elastic deflection of a beam under constant moment. Note poor results of 8-noded distorted element [ZTZ]

The computation of the cartesian derivatives of N_i and the Jacobian matrix is inmediate for straight-sided triangles (Example 6.2) giving

$$\mathbf{J}^{(e)} = \begin{bmatrix} x_2 - x_1 & y_2 - y_1 \\ x_3 - x_1 & y_3 - y_1 \end{bmatrix}^{(e)} \quad \text{and} \quad |\mathbf{J}^{(e)}| = 2A^{(e)} \tag{6.28}$$

The computation of the element integrals in this case is simple via the analytic expressions of Section 5.7.

For curve-sided triangles it is convenient to use the natural coordinates α and β defined in Section 5.5.6. This implies that L_2 and L_3 are replaced by α and β, respectively, and L_1 by $1-\alpha-\beta$ in all expressions. The computation of the cartesian derivatives of N_i follows precisely the steps described in the previous section, simply changing the coordinates ξ and η for α and β, respectively. For instance

$$\left\{\begin{array}{c}\dfrac{\partial N_i}{\partial x}\\ \dfrac{\partial N_i}{\partial y}\end{array}\right\} = \frac{1}{|\mathbf{J}^{(e)}|}\begin{bmatrix}\dfrac{\partial y}{\partial \beta} & -\dfrac{\partial y}{\partial \alpha}\\ -\dfrac{\partial x}{\partial \beta} & \dfrac{\partial x}{\partial \alpha}\end{bmatrix}\left\{\begin{array}{c}\dfrac{\partial N_i}{\partial \alpha}\\ \dfrac{\partial N_i}{\partial \beta}\end{array}\right\} \tag{6.29}$$

$$\mathbf{J}^{(e)} = \sum_{i=1}^{n}\begin{bmatrix}\dfrac{\partial N_i}{\partial \alpha}x_i & \dfrac{\partial N_i}{\partial \alpha}y_i\\ \dfrac{\partial N_i}{\partial \beta}x_i & \dfrac{\partial N_i}{\partial \beta}y_i\end{bmatrix} \tag{6.30}$$

$$\frac{\partial x}{\partial \alpha} = \sum_{i=1}^{n}\frac{\partial N_i(\alpha,\beta)}{\partial \alpha}x_i \quad ; \quad \frac{\partial x}{\partial \beta} = \sum_{i=1}^{n}\frac{\partial N_i(\alpha,\beta)}{\partial \beta}x_i \quad ; \quad \text{etc.} \tag{6.31}$$

The element stiffness matrix is obtained by an expression analogous to Eq.(6.11) by

$$\mathbf{K}_{ij}^{(e)} = \int_0^1\int_0^{1-\beta}\mathbf{B}_i^T\mathbf{D}\mathbf{B}_j\left|\mathbf{J}^{(e)}\right|t\,d\alpha\,d\beta = \int_0^1\int_0^{1-\beta}\mathbf{G}_{ij}\,(\alpha,\beta)\frac{t}{\left|\mathbf{J}^{(e)}\right|}\,d\alpha\,d\beta \tag{6.32}$$

where all the terms in $\mathbf{B}_i$ and $\mathbf{G}_{ij}$ are deduced from Eqs. (6.6)-(6.11) simply substituting ξ and η for α and β, respectively.

For curve-sided triangles the integrand of Eq.(6.32) is a rational polynomial and numerical integration is needed.

Example 6.2: Derive the expression of the Jacobian matrix for an isoparametric triangle with straight sides.

- ***Solution***

As the element sides are straight, a linear interpolation for the geometry suffices, i.e.

$$\begin{aligned}x &= L_1x_1 + L_2x_2 + L_3x_3\\ y &= L_1y_1 + L_2y_2 + L_3y_3\end{aligned} \tag{6.33}$$

where x_i, y_i $i=1,2,3$ are the coordinates of the three vertex nodes.

From Eq.(5.26) we deduce $L_1 = 1 - L_2 - L_3$. Substituting this into the above gives

$$\begin{aligned} x &= x_1 + (x_2 - x_1)L_2 + (x_3 - x_1)L_3 \\ y &= y_1 + (y_2 - y_1)L_2 + (y_3 - y_1)L_3 \end{aligned} \tag{6.34}$$

The derivatives of N_i with respect to L_2 and L_3 are obtained from

$$\begin{Bmatrix} \frac{\partial N_i}{\partial L_2} \\ \frac{\partial N_i}{\partial L_3} \end{Bmatrix} = \begin{bmatrix} \frac{\partial x}{\partial L_2} & \frac{\partial y}{\partial L_2} \\ \frac{\partial x}{\partial L_3} & \frac{\partial y}{\partial L_3} \end{bmatrix} \begin{Bmatrix} \frac{\partial N_i}{\partial x} \\ \frac{\partial N_i}{\partial y} \end{Bmatrix} = \mathbf{J}^{(e)} \begin{Bmatrix} \frac{\partial N_i}{\partial x} \\ \frac{\partial N_i}{\partial y} \end{Bmatrix} \tag{6.35}$$

The Jacobian matrix is deduced from the above two equations as

$$\mathbf{J}^{(e)} = \begin{bmatrix} \frac{\partial x}{\partial L_2} & \frac{\partial y}{\partial L_2} \\ \frac{\partial x}{\partial L_3} & \frac{\partial y}{\partial L_3} \end{bmatrix} = \begin{bmatrix} x_2 - x_1 & y_2 - y_1 \\ x_3 - x_1 & y_3 - y_1 \end{bmatrix} \tag{6.36}$$

The cartesian derivatives are obtained by (noting that $|\mathbf{J}^{(e)}| = 2A^{(e)}$)

$$\begin{Bmatrix} \frac{\partial N_i}{\partial x} \\ \frac{\partial N_i}{\partial y} \end{Bmatrix} = \frac{1}{2A^{(e)}} \begin{bmatrix} y_3 - y_1 & y_1 - y_2 \\ x_1 - x_3 & x_2 - x_1 \end{bmatrix} \begin{Bmatrix} \frac{\partial N_i}{\partial L_2} \\ \frac{\partial N_i}{\partial L_3} \end{Bmatrix} \tag{6.37}$$

Let us verify the above expression for the simple 3-noded triangle. For instance, for $i = 1$, $\frac{\partial N_1}{\partial L_2} = \frac{\partial N_1}{\partial L_3} = -1$ (as $N_1 = L_1 = 1 - L_2 - L_3$) and

$$\begin{Bmatrix} \frac{\partial N_1}{\partial x} \\ \frac{\partial N_1}{\partial y} \end{Bmatrix} = \frac{1}{2A^{(e)}} \begin{Bmatrix} y_2 - y_3 \\ x_3 - x_2 \end{Bmatrix} \tag{6.38}$$

Note the coincidence with the expressions obtained using Eqs.(4.33) directly. The expression for the cartesian derivatives obtained is completely general for straight-sided triangles of any approximation order (i.e. quadratic, cubic, etc.).

The reader is encouraged to repeat the process using the natural coordinates α and β.

6.4 NUMERICAL INTEGRATION IN TWO DIMENSIONS

It has been shown in the previous section that all the element integrals can be written in the natural coordinate space making use of the isoparametric formulation. The numerical integration by a Gauss quadrature will be considered next.

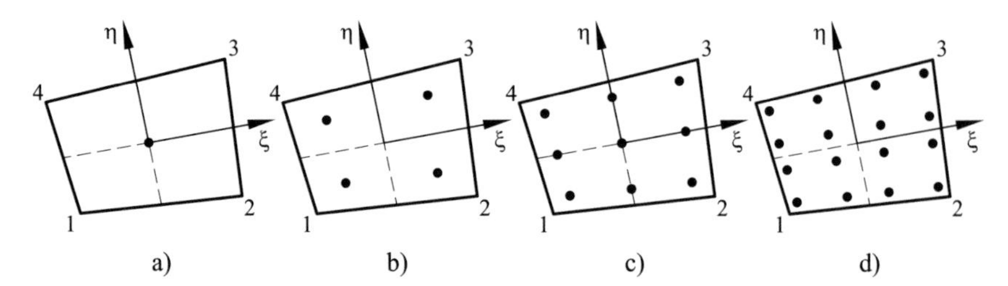

Fig. 6.4 Gauss quadratures over quadrilateral elements, a) 1×1, b) 2×2, c) 3×3, d) 4×4 integration points

6.4.1 Numerical integration in quadrilateral domains

The integral of a term $g(\xi, \eta)$ over the normalized isoparametric quadrilateral domain can be evaluated using a 2D Gauss quadrature by

$$\int_{-1}^{+1}\int_{-1}^{+1} g(\xi,\eta)\, d\xi\, d\eta = \int_{-1}^{+1} d\xi \left[\sum_{q=1}^{n_q} g(\xi,\eta_q) W_q\right] = \sum_{p=1}^{n_p}\sum_{q=1}^{n_q} g(\xi_p,\eta_q) W_p W_q \tag{6.39}$$

where n_p and n_q are the number of integration points along each natural coordinate ξ and η respectively; ξ_p and η_q are the natural coordinates of the pth integration point and W_p, W_q are the corresponding weights.

The coordinates and weights for each natural direction are directly deduced from those given in Table 3.1 for the 1D case. Let us recall that a 1D quadrature of qth order integrates *exactly* a polynomial of degree $q \leq 2n - 1$ (Section 3.4). Figure 6.4 shows the more usual quadratures for quadrilateral elements.

Example 6.3: Integrate numerically the function $f(\xi, \eta) = \xi^2 \eta^2$ over a rectangular element with dimensions $2a \times 2b$.

- ***Solution***

Since the element is rectangular $|\mathbf{J}^{(e)}| = ab$ (see Eq.(6.8)).
The integrand is a quadratic function in ξ and η and hence a 2×2 quadrature is needed (Figure 6.4b). Thus

$$I = \iint_A \xi^2\eta^2 dA = ab \int_{-1}^{+1}\int_{-1}^{+1} \xi^2\eta^2 = ab \sum_{p=1}^{2}\sum_{q=1}^{2} (\xi^2\eta^2)_{p,q} W_p W_q =$$

$$= ab\left[(-\frac{\sqrt{3}}{3})^2(-\frac{\sqrt{3}}{3})^2 + (-\frac{\sqrt{3}}{3})^2(\frac{\sqrt{3}}{3})^2 + (\frac{\sqrt{3}}{3})^2(-\frac{\sqrt{3}}{3})^2 + (\frac{\sqrt{3}}{3})^2(\frac{\sqrt{3}}{3})^2\right] = \frac{4}{9}ab$$

which is the exact solution.

6.4.2 Numerical integration over triangles

The Gauss quadrature for triangles is written as

$$\int_0^1 \int_0^{1-L_3} f(L_1, L_2, L_3)\, dL_2\, dL_3 = \sum_{p=1}^{n_p} f(L_{1_p}, L_{2_p}, L_{3_p})\, W_p \qquad (6.40)$$

where n_p is the number of integration points: $L_{1_p}, L_{2_p}, L_{3_p}$ and W_p are the area coordinates and the corresponding weight for the pth integration point.

Figure 6.5 shows the more usual coordinates and weights; the term "accuracy" in the figure refers to the highest degree polynomial which is exactly integrated by each quadrature. Figure 6.5 is also of direct application for computing the integrals defined in terms of the natural coordinates α and β, simply recalling that $L_2 = \alpha, L_3 = \beta$ and $L_1 = 1 - \alpha - \beta$.

The weights in Figure 6.5 are normalized so that their sum is 1/2. In many references this value is changed to the unity and this requires the sum of Eq.(6.40) to be multiplied by 1/2 so that the element area is correctly computed in those cases.

Example 6.4: Compute the area of a triangular element with straight sides by numerical integration.

- ***Solution***

$$A^{(e)} = \iint_{A^{(e)}} dx\, dy = \int_0^1 \int_0^{1-\beta} |\mathbf{J}^{(e)}|\, d\alpha\, d\beta = |\mathbf{J}^{(e)}| \sum_p W_p = \frac{|\mathbf{J}^{(e)}|}{2}$$

which corresponds with the value obtained in Eq.(6.28).

For further information on numerical integration over triangular domains see [CMPW,Cow,Du,ZTZ].

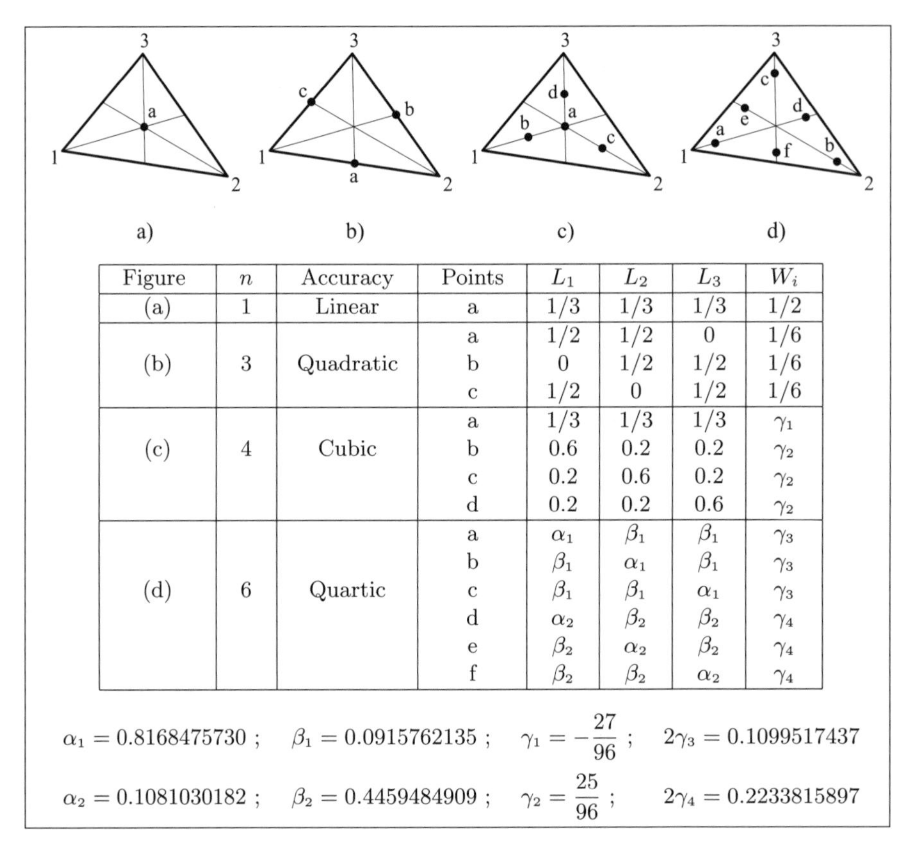

Figure	n	Accuracy	Points	L_1	L_2	L_3	W_i
(a)	1	Linear	a	1/3	1/3	1/3	1/2
(b)	3	Quadratic	a	1/2	1/2	0	1/6
			b	0	1/2	1/2	1/6
			c	1/2	0	1/2	1/6
(c)	4	Cubic	a	1/3	1/3	1/3	γ_1
			b	0.6	0.2	0.2	γ_2
			c	0.2	0.6	0.2	γ_2
			d	0.2	0.2	0.6	γ_2
(d)	6	Quartic	a	α_1	β_1	β_1	γ_3
			b	β_1	α_1	β_1	γ_3
			c	β_1	β_1	α_1	γ_3
			d	α_2	β_2	β_2	γ_4
			e	β_2	α_2	β_2	γ_4
			f	β_2	β_2	α_2	γ_4

$$\alpha_1 = 0.8168475730 \; ; \quad \beta_1 = 0.0915762135 \; ; \quad \gamma_1 = -\frac{27}{96} \; ; \quad 2\gamma_3 = 0.1099517437$$

$$\alpha_2 = 0.1081030182 \; ; \quad \beta_2 = 0.4459484909 \; ; \quad \gamma_2 = \frac{25}{96} \; ; \quad 2\gamma_4 = 0.2233815897$$

Fig. 6.5 Coordinates and weights for the Gauss quadrature in triangular elements

6.5 NUMERICAL INTEGRATION OF THE ELEMENT MATRICES AND VECTORS

6.5.1 Numerical integration of the stiffness matrix

The stiffness matrix for an isoparametric quadrilateral element is computed using numerical integration in the natural coordinate system as

$$\begin{aligned}\mathbf{K}_{ij}^{(e)} &= \iint_{A^{(e)}} \mathbf{B}_i^T \mathbf{D} \mathbf{B}_j t \, dx \, dy = \int_{-1}^{+1} \int_{-1}^{+1} \mathbf{B}_i^T \mathbf{D} \, \mathbf{B}_j \left| \mathbf{J}^{(e)} \right| t \, d\xi \, d\eta = \\ &= \sum_{p=1}^{n_p} \sum_{q=1}^{n_q} \left[\mathbf{B}_i^T \mathbf{D} \mathbf{B}_j \left| \mathbf{J}^{(e)} \right| t \right]_{p,q} W_{\dot{p}} W_{\dot{q}} = \sum_{p=1}^{n_p} \sum_{q=1}^{n_q} \left[\frac{t}{\left| \mathbf{J}^{(e)} \right|} \mathbf{G}_{ij} \right]_{p,q} W_p W_q\end{aligned} \tag{6.41}$$

where $\mathbf{G}_{ij}$ is the matrix deduced from Eq.(6.11).

For a triangular element we obtain from Eqs.(6.32) and (6.40)

$$\mathbf{K}_{ij}^{(e)} = \int_0^1 \int_0^{1-\beta} \mathbf{B}_i^T \mathbf{D} \mathbf{B}_j |\mathbf{J}^{(e)}| t \, d\alpha \, d\beta =$$

$$= \sum_{p=1}^{n_p} [\mathbf{B}_i^T \mathbf{D} \mathbf{B}_j |\mathbf{J}^{(e)}| t]_p W_p = \sum_{p=1}^{n_p} [\frac{t}{|\mathbf{J}^{(e)}|} \mathbf{G}_{ij}]_p W_p \tag{6.42}$$

The position of the integration points and the corresponding weights for Eqs.(6.41) and (6.42) are obtained from Figures 6.4 and 6.5, respectively.

It is important to note that the computation of matrix $\mathbf{G}_{ij}$ in the previous expressions is not necessary in practice. This is so because the numerical integration of the stiffness matrix just requires the evaluation of the Jacobian matrix $\mathbf{J}^{(e)}$ and its determinant, the strain matrix $\mathbf{B}_i$ and the constitutive matrix $\mathbf{D}$ *at each integration point* in the natural coordinate system. These computations can be performed in a sequential order which facilitates the implementation of Eqs.(6.41) and (6.42) in a computer program (Section 6.6 and Figure 6.8).

6.5.2 Numerical integration of the equivalent nodal force vector

The numerical integration of the equivalent nodal force vector due to body forces for isoparametric quadrilateral elements (Eq.(6.12)) gives

$$\mathbf{f}_{b_i}^{(e)} = \int_{-1}^{+1} \int_{-1}^{+1} \mathbf{N}_i^T \mathbf{b} \left|\mathbf{J}^{(e)}\right| t \, d\xi \, d\eta = \sum_{p=1}^{n_p} \sum_{q=1}^{n_q} \left(\mathbf{N}_i^T \mathbf{b} \left|\mathbf{J}^{(e)}\right| t\right)_{p,q} W_{\dot{p}} W_{\dot{q}} \tag{6.43}$$

For triangular elements the double summation is replaced by the single summation of Eq.(6.40).

The computation of the equivalent nodal force vector due to surface tractions deserves a special comment. Let us recall that this vector has the following expression (Eq.(4.68))

$$\mathbf{f}_{t_i}^{(e)} = \oint_{l^{(e)}} \mathbf{N}_i^T \mathbf{t} t \, ds \tag{6.44}$$

where $l^{(e)}$ is the loaded element boundary. In general this boundary represents a line ξ = constant or η = constant in the natural coordinate space. Therefore, the differential of length over the side $\eta = 1$ for the

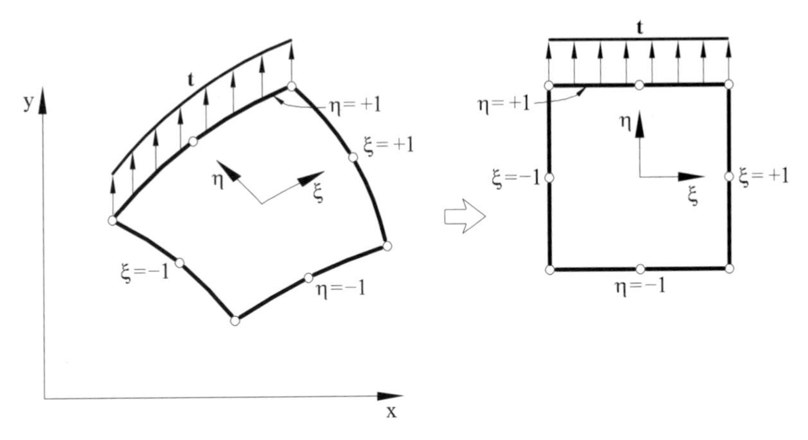

Fig. 6.6 Surface traction acting on the side $\eta = +1$

isoparametric quadrilateral element of Figure 6.6 is computed by

$$
\begin{aligned}
(ds)_{\eta=1} = (dx^2+dy^2)_{\eta=1} &= \left[\sqrt{\left(\frac{dx}{d\xi}\right)^2_{\eta=1}+\left(\frac{dy}{d\xi}\right)^2_{\eta=1}}\right]d\xi = \\
&= \left[\sqrt{\left(\sum_{i=1}^{n}\frac{dN_i}{d\xi}x_i\right)^2_{\eta=1}+\left(\sum_{i=1}^{n}\frac{dN_i}{d\xi}y_i\right)^2_{\eta=1}}\right]d\xi = c(\xi)\,d\xi
\end{aligned}
\tag{6.45}
$$

Substituting Eq.(6.45) into (6.44) yields a line integral which is a function of the natural coordinate ξ only. This is computed using a 1D quadrature by

$$
\mathbf{f}_{t_i}^{(e)} = \oint_{l^{(e)}} \left(\mathbf{N}_i^T\right)_{\eta=1} \mathbf{t} t c(\xi)\,d\xi = \int_{-1}^{+1} \mathbf{g}(\xi)\,d\xi = \sum_{p=1}^{n_p} \mathbf{g}(\xi_p)\,W_p \tag{6.46}
$$

Surface tractions act very frequently along the tangential or normal directions to the boundary (Figure 6.7) and this can simplify the computations. Transforming these forces to global axes gives

$$
\mathbf{t} = \begin{Bmatrix} t_x \\ t_y \end{Bmatrix} = \begin{Bmatrix} \tau\cos\,\beta - \sigma\sin\,\beta \\ \sigma\cos\,\beta + \tau\sin\,\beta \end{Bmatrix} \tag{6.47}
$$

where σ and τ are, respectively, the normal and tangential components of the surface traction, and β is the angle between the tangent to the boundary and the global x axis. Substituting Eq.(6.47) into the expression

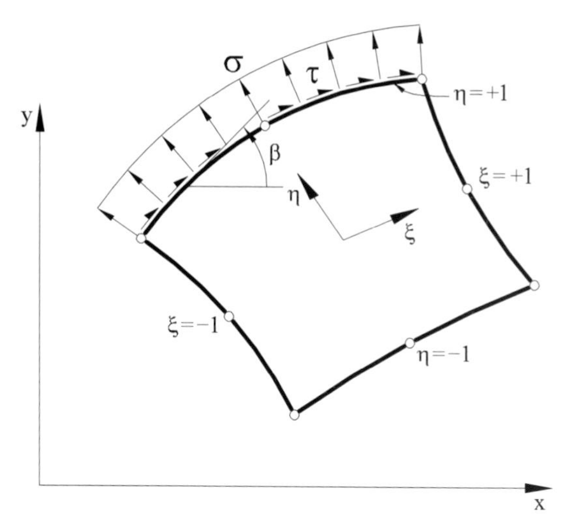

Fig. 6.7 Tangential and normal tractions acting on the side $\eta = +1$

of $\mathbf{t}$ of vector $\mathbf{f}_{t_i}^{(e)}$ in Eq.(6.44) yields

$$\mathbf{f}_{t_i}^{(e)} = \oint_{S^{(e)}} \mathbf{N}_i \begin{Bmatrix} \tau \cos\,\beta - \sigma \sin\,\beta \\ \sigma \cos\,\beta + \tau \sin\,\beta \end{Bmatrix} t\, ds = \oint_{s^{(e)}} \mathbf{N}_i \begin{Bmatrix} \tau\, dx - \sigma\, dy \\ \sigma\, dx + \tau\, dy \end{Bmatrix} t \quad (6.48)$$

On the loaded boundary

$$dx = \frac{\partial x}{\partial \xi}\, d\xi = J_{11}\, d\xi \quad ; \quad dy = \frac{\partial y}{\partial \xi}\, d\xi = J_{12}\, d\xi \qquad (6.49)$$

where J_{11} and J_{12} are obtained by sampling the Jacobian matrix at $\eta = 1$. Substituting Eq.(6.49) into (6.48) gives finally

$$\mathbf{f}_{t_i}^{(e)} = \int_{-1}^{+1} \mathbf{N}_i \begin{Bmatrix} \tau J_{11} - \sigma J_{12} \\ \sigma J_{11} + \tau J_{12} \end{Bmatrix} t\, d\xi = \sum_{p=1}^{n_p} \left[\mathbf{N}_i \begin{Bmatrix} \tau J_{11} - \sigma J_{12} \\ \sigma J_{11} + \tau J_{12} \end{Bmatrix} t \right]_p W_p \qquad (6.50)$$

6.6 COMPUTER PROGRAMMING OF $\mathbf{K}^{(e)}$ AND $\mathbf{f}^{(e)}$

Previous sections provided all the necessary expressions for programming the computation of the stiffness matrix and the equivalent nodal force vector for each element. The basic steps involved in programming the computation of $\mathbf{K}^{(e)}$ and $\mathbf{f}^{(e)}$ for quadrilateral elements are given next.

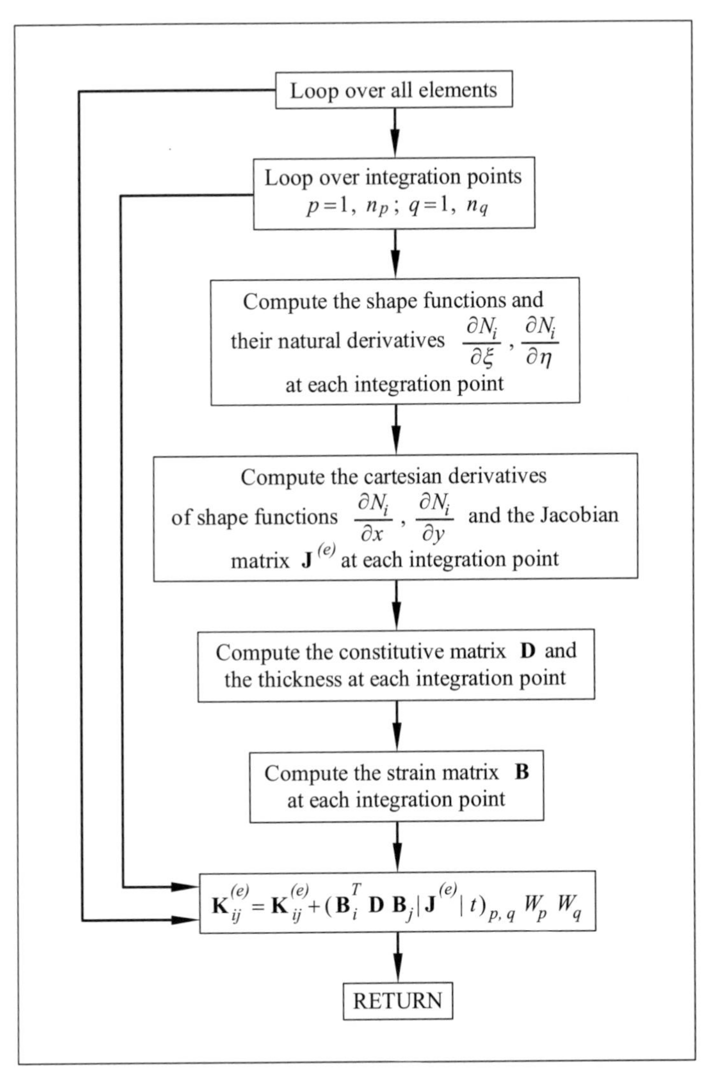

Fig. 6.8 Flow chart for the computation of $\mathbf{K}^{(e)}$

Figure 6.8 shows the flow diagram for computing $\mathbf{K}^{(e)}$ as deduced from Eq. (6.41).

The evaluation of the constitutive matrix $\mathbf{D}$ can be taken out of the numerical integration loop if the material properties are homogeneous over the element. The same applies for the thickness if this is constant over the

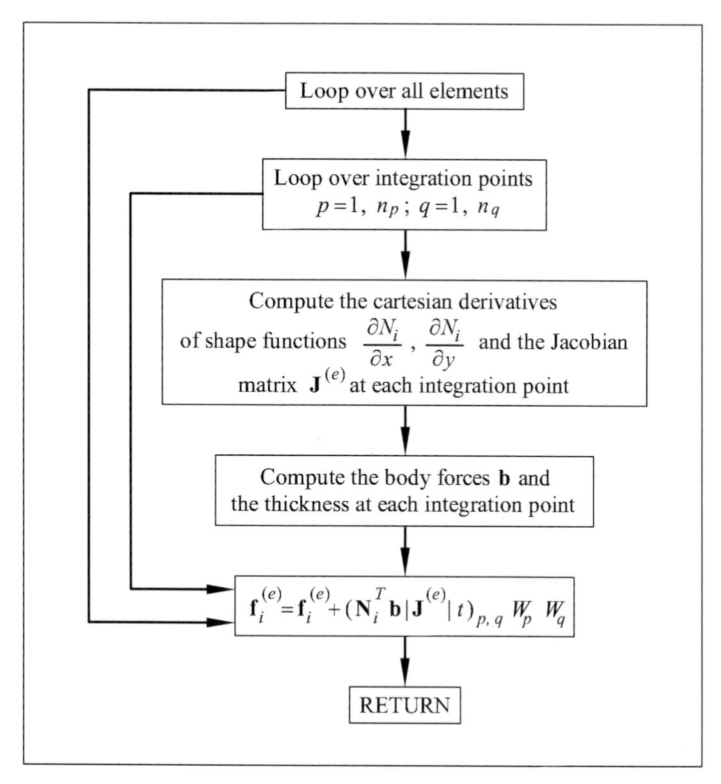

Fig. 6.9 Flow chart for computing $\mathbf{f}^{(e)}$ for body forces

element. The case of $\mathbf{D}$ (or t) variable is treated simply by using a standard interpolation within the element in terms of the nodal values as

$$\mathbf{D} = \sum_{i=1}^{n} N_i \mathbf{D}_i \quad ; \quad t = \sum_{i=1}^{n} N_i t_i \tag{6.51}$$

Eq.(6.51) is used to obtain the values for $\mathbf{D}$ and t at the Gauss points within the numerical integration loop.

Figure 6.9 shows the flow chart for computing vector $\mathbf{f}^{(e)}$ for the case of body forces as deduced from Eq.(6.43).

The computation of the body forces at each integration point can be taken out from the numerical integration loop if they are constant over the element. A variable body force can be accounted for by interpolating the known nodal values of the body force vector $\mathbf{b}$, in the same way as was done for the thickness in Eq. (6.51).

The flow charts in Figures 6.8 and 6.9 are completely general and applicable to all the structural problems treated in this book. Further details of the programming aspects of the FEM are given in Chapter 11. For more information see [Hu,HO,HO2].

6.7 OPTIMAL POINTS FOR COMPUTING STRAINS AND STRESSES

The strains (and the stresses) are obtained from the derivatives of the displacements. Therefore, their approximation is always of a lower order than that for the displacements. In general, if the shape functions are complete polynomials of pth degree the approximation for the stresses will be a polynomial of degree $p-1$ or $p-2$, depending if they are obtained as the first or second derivatives of the displacement field, respectively.

It can be proved that the stresses from the finite element solution can be considered a least square interpolation of the exact stress field [ZTZ]. Naturally, the exact stress field is unknown. However, an enhanced stress distribution can be found by the following property of the Gauss quadrature: A nth degree polynomial and a $n-1$th degree polynomial, obtained by least square fitting of the first one, take the same values at the points of the Gauss quadrature of order n. Hence, we can obtain an approximation of the stresses and strains that is one order higher by computing these at the Gauss points. This important property will be clarified with the following two examples.

Example 6.5: Verify that a second degree polynomial, and a first degree polynomial obtained by a least square smoothing of the former, take the same values at the points of the Gauss quadrature of order two.

- ***Solution***

Let us consider the second degree polynomial ($n=2$)

$$f(x) = 1 + x + x^2$$

We will now obtain a first degree polynomial ($n=1$) which approximates $f(x)$ in the least square sense; i.e. find a polynomial

$$g(x) = a + bx$$

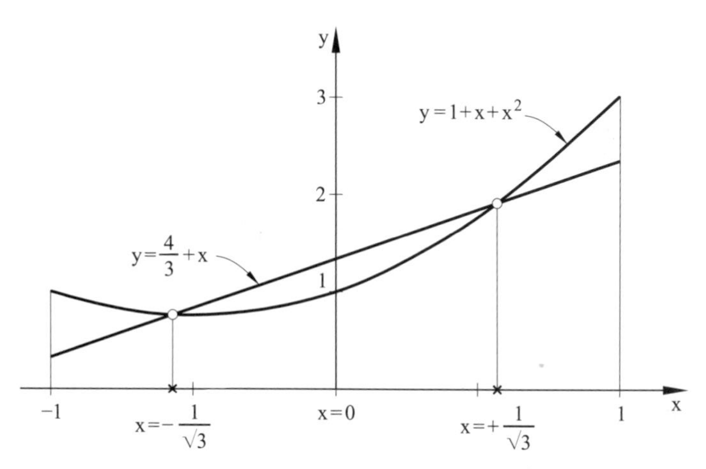

Fig. 6.10 Least square approximation of a quadratic polynomial by a linear one. Intersection of both polynomials at the points of the Gauss quadrature of order two

such that it minimizes the functional

$$M = \int_{-1}^{+1} \Big[f(x) - g(x)\Big]^2 dx = \int_{-1}^{+1} \Big[(1-a) + (1-b)x + x^2\Big]^2 dx$$

The parameters a and b are obtained by making

$$\frac{\partial M}{\partial a} = 0 \implies \int_{-1}^{+1} -2\Big[(1-a) \ + \ (1-b)x \ + \ x^2\Big]\, dx = 0$$
$$\frac{\partial M}{\partial b} = 0 \implies \int_{-1}^{+1} -2x\Big[(1-a) \ + \ (1-b)x \ + \ x^2\Big]\, dx = 0$$

which gives $a = \frac{4}{3}$ and $b = 1$ and hence $g(x) = \frac{4}{3} + x$.
Figure 6.10 shows the polynomials $f(x)$ and $g(x)$. Note that both take the same values at the points of the Gauss quadrature of order two.

Example 6.6: Verify that a cubic polynomial and a quadratic one obtained by least square smoothing of the former, take same values at the points of the third order Gauss quadrature.

- ***Solution***

Let us consider the third degree polynomial ($n = 3$)

$$f(x) = 1 + x + x^2 + x^3$$

We will obtain the second degree polynomial $g(x) = a + bx + cx^2$, such that

$$M = \int_{-1}^{+1} \Big[f(x) - g(x)\Big]^2 dx = \int_{-1}^{+1} \Big[(1-a) + (1-b)x + (1-c)x^2 + x^3\Big]^2 dx$$

is a minimum.
The three constants $a, b,$ and c are obtained by solving the following system

$$\begin{aligned}
\frac{\partial M}{\partial a} &= 0 \implies 1 - a + \frac{1-c}{3} = 0\\
\frac{\partial M}{\partial b} &= 0 \implies \frac{1-b}{3} + \frac{1}{5} = 0\\
\frac{\partial M}{\partial c} &= 0 \implies \frac{1-a}{3} + \frac{1-c}{5} = 0
\end{aligned}$$

which gives

$$a = 1 \quad , \quad b = \frac{8}{5} \quad , \quad c = 1 \quad \text{and} \quad g(x) = 1 + \frac{8}{5}x + x^2$$

Figure 6.11 shows that the exact and interpolating polynomials take the same values at the three points of the 3rd order Gauss quadrature (i,e. $\xi = 0$ and $\xi = \pm 0.774596$ (Table 3.1)).

The following conclusions can be drawn from what is explained:

1. If the *exact* distribution of the strain $\boldsymbol{\varepsilon}$ (or stress $\boldsymbol{\sigma}$) field is a polynomial of nth degree and the approximate finite element solution is a polynomial of $n-1$th degree, the computation of $\boldsymbol{\sigma}$ (or $\boldsymbol{\varepsilon}$) at the points of the Gauss quadrature of nth order gives the *exact* values.
2. The evaluation of $\boldsymbol{\sigma}$ or $\boldsymbol{\varepsilon}$ at the Gauss quadrature points chosen for the integration of $\mathbf{K}^{(e)}$ yields a solution of *one approximation order higher* than at any other point within the element.

The evaluation of stresses and strains at the so called "optimal" quadrature points is therefore of higher accuracy than at any other element point. The nodal stress values can subsequently be obtained from a local or global smoothing of the Gauss point values as detailed in Chapter 9.

The above concepts are rigorously true for 1D elements. For 2D and 3D elements the sampling of the strains and the stresses at the "optimal" Gauss quadrature points leads also to a substantial improvement in the results. Figure 6.12 shows the optimal points for computation of strains

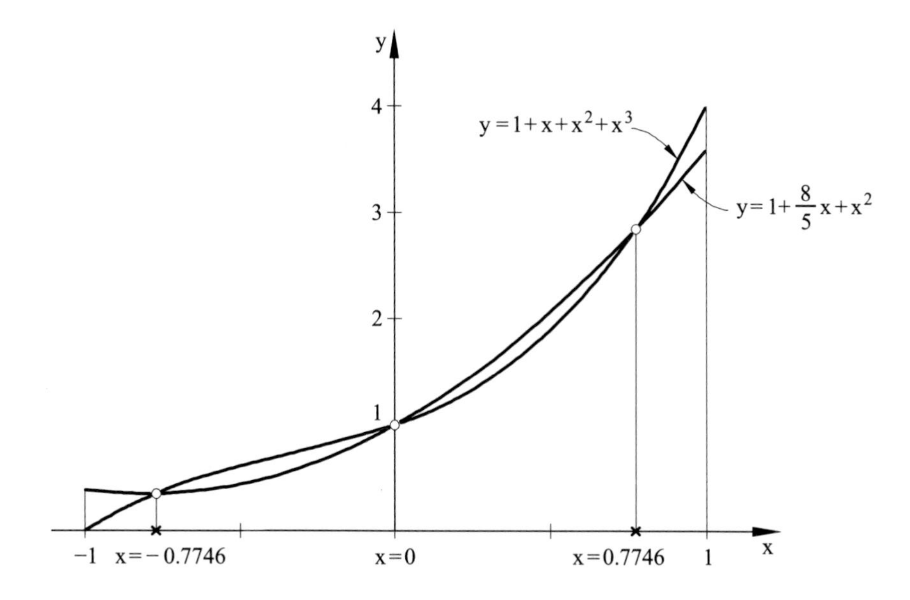

Fig. 6.11 Least square interpolation of a cubic polynomial by a quadratic one. Intersection of both polynomials at the points of the Gauss quadrature of order three

and stresses for some 1D and 2D elements. Extrapolation to the 3D case is simple.

Figure 6.15 shows an example of the analysis of a cantilever beam analyzed with 8-noded Serendipity rectangles. The accuracy of the shear stress value sampled at the 2×2 Gauss points is noticeable. This example is discussed further in the next section.

6.8 SELECTION OF THE QUADRATURE ORDER

The number of integration points is selected according to the degree of the polynomials appearing in the element integrals. Isoparametric elements contain rational terms within the integrals and exact integration is not longer possible. The alternative is to choose a quadrature order which integrates exactly the same expression for a rectangular or straight side triangular element. This quadrature is termed in practice *full integration* or *exact integration*. Remember that in these cases the Jacobian matrix is constant and the element integrals have a simple analytical form.

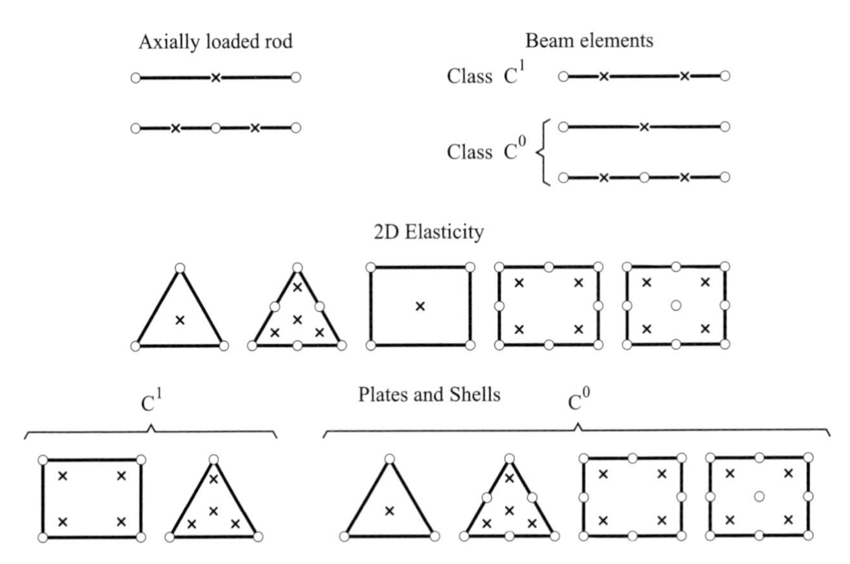

Fig. 6.12 Optimal points for computation of strains and stresses in some 1D and 2D elements

The minimum quadrature order for the stiffness matrix should preserve the convergence rate of the element. This is achieved by choosing a quadrature which integrates exactly all the complete polynomial terms contained in the shape functions. For rectangular elements this quadrature is of lower order than that required for the exact integration of the element stiffness matrix and, thus, some economy is obtained. Figure 6.13 shows the exact (full) and minimum quadratures for some popular rectangular and triangular elements. Note that both quadratures coincide for triangles.

Some authors associate the name "minimum quadrature" to that which guarantees that the element can reproduce in the limit a constant strain field [ZTZ]. This implies that the quadrature chosen should evaluate exactly the element area (or volume), which simply requires the exact computation of the following integral

$$A^{(e)} = \iint_{A^{(e)}} dA = \iint_{A^{(e)}} \left|\mathbf{J}^{(e)}\right| d\xi \, d\eta$$

In rectangles and straight side triangles this condition is too weak as it requires a single point quadrature which generally violates the minimum requirement for preserving the element convergence as described above (with the exception of the 3-noded triangle).

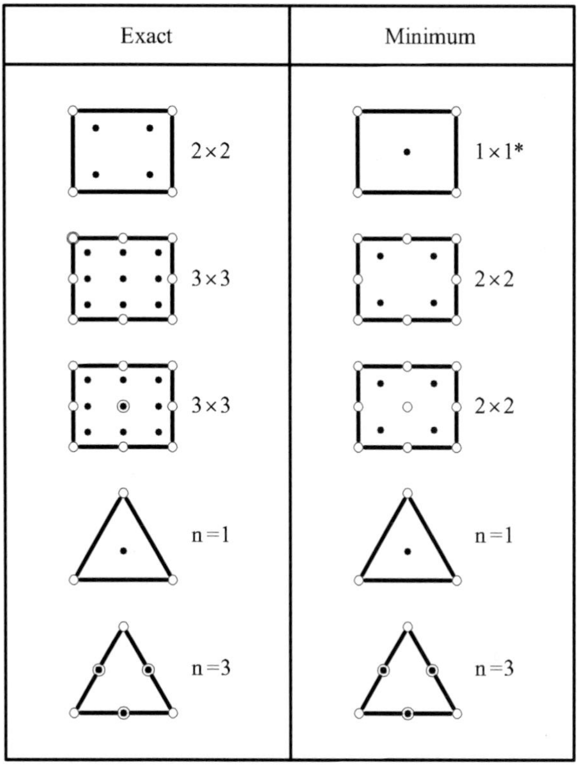

Exact	Minimum
2×2	1×1*
3×3	2×2
3×3	2×2
n=1	n=1
n=3	n=3

* Produces two propagable mechanims (Fig. 6.14a)

Fig. 6.13 Exact (full) and minimum quadrature rules for some rectangular and straight side triangular elements

Extreme care must also be taken so that a lower order quadrature does not introduce internal mechanisms in the element. These mechanisms appear when the displacement field generates a strain field which vanishes at the integration points, thus yielding a singular stiffness matrix. Sometimes these mechanisms are compatible between adjacent elements and lead to the singularity of the global stiffness matrix and, consequently, to an incorrect solution. A typical example are the two mechanisms induced by the reduced one point quadrature in the four-noded rectangle as shown in Figure 6.14a. These mechanisms invalidate the one point quadrature in this element for practical purposes unless some stabilization techniques are used to correct these deficiencies [BB,KZ,ZTZ]. See also Example 6.7.

In some cases, the mechanisms induced by the reduced integration of the stiffness matrix can not propagate in the mesh, and this preserves

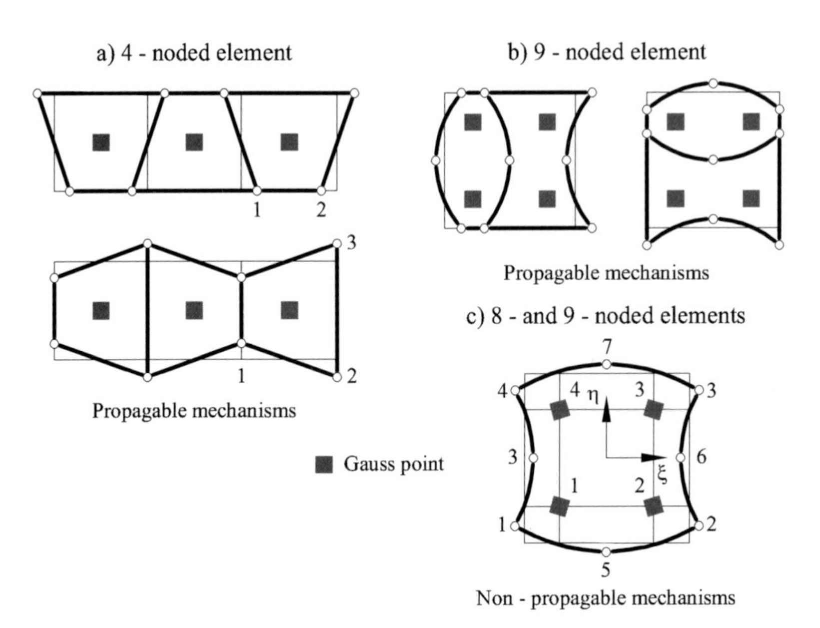

Fig. 6.14 a) Propagable mechanisms in the 4-noded rectangle with a single integration point, b) Mechanisms in the 8-noded and 9-noded quadrilaterals induced by reduced 2×2 quadrature. The mechanism in the 8-noded element is not propagable in a mesh

the correctness of the solution. This happens for the single mechanism originated by the 2×2 reduced quadrature in the 8-noded rectangle, as shown in Figure 6.14b. Unfortunately this is not the case for the 9-noded Lagrange element as the reduced integration introduces three mechanisms (Figure 6.14b,c) two of which are propagable and can pollute the solution and hence it is not recommended in practice.

The minimum quadrature points coincide in most cases with the optimum points for the computation of stresses. This can be easily verified by comparing the minimum and optimum quadratures shown in Figures 6.12 and 6.13, respectively. This important coincidence is shown in the analysis of a cantilever beam using 8-noded Serendipity rectangles. Figure 6.15 shows that the shear force distribution within each element computed by integrating the tangential stresses across the beam thickness is parabolic and, therefore, far from the correct linear solution. However, the tangential stresses at the 2×2 Gauss quadrature coincide with the exact values and the simple linear interpolation gives the exact shear force distribution.

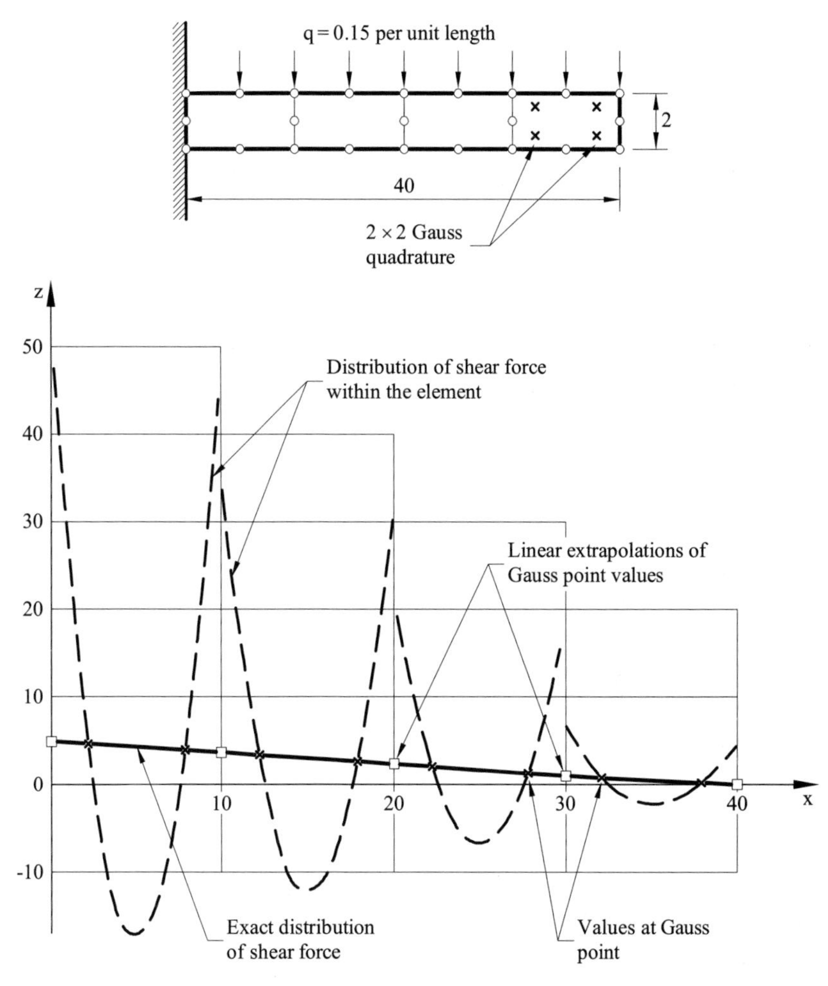

Fig. 6.15 Cantilever beam analyzed with four 8-noded Serendipity rectangles. Linear extrapolation of the shear force values from the transverse sections corresponding to the 2×2 Gauss quadrature

6.9 PERFORMANCE OF 2D ISOPARAMETRIC SOLID ELEMENTS

In general, the behaviour of 2D isoparametric solid elements is similar to that shown in the examples of Section 5.7 for rectangular and straight side triangles. Thus, isoparametric quadrilateral elements are usually more accurate than triangles of the same order. However, triangles are more convenient for modelling complex geometries. Also, the simplicity of lower order elements has a prize in the accuracy, in particular for problems with high stress gradients and a finer mesh is essential in the modelling of these zones. Here adaptive mesh refinement techniques are recommended. In practice, triangular elements are more versatile than quadrilaterals for mesh adaption (Chapter 9).

The distortion of rectangular elements to quadrilateral shapes leads to stiffer results [CMPW]. As mentioned earlier, extreme care should be taken when using reduced integration, as the relative improvement in the numerical results is sometimes at the expense of spurious mechanisms being introduced in the element which can pollute the solution. This issue is discussed further in the next section.

6.10 THE PATCH TEST FOR SOLID ELEMENTS

Application of the patch test to solid elements follows the general lines described in Section 3.9 for 1D elements. Here once again the test is a necessary and sufficient condition for the convergence of the element.

The application of the patch test to a solid element (either a 2D solid element, an axisymmetric solid element or a 3D solid element) can take the following three different modalities.

Patch test A. A known linear displacement field $\mathbf{a}^p$ is prescribed *at all nodes* of a patch of solid elements. For each internal node i in the patch we verify that (Figure 6.16a)

$$\mathbf{K}_{ij}\mathbf{a}_j^p - \mathbf{f}_i^p = \mathbf{0} \tag{6.52}$$

where $\mathbf{a}_j^p$ is the nodal displacement corresponding to the known field and $\mathbf{f}_i^p$ is a force vector resulting from any "body force" required to satisfy the governing differential equations of elasticity for the known solution. Generally, in problems expressed in cartesian coordinates $\mathbf{f}_i^p = 0$.

Patch test B. Only the values of $\mathbf{a}^p$ corresponding to the boundaries of the patch are inserted and $\mathbf{a}_i$ is found as (Figure 6.16b)

$$\mathbf{a}_i = \mathbf{K}_{ii}^{-1}(\mathbf{f}_i^p - \mathbf{K}_{ij}\mathbf{a}_j^p) \quad i \neq j \tag{6.53}$$

and compared against the exact value.

Patch tests A and B also involve the computation of the stresses within the elements and the comparison with the expected "exact" values.

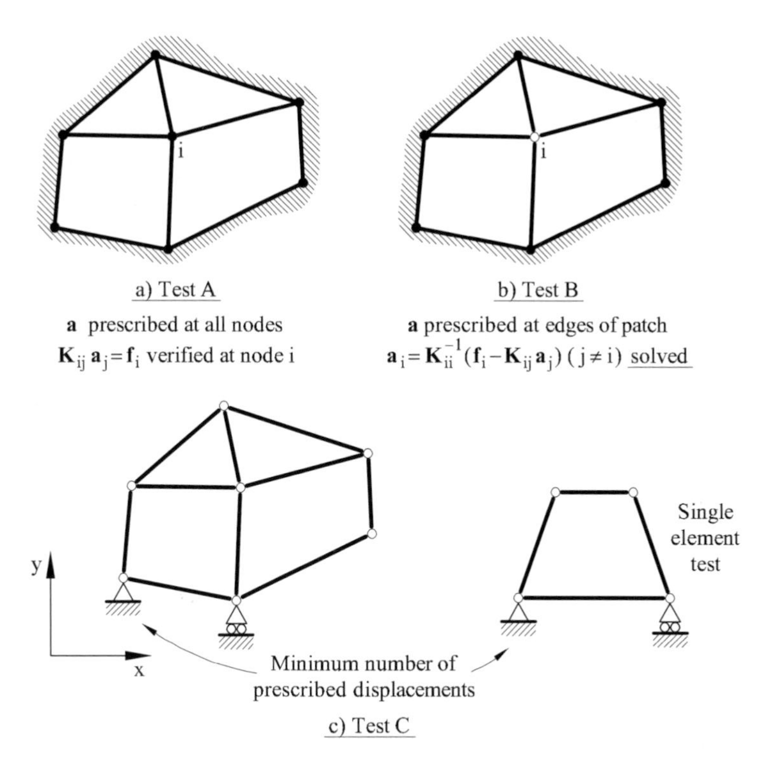

Fig. 6.16 Patch tests of form A, B and C

Satisfaction of patch tests A and B is a *necessary condition* for convergence of the element [ZTZ].

Patch test C. The assembled matrix system of the whole patch is written as

$$\mathbf{Ka} = \mathbf{f}^p \tag{6.54}$$

where $\mathbf{f}^p$ represents prescribed boundary forces corresponding to the known solution. The solution for $\mathbf{a}$ is sought after *fixing the minimum number of displacements* necessary to eliminate the rigid body motion, i.e. three displacements for 3D elasticity problems (Figure 6.16c) and it is compared with the known solution.

Patch test C allows us to detect any singularity in the stiffness matrix. This test is therefore an assessment of the *stability* of the finite element solution and hence provides not only a necessary but a sufficient condition for convergence.

When the patch is reduced to just one element the C test is termed the *single-element* test [ZTZ] (Figure 6.16c). This test is a requirement for a good finite element formulation as, on occasions, a larger patch may not reveal the intrinsic instabilities of a single element. A typical example is the 8-noded isoparametric solid element with reduced 2×2 Gauss quadrature. Here the singular deformation mode of a single element disappears when several elements are assembled. The satisfaction of the single-element test is not a sufficient condition for convergence and the test of at least one internal element boundary needs to be tested to assess sufficiency.

Example 6.7: Patch test for the 4-noded quadrilateral.

- ***Solution***

We consider a linear isotropic plane stress problem on the patch shown in Figure (a) below. The material properties are $E = 1000$ and $\nu = 0.3$.

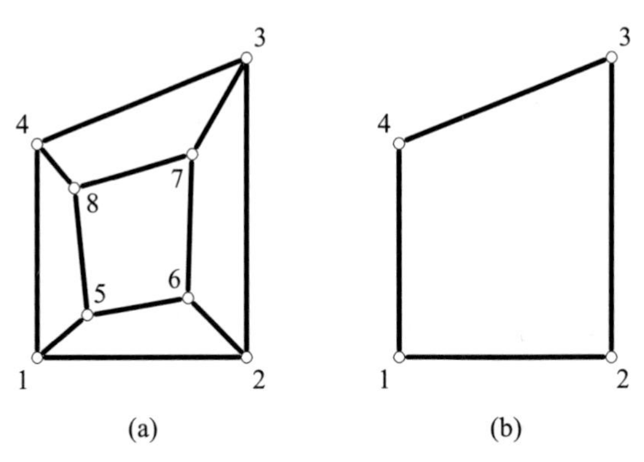

(a) (b)

We will verify that the patch test is satisfied for a linear displacement solution giving a constant stress field. This obviously satisfies the differential equation of equilibrium involving second derivatives of the displacement field.
The solution considered, taken from [ZTZ], is

$$u = 2 \times 10^{-3}x \quad , \quad v = -6 \times 10^{-4}y$$

which produces zero body forces and zero stresses except for $\sigma_x = 2$.
The patch test is first performed using a 2×2 Gauss quadrature. For patch test A all nodes are restrained and nodal displacements are specified according to the field chosen. Eq.(6.52) is satisfied at all nodes. Furthermore stresses at the Gauss points are all exact within round-off error. The reactions at the four boundary nodes are shown in Table 6.1.

	Coordinates		Computed displacements		Forces	
Node a	x_a	y_a	u_a	v_a	F_{x_a}	F_{y_a}
1	0.0	0.0	0.0	0.0	−2	0
2	2.0	0.0	0.0040	0.0	3	0
3	2.0	3.0	0.0040	−0.00180	2	0
4	0.0	2.0	0.0	−0.00120	−3	0
5	0.4	0.4	0.0008	−0.00024	0	0
6	1.4	0.6	0.0028	−0.00036	0	0
7	1.5	2.0	0.0030	−0.00120	0	0
8	0.3	1.6	0.0006	−0.00096	0	0

Table 6.1 Patch solution for figure of Example 6.7 [ZTZ]

Patch test B is verified by restraining only nodes 1 to 4 with their displacements specified according to Table 6.1. Exact results are once again recovered to within round-off errors.

Finally, patch test C is performed with node 1 fully restrained and node 4 restrained only in the x direction. Nodal forces are applied to nodes 2 and 3 with the values given in Table 6.1. This test also produced exact solutions for all other nodal quantities in Table 6.1 and recovered $\sigma_x = 2$ at all Gauss points in each element.

Test C was repeated using a 1×1 reduced Gauss quadrature to compute the element stiffness and nodal force quantities. Patch C indicated that the global stiffness matrix contained two global *zero energy nodes*, thus producing incorrect nodal displacements and, consequently, incorrect stresses, except at the 1×1 Gauss point used in each element to compute the stiffness and forces. Thus the 1×1 quadrature produces a failure in the patch test and therefore, its use is not recommended.

This deficiency can be corrected by a proper stabilization scheme [BB,KF,ZTZ] which provides a version of the one point reduced integrated 4-noded element useful for practical purposes.

The same conclusions are drawn after performing patch tests A, B and C on a one-element patch using the mesh shown in Figure (b) of previous page.

Example 6.8: Patch test for quadratic elements [ZTZ].

- ***Solution***

Figure 6.17 shows a two-element patch of quadratic isoparametric quadrilaterals. Both 8-noded Serendipity and 9-noded Lagrange types are considered. Patch test C is performed first for load case 1 (pure tension). For the 8-noded element both 2×2 (reduced) and (3×3) full gaussian quadrature satisfy the

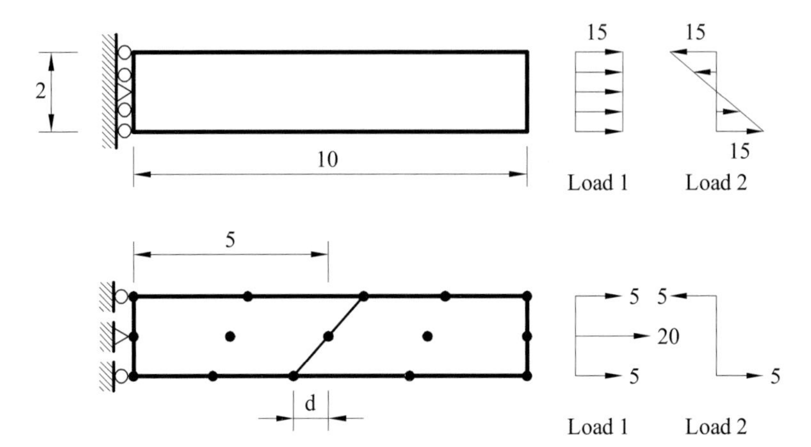

Fig. 6.17 Patch test for 8- and 9-noded isoparametric quadrilaterals

patch test, and for the 9-noded element only the 3×3 quadrature gives good results, whereas the 2×2 reduced quadrature leads to a mechanism due to failure in the rank of the stiffness matrix [ZTZ].

However, if we perform a one-element test for the 8-noded element with 2×2 quadrature, the spurious zero-energy mode shown in Figure 6.14 is found and thus the one-element test fails. As a conclusion both the 8-noded and the 9-noded elements with 2×2 quadrature are "suspect" and are to be used with great caution.

The same plane stress problem was next solved subjected to the bending loading shown as load 2 in Figure 6.17. The pure bending solution in elasticity is quadratic and no body forces are needed to satisfy the equilibrium equations. The equilibrating surface loads are two point loads equal and opposite acting on the top and bottom nodes. Table 6.2 shows results for the 8-noded and 9-noded elements for the indicated quadrature with $E = 100$ and $\nu = 0.3$. Results taken from [ZTZ] are given for a rectangular mesh ($d = 0$) and two distorted meshes ($d = 1$ and $d = 2$).

We observe that the 9-noded element with 3×3 quadrature passes patch test C for all meshes. On the other hand, the 8-noded element with 3×3 quadrature passes the test only for the rectangular mesh and its accuracy deteriorates very rapidly with increased mesh distorsion. This result confirms the failure of the 8-noded element to approximate a complete quadratic displacement function on linearly distorted meshes and the good performance of the 9-noded element in these cases.

The 2×2 quadrature improves results for the 8-noded element. However, the use of the 2×2 quadrature should be considered with great care.

Figure 6.18 shows an example of the dangers of 2×2 integration for the 8-noded element [ZTZ]. Here the structure is modelled by a single quasi-rigid element as the interest is centred in the "foundation" response. Use of 2×2

Element	Quadrature	d	v_A	u_B	v_B
8-noded	3×3		0.750	0.150	0.75225
8-noded	2×2	0	0.750	0.150	0.75225
9-noded	3×3		0.750	0.150	0.75225
8-noded	3×3		0.7448	0.1490	0.74572
8-noded	2×2	1	0.750	0.150	0.75100
9-noded	3×3		0.750	0.150	0.75225
8-noded	3×3		0.6684	0.1333	0.66364
8-noded	2×2	2	0.750	0.150	0.75225
9-noded	3×3		0.750	0.150	0.75225
Exact	—	—	0.750	0.150	0.75225

Table 6.2 Patch test of Figure 6.17. Bending load case ($E = 100, \nu = 0.3$) [ZTZ]

quadrature throughout leads to the spurious answers shown in Figure 6.18b, while the correct results for the 3×3 quadrature are shown in Figure 6.18c. We note that no zero-energy mode exists since more than one element is used. The spurious response is due to the larger variation in the elastic parameters between structure and foundation. This situation can easily occur in other structural problems and, therefore, use of the 8-noded 2×2 integrated element should be closely monitored and avoided for problems where anomalous behaviour is suspected.

6.11 APPLICATIONS

Some applications of plane elasticity elements to practical problems are presented next.

6.11.1 Analysis of concrete dams

Figure 6.19a shows the geometry of the transverse section of the Mequinenza gravity dam in Spain analyzed under the assumption of plane strain. A very coarse mesh of 140 8-noded Serendipity elements was used for the analysis. Full slip conditions were assumed in the dam base and only the horizontal displacements of the lower node of the downstream face were prescribed.

Figures 6.19b and c show the principal stress distribution obtained for self-weight plus hydrostatic loading with and without the effect of a gallery and a vertical joint, respectively. Figure 6.19b shows the stresses over the deformed shape of the dam amplified 200 times. Note that the horizontal constraint in just a point in the dam base leads to high stress values in

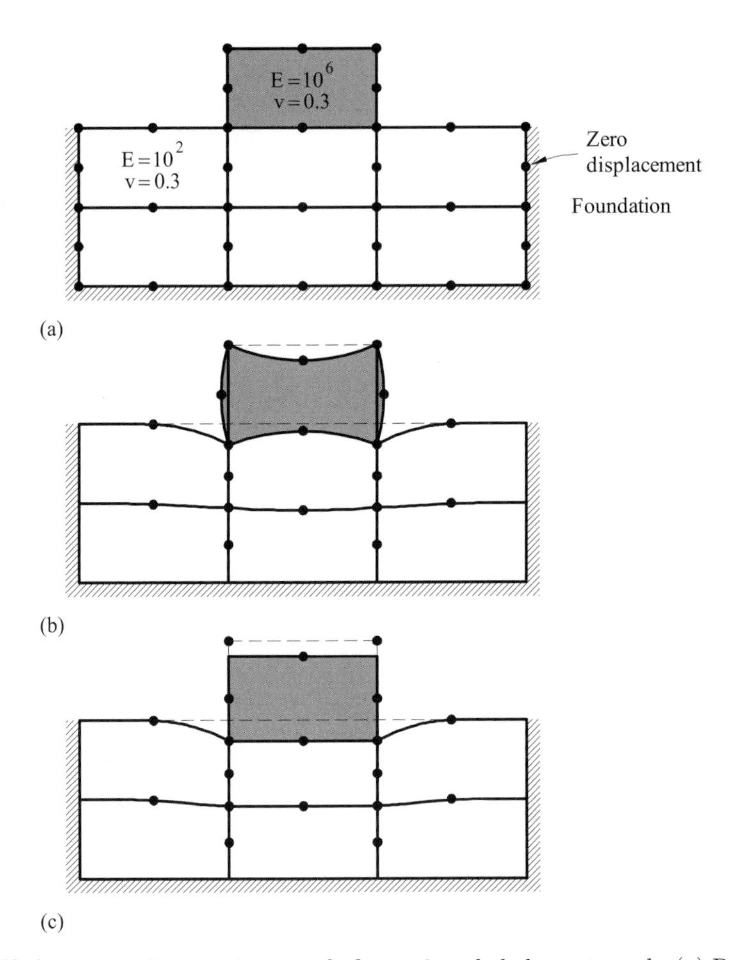

Fig. 6.18 A propagating spurious mode for an 8-noded element mesh. (a) Problem description. (b) 2×2 integration. (c) 3×3 integration

that zone. However, at a short distance from the prescribed point the stress distribution is correct, in accordance with Saint-Venant's principle [Ti]. Figure 6.19d shows the stress distribution around the gallery. More information on this study is reported in [OCOH,OOB].

Figure 6.20a shows a second gravity dam studied in Spain (Santa Coloma dam). A mesh of 651 plane strain eight-noded Serendipity quadrilaterals was used for discretizing the dam and the underlying foundation. Figure 6.20b shows the deformed shape of the dam under self-weight and hydrostatic loading. Contours of the displacement module and the stresses are shown in Figure 6.20c [SM].

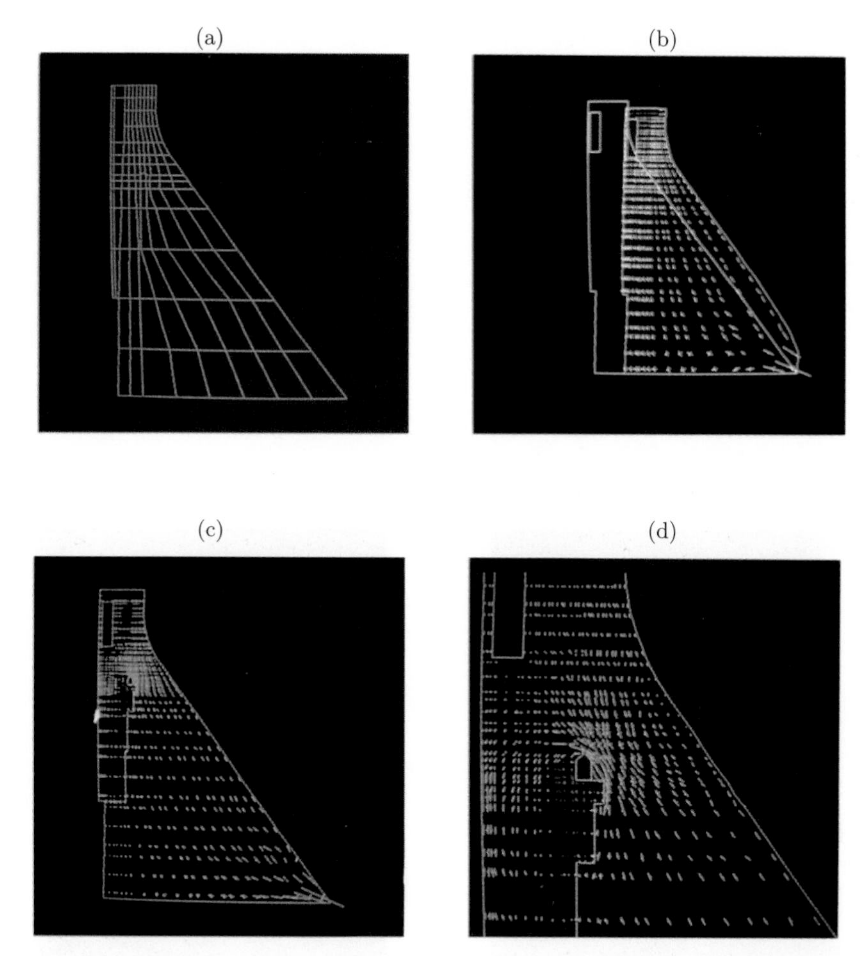

Fig. 6.19 Mequinenza gravity dam. a) Mesh of 140 8-noded Serendipity quadrilaterals, b) Deformed shape and principal stresses distribution under self-weight and hydrostatic loading, c) Idem including the effect of a gallery and a vertical joint, d) Detail of the stresses in the vicinity of the gallery

6.11.2 Analysis of an earth dam

Figure 6.21a shows the geometry of the Limonero earth dam in Málaga, Spain analyzed including the effect of the foundation and the mesh of 82 8-noded plane strain Serendipity quadrilaterals used for the analysis.

A non-tension material model was assumed. This requires the elimination of the tension stresses which excede a prescribed threshold value until a compression stress dominated equilibrium state is obtained via an iterative process [ZVK].

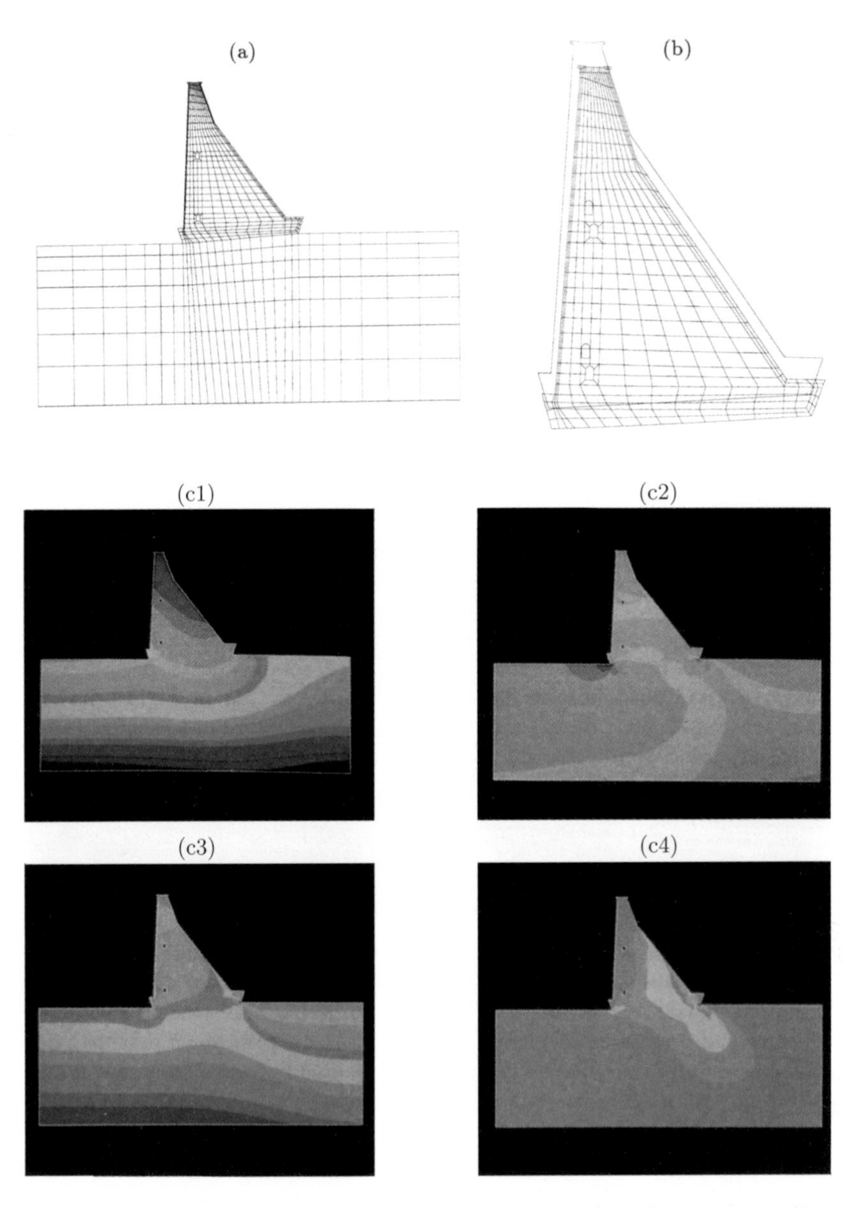

Fig. 6.20 Santa Coloma gravity dam. a) Mesh of 651 8-noded Serendipity quadrilaterals discretizing the dam and the foundation, b) Deformed shape for self-weight and hydrostatic loading, c) Contours of the displacement module (c1) and the stresses σ_x, σ_y and τ_{xy} (c2-c4) under the same loading

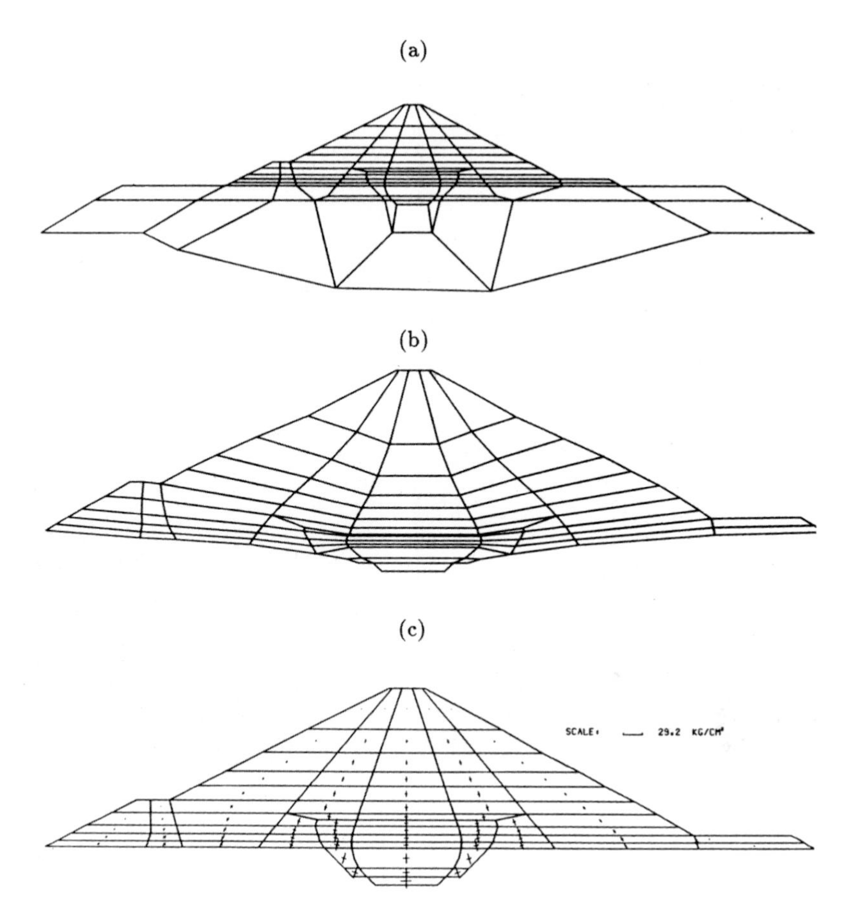

Fig. 6.21 Limonero earth dam. a) Mesh of 82 8-noded Serendipity quadrilaterals, b) and c) Deformed shape and principal stresses under self-weight

Figures 6.20b and c show the deformed geometry of the dam body under self-weight and the final stress distribution for an admissible threshold value of the tension stresses of 0.5 Mpa. Further information on this example can be found in [COHR+,HCA].

6.11.3 Analysis of an underground tunnel

This example shows some results of the analysis of an underground tunnel in the city of Barcelona in Spain. Plane strain conditions were assumed. Figure 6.22 shows the discretization of the transverse section analyzed using 8-noded quadrilaterals and the contours of the displacement modu-

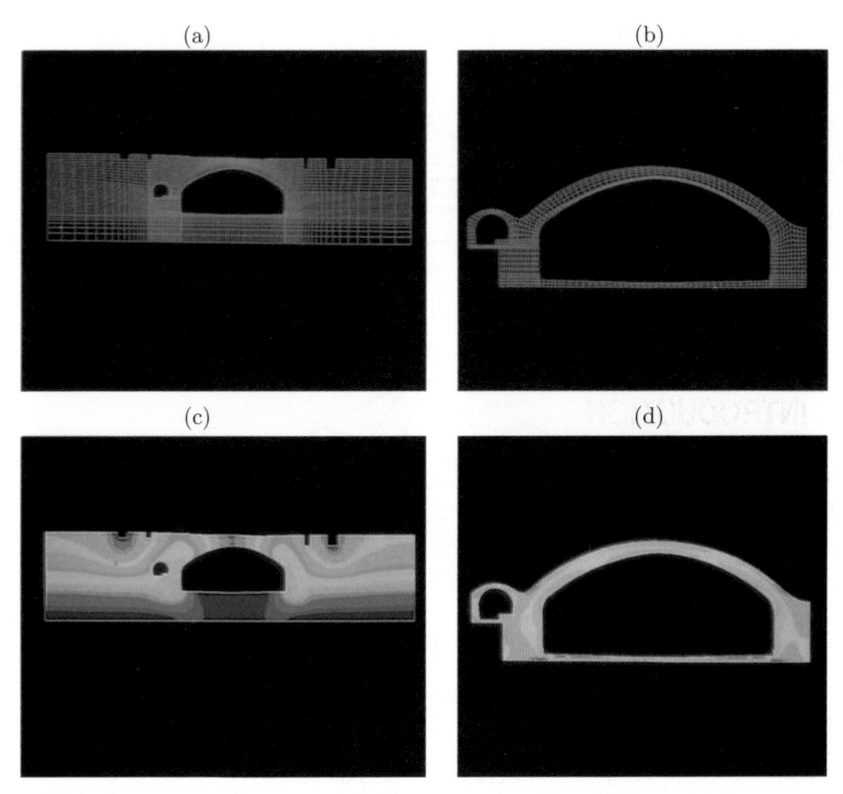

Fig. 6.22 Underground tunnel. a) Mesh of 8-noded Serendipity quadrilaterals, b) Detail of the mesh, c) Contours of total displacement for self-weight loading plus the weight of adjacent buildings, d) Detail of minor principal stress contours

lus and the minor principal stresses for self-weight plus a vertical loading equal to the weight of the adjacent buildings.

6.12 CONCLUDING REMARKS

This chapter has introduced most of the concepts necessary for the analysis of continuous structures with the FEM using elements of arbitrary shape. We point out once more that the underlying ideas are valid for other problems different of plane elasticity. Thus, the concepts and methodology explained for the isoparametric formulation, the numerical integration and the computation of the element stiffness matrix and the equivalent nodal force vector are applicable to other structural finite element models as a straightforward extension of the ideas studied in this chapter.

7
AXISYMMETRIC SOLIDS

7.1 INTRODUCTION

This chapter treats the analysis of solids with axial symmetry. Thus, only solids with geometrical and material properties independent of the circumferential coordinate θ are considered (Figure 7.1). This property allows the inherent 3D behaviour of a solid to be expressed with a much simpler 2D model.

If the loading is also axisymmetric, the displacement vector has only two components in the radial and axial directions. The analysis of axisymmetric solid structures by the FEM is not difficult and follows very similar steps to those explained in the previous chapters for plane elasticity problems. For arbitrary *non–axisymmetric loading* a full 3D analysis is needed. However, even in these cases the axial symmetry of the structure allows important simplifications to be introduced. For instance, the loading can be expanded in Fourier series and the effect of each harmonic term can be evaluated by a simpler 2D analysis. The final result is obtained by adding the contributions from the different 2D solutions. This avoids costly 3D computations. This chapter focuses only on the analysis of axisymmetric solids under axisymmetric loading. Axisymmetric solids under arbitrary loading will be studied in Volume 2 [On].

A thin-wall axisymmetric solid is usually termed an *axisymmetric shell.* The study of these structures is also covered in Volume 2 [On].

Axisymmetric solids represent a substantial percentage of engineering structures. Examples include water and oil tanks, cooling towers, silos, domes, cylindrical containment structures, chimneys, pressure vessels, etc. Also, some soil mechanics problems such as the analysis of foundations under vertical loads, can be solved using the methods explained in this chapter. Figure 7.2 shows examples of some typical axisymmetric structures.

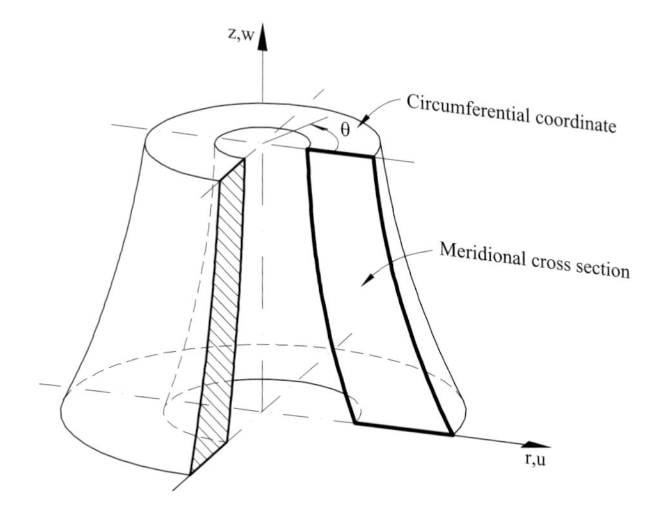

Fig. 7.1 Axisymmetric solid

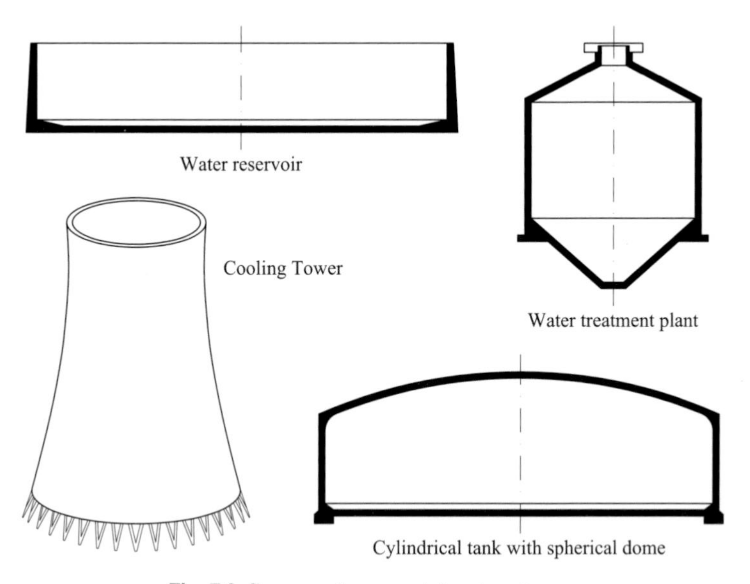

Fig. 7.2 Some axisymmetric structures

7.2 BASIC FORMULATION

7.2.1 Displacement field

Let us consider the axisymmetric solid under axisymmetric loading shown in Figure 7.1. The movement of a point is defined by the radial (u) and

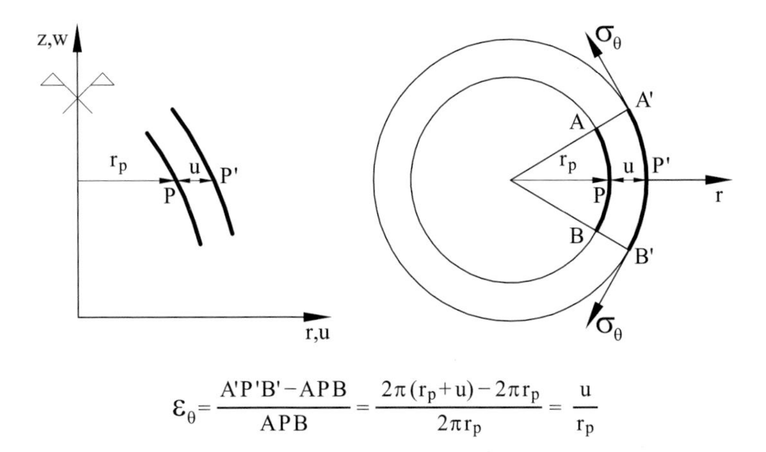

Fig. 7.3 Derivation of the circumferential strain ε_θ

axial (w) displacements, whereas the circumferential displacement (v) is zero due to the axial symmetry. The displacement vector is therefore

$$\mathbf{u} = \begin{Bmatrix} u(r,z) \\ w(r,z) \end{Bmatrix} \tag{7.1}$$

7.2.2 Strain field

Due to the axial symmetry, the displacements u and w are independent of the circumferential coordinate θ. Consequently, the tangential strains $\gamma_{r\theta}$ and $\gamma_{z\theta}$ are zero. Also, 3D elasticity theory gives [TG]

$$\varepsilon_r = \frac{\partial u}{\partial r} \quad ; \quad \varepsilon_z = \frac{\partial w}{\partial z} \quad ; \quad \gamma_{rz} = \frac{\partial u}{\partial x} + \frac{\partial w}{\partial r} \tag{7.2}$$

where $\varepsilon_r, \varepsilon_z$ and γ_{rz} are the radial, axial and tangential strains, respectively.

On the other hand, the points located on a circumference of radius r move, due to the axial deformation, to a circumference of radius $r + u$. This originates a circumferential strain which is defined as the relative elongation between these two circumferences (Figure 7.3); i.e.

$$\varepsilon_\theta = \frac{2\pi(r+u) - 2\pi r}{2\pi r} = \frac{u}{r} \tag{7.3}$$

The strain vector of a point has, therefore, the four components

$$\boldsymbol{\varepsilon} = [\varepsilon_r, \varepsilon_z, \varepsilon_\theta, \gamma_{rz}]^T = \left[\frac{\partial u}{\partial r}, \frac{\partial w}{\partial z}, \frac{u}{r}, \frac{\partial u}{\partial z} + \frac{\partial w}{\partial r}\right]^T \tag{7.4}$$

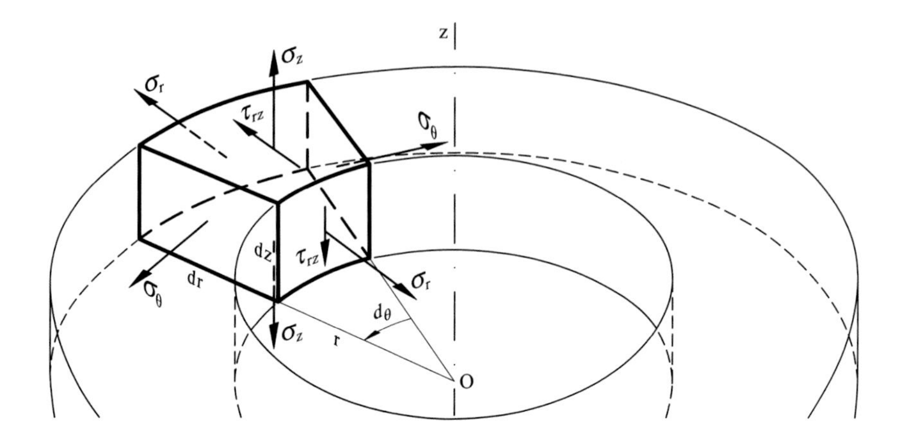

Fig. 7.4 Stresses acting on a differential volume of an axisymmetric solid under axisymmetric loading

7.2.3 Stress field

The stresses conjugate to the strains of Eq.(7.4) are (in vector form)

$$\boldsymbol{\sigma} = [\sigma_r, \sigma_z, \sigma_\theta, \tau_{rz}]^T \tag{7.5}$$

where σ_r, σ_z and σ_θ are, respectively, the radial, axial and circumferential stresses and τ_{rz} is the tangential stress. The rest of the stresses are zero. The sign convention for the stresses is shown in Figure 7.4.

The representation of the two principal stresses in the plane rz follows precisely that explained for 2D solids in Section 4.2.5.

7.2.4 Constitutive equation

The relationship between stresses and strains is deduced from 3D elasticity theory as for the plane elasticity case. The constitutive equation is written (in the presence of initial strains and initial stresses) as

$$\boldsymbol{\sigma} = \mathbf{D}\ (\boldsymbol{\varepsilon} - \boldsymbol{\varepsilon}^0) + \boldsymbol{\sigma}^0 \tag{7.6}$$

Matrix **D** for an isotropic material (recall that the axial symmetry of the material properties is needed) is written as

$$\mathbf{D} = \frac{E}{(1+\nu)\ (1-2\nu)} \begin{bmatrix} 1-\nu & \nu & \nu & 0 \\ \nu & 1-\nu & \nu & 0 \\ \nu & \nu & 1-\nu & 0 \\ 0 & 0 & 0 & \frac{1-2\nu}{2} \end{bmatrix} \tag{7.7}$$

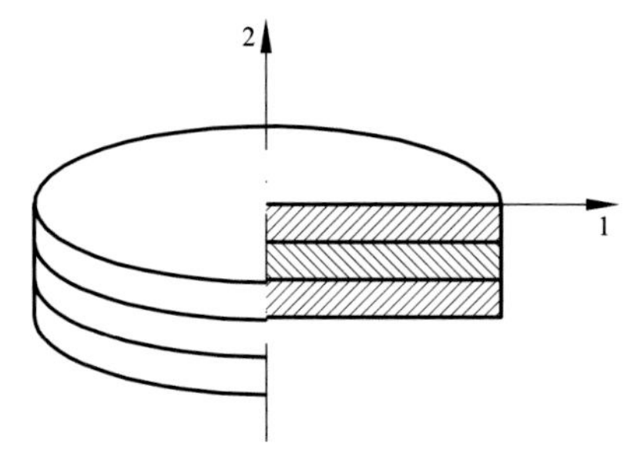

Fig. 7.5 Axisymmetric stratified solid

The initial strain vector for the thermal isotropic case is [TG]

$$\boldsymbol{\varepsilon}^0 = \frac{\alpha \; E \; (\Delta T)}{(1-2\nu)} \; [1,1,1,0]^T \tag{7.8}$$

where all the terms were defined in Chapter 4.

Stratified material

Stratified materials in axisymmetric solids are typically distributed in orthotropic layers with principal directions of orthotropy along axes 1 and 2 (Figure 7.5). The problem remains axisymmetric and $\mathbf{D}$ is now a function of five parameters E_1, E_2, G_2, ν_1 and ν_2 [TG,ZTZ], i.e.

$$\mathbf{D} = \frac{E_2}{d} \begin{bmatrix} n(1-\nu_2^2) & n\nu_2(1+\nu_1) & n(\nu_1+n\nu_2^2) & 0 \\ & 1-\nu_1^2 & 1-\nu_1^2 & 0 \\ & & n(1-n\nu_2^2) & 0 \\ \text{Sym.} & & & md \end{bmatrix} \tag{7.9}$$

where

$$n = \frac{E_1}{E_2} \quad , \quad m = \frac{G_2}{E_2} \quad \text{and} \quad d = (1+\nu_1)(1-\nu_1-2n\nu_2^2) \tag{7.10}$$

The initial thermal strains are modified due to the existence of two thermal expansion parameters α_z and α_r, corresponding to the directions normal and parallel to the layers plane. Vector $\boldsymbol{\varepsilon}^0$ of Eq.(7.8) is modified accordingly as

$$\boldsymbol{\varepsilon}^0 = \Delta T[\alpha_r, \alpha_r, \alpha_z, 0]^T \tag{7.11}$$

The constitutive matrix $\mathbf{D}$ and vector $\boldsymbol{\varepsilon}^0$ of Eqs.(7.7) and (7.8) for an isotropic material are obtained as a particular case of the above expressions for

$$n=1 \quad ; \quad m=\frac{1}{2(1+\nu)} \quad ; \quad \nu_1=\nu_2 \quad \text{and} \quad \alpha_r=\alpha_z=\frac{\alpha E}{2(1-\nu)} \tag{7.12}$$

7.2.5 Principle of virtual work

The virtual work expression is analogous to Eq.(4.22) for 2D elasticity with all the integrals now referring to the volume and the surface of the axisymmetric solid. The differential of volume can be expressed as (Figure 7.4)

$$dV=(rd\theta)\,dr\,dz=r\;d\theta\,dA \tag{7.13}$$

The PVW is therefore written as

$$\iint_A\int_0^{2\pi}\delta\boldsymbol{\varepsilon}^T\boldsymbol{\sigma}\;r\;d\theta\,dA=\iint_A\int_0^{2\pi}\delta\mathbf{u}^T\mathbf{b}r\;d\theta\,dA+ \\ +\oint_l\int_0^{2\pi}\delta\mathbf{u}^T\mathbf{t}r\;d\theta\,ds+\sum_i\int_0^{2\pi}\delta\mathbf{u}_i\mathbf{p}_i r_i\;d\theta \tag{7.14}$$

In the above, l and A are the boundary and area of the meridional section and

$$\mathbf{b}=\begin{Bmatrix}b_r\\b_z\end{Bmatrix} \quad ; \quad \mathbf{t}=\begin{Bmatrix}t_r\\t_z\end{Bmatrix} \quad ; \quad \mathbf{p}_i=\begin{Bmatrix}P_{r_i}\\P_{z_i}\end{Bmatrix} \tag{7.15}$$

are the vectors of body forces, surface tractions and point loads, respectively. We recall once again that all loads have axial symmetry (Figure 7.6).

Eq.(7.14) is simplified by integrating over each circumferential line

$$2\pi\iint_A\delta\boldsymbol{\varepsilon}^T\boldsymbol{\sigma}r\,dA=2\pi\iint_A\delta\mathbf{u}^T\mathbf{b}r\,dA+2\pi\oint_l\delta\mathbf{u}^T\mathbf{t}rds+2\pi\sum_i\delta\mathbf{u}_i\mathbf{p}_i r_i \tag{7.16}$$

Note that all the terms of Eq.(7.16) are multiplied by 2π and this number can therefore be simplified. However, it is very instructive to keep 2π in front of all the terms of Eq.(7.16) as a reminder of the axial symmetry of the problem and also to remind us that the point loads $\mathbf{p}_i$ refer to load intensities *per unit circumferential length* (Figure 7.6).

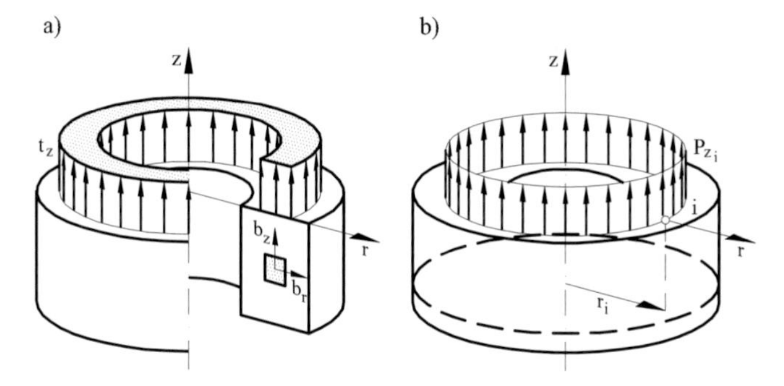

Fig. 7.6 Axisymmetric forces acting on an axisymmetric solid

7.3 FINITE ELEMENT FORMULATION. THREE-NODED AXISYMMETRIC TRIANGLE

Next, we will explain the details of the finite element analysis of axisymmetric solids. For the purpose of explanation, we will develop first the formulation for the simple linear 3-noded axisymmetric triangle. Figure 7.7 shows that the element is a ring with triangular cross-section. We note that all axisymmetric solid elements have an annular shape although the element integrals are computed over the 2D meridional section as clearly seen in Eq.(7.16).

7.3.1 Discretization of the displacement field

The displacement field is interpolated within each meridional section in a similar way to what we did for 2D elasticity. Thus, for the 3-noded triangle we write

$$\mathbf{u}=\begin{Bmatrix}u\\w\end{Bmatrix}=\begin{Bmatrix}N_1\ u_1\ +\ N_2\ u_2\ +\ N_3\ u_3\\N_1\ w_1\ +\ N_2\ w_2\ +\ N_3\ w_3\end{Bmatrix}=$$
$$=\sum_{i=1}^{3}\begin{bmatrix}N_i & 0\\0 & N_i\end{bmatrix}\begin{Bmatrix}u_i\\w_i\end{Bmatrix}=\sum_{i=1}^{3}\mathbf{N}_i\ \mathbf{a}_i^{(e)}=\mathbf{N}\ \mathbf{a}^{(e)} \tag{7.17}$$

where

$$\mathbf{N}=[\mathbf{N}_1,\mathbf{N}_2,\mathbf{N}_3]=\left[\begin{array}{cc|cc|cc}N_1 & 0 & N_2 & 0 & N_3 & 0\\ & & & & & \\0 & N_1 & 0 & N_2 & 0 & N_3\end{array}\right] \tag{7.18}$$

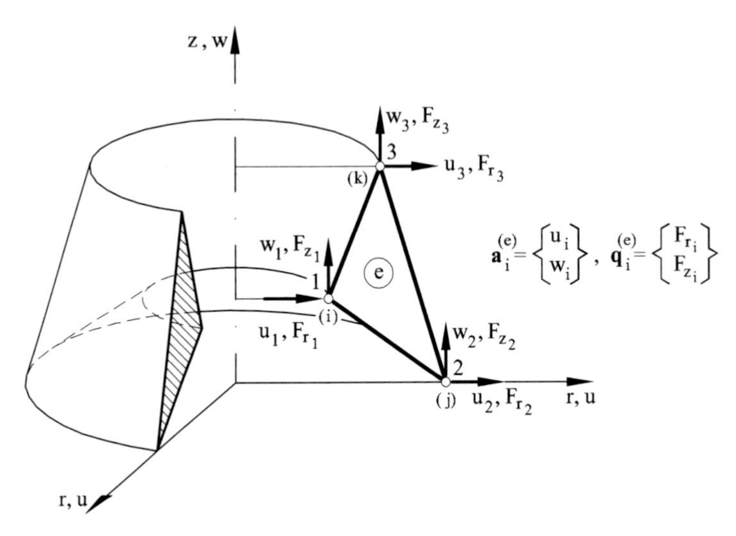

Fig. 7.7 Axisymmetric 3-noded triangle. Nodal displacements (u_i, w_i) and equilibrating nodal forces (F_{r_i}, F_{z_i}). Numbers in brackets denote global node numbers

and

$$\mathbf{a}^{(e)} = \begin{Bmatrix} \mathbf{a}_1^{(e)} \\ \mathbf{a}_2^{(e)} \\ \mathbf{a}_3^{(e)} \end{Bmatrix} \quad , \quad \mathbf{a}_i^{(e)} = \begin{Bmatrix} u_i \\ w_i \end{Bmatrix} \tag{7.19}$$

Note the coincidence of Eqs.(7.17), (7.18) and (7.19) with Eqs.(4.25), (4.28) and (4.29) for the analogous case of a 2D elasticity element. The shape functions N_i coincide with those of Eq.(4.33) simply substituting the coordinates x, y by r, z. The extension of Eqs. (7.17)–(7.19) to a n-noded element is straightforward as it only implies extending the summation of Eq.(7.17) from 3 to n.

7.3.2 Discretization of the strain and stress fields

Substituting the interpolation (7.17) into the strain vector of Eq.(7.4) gives

$$\boldsymbol{\varepsilon} = \sum_{i=1}^{3} \begin{bmatrix} \frac{\partial N_i}{\partial r} & 0 \\ 0 & \frac{\partial N_i}{\partial z} \\ \frac{N_i}{r} & 0 \\ \frac{\partial N_i}{\partial z} & \frac{\partial N_i}{\partial r} \end{bmatrix} \begin{Bmatrix} u_i \\ w_i \end{Bmatrix} = \Big[\mathbf{B}_1, \mathbf{B}_2, \mathbf{B}_3\Big] \begin{Bmatrix} \mathbf{a}_1^{(e)} \\ \mathbf{a}_2^{(e)} \\ \mathbf{a}_3^{(e)} \end{Bmatrix} = \mathbf{B}\ \mathbf{a}^{(e)} \tag{7.20}$$

where $\mathbf{B}$ is the strain matrix for the element and

$$\mathbf{B}_i = \begin{bmatrix} \frac{\partial N_i}{\partial r} & 0 \\ 0 & \frac{\partial N_i}{\partial z} \\ \frac{N_i}{r} & 0 \\ \frac{\partial N_i}{\partial z} & \frac{\partial N_i}{\partial r} \end{bmatrix} \tag{7.21}$$

is the strain matrix for the ith node.

The number of $\mathbf{B}_i$ matrices in Eq.(7.20) is extended from 3 to n for an element with n nodes.

Note that the strain matrix contains the term $\frac{N_i}{r}$ which is singular for $r = 0$. The way to overcome this problem is explained below.

The explicit form of $\mathbf{B}_i$ for the 3-noded triangle is obtained substituting Eq.(5.33) into (7.21) giving

$$\mathbf{B}_i = \frac{1}{2A^{(e)}} \begin{bmatrix} b_i & 0 \\ 0 & c_i \\ \frac{(a_i + b_i r + c_i z)}{r} & 0 \\ c_i & b_i \end{bmatrix} \tag{7.22}$$

The stresses are obtained in terms of the displacements by substituting Eq.(7.20) into (7.6) as

$$\boldsymbol{\sigma} = \sum_{i=1}^{3} \mathbf{B}_i \ \mathbf{D} \ \mathbf{a}^{(e)} - \mathbf{D} \ \boldsymbol{\varepsilon}^0 + \boldsymbol{\sigma}^0 \tag{7.23}$$

The computation of the circumferential strains ε_θ along the axis of symmetry presents us with a problem because the term $\frac{u}{r}$ leads to the undetermined value $\frac{0}{0}$. This difficulty can be overcome by computing ε_θ in the vicinity of the axis or, as it is more usual, by extrapolating the circumferential strain from the integration points located within the element to the symmetry axis. An alternative is to take advantage of the fact that $\varepsilon_r = \varepsilon_\theta$ at the axis of symmetry and simply compute the circumferential strain at those points using the first row of Eq.(7.20).

7.3.3 Equilibrium equations

The PVW particularized for a single element is (see Eq.(7.16))

$$2\pi \iint_{A^{(e)}} \delta \boldsymbol{\varepsilon}^T \boldsymbol{\sigma} r \, dA = 2\pi \iint_{A^{(e)}} \delta \mathbf{u}^T \mathbf{b} r \, dA + 2\pi \oint_{l^{(e)}} \delta \mathbf{u}^T \mathbf{t} r ds + 2\pi \sum_i [\delta \mathbf{a}_i^{(e)}]^T \mathbf{q}_i^{(e)} r_i \tag{7.24a}$$

where $\mathbf{q}_i^{(e)}$ is the vector of equilibrating nodal forces given by (Figure 7.7)

$$\mathbf{q}_i^{(e)} = [F_{r_i}, F_{z_i}]^T \tag{7.24b}$$

Assuming the standard interpolation for the virtual displacements and the virtual strains

$$\delta\mathbf{u} = \mathbf{N}\delta\mathbf{a} \quad , \quad \delta\boldsymbol{\varepsilon} = \mathbf{B}\delta\mathbf{a} \tag{7.24c}$$

and substituting these into Eq.(7.24a) gives after simplification of the virtual displacement field

$$2\pi \iint_{A^{(e)}} \mathbf{B}^T \boldsymbol{\sigma} r\, dA - 2\pi \iint_{A^{(e)}} \mathbf{N}^T \mathbf{b} r\, dA - 2\pi \oint_{l^{(e)}} \mathbf{N}^T \mathbf{t} r\, ds = \mathbf{q}^{(e)} \tag{7.25a}$$

where the equilibrating nodal force vector for the element is

$$\mathbf{q}^{(e)} = 2\pi \begin{Bmatrix} \mathbf{q}_1^{(e)} r_1 \\ \mathbf{q}_2^{(e)} r_2 \\ \mathbf{q}_3^{(e)} r_3 \end{Bmatrix} \tag{7.25b}$$

Substituting the constitutive equation for the stresses (Eq.(7.6)) into (7.25a) and using Eq.(7.20) gives

$$\begin{aligned} 2\pi \left(\iint_{A^{(e)}} \mathbf{B}^T \mathbf{D} \mathbf{B} r\, dA \right) \mathbf{a}^{(e)} - 2\pi \iint_{A^{(e)}} \mathbf{B}^T \mathbf{D} \boldsymbol{\varepsilon}_0 r\, dA + 2\pi \iint_{A^{(e)}} \mathbf{B}^T \boldsymbol{\sigma}_0 r\, dA - \\ - 2\pi \iint_{A^{(e)}} \mathbf{N}^T \mathbf{b} r\, dA - 2\pi \oint_{l^{(e)}} \mathbf{N}^T \mathbf{t} r\, ds = \mathbf{q}^{(e)} \end{aligned} \tag{7.25c}$$

or

$$\mathbf{K}^{(e)} \mathbf{a}^{(e)} - \mathbf{f}^{(e)} = \mathbf{q}^{(e)} \tag{7.26}$$

where $\mathbf{K}^{(e)}$ is the element stiffness matrix and $\mathbf{a}^{(e)}$ and $\mathbf{f}^{(e)}$ are the displacement vector and the equivalent nodal force vector for the element. These vectors have the following expressions

$$\mathbf{a}^{(e)} = \begin{Bmatrix} \mathbf{a}_1^{(e)} \\ \mathbf{a}_2^{(e)} \\ \mathbf{a}_3^{(e)} \end{Bmatrix} \quad , \quad \mathbf{a}_i^{(e)} = \begin{Bmatrix} u_i \\ w_i \end{Bmatrix} \quad , \quad \mathbf{f}^{(e)} = \begin{Bmatrix} \mathbf{f}_1^{(e)} \\ \mathbf{f}_2^{(e)} \\ \mathbf{f}_3^{(e)} \end{Bmatrix} \quad , \quad \mathbf{f}_i^{(e)} = \begin{Bmatrix} f_{r_i} \\ f_{z_i} \end{Bmatrix} \tag{7.27}$$

Vector $\mathbf{q}^{(e)}$ in Eqs.(7.25) is the total equilibrating force vector for the element whose components are obtained by integrating the individual force intensities $\mathbf{q}_i^{(e)}$ along a circumference with radius r_i.

The stiffness matrix and the equivalent nodal force vector for the element are

$$\underset{2n\times 2n}{\mathbf{K}^{(e)}} = 2\pi \iint_{A^{(e)}} \mathbf{B}^T \mathbf{D} \mathbf{B} r \, dA \tag{7.28a}$$

with

$$\underset{2\times 2}{\mathbf{K}_{ij}^{(e)}} = 2\pi \iint_{A^{(e)}} \underset{2\times 4}{\mathbf{B}_i^T} \, \underset{4\times 4}{\mathbf{D}} \, \underset{4\times 2}{\mathbf{B}_j} \, r \, drdz \tag{7.28b}$$

and

$$\mathbf{f}^{(e)} = 2\pi \iint_{A^{(e)}} \mathbf{N}^T \mathbf{b} r \, dA + 2\pi \oint_{l^{(e)}} \mathbf{N}^T \mathbf{t} r \, ds + 2\pi \iint_{A^{(e)}} \mathbf{B}^T \mathbf{D} \boldsymbol{\varepsilon}^0 r \, dA -$$
$$-2\pi \iint_{A^{(e)}} \mathbf{B}^T \boldsymbol{\sigma}^0 r \, dA = \mathbf{f}_b^{(e)} + \mathbf{f}_t^{(e)} + \mathbf{f}_\varepsilon^{(e)} + \mathbf{f}_\sigma^{(e)} \tag{7.29}$$

The first integral in Eq.(7.29) corresponds to the body force vector $\mathbf{f}_b^{(e)}$; the second one to the surface traction vector $\mathbf{f}_t^{(e)}$; the third one to the initial strain force vector $\mathbf{f}_\varepsilon^{(e)}$; and the fourth one to the initial stress force vector $\mathbf{f}_\sigma^{(e)}$. Vector $\mathbf{f}_i^{(e)}$ is obtained by substituting $\mathbf{N}$ and $\mathbf{B}$ by $\mathbf{N}_i$ and $\mathbf{B}_i$, respectively in above expressions.

The components of $\mathbf{f}^{(e)}$ are the total force acting at the element nodes, after integration over the circumferential direction at each node.

The expression of $\mathbf{K}^{(e)}$ and $\mathbf{f}^{(e)}$ of Eqs.(7.28) and (7.29) holds for any axisymmetric solid element with n nodes.

The global stiffness matrix $\mathbf{K}$ and the global equivalent nodal force vector $\mathbf{f}$ are obtained by assembling the element contributions in the standard manner. Once the global system of equation $\mathbf{Ka} = \mathbf{f}$ has been solved for the nodal displacements $\mathbf{a}$, the nodal reactions are computed via Eq.(2.25), or by Eq.(4.60a) with the following expression for the internal nodal force vector for the element

$$\mathbf{f}_{\text{int}}^{(e)} = 2\pi \iint_{A^{(e)}} \mathbf{B}^T \boldsymbol{\sigma} r \, dA \tag{7.30}$$

Eq.(7.30) is deduced from the first integral of Eq.(7.25a).

7.3.4 The stiffness matrix for the 3-noded triangle

Box 7.1 shows the integral expression of $\mathbf{K}_{ij}^{(e)}$ for the 3-noded triangle. Note that the integrand contains the terms $r, z, \frac{1}{r}, \frac{z}{r}$ and $\frac{z^2}{r}$. The exact integration of these terms over the element area is given in Box 7.2 [Ya,Ut].

$$\mathbf{K}_{ij}^{(e)} = \frac{\pi}{2(A^{(e)})^2} \iint_{A^{(e)}} \begin{bmatrix} (d_{11}b_ib_j + d_{44}c_ic_j)r+ & (d_{12}b_ic_j + d_{44}c_ib_j)r+ \\ +2A^{(e)}(d_{13}b_iN_j + d_{31}b_jN_i)+ & +2A^{(e)}d_{32}c_jN_i \\ +4(A^{(e)})^2 d_{33}\frac{N_iN_j}{r} & \\ (d_{21}c_ib_j + d_{44}b_ic_j)r+ & (d_{22}c_ic_j + d_{44}b_ib_j)r \\ +2A^{(e)}d_{23}c_iN_j & \end{bmatrix} drdz$$

d_{ij}: Element ij of the constitutive matrix $\mathbf{D}$
b_i, c_i: Coefficients of the shape function N_i

Box 7.1 Stiffness matrix for a 3-noded axisymmetric triangular element

The parameters C_{ij}, D_{ij} and E_{ij} of Box 7.2 are indeterminate for nodes belonging to the axis of symmetry or to an element side parallel to it. This problem can be overcome as follows.

Case 1. $r_i = 0$ and $r_i, r_j \neq 0$. In this case the axial displacement $u_i = 0$ and the element stiffness matrix can be reduced to a 5×5 matrix by eliminating the row and column corresponding to u_i. The I_4, I_5 and I_6 parameters of Box 7.2 are now:

$$I_4 = C_{jk} - z_i \ln \frac{r_j}{r_k}$$

$$I_5 = D_{jk} - \frac{1}{4}(z_j - z_i)(3z_i + z_j) - \frac{1}{4}(z_i - z_4)(3z_i + z_k) - \frac{1}{2}z_i^2 \ln \frac{r_j}{r_k}$$

$$I_6 = E_{jk} - \frac{1}{18}(z_j - z_i)(11z_i^2 + 5z_iz_j + 2z_j^2) - \\ -\frac{1}{18}(z_i - z_k)(11z_i^2 + 5z_iz_k + 2z_k^2) - \frac{1}{3}z_i^3 \ln \frac{r_j}{r_k} \tag{7.31}$$

Case 2 $r_i = r_j = 0, \quad r_k \neq 0$. Now u_i and u_j vanish and the stiffness matrix dimension can be reduced to 4×4 by eliminating the rows and columns corresponding to u_i and u_j. In this case $N_k = \frac{r}{r_k}$ and the integral terms I_4, I_5 and I_6 of Box 7.2 do not appear in $\mathbf{K}_{ij}^{(e)}$ [Ut].

Case 3. Side ij parallel to the symmetry axis $z \; : \; r_i - r_j = 0$. This situation can be overcome simply by using the L'Hopital rule for the

$$I_1 = \iint_{A^{(e)}} drdz = \frac{1}{2}\begin{vmatrix} 1 & 1 & 1 \\ r_1 & r_2 & r_3 \\ z_1 & z_2 & z_3 \end{vmatrix} \quad ; \quad I_4 = \iint_{A^{(e)}} \frac{drdz}{r} = C_{12} + C_{23} + C_{31}$$

$$I_2 = \iint_{A^{(e)}} rdrdz = \frac{r_1 + r_2 + r_3}{3} I_2 \quad ; \quad I_5 = \iint_{A^{(e)}} \frac{z}{r} drdz = D_{12} + D_{23} + D_{31}$$

$$I_3 = \iint_{A^{(e)}} zdrdz = \frac{z_1 + z_2 + z_3}{3} I_1 \quad ; \quad I_6 = \iint_{A^{(e)}} \frac{z^2}{r} drdz = E_{12} + E_{23} + E_{31}$$

with

$$C_{ij} = \frac{r_i z_j - r_j z_i}{r_i - r_j} \ln \frac{r_i}{r_j}$$

$$D_{ij} = \frac{z_j - z_i}{4(r_i - r_j)} \left[z_i(3r_j - r_i) - z_j(3r_i - r_j) \right] + \frac{1}{2} \left(\frac{r_i z_j - r_j z_i}{r_i - r_j} \right)^2 \ln \frac{r_i}{r_j}$$

$$E_{ij} = \frac{z_i - z_j}{18(r_i - r_j)^2} \left[z_j^2(11r_i^2 - 7r_i r_j + 2r_j^2) + z_i z_j (5r_i^2 - 22r_i r_j + 5r_j^2) + \right.$$

$$\left. + z_i^2(11r_j^2 - 7r_i r_j + 2r_i^2) \right] + \frac{1}{3} \left(\frac{r_i z_j - r_j z_i}{r_i - r_j} \right)^3 \ln \frac{r_i}{r_j}$$

Box 7.2 Exact values of some integral terms in axisymmetric straight side triangles with vertices i, j, k

computation of C_{ij}, D_{ij} and E_{ij} to give [Ya]

$$C_{ij} = z_j - z_i \quad ; \quad D_{ij} = E_{ij} = 0$$

A practical rule for cases 1 and 2 is to keep the original size of $\mathbf{K}^{(e)}$ and simply make equal to zero the terms of the rows and columns corresponding to the prescribed values of u_i along the axis of symmetry. An arbitrary non-zero value must be then assigned to the corresponding main diagonal terms to avoid the singularity of the stiffness matrix.

The element integrals can also be computed numerically using a Gauss quadrature. The singularity of the term $\frac{u}{r}$ is avoided by choosing a quadrature not containing points along the axis of symmetry. The approximation of the integrals I_4, I_5 and I_6 of Box 7.2 also requires a larger number of integration points. Excellent results can however be obtained for the 3-noded axisymmetric triangle (with homogeneous material properties) by computing the element integrals with a *single integration point.*

The reduced one-point integration allows us to derive a simple explicit form of the stiffness matrix for the 3-noded triangle as

$$\mathbf{K}^{(e)} = \overline{\mathbf{B}}^T \, \overline{\mathbf{D}} \, \overline{\mathbf{B}} \, \bar{r} \, A^{(e)} \tag{7.32}$$

where $\overline{(\cdot)}$ denotes values computed at the element centroid. The analytical expression of $\mathbf{K}_{ij}^{(e)}$ for this case can be deduced directly from Box 7.1 by substituting r for $\bar{r}$, N_i for $\frac{1}{3}$, and the integral for the integrand multiplied by the element area. The one point quadrature is exact for all terms of $\mathbf{K}_{ij}^{(e)}$ with the exception of $\frac{N_i N_j}{r}$. The error in the evaluation of this term does not really affect the behaviour of the element and very good results are obtained using a finer mesh in zones where higher stresses are expected [ZTZ].

The reduced one-point quadrature preserves the correct rank of the stiffness matrix and hence the element is free from spurious mechanisms.

7.3.5 Equivalent nodal force vectors for the 3-noded triangle

Body forces and surface tractions

From Eq.(7.29) we deduce

$$\mathbf{f}_b = 2\pi \iint_{A^{(e)}} \mathbf{N}^T \mathbf{b} r \, dA \tag{7.33a}$$

$$\mathbf{f}_t = 2\pi \oint_{l^{(e)}} \mathbf{N}^T \mathbf{t} r \, ds \tag{7.33b}$$

The expressions for $\mathbf{f}_b$ and $\mathbf{f}_t$ for the 3-noded triangle with nodes i, j, k of Figure 7.7 simplify if the acting forces are constant over the element. Substituting the shape functions (Eqs.(4.33) and (7.18)) into (7.31) and assuming that the surface load $\mathbf{t}$ acts on the side ij gives

$$\mathbf{f}_b^{(e)} = \frac{\pi A^{(e)}}{6} \begin{Bmatrix} (2r_i + r_j + r_k) \, b_r \\ (2r_i + r_j + r_k) \, b_z \\ (r_i + 2r_j + r_k) \, b_r \\ (r_i + 2r_j + r_k) \, b_z \\ (r_i + r_j + 2r_k) \, b_r \\ (r_i + r_j + 2r_k) \, b_z \end{Bmatrix} \quad ; \quad \mathbf{f}_t^{(e)} = \frac{\pi l_{ij}^{(e)}}{3} \begin{Bmatrix} (2r_i + r_j) \, t_r \\ (2r_i + r_j) \, t_z \\ (r_i + 2r_j) \, t_r \\ (r_i + 2r_j) \, t_z \\ 0 \\ 0 \end{Bmatrix} \tag{7.34a}$$

where $l_{ij}^{(e)}$ is the length of side ij subjected to the surface tractions. Note that the nodal forces due to body forces and surface tractions are not distributed in equal parts between the nodes as it occurs for the plane elasticity triangle. The highest load intensity now corresponds to the node situated further from the symmetry axis.

The expressions for $\mathbf{f}_t^{(e)}$ due to surface tractions acting over the sides jk and ik are respectively

$$\mathbf{f}_t^{(e)} = \frac{\pi l_{jk}^{(e)}}{3} \begin{Bmatrix} 0 \\ 0 \\ (2r_j + t_k)\ t_r \\ (2r_j + r_k)\ t_z \\ (r_j + 2r_k)\ t_r \\ (r_j + 2r_k)\ t_z \end{Bmatrix} \quad ; \quad \mathbf{f}_t^{(e)} = \frac{\pi l_{ik}^{(e)}}{3} \begin{Bmatrix} (2r_i + r_k)\ t_r \\ (2r_i + r_k)\ t_z \\ 0 \\ 0 \\ (r_i + 2r_k)\ t_r \\ (r_i + 2r_k)\ t_z \end{Bmatrix} \tag{7.34b}$$

Initial strain force vector

From Eq.(7.29) we deduce

$$\mathbf{f}_\varepsilon^{(e)} = 2\pi \iint_{A^{(e)}} \mathbf{B}^T\ \mathbf{D}\ \boldsymbol{\varepsilon}^0\ r\ dA \tag{7.35}$$

The integral in Eq.(7.35) for the 3-noded triangle can be computed using the expressions for I_1, I_2 and I_3 from Box 7.2. For thermal initial strains and homogeneous material properties we find that

$$\mathbf{f}_{\varepsilon_i}^{(e)} = \pi\ \alpha \Delta T \begin{Bmatrix} (d_{11} + d_{12} + d_{13}) b_i\ \bar{r} + \dfrac{2A^{(e)}}{3}(d_{31} + d_{32} + d_{33}) \\ (d_{21} + d_{22} + d_{23}) c_i\ \bar{r} \end{Bmatrix} \tag{7.36}$$

where $\bar{r}$ is the radial coordinate of the element centroid and d_{ij} are the terms of the constitutive matrix of Eqs.(7.7) or (7.9).

Initial stress force vector

Eq.(7.29) gives

$$\mathbf{f}_\sigma^{(e)} = -2\pi \int_{A^{(e)}} \mathbf{B}^T\ \boldsymbol{\sigma}^0\ r\ dA \tag{7.37a}$$

The exact form of $\mathbf{f}_{\sigma_i}^{(e)}$ for the 3-noded triangle under a *constant* initial stress field is

$$\mathbf{f}_{\sigma_i}^{(e)} = -\pi \begin{Bmatrix} (b_i\ \sigma_r^0 + c_i\ \tau_{rz}^0)\bar{r} + \frac{2}{3} A^{(e)}\ \sigma_\theta^0 \\ (c_i\ \sigma_z^0 + b_i\ \tau_{rz}^0)\bar{r} \end{Bmatrix} \tag{7.37b}$$

where again $\bar{r}$ is the radial coordinate of the element centroid.

Circular point loads

The nodal point vector is deduced from the PVW (Eq.(7.6)) as

$$\mathbf{p}_i = 2\pi r_i [P_{r_i}, P_{z_i}]^T \tag{7.38}$$

where P_{r_i} and P_{z_i} are the intensities of the radial and vertical point loads acting uniformly along a circumferential line at node i. Consequently, the nodal reactions per unit circumferential length corresponding to the prescribed displacements are computed by dividing the value of the total reactions by $2\pi r_i$, where r_i is the radial coordinate of the node.

7.4 OTHER RECTANGULAR OR STRAIGHT-SIDED TRIANGULAR AXISYMMETRIC SOLID ELEMENTS

Eqs.(7.28) and (7.29) provide the general expressions for the stiffness matrix and the equivalent nodal force vector for any axisymmetric solid element. The computation of these expressions for rectangular and straight-sided triangles is simple and analytical forms can be obtained in most cases as shown in the following examples. However, it might become more complex when integrating terms such as $\frac{z^m}{r}$. In practice it is more convenient to define the element integrals in terms of the natural coordinate system and to use numerical integration.

Example 7.1: Obtain the stiffness matrix and the equivalent nodal force vector for a 4-noded axisymmetric rectangle.

- ***Solution***

Eq.(4.80) for the shape functions in cartesian coordinates will be used, i.e.

$$N_1 = \frac{1}{4}(1 - \frac{r}{a})(1 - \frac{z}{b}) \quad ; \quad N_2 = \frac{1}{4}(1 + \frac{r}{a})(1 - \frac{z}{b})$$
$$N_3 = \frac{1}{4}(1 + \frac{r}{a})(1 + \frac{z}{b}) \quad ; \quad N_4 = \frac{1}{4}(1 - \frac{r}{a})(1 + \frac{z}{b})$$

The strain matrix of Eq.(7.20) is

$$\mathbf{B} = \left[\mathbf{B}_1, \mathbf{B}_2, \mathbf{B}_3, \mathbf{B}_4\right]$$

where, for instance,

$$\mathbf{B}_1 = \begin{bmatrix} -\frac{1}{4a}(1-\frac{z}{b}) & 0 \\ 0 & -\frac{1}{4b}(1-\frac{r}{a}) \\ \frac{(1-\frac{r}{a})(1-\frac{z}{b})}{4r} & 0 \\ -\frac{1}{4b}(1-\frac{r}{a}) & -\frac{1}{4a}(1-\frac{z}{b}) \end{bmatrix}$$

with similar expressions for $\mathbf{B}_2, \mathbf{B}_3$ and $\mathbf{B}_4$.
The element stiffness matrix is obtained from Eq.(7.28). The expression of $\mathbf{B}_i$ shows that the stiffness matrix integrals contain terms like $r, z, z^2, \frac{1}{r}, \frac{z}{r}, \frac{z^2}{r}$. The computation of the integrals over the element area can be performed analytically as explained in Section 5.6. The expressions from Box 7.2 are also applicable after splitting the element into two triangular regions.
The equivalent nodal force vector due to body forces is deduced from Eq.(7.33a) as

$$\mathbf{f}_b^{(e)} = 2\pi \iint_{A^{(e)}} [\mathbf{N}_1, \mathbf{N}_2, \mathbf{N}_3, \mathbf{N}_4]^T \; \mathbf{b} \; r \; dr \; dz$$

with

$$\mathbf{f}_{b_i}^{(e)} = 2\pi \iint_{A^{(e)}} \mathbf{N}_i^T \; \mathbf{b} \; r \; dr \; dz$$

The integration of the terms in r, r^2 and rz in above expression can be obtained analytically as explained in Section 5.6.

Example 7.2: Compute the stiffness matrix and the equivalent nodal body force vector for a straight-sided 6-noded axisymmetric triangle.

- ***Solution***

The shape functions for the 6-noded triangle written in area coordinates are (Eq.(5.34))

$$N_1 = 2(L_1 - 1)L_1 \quad ; \quad N_2 = 2(L_2 - 1)L_2 \quad ; \quad N_3 = 2(L_3 - 1)L_3$$
$$N_4 = 4L_1L_2 \quad ; \quad N_5 = 4L_2L_3 \quad ; \quad N_6 = 4L_1L_3$$

where

$$L_i = \frac{1}{2A^{(e)}}(a_i + b_i \; r + c_i \; z) \quad ; \quad i = 1, 2, 3$$

and a_i, b_i, c_i are the parameters given in Eq.(4.32b) where the x, y coordinates are replaced now by r, z respectively.

The element strain matrix is

$$\mathbf{B} = [\mathbf{B}_1, \mathbf{B}_2, \mathbf{B}_3, \mathbf{B}_4, \mathbf{B}_5, \mathbf{B}_6]$$

The cartesian derivatives of the shape functions are computed by

$$\begin{aligned}\frac{\partial N_i}{\partial r} &= \frac{\partial N_i}{\partial L_1}\frac{\partial L_1}{\partial r} + \frac{\partial N_i}{\partial L_2}\frac{\partial L_2}{\partial r} + \frac{\partial N_i}{\partial L_3}\frac{\partial L_3}{\partial r} = \\ &= \frac{1}{2A^{(e)}}\left[b_1 \frac{\partial N_i}{\partial L_1} + b_2 \frac{\partial N_i}{\partial L_2} + b_3 \frac{\partial N_i}{\partial L_3}\right] = B_i\end{aligned}$$

Similarly we obtain

$$\frac{\partial N_i}{\partial z} = \frac{1}{2A^{(e)}}\left[c_1 \frac{\partial N_i}{\partial L_1} + c_2 \frac{\partial N_i}{\partial L_2} + c_3 \frac{\partial N_i}{\partial L_3}\right] = C_i$$

The nodal strain matrices are given by

$$\mathbf{B}_i = \begin{bmatrix} \frac{\partial N_i}{\partial r} & 0 \\ 0 & \frac{\partial N_i}{\partial z} \\ \frac{N_i}{r} & \\ \frac{\partial N_i}{\partial z} & \frac{\partial N_i}{\partial r} \end{bmatrix} = \begin{bmatrix} B_i & 0 \\ 0 & C_i \\ \frac{N_i}{r} & 0 \\ C_i & B_i \end{bmatrix}$$

For instance

$$\mathbf{B}_1 = \begin{bmatrix} \frac{b_1(2L_1-1)}{A^{(e)}} & 0 \\ 0 & \frac{c_1(2L_1-1)}{A^{e)}} \\ \frac{2(L_1-1)L_1}{r} & 0 \\ \frac{c_1(2L_1-1)}{A^{(e)}} & \frac{b_1(2L_1-1)}{A^{(e)}} \end{bmatrix} \qquad \text{etc.}$$

The element stiffness matrix is obtained by Eq.(7.28). The analytical computation of the integrals can be simplified by expressing the integrands in terms of the cartesian coordinates. The terms appearing in the integrals are of the type $r^m z^n$ and $\frac{z^n}{r}$, with $m = 0, 1, 2$ and $n = 0, 1, 2, 3, 4$. The integral of the terms $r^m z^n$ is directly obtained by Eq.(5.40). The exact integration of the terms $\frac{z^n}{r}$, $n = 0, 1, 2$ is given in Box 7.2. The analytical integration of the terms $\frac{z^3}{r}$ and $\frac{z^4}{r}$ is more complicated and it is simpler to use numerical integration.
The equivalent nodal force vector for body forces is given by

$$\mathbf{f}_b^{(e)} = 2\pi \iint_{A^{(e)}} [N_1, N_2, N_3, N_4, N_5, N_6]^T \, \mathbf{b} \, r \, dr \, dz$$

The integrals of $\mathbf{f}_{b_i}^{(e)}$ contain terms such as $r^m z^n$ with $m, n = 0, 1, 2$ which can be computed exactly by Eq.(5.40).

We conclude from the above examples that the integration of the matrices and vectors for rectangular and straight-sided triangular axisymmetric elements is simple and involves the integration of terms like $r^m z^n$ and $\frac{z^n}{r}$ with $m, n = 0, 1, 2 \ldots p$ $(m + n \leq p)$, where p is the highest degree polynomial contained in the shape functions. The analytical integration of the terms $r^m\ z^n$ can be performed using the expressions given in Section 5.6. The exact integration of the terms $\frac{z^n}{r}$ over straight side triangles is given in Box 7.2 for $n \leq 2$. For $n > 2$ it is recommended that the integrals are computed approximately using numerical integration.

Numerical integration is also mandatory for arbitrary quadrilateral shapes and curve-sided elements.

7.5 ISOPARAMETRIC AXISYMMETRIC SOLID ELEMENTS

The formulation of isoparametric axisymmetric solid elements follows precisely the lines explained for 2D solid elements (Chapter 6). The axial and vertical coordinates of an isoparametric axisymmetric solid element with n nodes are expressed in terms of the nodal values as

$$\mathbf{x} = \begin{Bmatrix} r \\ z \end{Bmatrix} = \sum_{i=1}^{n} \mathbf{N}_i\ \mathbf{x}_i^{(e)} \tag{7.39a}$$

with

$$\mathbf{N}_i = \begin{bmatrix} N_i & 0 \\ 0 & N_i \end{bmatrix} \quad ; \quad \mathbf{x}_i^{(e)} = \begin{Bmatrix} r_i \\ z_i \end{Bmatrix} \tag{7.39b}$$

The cartesian derivatives of the shape functions are obtained in terms of the natural coordinates ξ, η as explained for 2D solid elements simply replacing the coordinates x, y by r, z, respectively. The integrals in the element stiffness matrix are expressed in the natural coordinate system as

$$\begin{aligned} \mathbf{K}_{ij}^{(e)} &= 2\pi \iint_{A^{(e)}} \mathbf{B}_i^T\ \mathbf{D}\ \mathbf{B}_j\ r\ drdz = \\ &= 2\pi \int_{-1}^{+1} \int_{-1}^{+1} \mathbf{B}_i^T\ \mathbf{D}\ \mathbf{B}_j \left(\sum_{k=1}^{n} N_k\ r_k \right) \left| \mathbf{J}^{(e)} \right| d\xi\ d\eta \end{aligned} \tag{7.40}$$

Similarly, for the equivalent nodal force vector we have

$$\begin{aligned}\mathbf{f}_{b_i}^{(e)} &= 2\pi \iint_{A^{(e)}} \mathbf{N}_i^T \; \mathbf{b} \; r \; drdz = \\ &= 2\pi \int_{-1}^{+1} \int_{-1}^{+1} \mathbf{N}_i^T \; \mathbf{b} \left(\sum_{k=1}^{n} N_k r_k \right) \left| \mathbf{J}^{(e)} \right| \, d\xi \; d\eta \end{aligned} \tag{7.41}$$

The expression of the Jacobian matrix $\mathbf{J}^{(e)}$ is deduced from Eq.(6.3) simply changing the coordinates x and y by r and z, respectively.

As explained for 2D solid elements the integrals in Eqs.(7.40) and (7.41) contain rational algebraic functions in the natural coordinates. An exception is the case of rectangles and straight-sided triangles discussed previously. In general, the element integrals are computed numerically.

A $n_p \times n_q$ quadrature over quadrilaterals gives (Sections 6.4.1 and 6.5.1)

$$\begin{aligned}\mathbf{K}_{ij}^{(e)} &= 2\pi \sum_{p=1}^{n_p} \sum_{q=1}^{n_q} \left[\mathbf{B}_i^T \; \mathbf{D} \; \mathbf{B}_j \Big(\sum_{k=1}^{n} N_k r_k \Big) |\mathbf{J}^{(e)}| \right]_{p,q} W_p W_q \\ \mathbf{f}_{b_i}^{(e)} &= 2\pi \sum_{p=1}^{n_p} \sum_{q=1}^{n_q} \left[\mathbf{N}_i^T \; \mathbf{b} \Big(\sum_{k=1}^{n} N_k r_k \Big) |\mathbf{J}^{(e)}| \right]_{p,q} W_p \; W_q \end{aligned} \tag{7.42}$$

For triangular elements the Gauss quadrature is (Sections 6.4.2 and 6.5.2)

$$\begin{aligned}\mathbf{K}_{ij}^{(e)} &= 2\pi \sum_{p=1}^{n_p} \left[\mathbf{B}_i^T \mathbf{D} \mathbf{B}_j \left(\sum_{k=1}^{n} N_k r_k \right) |\mathbf{J}^{(e)}| \right]_p W_p \\ \mathbf{f}_{b_i}^{(e)} &= 2\pi \sum_{p=1}^{n_p} \left[\mathbf{N}_i^T \mathbf{b} \left(\sum_{k=1}^{n} N_k r_k \right) |\mathbf{J}^{(e)}| \right]_p W_p \end{aligned} \tag{7.43}$$

The only difference between the Gauss quadrature for axisymmetric solid elements and plane elasticity elements is the presence of the radial coordinate within the integrals in the former.

7.6 ANALOGIES BETWEEN THE FINITE ELEMENT FORMULATIONS FOR PLANE ELASTICITY AND AXISYMMETRIC SOLIDS

Axisymmetric solids are conceptually different from plane elasticity structures. However, the finite element methodology for both cases shares many common features. For instance, the strain matrices are very similar as shown below.

$\boxed{\mathbf{B}_i \ matrix}$

Plane elasticity **Axisymmetric solids**

$$\begin{bmatrix} \frac{\partial N_i}{\partial x} & 0 \\ 0 & \frac{\partial N_i}{\partial y} \\ \frac{\partial N_i}{\partial y} & \frac{\partial N_i}{\partial x} \end{bmatrix} \begin{matrix} \varepsilon_x \\ \varepsilon_y \\ \gamma_{xy} \end{matrix} \qquad \begin{bmatrix} \frac{\partial N_i}{\partial r} & 0 \\ 0 & \frac{\partial N_i}{\partial z} \\ \frac{\partial N_i}{\partial z} & \frac{\partial N_i}{\partial r} \\ \cdots\cdots & \cdots\cdots \\ \frac{N_i}{r} & 0 \end{bmatrix} \begin{matrix} \varepsilon_r \\ \varepsilon_z \\ \gamma_{rz} \\ \\ \varepsilon_\theta \end{matrix} \tag{7.45}$$

The first three rows of both matrices coincide if the coordinates r, z are replaced by x, y. The fourth row term $\frac{N_i}{r}$ corresponding to the circumferential strain in axisymmetric solids tends to zero for large values of r and thus in the limit case $(r \rightarrow \infty)$ the non-zero terms of both matrices coincide. This coincidence has a clear physical meaning as the behaviour of an axisymmetric solid with a large radius resembles to that of a prismatic solid under plane strain conditions. In fact, a prismatic solid can always be considered as part of an axisymmetric solid with an infinite radius and the analogy between the two problems is clear in this case.

These analogies extend to the element integrals. It is deduced from Eqs.(4.61) and (7.28) that both expressions are identical if the term $2\pi r dr dz$, expressing the area differential in axisymmetric solids, is replaced by $t dx dy$ for plane elasticity. It is therefore very simple to write a computer program for solving both types of problems in a unified manner. See Chapter 11 and [Hu,HO,HO2].

7.7 EXAMPLES OF APPLICATION

7.7.1 Infinitely long cylinder under external pressure

This example shows the analysis of an infinitely long thick cylinder under external pressure. This is a plane strain problem and, thus, only a single slice needs to be analyzed as shown in Figure 7.8. The study has been performed using twenty 3-noded axisymmetric triangles with the mesh and boundary conditions shown in the figure.

Figure 7.8 shows the results obtained for the radial circumferential stress and the axial stress distributions along a radial line. Excellent agreement with the exact solution [CR] is obtained in all cases.

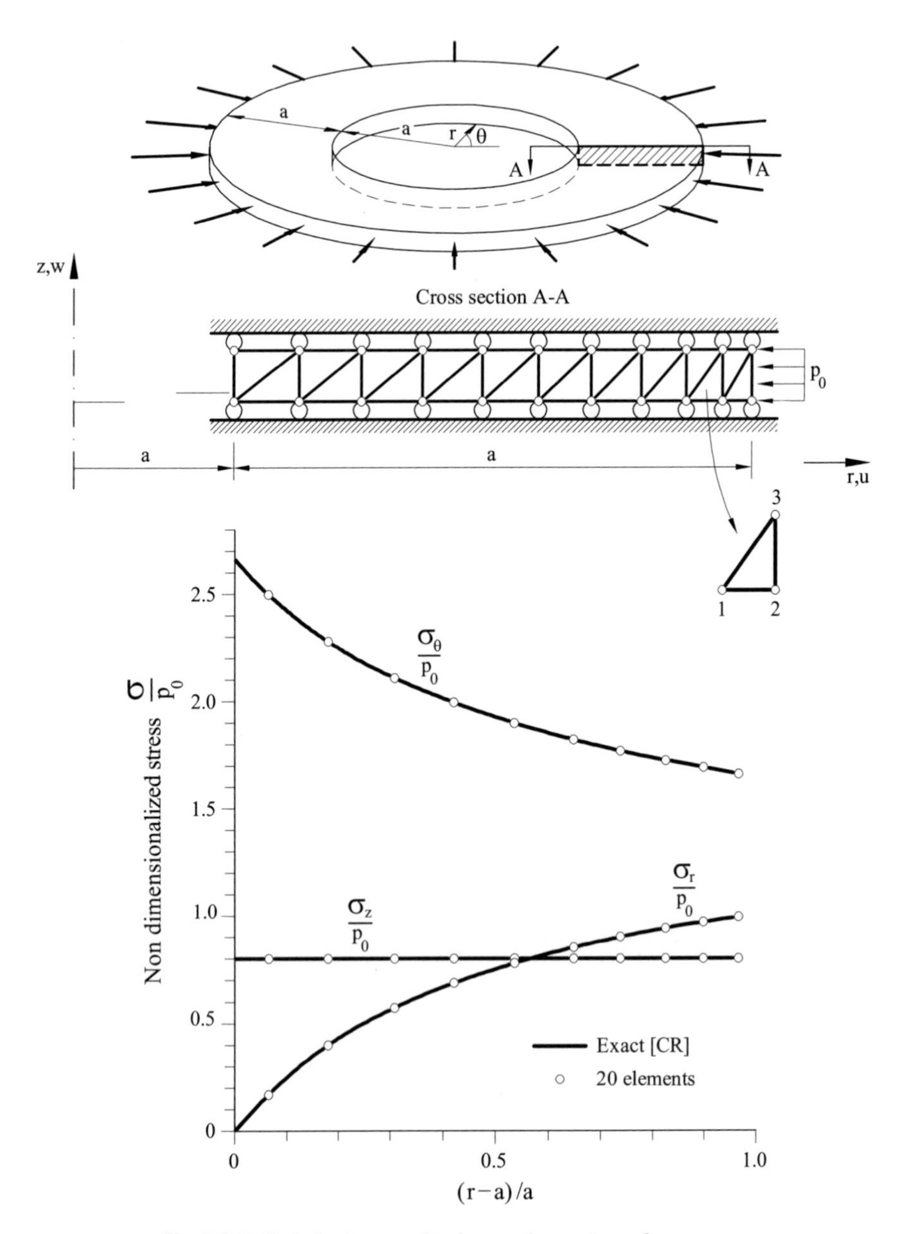

Fig. 7.8 Infinitely long cylinder under external pressure

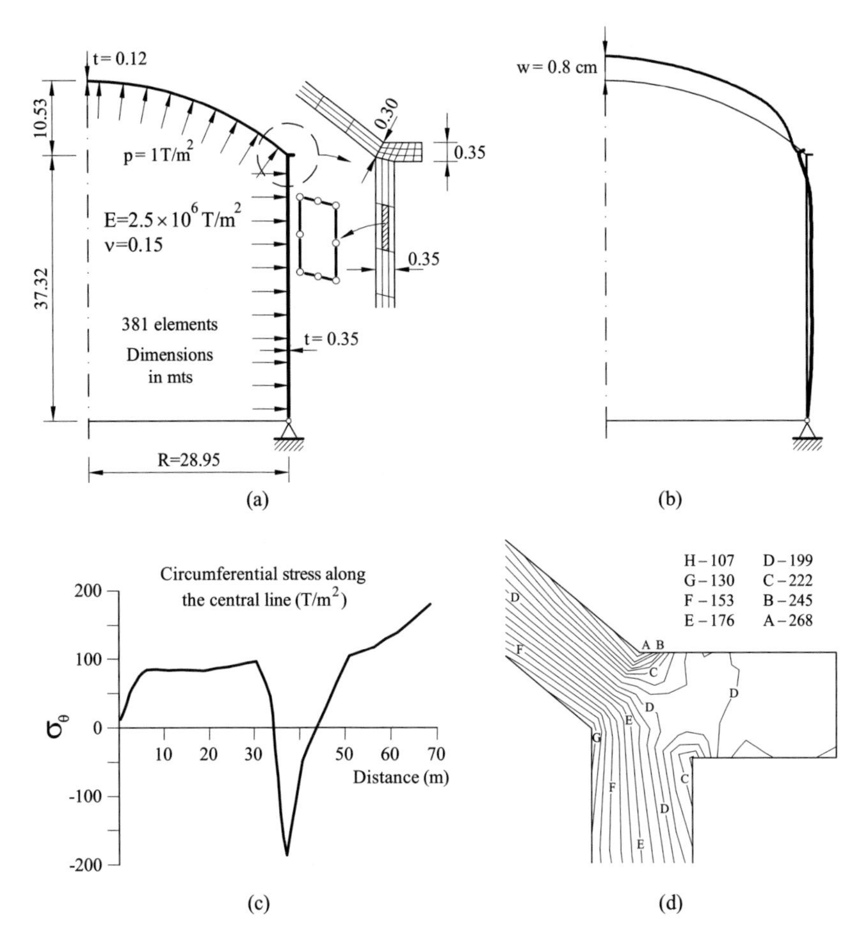

Fig. 7.9 Cylindrical tank with spherical dome under internal pressure. (a) Geometry and mesh. (b) Deformed mesh. (c) Distribution of the circumferential stress σ_θ along the central line. (d) Isolines for σ_θ in the wall-dome joint region

7.7.2 Cylindrical tank with spherical dome under internal pressure

Figure 7.9a shows the geometry of the tank and the mechanical properties. An internal pressure of $1\mathrm{T/m}^2$ acts along the inner wall. Clamped boundary conditions ($u = w = 0$) have been imposed on the base nodes. A regular mesh of 381 8-noded Serendipity axisymmetric quadrilaterals has been used. Three elements have been used to discretize the wall thickness.

The deformed mesh is plotted in Figure 7.9b. Figure 7.9c shows the distribution of the circumferential stress along the central line of the meridional section. Note that most of the cylindrical wall is subjected to a uniform tensile circumferential stress state, whereas high compression stress gradients occur in the vicinity of the wall-dome joint. Figure 7.9d shows a detail of the isolines for the circumferential stress in that region.

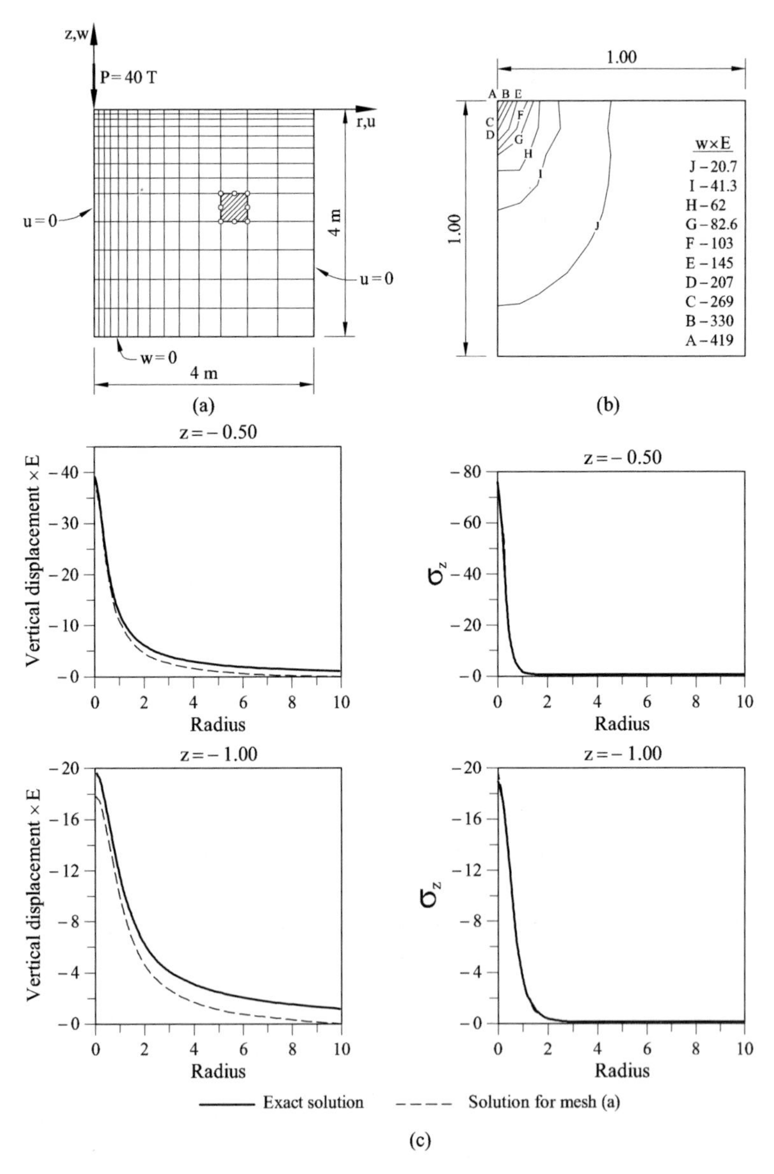

Fig. 7.10 Boussinesq problem. (a) Geometry and mesh. (b) Isolines for the vertical displacement ($w \times E$) in the vicinity of the load. (c) Radial distributions for σ_z and the vertical displacement for $z = -0.50$ and $z = -1.00$ ($\nu = 0.20$)

7.7.3 Semi-infinite elastic space under point load

This example corresponds to the well known Boussinesq problem of a semi-infinite elastic space under a point load. The analysis has been performed for a finite domain of 4×4 mts. discretized with a mesh of 8-noded Serendipity rectangles as shown in Figure 7.10a. The isolines for the vertical displacement in the vicinity of the load are plotted in Figure 7.10b. High displacement gradients are obtained in this region as expected, since the theoretical value of the vertical displacement under the load is infinity. Figure 7.10c shows the distributions of the stress σ_z and the vertical displacement w along two horizontal lines corresponding to the distances $z = -0.50$ and $-1.0m$. respectively. The comparison of the numerical results with the theoretical solution [TG] is very good in both cases.

7.8 CONCLUDING REMARKS

The finite element analysis of axisymmetric solids shares many features with that of the plane elasticity solids studied in Chapters 4–6. In particular, the displacement interpolation and the derivation of the element stiffness matrix and the equivalent nodal force vector are very similar in both cases. This makes it simple to organize a computer program that is valid for both problems.

Axisymmetric solid elements behave very similarly to 2D elasticity elements: quadrilateral elements are generally more precise than triangles, and the quadratic elements perform better than the linear ones. Here again the simple 3-noded axisymmetric triangle is the more versatile element for the discretization of complex axisymmetric geometries using unstructured meshes, as well as for using adaptive mesh refinement techniques. Moreover, the linear triangle is highly accurate in tension or compression dominated problems (Example 7.1). All this has contributed to the popularity of the 3-noded triangle for the analysis of axisymmetric solids.

8

THREE DIMENSIONAL SOLIDS

8.1 INTRODUCTION

Many structures have geometrical, mechanical or loading features which make it impossible to use the simple plane stress/plane strain and axi-symmetric models studied in previous chapters; or even the plate and shell models to be described in the second volume of the book [On]. The only alternative is to perform a full three dimensional (3D) analysis based on general 3D elasticity theory [TG].

Examples of these situations are found in solids with irregular shapes and in the study of prismatic solids with heterogeneous material properties or arbitrary loading. Figure 8.1 shows some examples of typical structures requiring a full 3D analysis.

Despite its apparent complexity, the analysis of a 3D solid with the FEM does not introduce major conceptual problems. 3D elasticity theory is a straightforward extension of the 2D case and the steps involved in the 3D finite element analysis of a structure are a repetition of those studied in Chapters 4–7. In that respect, this chapter closes the cycle of structural problems which can be analyzed using elasticity theory, either by the general 3D form or by any of the simplified 2D cases previously studied.

Although conceptually simple, 3D finite element computations involve a considerable amount of work in comparison with 2D analyses. The principal reason is the introduction of an additional space dimension, for this leads to greater computational time as well as requiring more effort to input data and visualize the results. Consequently, 3D analyses tend to be avoided in practice whenever possible in favour of simpler 2D solutions. Unfortunately this is not possible in many practical situations which require a 3D analysis.

The first part of the chapter introduces the basic concepts of 3D elasticity theory necessary for application of the FEM. Details of the derivation of the element stiffness matrix and the nodal load vectors are given for the general case and are particularized for the simple 4-noded tetrahedral

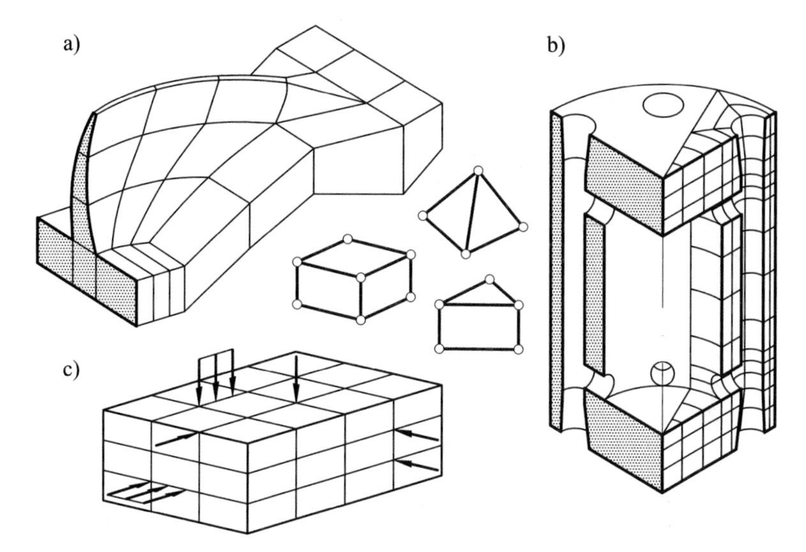

Fig. 8.1 Structures which require a 3D analysis: (a) Double arch dam including foundation effects. (b) Pressure vessel. (c) Prismatic solid under arbitrary loading

element. The derivation of the shape functions for hexahedral and tetrahedral elements is then explained. Next, the formulation of 3D isoparametric elements is detailed. Finally, some examples of practical application of 3D finite element analysis are given.

8.2 BASIC THEORY

8.2.1 Displacement field

Let us consider the 3D solid shown in Figure 8.2. The movement of a point is defined by the three components of the displacement vector, i.e.

$$\mathbf{u} = [u, v, w]^T \tag{8.1}$$

where u, v, w are the displacements of the point in the directions of the cartesian axes x, y, z, respectively.

8.2.2 Strain field

The strain field is defined by the standard six strain components of 3D elasticity [TG]. The strain vector is written as

$$\boldsymbol{\varepsilon} = \left[\varepsilon_x, \varepsilon_y, \varepsilon_z, \gamma_{xy}, \gamma_{xz}, \gamma_{yz}\right]^T \tag{8.2}$$

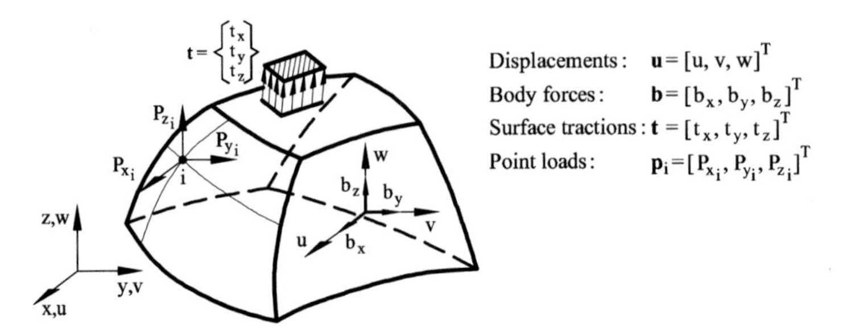

Fig. 8.2 3D solid. Displacements and loads

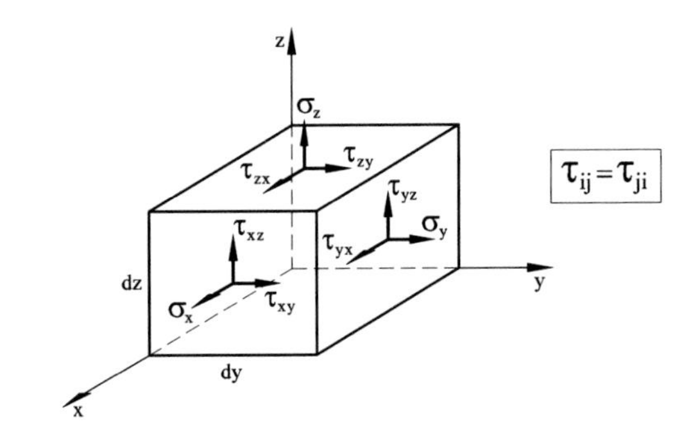

Fig. 8.3 Sign criterion for the stresses in a 3D solid

with

$$\begin{aligned} \varepsilon_x &= \frac{\partial u}{\partial x} \quad ; \quad \varepsilon_y = \frac{\partial v}{\partial y} \quad ; \quad \varepsilon_z = \frac{\partial w}{\partial z} \\ \gamma_{xy} &= \frac{\partial u}{\partial y} + \frac{\partial v}{\partial x} \quad ; \quad \gamma_{xz} = \frac{\partial u}{\partial z} + \frac{\partial w}{\partial x} \quad ; \quad \gamma_{yz} = \frac{\partial v}{\partial z} + \frac{\partial w}{\partial y} \end{aligned} \tag{8.3}$$

where $\varepsilon_x, \varepsilon_y, \varepsilon_z$ are the normal strains and $\gamma_{xy}, \gamma_{xz}, \gamma_{yz}$ are the tangential strains.

8.2.3 Stress field

The stress field is defined by the six stress components which are conjugate of the six non-zero strains of Eq.(8.2). The stress vector is

$$\boldsymbol{\sigma} = \left[\sigma_x, \sigma_y, \sigma_z, \tau_{xy}, \tau_{xz}, \tau_{yz}\right]^T \tag{8.4}$$

where $\sigma_x, \sigma_y, \sigma_z$ are the normal stresses and $\tau_{xy}, \tau_{xz}, \tau_{yz}$ are the tangential stresses. Note that $\tau_{ij} = \tau_{ji}$. For the sign criteria see Figure 8.3.

8.2.4 Stress-strain relationship

The stress-strain relationship is expressed for the general case of anisotropic elasticity by a 6×6 symmetric constitutive matrix with 21 independent parameters. For *orthotropic materials* with principal ortrotopy directions x', y', z' the constitutive equation is written as

$$\varepsilon_1' = \frac{1}{E_1}\sigma_1' - \frac{\nu_{21}}{E_2}\sigma_2' - \frac{\nu_{31}}{E_3}\sigma_3' \quad , \quad \varepsilon_2' = \frac{1}{E_2}\sigma_2' - \frac{\nu_{12}}{E_1}\sigma_1' - \frac{\nu_{32}}{E_3}\sigma_3'$$

$$\varepsilon_3' = \frac{1}{E_3}\sigma_3' - \frac{\nu_{13}}{E_1}\sigma_1' - \frac{\nu_{23}}{E_2}\sigma_2' \quad , \quad \gamma_{12} = \frac{\tau_{12}}{G_{12}} \; ; \; \gamma_{13} = \frac{\tau_{13}}{G_{13}} \; ; \; \gamma_{23} = \frac{\tau_{23}}{G_{23}} \tag{8.5a}$$

The symmetry of the constitutive matrix requires

$$E_1\nu_{21} = E_2\nu_{12} \quad ; \quad E_2\nu_{32} = E_3\nu_{23} \quad ; \quad E_3\nu_{13} = E_1\nu_{31} \tag{8.5b}$$

and the total number of independent material parameters reduce to nine for this case.

The number of material parameters reduces further to five for an orthotropic material in the plane 1-2 and isotropic in the plane 23. This situation is typical of fiber-reinforced composite materials. The transformation of the local constitutive matrix from the principal orthotropy axes to the global cartesian axes x, y, z follows the procedure explained in Section 4.2.4 for plane problems.

Isotropic materials require two material parameters only, the Young modulus E and the Poisson's ratio ν. The constitutive matrix for isotropic materials can be directly written in global cartesian axes. If initial strains and stresses are taken into account we can write

$$\boldsymbol{\sigma} = \mathbf{D}\,(\boldsymbol{\varepsilon} - \boldsymbol{\varepsilon}^0) + \boldsymbol{\sigma}^0 \tag{8.6}$$

where the isotropic constitutive matrix $\mathbf{D}$ is given by

$$\mathbf{D} = \frac{E(1-\nu)}{(1+\nu)(1-2\nu)} \begin{bmatrix} 1 & \frac{\nu}{1-\nu} & \frac{\nu}{1-\nu} & 0 & 0 & 0 \\ & 1 & \frac{\nu}{1-\nu} & 0 & 0 & 0 \\ & & 1 & 0 & 0 & 0 \\ & & & \frac{1-2\nu}{2(1-\nu)} & 0 & 0 \\ & \text{Symmetrical} & & & \frac{1-2\nu}{2(1-\nu)} & 0 \\ & & & & & \frac{1-2\nu}{2(1-\nu)} \end{bmatrix} \tag{8.7}$$

The initial strain vector due to thermal strains is

$$\boldsymbol{\varepsilon}^0 = \alpha(\Delta T)\ [1, 1, 1, 0, 0, 0]^T \tag{8.8}$$

8.2.5 Principal stresses, stress invariants and failure criteria

The principal stresses $\sigma_I, \sigma_{II}, \sigma_{III}$ for 3D solids are the three roots of the characteristic polynomial

$$\det([\sigma] - \lambda \mathbf{I}_3) = 0 \tag{8.9a}$$

where

$$[\sigma] = \begin{bmatrix} \sigma_x & \tau_{xy} & \tau_{xz} \\ \tau_{xy} & \sigma_y & \tau_{yz} \\ \tau_{xz} & \tau_{yz} & \sigma_z \end{bmatrix} \quad , \quad \mathbf{I}_3 = \begin{bmatrix} 1 & 0 & 0 \\ 0 & 1 & 0 \\ 0 & 0 & 1 \end{bmatrix} \tag{8.9b}$$

Eq.(8.9a) can be expressed as

$$\lambda^3 - I_1\lambda^2 + I_2\lambda - I_3 = 0 \tag{8.10a}$$

where

$$\begin{aligned} I_1 &= \sigma_I + \sigma_{II} + \sigma_{III} = \sigma_x + \sigma_y + \sigma_z = t_r[\sigma] \\ I_2 &= \sigma_I\sigma_{II} + \sigma_I\sigma_{III} + \sigma_{II}\sigma_{III} = \frac{1}{2}[(t_r[\sigma])^2 - t_r[\sigma]^2] = \\ &= \sigma_x\sigma_y + \sigma_x\sigma_z + \sigma_y\sigma_z - \tau_{xy}^2 - \tau_{xz}^2 - \tau_{yz}^2 \\ I_3 &= \sigma_I\sigma_{II}\sigma_{III} = \det[\sigma] = \sigma_x\sigma_y\sigma_z + 2\tau_{xy}\tau_{xz}\tau_{yz} - \sigma_x\tau_{yz}^2 - \sigma_y\tau_{xz}^2 - \sigma_z - \tau_{xy}^2 \end{aligned} \tag{8.10b}$$

The three principal stresses are associated to the unit normal vectors $\mathbf{n}_I, \mathbf{n}_{II}, \mathbf{n}_{III}$ defining the principal stress directions (Figure 8.4). These vectors are found solving the system

$$[[\sigma] - \lambda_i \mathbf{I}_3]\,\mathbf{n}_i = 0 \quad \text{with} \quad \mathbf{n}_i^T\mathbf{n}_i = 1 \quad , \quad i = I, II, III \tag{8.11}$$

The principal stresses are independent of the coordinate system. Hence, I_1, I_2 and I_3 are the invariants of tensor $[\sigma]$ for an orthogonal transformation of coordinates.

The mean stress, or hydrostatic stress σ_n is also an invariant

$$\sigma_n = \frac{1}{3}(\sigma_x + \sigma_y + \sigma_z) = \frac{1}{3}(\sigma_I + \sigma_{II} + \sigma_{III}) = \frac{1}{3}I_1 \tag{8.12}$$

Let us introduce now the deviatoric stress tensor

$$[s] = [\sigma] - \sigma_n \mathbf{I}_3 \tag{8.13}$$

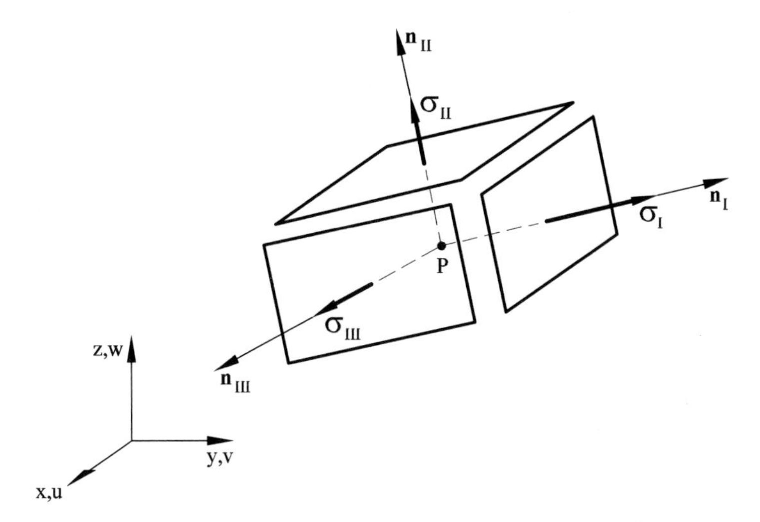

Fig. 8.4 Principal stresses in a 3D solid

Eq.(8.10a) is written as

$$\lambda_s^3 - J_2\lambda_s - J_3 = 0 \quad \text{with} \quad \lambda_s = \lambda - \sigma_n \tag{8.14a}$$

where J_1, J_2, J_3 are the deviatoric stress invariants

$$J_1 = t_r[s] = 0$$

$$J_2 = \frac{1}{6}\left[(\sigma_I - \sigma_{II})^2 + (\sigma_I - \sigma_{III})^2 + (\sigma_{II} - \sigma_{III})^2\right] = \tag{8.14b}$$

$$= \frac{1}{6}\left[(\sigma_x - \sigma_y)^2 + (\sigma_x - \sigma_z)^2 + (\sigma_y - \sigma_z)^2 + 6(\tau_{xy}^2 + \tau_{yz}^2 + \tau_{xz}^2)\right]$$

$$J_3 = \det[s] = \det([\sigma] - \sigma_n \mathbf{I}_3)$$

The quantities J_2 and J_3 can be expressed in terms of I_1, I_2, I_3 as

$$J_2 = -I_2 + \frac{1}{3}I_1^2 \quad , \quad J_3 = I_3 - \frac{1}{3}I_1 I_2 + \frac{2}{27}I_1^3 \tag{8.15}$$

The octaedric shear stress τ_0 is

$$\tau_0^2 = \frac{2}{3}J_2 \tag{8.16}$$

This stress is the same for all the (eight) planes inclined the same angle with respect to the principal directions $\mathbf{n}_I, \mathbf{n}_{II}, \mathbf{n}_{III}$ (i.e. planes defined by the normal vectors $\mathbf{n} = \frac{1}{\sqrt{3}}(\pm 1, \pm 1, \pm 1))$. The normal stress acting on each of these planes is the hydrostatic stress σ_n [ZT].

The three real roots of Eq.(8.14a) are given by [Bey]

$$s_I = \sqrt{2}\tau_0 \cos\left(w - \frac{2\pi}{3}\right) \; ; \; s_{II} = \sqrt{2}\tau_0 \cos\left(w + \frac{2\pi}{3}\right) \; ; \; s_{III} = \sqrt{2}\tau_0 \cos w \tag{8.17a}$$

with

$$\cos 3w = \sqrt{2}\frac{J_3}{\tau_0^3} \quad \left(0 \leq w \leq \frac{\pi}{3}\right) \tag{8.17b}$$

The principal stresses therefore are

$$\sigma_I = s_I + \sigma_n \quad ; \quad \sigma_{II} = s_{II} + \sigma_n \quad ; \quad \sigma_{III} = s_{III} + \sigma_n \tag{8.18}$$

As mentioned in previous chapters, onset of failure at a point can be monitored by the maximum principal stress reaching a prescribed limit value. This a usual procedure to detect fracture in fragile materials. For most materials however, the initiation of failure at a point is governed by the so called yield rule expressed in terms of the stress invariants. Here we simply note that both the octaedric shear stress τ_0 and the hydrostatic stress σ_n play an important role in the definition of yield surfaces in elasto-plasticity theories [ZT].

For example the equivalent stress (also called von Mises stress) σ_{eq} used in classical Hencky-Mises elastoplasticity is defined as [ZT]

$$\sigma_{\text{eq}} = (3J_2)^{1/2} = \frac{3}{\sqrt{2}}\tau_0 \tag{8.19}$$

8.2.6 Virtual work principle

The expression of the PVW for 3D solids is

$$\iiint_V \delta\boldsymbol{\varepsilon}^T \boldsymbol{\sigma} \, dV = \iiint_V \delta\mathbf{u}^T \, \mathbf{b} \, dV + \iint_A \delta\mathbf{u}^T \, \mathbf{t} \, dA + \sum_i \delta\mathbf{a}_i^T \, \mathbf{p}_i \tag{8.20}$$

where V and A are respectively the volume and the surface of the solid over which the body forces $\mathbf{b} = [b_x, b_y, b_z]^T$ and the surface tractions $\mathbf{t} = [t_x, t_y, t_z]^T$ act, and $\mathbf{p}_i = [P_{x_i}, P_{y_i}, P_{z_i}]^T$ are the point loads acting at node i. Eq.(8.20) is an extension of the PVW for 2D solids (Eq.(4.23)).

The PVW integrals involve first derivatives of the displacements only. Thus C^o continuity is required for the finite element approximation as for 2D elasticity and axisymmetric problems.

8.3 FINITE ELEMENT FORMULATION. THE FOUR-NODED TETRAHEDRON

The finite element formulation for 3D solid elements will be detailed next. It is important to note that most of the expressions are general and applicable to any 3D element with n nodes. However we will particularize all matrices and vectors for the simple 4-noded tetrahedron as an example. This element is a natural extension of the 3-noded triangle in Chapter 4.

8.3.1 Discretization of the displacement field

Let us consider a 3D solid discretized into 4-noded tetrahedra as that of Figure 8.5. The displacement field within each element is interpolated as

$$\mathbf{u} = \begin{Bmatrix} u \\ v \\ w \end{Bmatrix} = \begin{Bmatrix} N_1\, u_1 + N_2\, u_2 + N_3\, u_3 + N_4\, u_4 \\ N_1\, v_1 + N_2\, v_2 + N_3\, v_3 + N_4\, v_4 \\ N_1\, w_1 + N_2\, w_2 + N_3\, w_3 + N_4\, w_4 \end{Bmatrix} = \sum_{i=1}^{4} \mathbf{N}_i\, \mathbf{a}_i^{(e)} = \mathbf{N}\, \mathbf{a}^{(e)} \tag{8.21a}$$

where

$$\mathbf{N} = [\mathbf{N}_1, \mathbf{N}_2, \mathbf{N}_3, \mathbf{N}_4] \quad ; \quad \mathbf{N}_i = \begin{bmatrix} N_i & 0 & 0 \\ 0 & N_i & 0 \\ 0 & 0 & N_i \end{bmatrix} \tag{8.21b}$$

and

$$\mathbf{a}^{(e)} = \begin{Bmatrix} \mathbf{a}_1^{(e)} \\ \mathbf{a}_2^{(e)} \\ \mathbf{a}_3^{(e)} \\ \mathbf{a}_4^{(e)} \end{Bmatrix} \quad ; \quad \mathbf{a}_i^{(e)} = \begin{Bmatrix} u_i \\ v_i \\ w_i \end{Bmatrix} \tag{8.21c}$$

are the shape function matrix and the displacement vector for the element and a node i, respectively. As usual, the same interpolation has been used for the three displacement components. The shape functions are therefore the same for the three displacements.

The extension of above expressions for an element with n nodes simply involves changing from 4 to n the number of matrices $\mathbf{N}_i$ and vectors $\mathbf{a}_i^{(e)}$ in Eqs.(8.21a).

The analytical form of the shape functions N_i is obtained in a similar way as for the 3-noded triangle (Section 4.3.1). The four nodes define a linear displacement field in 3D. Choosing the u displacement we can write

$$u = \alpha_1 + \alpha_2 x + \alpha_3 y + \alpha_4 z \tag{8.22}$$

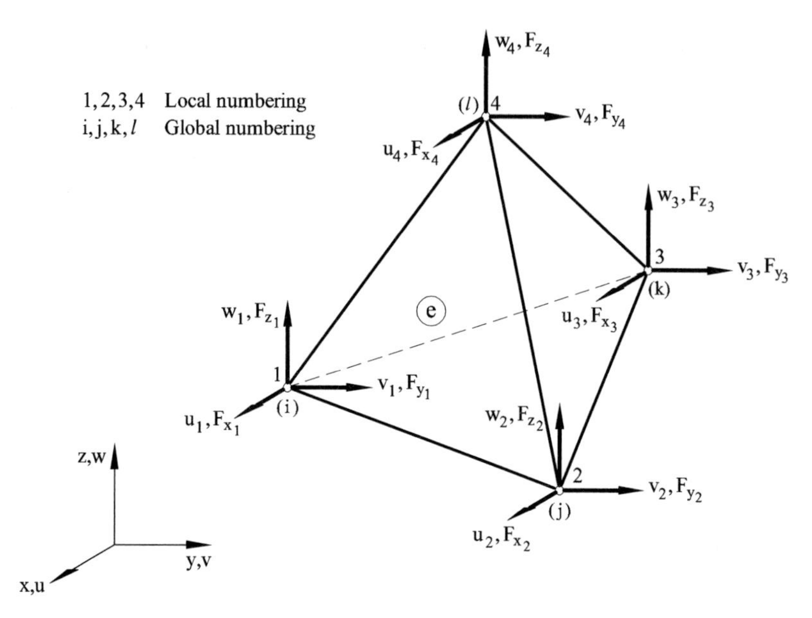

Fig. 8.5 Four-noded tetrahedral element. Nodal displacements (u_i, v_i, w_i) and equilibrating nodal forces $(F_{x_i}, F_{y_i}, F_{z_i})$

The α_i parameters are obtained by substituting the nodal coordinates into Eq.(8.22) and then making the displacements equal to their nodal values, i.e.

$$
\begin{aligned}
u_1 &= \alpha_1 + \alpha_2\, x_1 + \alpha_3\, y_1 + \alpha_4\, z_1 \\
u_2 &= \alpha_1 + \alpha_2\, x_2 + \alpha_3\, y_2 + \alpha_4\, z_2 \\
u_3 &= \alpha_1 + \alpha_2\, x_3 + \alpha_3\, y_3 + \alpha_4\, z_3 \\
u_4 &= \alpha_1 + \alpha_2\, x_4 + \alpha_3\, y_4 + \alpha_4\, z_4
\end{aligned}
\tag{8.23}
$$

Eq. (8.23) is used to solve for $\alpha_1, \alpha_2, \alpha_3$ and α_4. Substituting these values into (8.22) yields, after rearranging the terms,

$$
u = \sum_{i=1}^{4} \frac{1}{6V^{(e)}} (a_i + b_i x + c_i y + d_i z) u_i \tag{8.24}
$$

The nodal shape function N_i is obtained by comparing Eqs.(8.24) and (8.21a) as

$$
N_i = \frac{1}{6V^{(e)}} (a_i + b_i x + c_i y + d_i z) \tag{8.25a}
$$

where $V^{(e)}$ is the element volume and

$$a_i = \det \begin{vmatrix} x_j & y_j & z_j \\ x_k & y_k & z_k \\ x_l & y_l & z_l \end{vmatrix} \quad ; \quad b_i = -\det \begin{vmatrix} 1 & y_j & z_j \\ 1 & y_k & z_k \\ 1 & y_l & z_l \end{vmatrix}$$

$$c_i = \det \begin{vmatrix} x_j & 1 & z_j \\ x_k & 1 & z_k \\ x_l & 1 & z_l \end{vmatrix} \quad ; \quad d_i = -\det \begin{vmatrix} x_j & y_j & 1 \\ x_k & y_k & 1 \\ x_l & y_l & 1 \end{vmatrix} \tag{8.25b}$$

The different parameters for $i = 1, 2, 3, 4$, are obtained by adequate cyclic permutation of the indexes i, j, k, l.

As usual, the shape function N_i has the same expression as in Eqs.(8.25) for the other two displacement components v and w.

The graphic form of the shape functions for 3D elements is not straightforward since they are a functions of three variables. Note, however, that the expression of N_i over an element face coincides precisely with the shape function of the 2D element corresponding to that face. Thus, the shape functions for the 4-noded tetrahedron take forms over each face that are identical to those for the 3-noded triangle shown in Figure 4.8.

8.3.2 Strain matrix

Substituting Eq.(8.21a) into (8.2) gives for a 3D element with n nodes

$$\boldsymbol{\varepsilon} = \sum_{i=1}^{n} \left\{ \begin{array}{c} \dfrac{\partial N_i}{\partial x} u_i \\ \dfrac{\partial N_i}{\partial y} v_i \\ \dfrac{\partial N_i}{\partial z} w_i \\ \dfrac{\partial N_i}{\partial y} u_i + \dfrac{\partial N_i}{\partial x} v_i \\ \dfrac{\partial N_i}{\partial z} u_i + \dfrac{\partial N_i}{\partial x} w_i \\ \dfrac{\partial N_i}{\partial z} v_i + \dfrac{\partial N_i}{\partial y} w_i \end{array} \right\} = \sum_{i=1}^{n} \mathbf{B}_i \, \mathbf{a}_i^{(e)} = \mathbf{B} \, \mathbf{a}^{(e)} \tag{8.26}$$

where $\mathbf{B}$ is the element strain matrix given by

$$\mathbf{B} = [\mathbf{B}_1, \mathbf{B}_2, \mathbf{B}_3, \ldots, \mathbf{B}_n] \tag{8.27a}$$

and $\mathbf{B}_i$ is the strain matrix of node i, with

$$\mathbf{B}_i = \begin{bmatrix} \frac{\partial N_i}{\partial x} & 0 & 0 \\ 0 & \frac{\partial N_i}{\partial y} & 0 \\ 0 & 0 & \frac{\partial N_i}{\partial z} \\ \frac{\partial N_i}{\partial y} & \frac{\partial N_i}{\partial x} & 0 \\ \frac{\partial N_i}{\partial z} & 0 & \frac{\partial N_i}{\partial x} \\ 0 & \frac{\partial N_i}{\partial z} & \frac{\partial N_i}{\partial y} \end{bmatrix} \quad (8.27b)$$

As usual matrix $\mathbf{B}$ contains as many $\mathbf{B}_i$ matrices as element nodes. For the 4-noded tetrahedron

$$\mathbf{B} = [\mathbf{B}_1, \mathbf{B}_2, \mathbf{B}_3, \mathbf{B}_4] \quad (8.28a)$$

Making use of Eqs.(8.25a) gives for the 4-noded tetrahedron

$$\mathbf{B}_i = \frac{1}{6V^{(e)}} \begin{bmatrix} b_i & 0 & 0 \\ 0 & c_i & 0 \\ 0 & 0 & d_i \\ c_i & b_i & 0 \\ d_i & 0 & b_i \\ 0 & d_i & c_i \end{bmatrix} \quad (8.28b)$$

The strain matrix is constant, as it is for the 3-noded triangle for plane elasticity.

8.3.3 Equilibrium equations

The PVW for a single element is (Eq.(8.20))

$$\iiint_{V^{(e)}} \delta\boldsymbol{\varepsilon}^T \boldsymbol{\sigma}\, dV = \iiint_V \delta\mathbf{u}^T \mathbf{b}\, dV + \iint_A \delta\mathbf{u}^T \mathbf{t}\, dA + [\delta\mathbf{a}^{(e)}]^T \mathbf{q}^{(e)} \quad (8.29)$$

where, as usual, $\mathbf{q}^{(e)}$ is the vector of equilibrating nodal forces for the element acting on the virtual nodal displacements $\delta\mathbf{a}^{(e)}$ with

$$\delta\mathbf{a}^{(e)} = \begin{Bmatrix} \delta\mathbf{a}_1^{(e)} \\ \delta\mathbf{a}_2^{(e)} \\ \delta\mathbf{a}_3^{(e)} \end{Bmatrix} \quad , \quad \delta\mathbf{a}_i^{(e)} = \begin{Bmatrix} \delta u_i \\ \delta v_i \\ \delta w_i \end{Bmatrix} \quad , \quad \mathbf{q}^{(e)} = \begin{Bmatrix} \delta\mathbf{q}_1^{(e)} \\ \delta\mathbf{q}_2^{(e)} \\ \delta\mathbf{q}_3^{(e)} \end{Bmatrix} \quad , \quad \mathbf{q}_i^{(e)} = \begin{Bmatrix} F_{x_i} \\ F_{y_i} \\ F_{z_i} \end{Bmatrix} \tag{8.30a}$$

The virtual displacements and the virtual strains are interpolated in terms of the virtual displacement values in the standard form, i.e.

$$\delta\mathbf{u} = \mathbf{N}\delta\mathbf{a} \quad , \quad \delta\boldsymbol{\varepsilon} = \mathbf{B}\delta\mathbf{a} \tag{8.30b}$$

Substituting Eqs.(8.30b) into (8.29) gives after simplification of the virtual displacements

$$\iiint_{V^{(e)}} \mathbf{B}^T\boldsymbol{\sigma}\,dV - \iiint_{V^{(e)}} \mathbf{N}^T\mathbf{b}\,dV - \iiint_{A^{(e)}} \mathbf{N}^T\mathbf{t}\,dA = \mathbf{q}^{(e)} \tag{8.31}$$

Substituting the constitutive equation for the stresses (Eq.(8.6)) into (8.31) gives the equilibrium equation for the element in the standard matrix form

$$\begin{aligned} \left(\iiint_{V^{(e)}} \mathbf{B}^T\mathbf{D}\mathbf{B}\,dV\right)\mathbf{a}^{(e)} - \iiint_{V^{(e)}} \mathbf{B}^T\mathbf{D}\boldsymbol{\varepsilon}^0\,dV + \iiint_{V^{(e)}} \mathbf{B}^T\boldsymbol{\sigma}^0\,dV \\ - \iiint_{V^{(e)}} \mathbf{N}^T\mathbf{b}\,dV - \iiint_{A^{(e)}} \mathbf{N}^T\mathbf{t}\,dA = \mathbf{q}^{(e)} \end{aligned} \tag{8.32a}$$

or

$$\mathbf{K}^{(e)}\,\mathbf{a}^{(e)} - \mathbf{f}^{(e)} = \mathbf{q}^{(e)} \tag{8.32b}$$

The global system of equations $\mathbf{K}\mathbf{a} = \mathbf{f}$ is obtained by assembling the contributions of $\mathbf{K}^{(e)}$ and $\mathbf{f}^{(e)}$ for each element in the usual manner.

The reactions at the prescribed nodes can be obtained once the nodal displacements have been found via Eq.(2.26a), or by using Eq.(4.60a) with the vector of internal nodal forces for each element given by

$$\mathbf{f}_{\text{int}}^{(e)} = \iiint_{V^{(e)}} \mathbf{B}^T\boldsymbol{\sigma}\,dV \tag{8.33}$$

This expression is deduced from the first integral in the l.h.s. of Eq.(8.31).

8.3.4 Stiffness matrix for the element

The element stiffness matrix is deduced from Eq.(8.32a) as

$$\underset{3n\times 3n}{\mathbf{K}^{(e)}} = \iiint_{V^{(e)}} \underset{3n\times 6}{\mathbf{B}^T} \; \underset{6\times 6}{\mathbf{D}} \; \underset{6\times 3n}{\mathbf{B}} \, dV \tag{8.34a}$$

with

$$\underset{3\times 3}{\mathbf{K}_{ij}^{(e)}} = \iiint_{V^{(e)}} \underset{3\times 6}{\mathbf{B}_i^T} \; \underset{6\times 6}{\mathbf{D}} \; \underset{6\times 3}{\mathbf{B}_j} \, dV \tag{8.34b}$$

The form of matrices $\mathbf{K}$ *and* $\mathbf{K}_{ij}^{(e)}$ *of Eqs.(8.34) is completely general and applicable to any 3D solid element with n nodes.*

The expression of $\mathbf{K}_{ij}^{(e)}$ for the 4-noded tetrahedron with homogeneous material properties is simple since the strain matrix is constant. This gives

$$\mathbf{K}_{ij}^{(e)} = \mathbf{B}_i^T \mathbf{D} \mathbf{B}_j V^{(e)} \tag{8.35}$$

The explicit form of $\mathbf{K}_{ij}^{(e)}$ for this element is shown in Box 8.1.

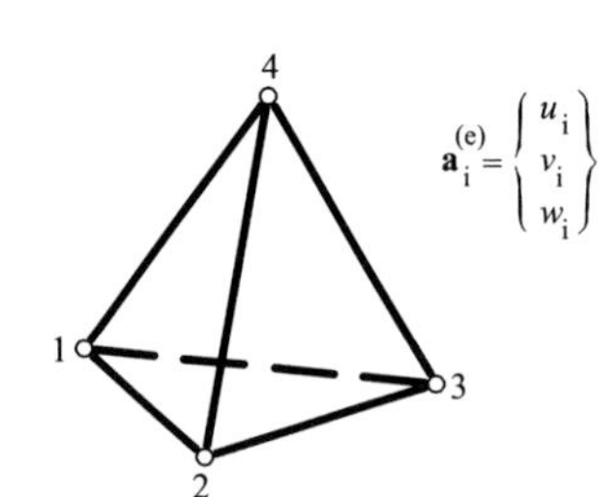

$$\mathbf{a}_i^{(e)} = \begin{Bmatrix} u_i \\ v_i \\ w_i \end{Bmatrix}$$

$$\mathbf{K}^{(e)} = \begin{bmatrix} \mathbf{K}_{11}^{(e)} & \mathbf{K}_{12}^{(e)} & \mathbf{K}_{13}^{(e)} & \mathbf{K}_{14}^{(e)} \\ & \mathbf{K}_{22}^{(e)} & \mathbf{K}_{23}^{(e)} & \mathbf{K}_{24}^{(e)} \\ & & \mathbf{K}_{33}^{(e)} & \mathbf{K}_{34}^{(e)} \\ \text{Symmetrical} & & & \mathbf{K}_{44}^{(e)} \end{bmatrix}$$

$$\mathbf{K}_{ij}^{(e)} = \frac{1}{36V^{(e)}} \begin{bmatrix} (d_{11}b_ib_j + d_{44}c_ic_j + d_{55}d_id_j) & (d_{12}b_ic_j + d_{44}c_ib_j) & (d_{13}b_id_j + d_{55}d_ib_j) \\ (d_{21}c_ib_j + d_{44}b_ic_j) & (d_{22}c_ic_j + d_{44}b_ib_j + d_{66}d_id_j) & (d_{23}c_id_j + d_{66}d_ic_j) \\ (d_{31}d_ib_j + d_{55}b_id_j) & (d_{32}d_ic_j + d_{66}c_id_j) & (d_{33}d_id_j + d_{55}b_ib_j + d_{66}c_ic_j) \end{bmatrix}$$

d_{ij} are the elements of the constitutive matrix $\mathbf{D}$

b_i, c_i, d_i are the parameters of the shape function N_i

Box 8.1 Stiffness matrix $\mathbf{K}_{ij}^{(e)}$ for the 4-noded linear tetrahedral element

8.3.5 Equivalent nodal force vector for the element

The equivalent nodal force vector of Eq.(8.32b) is

$$\mathbf{f}^{(e)} = \iiint_{V^{(e)}} \mathbf{N}^T \, \mathbf{b} \, dV + \iint_{A^{(e)}} \mathbf{N}^T \mathbf{t} \, dA + \iiint_{V^{(e)}} \mathbf{B}^T \mathbf{D} \boldsymbol{\varepsilon}^0 \, dV - \\ - \iiint_{V^{(e)}} \mathbf{B}^T \boldsymbol{\sigma}^0 \, dV = \mathbf{f}_b^{(e)} + \mathbf{f}_t^{(e)} + \mathbf{f}_\varepsilon^{(e)} + \mathbf{f}_\sigma^{(e)} \tag{8.36}$$

The first two integrals are contributed by the body forces and the surface tractions, respectively, and the last two ones are due to the initial strains and the initial stresses.

The general expression for the equivalent force vectors for an arbitrary 3D solid element is given below. The particular form for the 4-noded tetrahedron is detailed.

Body forces

$$\underset{3n \times 1}{\mathbf{f}_b^{(e)}} = \iiint_{V^{(e)}} \mathbf{N}^T \mathbf{b} dV \tag{8.37a}$$

For the 4-noded tetrahedron:

$$\underset{12 \times 1}{\mathbf{f}_b^{(e)}} = \begin{Bmatrix} \mathbf{f}_{b_1}^{(e)} \\ \mathbf{f}_{b_2}^{(e)} \\ \mathbf{f}_{b_3}^{(e)} \\ \mathbf{f}_{b_4}^{(e)} \end{Bmatrix} \quad \text{with} \quad \mathbf{f}_{b_i}^{(e)} = \iiint_{V^{(e)}} \mathbf{N}_i^T \mathbf{b} \, dV \tag{8.37b}$$

For a uniform body force:

$$\mathbf{f}_{b_i}^{(e)} = \frac{V^{(e)}}{4} \left[b_x, b_y, b_z \right]^T \tag{8.37c}$$

i.e. the total body force is distributed in equal parts between the four nodes, as expected.

Surface tractions

$$\underset{3n \times 1}{\mathbf{f}_t^{(e)}} = \iint_{A^{(e)}} \mathbf{N}^T \mathbf{t} dA \tag{8.38a}$$

For the 4-noded tetrahedron:

$$\underset{12 \times 1}{\mathbf{f}_t^{(e)}} = \begin{Bmatrix} \mathbf{f}_{t_1} \\ \mathbf{f}_{t_2}^{(e)} \\ \mathbf{f}_{t_3}^{(e)} \\ \mathbf{f}_{t_4}^{(e)} \end{Bmatrix} \quad \text{with} \quad \mathbf{f}_{t_i}^{(e)} = \iint_{A^{(e)}} \mathbf{N}_i^T \mathbf{t} \, dA \tag{8.38b}$$

The components of the surface tractions vector depend on the element face over which the external force acts; i.e.

Uniform force acting on the face defined by nodes 1-2-3

$$\mathbf{f}_t^{(e)} = \frac{A_{123}^{(e)}}{3}\left[t_x, t_y, t_z, t_x, t_y, t_z . t_x, t_y, t_z, 0, 0, 0\right]^T \tag{8.39}$$

where $A_{123}^{(e)}$ is the area of the face. Note that the last three terms of Eq.(8.39) are zero as the shape function N_4 vanishes on that face.

Uniform force acting on the face defined by nodes 1-2-4

$$\mathbf{f}_t^{(e)} = \frac{A_{124}^{(e)}}{3}\,[t_x, t_y, t_z, t_x, t_y, t_z, 0, 0, 0, t_x, t_y, t_z]^T \tag{8.40}$$

Uniform force acting on the face defined by nodes 2-3-4

$$\mathbf{f}_t^{(e)} = \frac{A_{234}^{(e)}}{3}[0, 0, 0, t_x, t_y, t_z, t_x, t_y, t_z, t_x, t_y, t_z]^T \tag{8.41}$$

and

Uniform force acting on the face defined by nodes 1-3-4

$$\mathbf{f}_t^{(e)} = \frac{A_{134}^{(e)}}{3}[t_x, t_y, t_z, 0, 0, 0, t_x, t_y, t_z, t_x, t_y, t_z]^T \tag{8.42}$$

Uniform surface tractions are distributed in equal parts between the three nodes of the linear tetrahedron face affected by the loading.

Forces due to initial strains

$$\underset{3n\times 1}{\mathbf{f}_\varepsilon^{(e)}} = \iiint_{V^{(e)}} \mathbf{B}^T \mathbf{D} \boldsymbol{\varepsilon}^0 \, dV \tag{8.43a}$$

For the 4-noded tetrahedron:

$$\underset{12\times 1}{\mathbf{f}_\varepsilon^{(e)}} = \begin{Bmatrix} \mathbf{f}_{\varepsilon_1}^{(e)} \\ \mathbf{f}_{\varepsilon_2}^{(e)} \\ \mathbf{f}_{\varepsilon_3}^{(e)} \\ \mathbf{f}_{\varepsilon_4}^{(e)} \end{Bmatrix} ; \; \mathbf{f}_{\varepsilon_i}^{(e)} = \iiint_{V^{(e)}} \mathbf{B}_i^T \mathbf{D} \boldsymbol{\varepsilon}^0 \, dV = \frac{1}{6} \begin{Bmatrix} (d_{11}\varepsilon_x^0 + d_{12}\varepsilon_y^0 + d_{13}\varepsilon_z^0) b_i \\ (d_{21}\varepsilon_x^0 + d_{22}\varepsilon_y^0 + d_{23}\varepsilon_z^0) c_i \\ (d_{31}\varepsilon_x^0 + d_{32}\varepsilon_y^0 + d_{33}\varepsilon_z^0) d_i \\ 0 \\ 0 \\ 0 \end{Bmatrix} \tag{8.43b}$$

where d_{ij} is the element ij of matrix $\mathbf{D}$. For initial strains due to thermal effects and isotropic material we have

$$\mathbf{f}_{\varepsilon_i}^{(e)} = \frac{E\,\alpha(\Delta T)}{6(1-2\nu)}[b_i, c_i, d_i, 0, 0, 0]^T \tag{8.43c}$$

Forces due to initial stresses

$$\underset{3n\times 1}{\mathbf{f}_{\sigma}^{(e)}} = \iiint_{V^{(e)}} \mathbf{B}^T \boldsymbol{\sigma}^0 \, dV \tag{8.44a}$$

For the 4-noded tetrahedron:

$$\underset{12\times 1}{\mathbf{f}_{\sigma}^{(e)}} = \begin{Bmatrix} \mathbf{f}_{\sigma_1}^{(e)} \\ \mathbf{f}_{\sigma_2}^{(e)} \\ \mathbf{f}_{\sigma_3}^{(e)} \\ \mathbf{f}_{\sigma_4}^{(e)} \end{Bmatrix} ; \; \mathbf{f}_{\sigma_i}^{(e)} = \iiint_{V^{(e)}} \mathbf{B}_i \boldsymbol{\sigma}^0 \, dV = \frac{1}{6} \begin{Bmatrix} b_i\sigma_x^0 + c_i\tau_{xy}^0 + d_i\tau_{xz}^0 \\ c_i\sigma_y^0 + b_i\tau_{xy}^0 + d_i\tau_{yz}^0 \\ d_i\sigma_z^0 + b_i\tau_{xz}^0 + c_i\tau_{yz}^0 \end{Bmatrix} \tag{8.44b}$$

8.3.6 The performance of the 4-noded tetrahedron

The 4-noded tetrahedron behaves similarly to the 3-noded linear triangle presented in Chapter 4. The element has a good ability to model uniform stress fields. However, its accuracy is poor for bending dominated problems, as well as in the presence of high stress gradients and finer meshes are needed in these zones. More details are given in Section 8.11.

We emphasize the intrinsic difficulty of discretizing a solid with tetrahedra. This is a serious problem for the application of the linear tetrahedron to arbitrary shaped solids and particularly when adaptive mesh refinement is used. Much work has been reported in recent years on the development of efficient mesh generators for 3D solids using linear tetrahedral elements (see Chapter 10, Annex D and [BP,GiD,Ng,Pe,Pi,PVMZ]).

8.4 OTHER 3D SOLID ELEMENTS

The formulation presented in Section 8.3 is completely general. This means that the expressions for the stiffness matrix and the equivalent nodal force vector *for any 3D solid element* coincide with Eqs.(8.34) and (8.36), respectively. The computation of the element integrals simply requires substituting the adequate element shape functions in the expressions for $\mathbf{B}_i$ and $\mathbf{f}^{(e)}$. In common with 2D solid elements, 3D hexahedra can be of Lagrange or Serendipity types. The shape functions for hexahedral elements

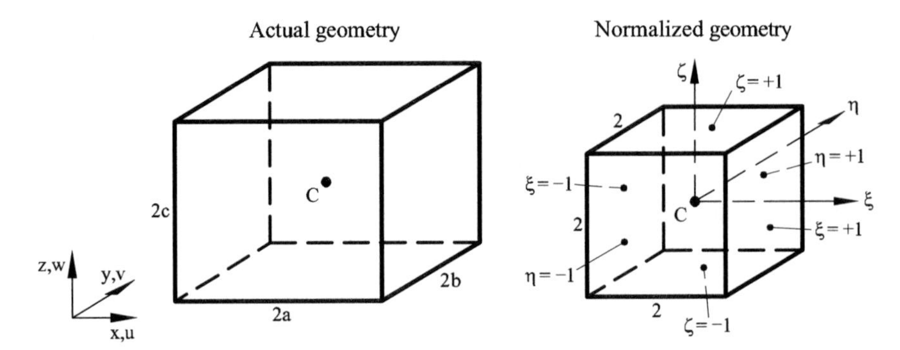

Fig. 8.6 Right prism. Actual and normalized geometries

are obtained for a right prism in the normalized geometry in terms of the natural coordinates (Figure 8.6) following similar rules as for 2D rectangles. The modeling of arbitrary 3D geometries of irregular shape with hexahedra is straightforward using an isoparametric formulation.

The shape functions for Lagrange right prisms are simply obtained by the product of three Lagrange polynomials in 1D. The derivation of the shape functions for 3D Serendipity prismatic elements is more cumbersome, although the same rules given for the 2D case apply.

Like triangles, the shape functions for tetrahedra are complete polynomials and they are more easily expressed in terms of volume and natural coordinates.

8.5 RIGHT PRISMS

Let us consider a right prism with edges $2a \times 2b \times 2c$, with the natural coordinates ξ, η, ζ defined as shown in Figure 8.6. We can write

$$\xi = \frac{(x - x_c)}{a} \quad ; \quad \eta = \frac{(y - y_c)}{b} \quad ; \quad \zeta = \frac{(z - z_c)}{c} \tag{8.45}$$

where (x_c, y_c, z_c) are the coordinates of the centroid. Note that the prism becomes a cube in the natural coordinate system (Figure 8.6). From Eq.(8.45)

$$\frac{d\xi}{dx} = \frac{1}{a} \quad ; \quad \frac{d\eta}{dy} = \frac{1}{b} \quad ; \quad \frac{d\zeta}{dz} = \frac{1}{c} \tag{8.46}$$

and a differential of volume is expressed by

$$dx\,dy\,dz = abc\;d\xi\,d\eta\,d\zeta \tag{8.47}$$

The integration of a function $f(x,y,z)$ over the element is expressed in the natural coordinate system as

$$\iiint_{V^{(e)}} f(x,y,z)\,dV = \int_{-1}^{+1}\int_{-1}^{+1}\int_{-1}^{+1} f(\xi,\eta,\zeta) abc\, d\xi\, d\eta\, d\zeta \qquad (8.48)$$

Since the element is a right prism, the cartesian derivatives of the shape functions are directly obtained by

$$\frac{\partial N_i}{\partial x} = \frac{1}{a}\frac{\partial N_i}{\partial \xi} \quad ; \quad \frac{\partial N_i}{\partial y} = \frac{1}{b}\frac{\partial N_i}{\partial \eta} \quad ; \quad \frac{\partial N_i}{\partial z} = \frac{1}{c}\frac{\partial N_i}{\partial \zeta} \qquad (8.49)$$

The shape functions must satisfy the standard conditions (Section 5.2.2)

$$N_i(\xi_j,\eta_j,\zeta_j) = \begin{cases} 1 & \text{if} \quad i=j \\ 0 & \text{if} \quad i \neq j \end{cases} \qquad (8.50a)$$

and

$$\sum_{i=1}^{n} N_i\ (\xi,\eta,\zeta) = 1 \qquad (8.50b)$$

8.5.1 Right prisms of the Lagrange family

The shape functions for Lagrange right prisms are obtained by multilying three 1D Lagrange polynomials as

$$N_i(\xi,\eta,\zeta) = \ l_I^i(\xi)\ l_I^i(\eta)\ l_I^i(\zeta) \qquad (8.51)$$

where $l_I^i(\xi)$ is the normalized Lagrange polynomial of Ith degree passing by node i, etc. For the same reasons mentioned in Section 5.3.4 it is usual to choose the same polynomial approximation in each of the three directions ξ,η and ζ.

The terms contained in the shape functions of prismatic elements are deduced from the Pascal tetrahedron. Figure 8.7 shows the linear and quadratic elements of this family whose shape functions are derived next.

8.5.1.1 Linear right prism of the Lagrange family

The simplest Lagrange prismatic element is the 8-noded linear prism shown in Figure 8.7.

The nodal shape function is obtained by multiplying the three normalized linear polynomials in ξ,η and ζ, corresponding to the node. In general form

$$N_i(\xi,\eta,\zeta) = \frac{1}{8}(1+\xi_i\xi)\ (1+\eta_i\eta)\ (1+\zeta_i\zeta) \qquad (8.52)$$

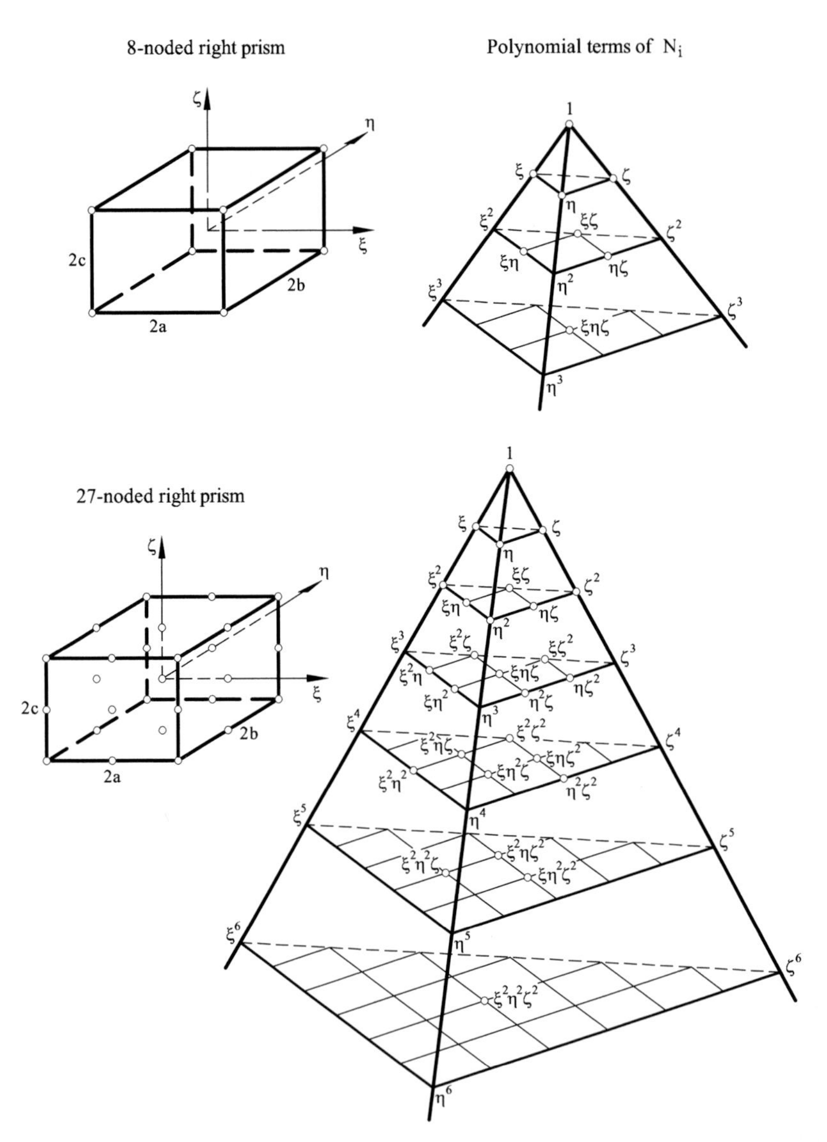

Fig. 8.7 Linear and quadratic right prisms of the Lagrange family. Polynomial terms contained in the shape functions deduced from the Pascal tetrahedron

Note that:

1. The shape functions contain a complete linear polynomial in ξ, η, ζ and the incomplete quadratic and cubic terms $\xi\eta$, $\xi\zeta$, $\eta\xi$ and $\xi\eta\zeta$, $\xi\eta\zeta$ (Figure 8.8).

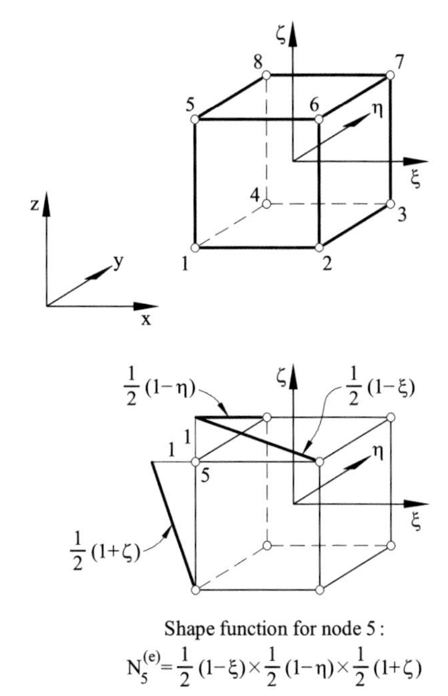

Local node number	Local coordinates ξ_i	η_i	ζ_i
1	−1	−1	−1
2	1	−1	−1
3	1	1	−1
4	−1	1	−1
5	−1	−1	1
6	1	−1	1
7	1	1	1
8	−1	1	1

In general :

$$N_i^{(e)}=\frac{1}{8}(1+\xi_i\xi)(1+\eta_i\eta)(1+\zeta_i\zeta)$$

Fig. 8.8 Shape functions for the linear right prism

2. The shape functions satisfy conditions (8.50).

The linear hexahedron is a popular element due to the small number of nodal variables, which is attractive for practical 3D analysis.

The linear hexahedron behaves for linear elasticity analysis similarly as the 4-noded rectangle of Chapter 4. Its performance is excellent for tension or compression dominated problems, whereas its accuracy is poor for bending dominated situations. This is due to its inability to follow curved deformation patterns and finer meshes are needed to obtain good solutions in these cases.

The behaviour of the 8-noded hexahedron can be improved by using reduced integration for the shear stiffness terms, by adding internal or incompatible modes, or by using an assumed strain field, in a similar way to the 4-noded quadrilateral (Section 4.4.2). A popular alternative is to add incompatible internal modes to the original displacement field as

$$\begin{Bmatrix} u \\ v \\ w \end{Bmatrix} = \sum_{i=1}^{8} N_i \begin{Bmatrix} u_i \\ v_i \\ w_i \end{Bmatrix}^{(e)} + \begin{Bmatrix} (1-\xi^2)a_1 + (1-\eta^2)a_2 + (1-\zeta^2)a_3 \\ (1-\xi^2)a_4 + (1-\eta^2)a_5 + (1-\zeta^2)a_6 \\ (1-\xi^2)a_7 + (1-\eta^2)a_8 + (1-\zeta)a_9 \end{Bmatrix} \quad (8.53)$$

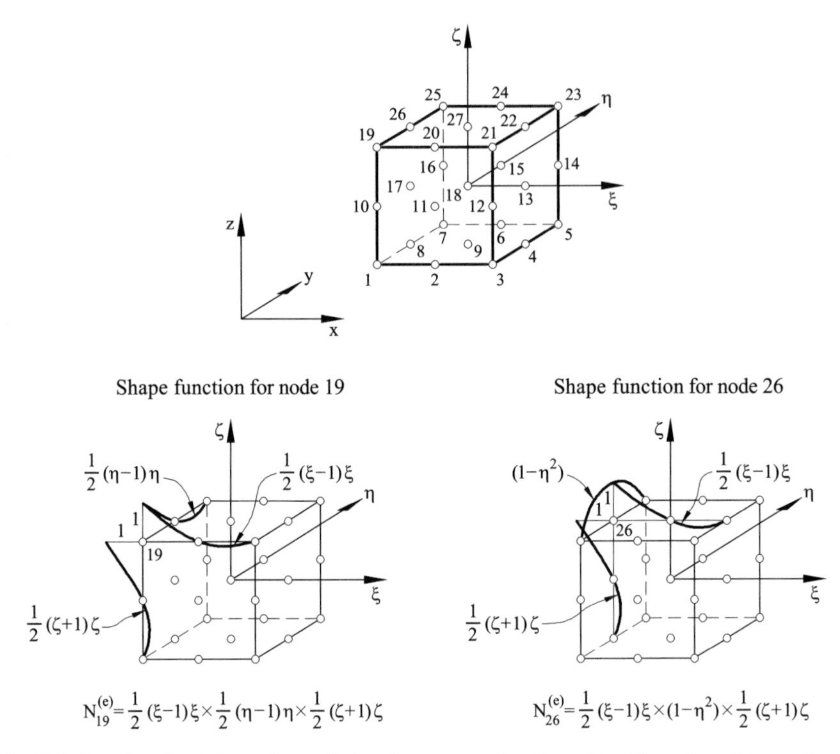

Fig. 8.9 Quadratic right prism of the Lagrange family with 27 nodes. Shape functions for a corner node and a mid-side node

The enhanced 8-noded hexahedron requires a one point reduced quadrature for the terms corresponding to the incompatible modes a_i, so that the patch test is satisfied. It can be proved that in its right form the enhanced element reproduces pure bending states exactly [CMPW]. The internal incompatible modes can be eliminated by static condensation to yield a 24×24 stiffness matrix. Hence, the size of the global stiffness matrix does not increase. An explicit form of the element stiffness matrix for right prisms can be obtained as for the 4-noded rectangle (Box 4.2).

8.5.1.2 Quadratic right prism of the Lagrange family

The quadratic Lagrange prism has 27 nodes (Figure 8.9). The shape functions are obtained by the product of three 1D normalized quadratic Lagrange polynomials. Figure 8.9 shows the derivation of the shape functions for a corner node and a mid-side node. It is simple to extrapolate this procedure to obtain the following general expressions

Corner nodes

$$N_i = \frac{1}{8}(\xi^2 + \xi\xi_i)(\eta^2 + \eta\eta_i)(\zeta^2 + \zeta\zeta_i) \quad ; \quad i = \begin{smallmatrix} 1,3,5,7 \\ 19,21,23,25 \end{smallmatrix} \tag{8.54}$$

Mid-side nodes

$$\begin{aligned} N_i = \frac{1}{4}\eta_i^2(\eta^2 - \eta\eta_i)\zeta^2(\zeta^2 - \zeta\zeta_i)(1-\xi^2) + \frac{1}{4}\zeta_i^2(\zeta^2 - \zeta\zeta_i) + \\ + \xi_i^2(\xi^2 - \xi\xi_i)(1-\eta^2) + \frac{1}{4}\xi_i^2(\xi^2 - \xi\xi_i)\eta_i^2(\eta^2 - \eta\eta_i)(1-\zeta^2) \end{aligned} \; ; \; i = \begin{smallmatrix} 2,4,6,8 \\ 10,12,14,16 \\ 20,22,24,26 \end{smallmatrix} \tag{8.55a}$$

Face nodes

$$\begin{aligned} N_i = \frac{1}{2}(1-\xi^2)(1-\eta^2)(\zeta + \zeta_i\zeta^2) + \frac{1}{2}(1-\eta^2)(1-\zeta^2)(\xi + \xi_i\xi^2) + \\ + \frac{1}{2}(1-\xi^2)(1-\zeta^2)(\eta + \eta_i\eta^2) \quad ; \quad i = \begin{smallmatrix} 9,11,13 \\ 15,17,27 \end{smallmatrix} \end{aligned} \tag{8.55b}$$

Central internal node

$$N_{18} = (1-\xi^2)(1-\eta^2)(1-\zeta^2) \tag{8.55c}$$

8.5.1.3 Other hexahedral elements of the Lagrange family

The next members of the Lagrange family are the 64-noded cubic prism (4 nodes along each edge) and the 125-noded quartic prism (5 nodes alone each edge). Their shape functions are obtained by the product of three 1D normalized cubic and quartic polynomials respectively. These elements in their right form are in general not competitive versus the analogous Serendipity elements which have less nodal variables.

8.5.2 Serendipity prisms

Serendipity prisms are obtained by extension of the corresponding 2D rectangular Serendipity elements. Figure 8.10 shows the first two elements in the family, i.e. the 8- and 20-noded right prisms. Note that the 8-noded prism is common to the Lagrange and Serendipity families and hence its shape functions coincide with those given in Section 8.5.1.1.

8.5.2.1 20-noded quadratic Serendipity prism

The shape functions are obtained using similar criteria as for the 8-noded rectangle (Section 5.4.1). The shape functions for the side nodes are ob-

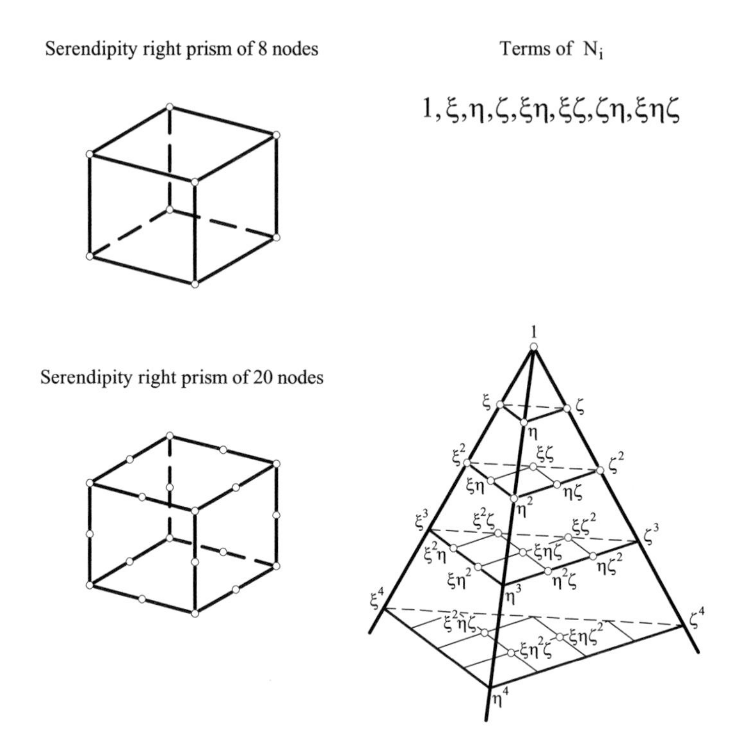

Fig. 8.10 8-noded and 20-noded Serendipity prisms. Polynomial terms contained in the shape functions deduced from the Pascal's tetrahedron

tained by multiplying a 1D normalized quadratic Lagrange polynomial and two 1D normalized linear polynomials expressed in the natural coordinates. For the corner nodes, a two step procedure is followed. The first step involves the derivation of the trilinear function corresponding to the node. This function is subsequently modified so that it takes a zero value at the side nodes. This is achieved by substracting one half of the values of the shape function of the side nodes adjacent to the relevant corner node under consideration. See Section 5.4.1 for details.

Figure 8.11 shows the derivation of the shape function for a side node (*20*) and a corner node (*13*). The element shape functions are written in compact form as

Corner nodes

$$N_i = \frac{1}{8}(1+\xi_i\xi)(1+\eta_i\eta)(1+\zeta_i\zeta)(\xi_i\xi+\eta_i\eta+\zeta_i\zeta-2) \quad ; \quad i = \begin{matrix}1,3,5,7\\ 13,15,17,19\end{matrix} \tag{8.56a}$$

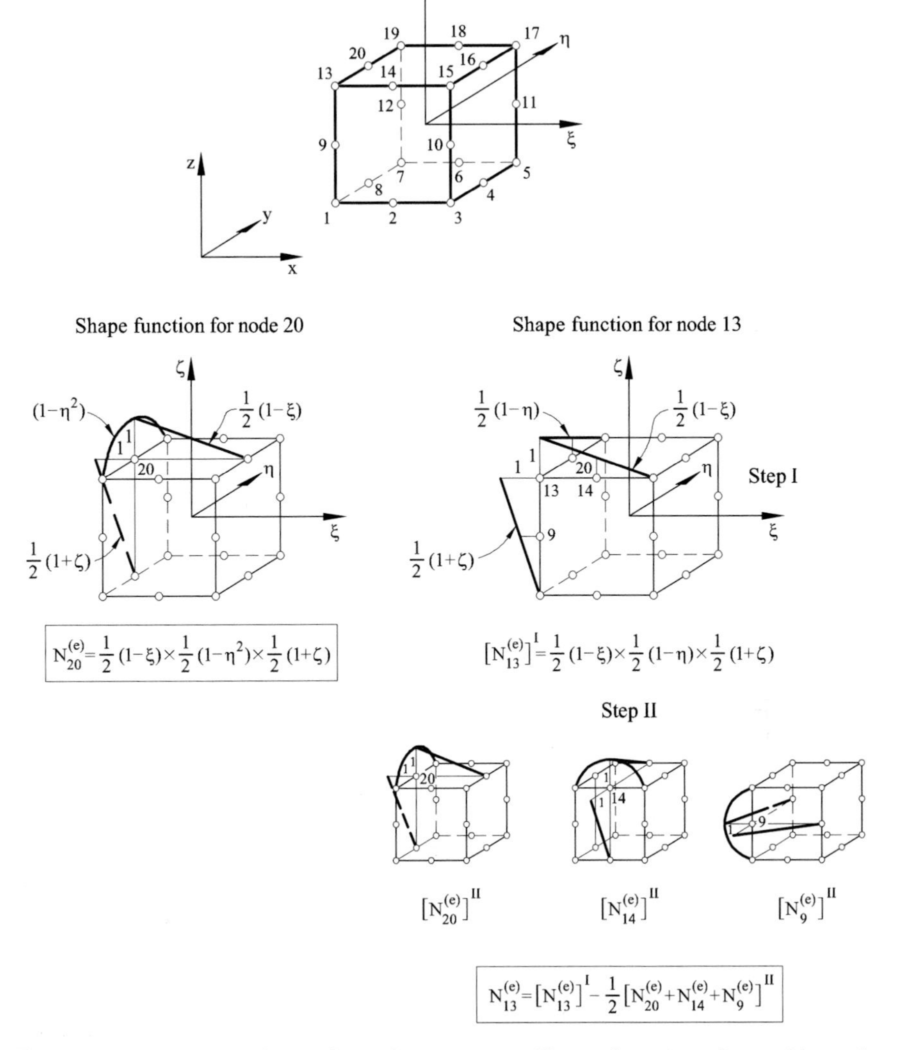

Fig. 8.11 20-noded quadratic Serendipity prism. Shape functions for a side node and a corner node

Side nodes

$$
\begin{aligned}
N_i &= \tfrac{1}{4}(1-\xi^2)(1+\eta_i\eta)(1+\zeta_i\zeta) \quad ; \quad i = 2,6,14,18 \\
&= \tfrac{1}{4}(1-\eta^2)(1+\zeta_i\zeta)(1+\xi_i\xi) \quad ; \quad i = 4,8,16,20 \qquad (8.56b) \\
&= \tfrac{1}{4}(1-\zeta^2)(1+\eta_i\eta)(1+\xi_i\xi) \quad ; \quad i = 9,10,11,12
\end{aligned}
$$

Note that:

1. The shape functions contain a complete quadratic polynomial plus the terms $\xi\eta^2$, $\xi^2\eta$, $\xi^2\zeta$, $\xi.\zeta^2$, $\zeta^2\eta$, $\eta^2\zeta$, $\xi\eta\zeta$, $\xi^2\eta\zeta$, $\xi\eta^2\zeta$ and $\xi\eta\zeta^2$. The shape functions satisfy the conditions (8.50).
2. The 20-noded Serendipity prism has the same quadratic approximation as the 27-noded Lagrange prism. This means savings of 21 nodal variables per element, which explains the popularity of the 20-noded right prism for practical applications.

8.5.2.2 32-noded cubic Serendipity prism

This element has 8 corner nodes and 24 nodes along the edges as shown in Figure 8.12. The 12 face nodes define a cubic polynomial over each face as for the corresponding quadrilateral element (Section 5.4.2). The shape functions for the side nodes are obtained by multiplying a 1D normalized cubic Lagrange polynomial and two 1D linear polynomials expressed in the natural coordinates. For the corner nodes the starting point is the trilinear shape function, from which a proportion of the shape functions of the side nodes is substracted so that the final shape function takes a zero value at these nodes. The reader is invited to derive the expressions of the shape functions for this element shown in Figure 8.12.

The shape functions contain a complete cubic polynomial (20 terms) plus the following twelve terms: $\xi^3\eta$, $\xi\eta^3$, $\eta^3\zeta$, $\eta\zeta^3$, $\xi\zeta^3$, $\zeta^3\xi$, $\xi^2\eta\zeta$, $\xi\eta^2\zeta$, $\xi\eta\zeta^2$, $\xi^3\eta\zeta$, $\xi\eta^3\zeta$, $\xi\eta\zeta^3$. Therefore, the cubic Serendipity prism has the same approximation as the analogous 64-noded Lagrangian prism, but with a substantial reduction in the number of nodal variables. However, this element is less popular than the 20-noded prism.

Example 8.1: Compute the matrix $\mathbf{K}_{11}^{(e)}$ for the 8-noded right prism of Figure 8.8 for homogeneous isotropic material.

- ***Solution***

The cartesian derivatives of the shape function $\mathbf{N}_1$ are obtained using Eqs.(8.46) and (8.49) as (note that the element sides are straight)

$$\frac{\partial N_1}{\partial x} = \frac{\partial N_1}{\partial \xi}\frac{\partial \xi}{\partial x} = \frac{1}{a}\frac{\partial}{\partial \xi}\left[\frac{1}{8}(1-\xi)(1-\eta)(1-\xi)\right] = -\frac{1}{8a}(1-\eta)(1-\zeta)$$

Serendipity right prism of 32 nodes

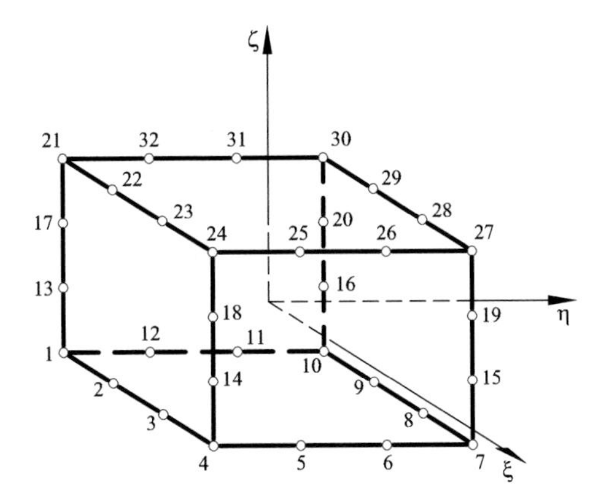

Shape functions

$$N_i = \frac{9}{64}(1+\xi\xi_i)(1+\eta\eta_i)(1+\zeta\zeta_i)\left(-\frac{19}{9}+\xi^2+\eta^2+\zeta^2\right) \qquad i=\begin{cases} 1,\ 4,\ 7,10 \\ 21,24,27,30 \end{cases}$$

$$N_i = \frac{81}{64}(1-\xi^2)\left(\frac{1}{9}+\xi\xi_i\right)(1+\eta\eta_i)(1+\zeta\zeta_i) \qquad i=\begin{cases} 2,\ 3,\ 8,\ 9 \\ 22,23,28,29 \end{cases}$$

$$N_i = \frac{81}{64}(1+\xi\xi_i)(1-\eta^2)\left(\frac{1}{9}+\eta\eta_i\right)(1+\zeta\zeta_i) \qquad i=\begin{cases} 5,\ 6,11,12 \\ 25,26,31,32 \end{cases}$$

$$N_i = \frac{81}{64}(1+\xi\xi_i)(1+\eta\eta_i)\left(\frac{1}{9}+\zeta\zeta_i\right)(1-\zeta^2) \qquad i=\begin{cases} 13,14,15,16 \\ 17,18,19,20 \end{cases}$$

Fig. 8.12 Shape functions for the 32-noded cubic Serendipity prism

and

$$\frac{\partial N_1}{\partial y} = \frac{\partial N_i}{\partial y}\frac{\partial y}{\partial \eta} = -\frac{1}{8b}(1-\xi)(1-\zeta)\ ,\ \frac{\partial N_1}{\partial z} = \frac{\partial N_i}{\partial z}\frac{\partial z}{\partial \zeta} = -\frac{1}{8c}(1-\xi)(1-\eta)$$

The strain matrix $\mathbf{B}_1$ is

$$\mathbf{B}_1 = \frac{1}{8}\begin{bmatrix} -\frac{1}{a}(1-\eta)(1-\zeta) & 0 & 0 \\ 0 & -\frac{1}{b}(1-\xi)(1-\zeta) & 0 \\ 0 & 0 & -\frac{1}{c}(1-\xi)(1-\eta) \\ -\frac{1}{b}(1-\xi)(1-\zeta) & -\frac{1}{a}(1-\eta)(1-\zeta) & 0 \\ -\frac{1}{c}(1-\xi)(1-\eta) & 0 & -\frac{1}{a}(1-\eta)(1-\zeta) \\ 0 & -\frac{1}{c}(1-\zeta)(1-\eta) & -\frac{1}{b}(1-\xi)(1-\zeta) \end{bmatrix}$$

Matrix $\mathbf{K}_{11}^{(e)}$ is obtained by

$$\mathbf{K}_{11}^{(e)} = \iiint_{V^{(e)}} \mathbf{B}_1^T \mathbf{D} \mathbf{B}_1 dV = \int_{-1}^{+1}\int_{-1}^{+1}\int_{-1}^{+1} \mathbf{B}_1^T \mathbf{D} \mathbf{B}_1 abc\, d\xi\, d\eta\, d\zeta$$

Denoting $\bar{\xi} = 1 - \xi$, $\overline{\eta} = 1 - \eta$ and $\bar{\zeta} = 1 - \zeta$ we can write

$$\mathbf{D}\,\mathbf{B}_1 = \frac{1}{8}\begin{bmatrix} d_{11} & d_{12} & d_{13} & & & \\ d_{12} & d_{22} & d_{23} & & \mathbf{0} & \\ d_{13} & d_{23} & d_{33} & & & \\ & & & d_{44} & & \\ & \mathbf{0} & & & d_{55} & \\ & & & & & d_{66} \end{bmatrix} \begin{bmatrix} -\frac{1}{a}\overline{\eta}\bar{\xi} & 0 & 0 \\ 0 & -\frac{1}{b}\overline{\xi\zeta} & 0 \\ 0 & 0 & -\frac{1}{c}\bar{\xi}\overline{\eta} \\ -\frac{1}{b}\overline{\xi\zeta} & -\frac{1}{a}\overline{\eta}\bar{\zeta} & 0 \\ -\frac{1}{c}\bar{\xi}\overline{\eta} & 0 & -\frac{1}{a}\overline{\eta}\bar{\zeta} \\ 0 & -\frac{1}{c}\bar{\xi}\overline{\eta} & -\frac{1}{b}\overline{\xi\zeta} \end{bmatrix} =$$

$$= \frac{1}{8}\begin{bmatrix} -\frac{d_{11}}{a}\overline{\eta}\bar{\zeta} & -\frac{d_{12}}{b}\overline{\xi\zeta} & -\frac{d_{13}}{c}\bar{\xi}\overline{\eta} \\ -\frac{d_{12}}{a}\overline{\eta}\bar{\zeta} & -\frac{d_{22}}{b}\overline{\xi\zeta} & -\frac{d_{23}}{c}\bar{\xi}\overline{\eta} \\ -\frac{d_{13}}{a}\overline{\eta}\bar{\zeta} & -\frac{d_{23}}{b}\overline{\xi\zeta} & -\frac{d_{33}}{c}\bar{\xi}\overline{\eta} \\ -\frac{d_{44}}{b}\overline{\xi\zeta} & -\frac{d_{44}}{a}\overline{\eta}\bar{\zeta} & 0 \\ -\frac{d_{55}}{c}\bar{\xi}\overline{\eta} & 0 & -\frac{d_{55}}{a}\overline{\eta}\bar{\zeta} \\ 0 & -\frac{d_{66}}{c}\bar{\zeta}\overline{\eta} & -\frac{d_{66}}{b}\overline{\xi\zeta} \end{bmatrix}$$

Multiplying the previous equation by $\mathbf{B}_1^T$ gives

$$\mathbf{B}_1^T \mathbf{D} \mathbf{B}_1 =$$

$$= \frac{1}{64}\begin{bmatrix} \left(\frac{d_{11}}{a^2}\overline{\eta}_1 + \frac{d_{44}}{b^2}\bar{\xi}_2 + \frac{d_{55}}{c^2}\bar{\xi}_1\right) & \left(d_{12} + d_{44}\right)\frac{\bar{\xi}_3}{ab} & \left(d_{13} + d_{55}\right)\frac{\bar{\xi}_4}{ac} \\ & \left(\frac{d_{22}}{b^2}\bar{\xi}_2 + \frac{d_{44}}{a^2}\overline{\eta}_1 + \frac{d_{66}}{c^2}\bar{\xi}_1\right) & \left(d_{23} + d_{66}\right)\frac{\bar{\xi}_5}{bc} \\ \text{Symmetrical} & & \left(\frac{d_{33}}{c^2}\bar{\xi}_1 + \frac{d_{55}}{a^2}\overline{\eta}_1 + \frac{d_{66}}{b^2}\bar{\xi}_2\right) \end{bmatrix}$$

with $\bar{\xi}_1 = \bar{\xi}^2\bar{\eta}^2$, $\bar{\xi}_2 = \bar{\xi}^2\bar{\zeta}^2$, $\bar{\xi}_3 = \bar{\xi}\bar{\eta}\bar{\zeta}^2$, $\bar{\xi}_4 = \bar{\xi}\bar{\eta}^2\bar{\zeta}$, $\bar{\xi}_5 = \bar{\xi}^2\bar{\eta}\bar{\zeta}$, $\bar{\eta}_1 = \bar{\eta}^2\bar{\zeta}^2$.
Taking into account that

$$\int_{-1}^{+1} \bar{\xi}^2 d\xi = \int_{-1}^{+1} \overline{\eta}^2 d\eta = \int_{-1}^{+1} \bar{\zeta}^2 = \frac{8}{3}$$

$$\int_{-1}^{+1} \bar{\xi}\, d\xi = \int_{-1}^{+1} \overline{\eta}\, d\eta = \int_{-1}^{+1} \bar{\zeta}\, d\zeta = 2$$

we finally obtain

$$\mathbf{K}_{11}^{(e)} = \frac{V^{(e)}}{8} \begin{bmatrix} \frac{2}{9}\left(\frac{d_{11}}{a^2}+\frac{d_{44}}{b^2}+\frac{d_{55}}{c^2}\right) & \frac{1}{6ab}\left(d_{12}+d_{44}\right) & \frac{1}{6ac}\left(d_{13}+d_{55}\right) \\ & \frac{2}{9}\left(\frac{d_{22}}{b^2}+\frac{d_{44}}{a^2}+\frac{d_{66}}{c^2}\right) & \frac{1}{6bc}\left(d_{23}+d_{66}\right) \\ \text{Symmetrical} & & \frac{2}{9}\left(\frac{d_{33}}{c^2}+\frac{d_{55}}{a^2}+\frac{d_{66}}{b^2}\right) \end{bmatrix}$$

The rest of the $\mathbf{K}_{ij}^{(e)}$ matrices are obtained following a similar procedure.

8.6 STRAIGHT-EDGED TETRAHEDRA

Straight-edged tetrahedral elements are a direct 3D extension of straight-sided triangles. Their shape functions are also complete polynomials whose terms can easily be deduced from the Pascal tetrahedron as shown in Figure 8.13.

The shape functions for tetrahedral elements can be written in terms of volume coordinates and/or natural coordinates. *The volume coordinates* are identified by L_1, L_2, L_3 and L_4 and have a similar meaning to the area coordinates in triangles. Each coordinate L_i is now defined as the ratio between the volume of the tetrahedron formed by a point inside the element P and the face opposite to node i, and the total volume (Figure 8.14). Thus

$$L_i = \frac{\text{Volume } Pjkl}{V^{(e)}} \quad ; \quad i = 1,2,3,4 \tag{8.57}$$

Obviously, the following expression holds

$$L_1 + L_2 + L_3 + L_4 = 1 \tag{8.58}$$

Volume coordinates can be used to define a linear interpolation of the element geometry as

$$x = \sum_{i=1}^{4} L_i\, x_i, \quad y = \sum_{i=1}^{4} L_i\, y_i, \quad z = \sum_{i=1}^{4} L_i\, z_i \tag{8.59}$$

Eqs. (8.58) and (8.59) allow us to eliminate L_i in terms of the cartesian coordinates as

$$L_i = \frac{l}{6V^{(e)}}(a_i + b_i x + c_i y + d_i z) \tag{8.60}$$

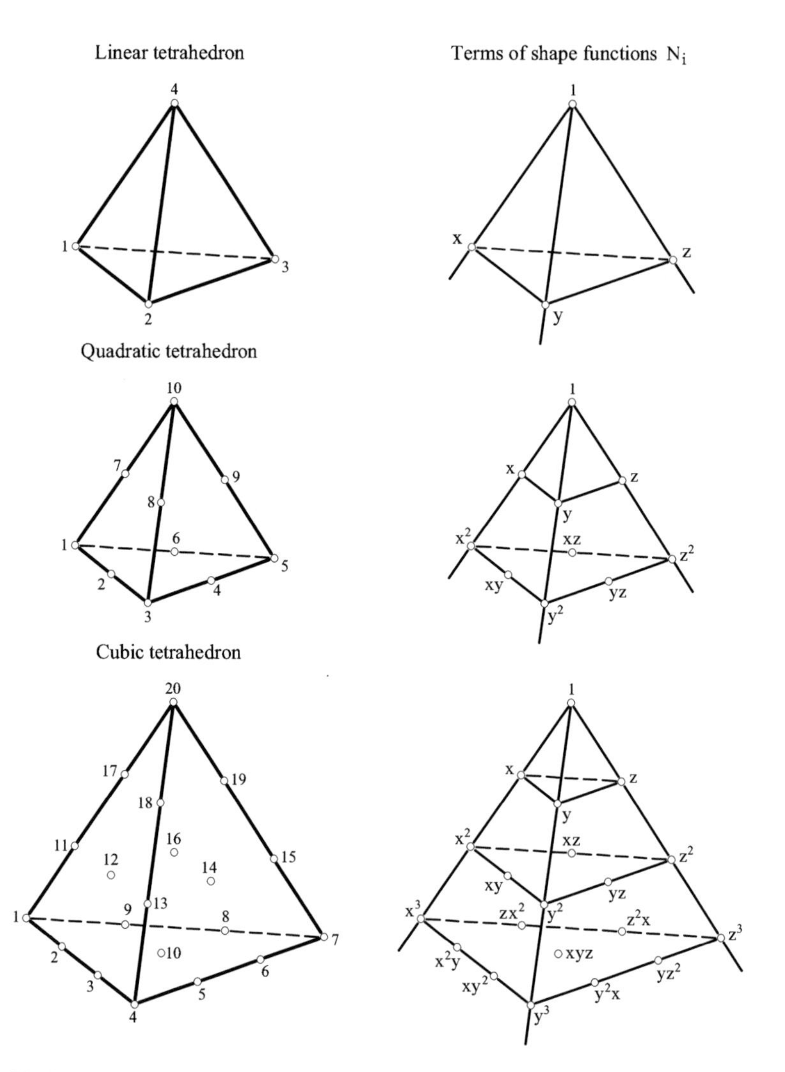

Fig. 8.13 Straight-edged tetrahedral elements: linear (4 nodes), quadratic (10 nodes), cubic (20 nodes). Polynomial terms contained in the shape functions

where a_i, b_i, c_i, d_i coincide with the values given in Eq.(8.18). Therefore, *the volume coordinates coincide with the shape functions for the 4-noded tetrahedron.*

Eq.(8.60) allows us to obtain the cartesian derivatives of the volume coordinates as

$$\frac{\partial L_i}{\partial x} = \frac{l}{6V^{(e)}}\, b_i \quad ; \quad \frac{\partial L_i}{\partial y} = \frac{l}{6V^{(e)}}\, c_i \quad ; \quad \frac{\partial L_i}{\partial z} = \frac{l}{6V^{(e)}}\, d_i \tag{8.61}$$

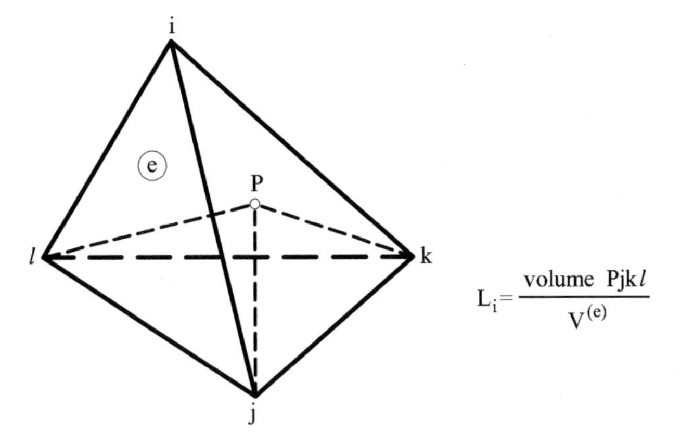

Fig. 8.14 Volume coordinates in a tetrahedron

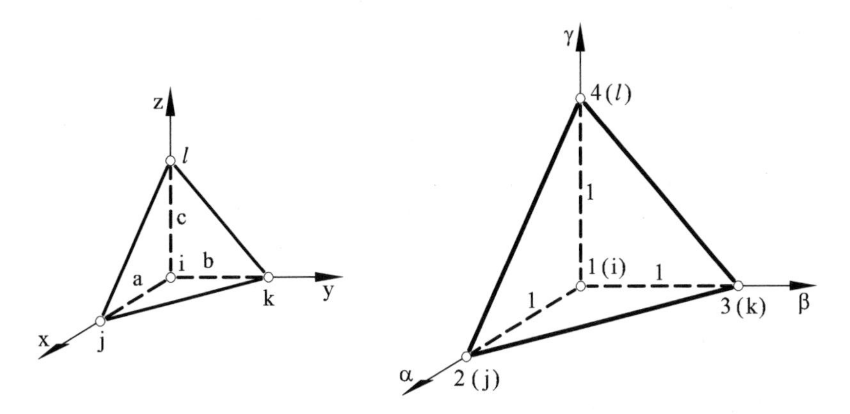

Fig. 8.15 Natural coordinate system α, β, γ in a tetrahedron

The *natural coordinates* α, β, γ define a normalized straight tetrahedron of unit right edges and faces for $\alpha = 0, \beta = 0, \gamma = 0$ and $1 - \alpha - \beta - \gamma = 0$ (Figure 8.15). For a tetrahedron with right edges a, b, c we have

$$\alpha = \frac{x - x_i}{a} \quad ; \quad \beta = \frac{y - y_i}{b} \quad ; \quad \gamma = \frac{z - z_i}{c} \tag{8.62}$$

where i is the node taken as origin of the natural coordinate system ($i = 1$ in the local numbering system). From Eq.(8.62) we deduce

$$\frac{d\alpha}{dx} = \frac{1}{a} \quad ; \quad \frac{d\beta}{dy} = \frac{1}{b} \quad ; \quad \frac{d\gamma}{dz} = \frac{1}{c} \tag{8.63a}$$

A differential of volume can be expressed as

$$dV = dx\ dy\ dz = abc\ d\alpha\ d\beta\ d\gamma \tag{8.63b}$$

The integral of a function $f(x, y, z)$ over the element can be written in the natural coordinate system as

$$\iiint_{V^{(e)}} f(x,y,z)\,dx\,dy\,dz = \int_0^1 \int_0^{1-\alpha} \int_0^{1-\beta-\gamma} f(\alpha,\beta,\gamma) abc\,d\alpha\,d\beta\,d\gamma \tag{8.64}$$

The shape functions for the linear tetrahedron can be expressed simply in terms of the natural coordinates α, β, γ as

$$N_1 = 1 - \alpha - \beta - \gamma \ \ ; \ \ N_2 = \alpha; \quad N_3 = \beta; \quad N_4 = \gamma \tag{8.65}$$

The shape functions in the natural coordinate system satisfy Eqs.(8.50). Also from Eqs.(8.63a) and (8.65) we obtain

$$\frac{\partial N_i}{\partial x} = \frac{1}{a}\frac{\partial N_i}{\partial \alpha} \ ; \quad \frac{\partial N_i}{\partial y} = \frac{1}{b}\frac{\partial N_i}{\partial \beta} \ ; \quad \frac{\partial N_i}{\partial z} = \frac{1}{c}\frac{\partial N_i}{\partial \gamma} \tag{8.66}$$

The relationship between the volume and natural coordinates is readily deduced from Eqs.(8.60) and (8.65) (recalling that $N_i = L_i$) as

$$L_1 = 1 - \alpha - \beta - \gamma; \quad L_2 = \alpha; \quad L_3 = \beta \quad L_4 = \gamma \tag{8.67}$$

Volume coordinates allow us to express the element shape functions for tetrahedral elements by the product of four 1D normalized Lagrange polynomials, in a similar way as explained in Section 5.5.2 for triangular elements. Thus the shape function of a node i with generalized coordinates (I, J, K, L) is given by

$$N_i = l_I^i(L_1)\ l_J^i(L_2)\ l_K^i(L_3)\ l_L^i(L_4) \tag{8.68}$$

where the value of the generalized coordinates I, J, K, and L coincides with the power of each volume coordinate in the expression of N_i. Hence $I + J + K + L = M$ where M is the degree of the complete polynomial contained in N_i. Also, $l_I^i(L_j)$ is the normalized Lagrange polynomial of Ith degree in L_j passing by node i (Eq.(3.5b)). Figures 8.16 and 8.17 show the values of the generalized coordinates I, J, K, L for two typical tetrahedral elements.

The expression of the shape functions in terms of natural coordinates α, β, γ is straightforward using the transformations (8.67).

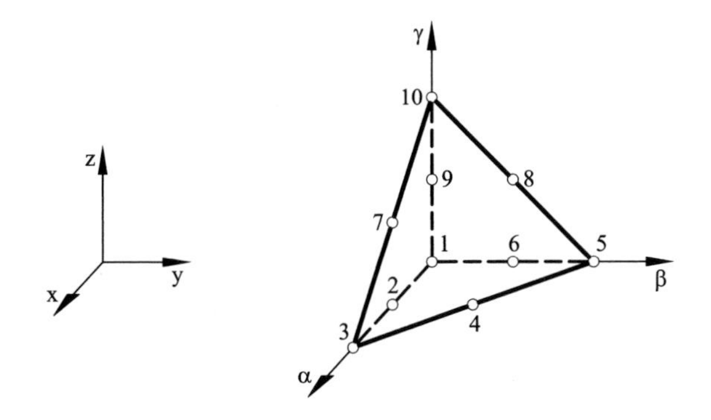

Local node numbers	Coordinates				Natural coordinates		
	I	J	K	L	α_i	β_i	γ_i
1	2	0	0	0	0	0	0
2	1	1	0	0	1/2	0	0
3	0	2	0	0	1	0	0
4	0	1	1	0	1/2	1/2	0
5	0	0	2	0	0	1	0
6	1	0	1	0	0	1/2	0
7	0	1	0	1	1/2	0	1/2
8	0	0	1	1	0	1/2	1/2
9	1	0	0	1	0	0	1/2
10	0	0	0	2	0	0	1

Fig. 8.16 10-noded quadratic tetrahedron. Generalized coordinates I, J, K, L and natural coordinates α, β, γ for each node

8.6.1 Shape functions for the 10-noded quadratic tetrahedron

The nodal values of the generalized coordinates I, J, K, L and of the natural coordinates are shown in Figure 8.16. The values of the volume coordinates L_i are deduced from Eq.(8.67). The shape functions are derived next using Eq.(8.68).

Node 1

Position (I, J, K, L) : $(2, 0, 0, 0)$. Volume coordinates : $(1, 0, 0, 0)$

$$N_1 = l_2^1(L_1) = \frac{\left(L_1 - \frac{1}{2}\right)L_1}{1 - \frac{1}{2}} = (2L_1 - 1)L_1 \tag{8.69a}$$

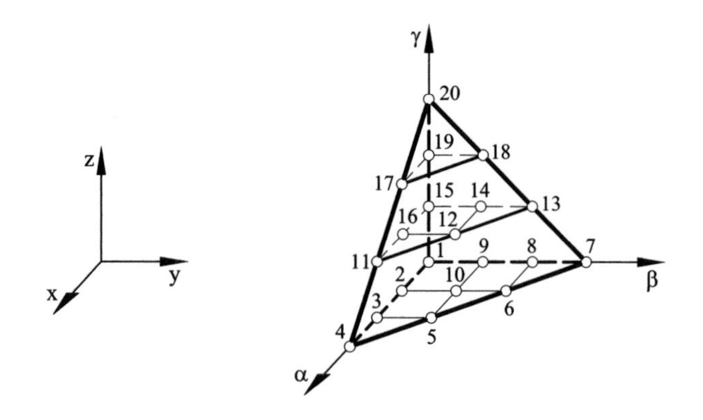

Local node numbers	Coordinates I	J	K	L	Natural Coordinates α_i	β_i	γ_i
1	3	0	0	0	0	0	0
2	2	1	0	0	1/3	0	0
3	1	2	0	0	2/3	0	0
4	0	3	0	0	1	0	0
5	0	2	1	0	2/3	1/3	0
6	0	1	2	0	1/2	2/3	0
7	0	0	3	0	0	1	0
8	1	0	2	0	0	2/3	0
9	2	0	1	0	0	1/3	0
10	1	1	1	0	1/3	1/3	0
11	0	2	0	1	2/3	0	1/3
12	0	1	1	1	1/3	1/3	1/3
13	0	0	2	1	0	2/3	1/3
14	1	0	1	1	0	1/3	1/3
15	2	0	0	1	0	0	1/3
16	1	1	0	1	1/3	0	1/3
17	0	1	0	2	1/3	0	2/3
18	0	0	1	2	0	1/3	2/3
19	1	0	0	2	0	0	2/3
20	0	0	0	3	0	0	1

Fig. 8.17 20-noded cubic tetrahedron. Generalized coordinates I, J, K, L and natural coordinates for each node

Node 2

Position (I, J, K, L) : $(1,1,0,0)$. Volume coordinates : $(\frac{1}{2}, \frac{1}{2}, 0, 0)$

$$N_2 = l_1^2(L_1)\ l_1^2(L_2) = \frac{L_1}{\frac{1}{2}}\ \frac{L_2}{\frac{1}{2}} = 4L_1\ L_2 \tag{8.69b}$$

The same procedure gives

$$\begin{aligned} N_3 &= (2L_2 - 1)L_2 \quad &;\quad N_7 &= 4\ L_2\ L_4 \\ N_4 &= 4L_2\ L_3 \quad &;\quad N_8 &= 4\ L_3\ L_4 \\ N_5 &= (2\ L_3 - 1)L_3 \quad &;\quad N_9 &= 4\ L_1\ L_4 \\ N_6 &= 4\ L_1\ L_3 \quad &;\quad N_{10} &= (2\ L_4 - 1)\ L_4 \end{aligned} \tag{8.69c}$$

The expression for N_i in terms of the natural coordinates is obtained using Eq.(8.67). The cartesian form of N_i for a straight-edged tetrahedron is obtained by substituting L_i in terms of x, y, z using Eq.(8.60).

The shape functions for this element contain all the terms of a quadratic polynomial (Figure 8.13) and they satisfy Eqs.(8.50).

8.6.2 Shape functions for the 20-noded quadratic tetrahedron

The nodal values of the generalized coordinates I, J, K, L and α, β, γ are shown in Figure 8.17. From Eqs.(8.68) we obtain:

Node 1

Position (I, J, K, L) : $(3,0,0,0)$. Volume coordinates : $(3,0,0,0)$

$$N_1 = l_3^1\ (L_1) = \frac{\left(L_1 - \frac{2}{3}\right)\left(L_1 - \frac{1}{3}\right)L_1}{\left(1 - \frac{2}{3}\right)\left(1 - \frac{1}{3}\right)1} = \frac{1}{2}L_1(3L_1 - 1)\ (3L_1 - 2) \tag{8.70a}$$

Node 2

Position (I, J, K, L) : $(2,1,0,0)$. Volume coordinates : $(\frac{2}{3}, \frac{1}{3}, 0, 0)$

$$N_2 = l_2^2(L_1)\ L_1^2(L_2) = \frac{\left(L_1 - \frac{1}{3}\right)\ L - 1}{\left(\frac{2}{3} - \frac{1}{3}\right)\frac{2}{3}}\ \frac{L_2}{\frac{1}{3}} = \frac{9}{2}(3L_1 - 1)\ L_1\ L_2 \tag{8.70b}$$

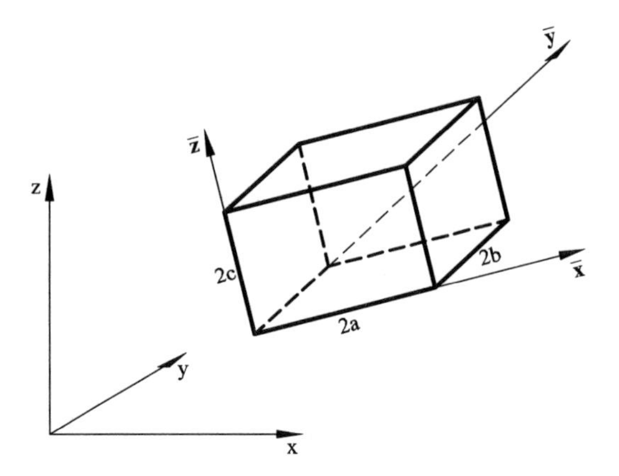

Fig. 8.18 Local coordinate system $\bar{x}\bar{y}\bar{z}$ for the analytical computation of volume integrals over right prisms

Following a similar procedure gives

$$
\begin{array}{lll}
N_3 = \frac{9}{2}(3L_2 - 1)L_1L_2 & ; \; N_9 = \frac{9}{2}(3L_1 - 1)L_1L_3 & ; \; N_{15} = \frac{9}{2}(3L_1 - 1)L_1L_4 \\
N_4 = \frac{1}{2}L_2(3L_2 - 1)(3L_2 - 2) & ; \; N_{10} = 27\ L_1\ L_2L_3 & ; \; N_{16} = 27L_1L_2L_4 \\
N_5 = \frac{9}{2}(3L_2 - 1)L_2\ {}_3 & ; \; N_{11} = \frac{9}{2}(3L_2 - 1)L_2\ L_4 & ; \; N_{17} = \frac{9}{2}(3L_4 - 1)L_4\ L_2 \\
N_6 = \frac{9}{2}(3\ L_3 - 1)L_3\ L_2 & ; \; N_{12} = 27\ L_2\ L_3\ L_4 & ; \; N_{18} = \frac{9}{2}(3L_4 - 1)L_4\ L_3 \\
N_7 = \frac{1}{2}L_3(3L_3 - 1)(3L_3 - 2) & ; \; N_{13} = \frac{9}{2}(3L_3 - 1)L_3\ L_4 & ; \; N_{19} = \frac{9}{2}(3L_4 - 1)L_4\ L_1 \\
N_8 = \frac{9}{2}(3L_3 - 1)L_3L_1 & ; \; N_{14} = 27L_1L_3L_4 & ; \; N_{20} = \frac{1}{2}L_4(3L_4 - 1)(3L_4 - 2)
\end{array}
\tag{8.70c}
$$

Eqs.(8.67) allow us to express N_i in terms of the natural coordinates. The cartesian form of N_i for a straight-edged tetrahedron is obtained substituting Eq.(8.60) into above expressions. The shape functions are complete cubic polynomials and they satisfy Eq.(8.50).

8.7 COMPUTATION OF ELEMENT INTEGRALS

8.7.1 Analytical computation of element integrals

In general, the computation of element integrals is carried out via numerical integration. Useful analytical rules can, however, be derived for straight-edged tetrahedra or right prisms as shown below.

Volume integrals over right prisms can be computed exactly using the local coordinate system $\bar{x}\bar{y}\bar{z}$ shown in Figure 8.18. For instance, the integral over a face with $\overline{z} = constant$, is first obtained by Eq.(5.41) and then a simple line integral in the $\overline{z}$ direction is performed. The resulting local stiffness matrix $\overline{\mathbf{K}}$ is next transformed to the global axes xyz by a transformation similar to Eq.(5.42) taking into account the third cartesian axis. This procedure is not applicable to irregular or curve-sided hexahedra for which the use of an isoparametric formulation and numerical integration is essential.

The integrals over straight-sided tetrahedra have simpler expressions. Thus, the volume integral of a polynomial term expressed in volume coordinates is given by

$$\iiint_{V^{(e)}} L_1^k L_2^l L_3^m L_4^n \, dV = 6V^{(e)} \frac{k!\; l!\; m!\; n!}{(k+l+m+n+3)!} \tag{8.71}$$

Similarly, the surface integrals over the element faces can be obtained by Eqs.(5.40) or (5.47a).

The use of natural coordinates does not introduce any additional difficulty. The volume integrals in tetrahedra are computed by

$$\iiint_{V^{(e)}} \alpha^k \beta^l \gamma^m \, dV = 6V^{(e)} \frac{k!\; l!\; m!}{(k+l+m+3)!} \tag{8.72}$$

and the surface integrals can be obtained by Eq.(5.48). Recall that if any of the coordinates is missing in the integrals of Eqs.(8.71) and (8.72) then the corresponding power is made equal to one in the numerator and to zero in the denominator of the corresponding right-hand side.

Curve-sided tetrahedra require an isoparametric formulation and numerical integration.

Example 8.2: Compute the submatrix $\mathbf{K}_{11}^{(e)}$ for a 20-noded quadratic tetrahedron with straight sides.

- ***Solution***

We compute first the cartesian derivatives of the shape function N_1 expressed in terms of volume coordinates. For instance,

$$\frac{\partial N_1}{\partial x} = \frac{\partial N_1}{\partial L_1}\frac{\partial L_1}{\partial x} + \frac{\partial N_1}{\partial L_2}\frac{\partial L_2}{\partial x} + \frac{\partial N_1}{\partial L_3}\frac{\partial L_3}{\partial x} + \frac{\partial N_1}{\partial L_4}\frac{\partial L_4}{\partial x}$$

Since $N_1 = L_1(2L_1 - 1)$ we have, using Eq.(8.60)

$$\frac{\partial N_1}{\partial x} = (4L_1 - 1)\,\frac{b_i}{6V^{(e)}}$$

Following a similar procedure gives

$$\frac{\partial N_1}{\partial y} = (4L_1 - 1)\,\frac{c_i}{6V^{(e)}} \qquad \text{and} \qquad \frac{\partial N_1}{\partial z} = (4L_1 - 1)\,\frac{d_i}{6V^{(e)}}$$

The strain matrix $\mathbf{B}_1$ is written as

$$\mathbf{B}_1 = \frac{(4L_1 - 1)}{6V^{(e)}} \begin{bmatrix} b_i & 0 & 0 \\ 0 & c_i & 0 \\ 0 & 0 & d_i \\ c_i & b_i & 0 \\ d_i & 0 & b_i \\ 0 & d_i & c_i \end{bmatrix} = \frac{(4L_1 - 1)}{6V^{(e)}}\,\overline{\mathbf{B}}_1$$

and matrix $\mathbf{K}_{11}^{(e)}$ is given by

$$\mathbf{K}_{11}^{(e)} = \iiint_{V^{(e)}} \mathbf{B}_1^T \mathbf{D} \mathbf{B}_1 dV = \frac{1}{36(V^{(e)})^2} \overline{\mathbf{B}}_1^T \mathbf{D} \overline{\mathbf{B}}_1^T \iiint_{V^{(e)}} (4L_1 - 1)^2\, dV$$

Making use of Eq.(8.72) gives finally

$$\mathbf{K}_{11}^{(e)} = \frac{1}{60V^{(e)}} \begin{bmatrix} (d_{11}b_1^2 + d_{44}c_1^2 + d_{55}d_1^2) & b_1c_1(d_{12} + d_{44}) & (d_{13}b_1d_1 + d_{55}d_1c_1) \\ & (d_{22}c_1^2 + d_{44}b_1^2 + d_{66}d_1^2) & c_1d_1(d_{23} + d_{66}) \\ \text{Symmetrical} & & (d_{33}d_1^2 + d_{55}b_1^2 + d_{66}c_1^2) \end{bmatrix}$$

The rest of the $\mathbf{K}_{ij}^{(e)}$ matrices are obtained in a similar manner.

8.8 3D ISOPARAMETRIC ELEMENTS

3D elements with arbitrary geometry can be derived using an isoparametric formulation. The element geometry is defined in the global cartesian coordinate system in terms of the nodal coordinates, whereas the element integrals are performed in the natural coordinate space over cubes of side length equal to two or straight tetrahedra with right edges of unit length using simple transformations (Figure 8.19).

Let us consider first hexahedral elements. The coordinates of a point within a n-noded element are expressed in isoparametric form as

$$\mathbf{x} = \begin{Bmatrix} x \\ y \\ z \end{Bmatrix} = \sum_{i=1}^{n} \mathbf{N}_i \begin{Bmatrix} x_i \\ y_i \\ z_i \end{Bmatrix} = \mathbf{N}\,\mathbf{x}^{(e)} \tag{8.73a}$$

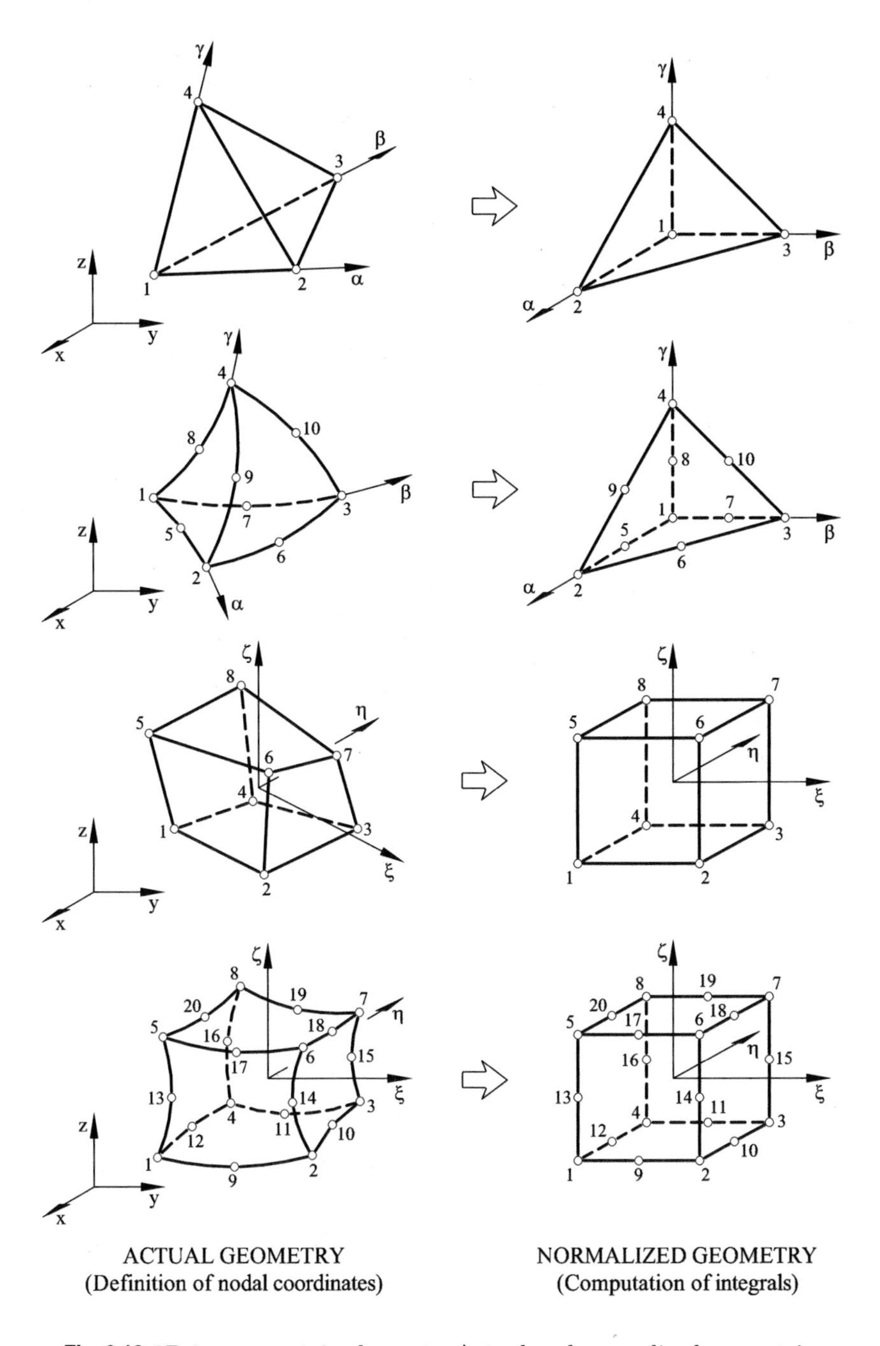

Fig. 8.19 3D isoparametric elements. Actual and normalized geometries

with

$$\mathbf{N} = [\mathbf{N}_1, \mathbf{N}_2, \ldots, \mathbf{N}_n] \quad ; \quad \mathbf{N}_i = \begin{bmatrix} N_i & & \\ & N_i & \\ & & N_i \end{bmatrix} \quad ; \quad N_i = f(\xi, \eta, \zeta) \quad (8.73b)$$

where N_i are the shape functions used for the displacement field.

Eq.(8.73a) relates the cartesian and the natural coordinates. This relationship is uniquely defined if the sign of the determinant of the Jacobian matrix is positive over the element. This is usually fulfilled except for very distorted element shapes. Typically the rules given in Section 6.2 for 2D quadrilateral elements also apply in this case.

The cartesian derivatives of the shape functions are computed following the same procedure as explained for 2D elements. The chain rule of derivation gives

$$\begin{Bmatrix} \frac{\partial N_i}{\partial \xi} \\ \frac{\partial N_i}{\partial \eta} \\ \frac{\partial N_i}{\partial \zeta} \end{Bmatrix} = \begin{bmatrix} \frac{\partial x}{\partial \xi} & \frac{\partial y}{\partial \xi} & \frac{\partial z}{\partial \xi} \\ \frac{\partial x}{\partial \eta} & \frac{\partial y}{\partial \eta} & \frac{\partial z}{\partial \eta} \\ \frac{\partial x}{\partial \zeta} & \frac{\partial y}{\partial \zeta} & \frac{\partial z}{\partial \zeta} \end{bmatrix} \begin{Bmatrix} \frac{\partial N_i}{\partial x} \\ \frac{\partial N_i}{\partial y} \\ \frac{\partial N_i}{\partial z} \end{Bmatrix} = \mathbf{J}^{(e)} \frac{\partial \mathbf{N}_i}{\partial \mathbf{x}} \quad (8.74)$$

where $\mathbf{J}^{(e)}$ is the Jacobian matrix. Using Eq.(8.73a) yields

$$\mathbf{J}^{(e)} = \sum_{i=1}^{n} \begin{bmatrix} \frac{\partial N_i}{\partial \xi} x_i & \frac{\partial N_i}{\partial \xi} y_i & \frac{\partial N_i}{\partial \xi} z_i \\ \frac{\partial N_i}{\partial \eta} x_i & \frac{\partial N_i}{\partial \eta} y_i & \frac{\partial N_i}{\partial \eta} z_i \\ \frac{\partial N_i}{\partial \zeta} x_i & \frac{\partial N_i}{\partial \zeta} y_i & \frac{\partial N_i}{\partial \zeta} z_i \end{bmatrix} \quad (8.75)$$

The cartesian derivatives of N_i are deduced from Eq.(8.74) as

$$\begin{Bmatrix} \frac{\partial N_i}{\partial x} \\ \frac{\partial N_i}{\partial y} \\ \frac{\partial N_i}{\partial z} \end{Bmatrix} = \left[\mathbf{J}^{(e)}\right]^{-1} \begin{Bmatrix} \frac{\partial N_i}{\partial \xi} \\ \frac{\partial N_i}{\partial \eta} \\ \frac{\partial N_i}{\partial \zeta} \end{Bmatrix} \quad (8.76)$$

Appendix A gives the analytical form of the inverse of the 3×3 Jacobian matrix. The volume differential is expressed as

$$dV = dx\ dy\ dz = \left|\mathbf{J}^{(e)}\right| d\xi\ d\eta\ d\zeta \tag{8.77}$$

The strain matrix for an isoparametric prismatic element can be expressed in terms of the natural coordinates using Eqs.(8.27b) and (8.76) as

$$\mathbf{B}_i(\xi,\eta,\zeta) = \begin{bmatrix} \bar{b}_i & 0 & 0 \\ 0 & \bar{c}_i & 0 \\ 0 & 0 & \bar{d}_i \\ \bar{c}_i & \bar{b}_i & 0 \\ \bar{d}_i & 0 & \bar{b}_i \\ 0 & \bar{d}_i & \bar{c}_i \end{bmatrix} \tag{8.78a}$$

with

$$\begin{Bmatrix} \bar{b}_i \\ \bar{c}_i \\ \bar{d}_i \end{Bmatrix} = \sum_{k=1}^{3} \begin{Bmatrix} \bar{J}_{1k}^{(e)} \\ \bar{J}_{2k}^{(e)} \\ \bar{J}_{3k}^{(e)} \end{Bmatrix} \frac{\partial N_i}{\partial \xi_k} \tag{8.78b}$$

where $\bar{J}_{ij}^{(e)}$ is the term ij of the inverse Jacobian matrix $[\mathbf{J}^{(e)}]^{-1}$, $\xi_1 = \xi, \xi_2 = \eta$ and $\xi_3 = \zeta$.

The stiffness matrix for a 3D isoparametric hexahedral can be computed over the unit cubic domain in the normalized space by the following expressions

$$\begin{aligned} \mathbf{K}_{ij}^{(e)} &= \iiint_{V^{(e)}} \mathbf{B}_i^T\ \mathbf{D}\ \mathbf{B}_j\ dV = \\ &= \int_{-1}^{+1}\int_{-1}^{+1}\int_{-1}^{+1} \mathbf{B}_i^T(\xi,\eta,\zeta)\mathbf{D}\ \mathbf{B}_j(\xi,\eta,\zeta)\left|\mathbf{J}^{(e)}\right| d\xi\, d\eta\, d\zeta = \\ &= \int_{-1}^{+1}\int_{-1}^{+1}\int_{-1}^{+1} \mathbf{G}_{ij}(\xi,\eta,\zeta)\, d\xi\, d\eta\, d\zeta \end{aligned} \tag{8.79a}$$

with

$$\mathbf{G}_{ij} = \begin{bmatrix} (d_{11}\bar{b}_{ij} + d_{44}\bar{c}_{ij} + d_{55}\bar{d}_{ij}) & (d_{12}\bar{b}_i\bar{c}_j + d_{44}\bar{c}_i\bar{b}_j) & (d_{13}\bar{b}_i\bar{d}_j + d_{55}\bar{d}_i\bar{b}_j) \\ (d_{21}\bar{c}_i\bar{b}_j + d_{44}\bar{b}_i\bar{c}_j) & (d_{22}\bar{c}_{ij} + d_{44}\bar{b}_{ij} + d_{66}\bar{d}_{ij}) & (d_{23}\bar{c}_i\bar{d}_j + d_{66}\bar{d}_i\bar{c}_j) \\ (d_{31}\bar{d}_i\bar{b}_j + d_{55}\bar{b}_i\bar{d}_j) & (d_{32}\bar{d}_i\bar{c}_j + d_{66}\bar{c}_i\bar{d}_j) & (d_{33}\bar{d}_{ij} + d_{55}\bar{b}_{ij} + d_{66}\bar{c}_{ij}) \end{bmatrix} \left|\mathbf{J}^{(e)}\right| \tag{8.79b}$$

where $\bar{b}_{ij} = \bar{b}_i\bar{b}_j$, $\bar{c}_{ij} = \bar{c}_i\bar{c}_j$ and $\bar{d}_{ij} = \bar{d}_i\bar{d}_j$, with $\bar{b}_i, \bar{c}_i, \bar{d}_i$ as given in Eq.(8.78b) and d_{ij} are the terms of matrix $\mathbf{D}$ of Eq.(8.8). Matrix $\mathbf{G}$ typically contains rational algebraic functions arising from the Jacobian inverse contributions. Numerical integration is mandatory in this case.

Isoparametric tetrahedral elements follow a similar procedure. The geometry interpolation is defined by Eq.(8.73a) with N_i expressed in terms of volume or natural coordinates. The computation of the cartesian derivatives of N_i for straight-side tetrahedra is immediate from Eqs.(8.61) and (8.66). The element integrals can be expressed in terms of volume or natural coordinates. Natural coordinates are more convenient for curved tetrahedra. The derivation of the stiffness matrix follows precisely the same steps as explained above for hexahedra simply substituting the coordinates ξ, η, ζ for α, β, γ, respectively. The element stiffness matrix is then computed in the normalized tetrahedron (Figure 8.15) as

$$\mathbf{K}_{ij}^{(e)} = \int_0^1 \int_0^{1-\alpha} \int_0^{1-\alpha-\beta} \mathbf{G}_{ij}(\alpha, \beta, \gamma)\, d\alpha\, d\beta\, d\gamma \tag{8.80}$$

where $\mathbf{G}(\alpha, \beta, \gamma)$ is deduced from Eq.(8.79b). Here again numerical integration is mandatory.

8.9 NUMERICAL INTEGRATION

8.9.1 Hexahedral elements

Let us consider the integration of a function $f(x, y, z)$ over a hexahedral isoparametric element. The following transformations are required

$$\iiint_{V^{(e)}} f(x,y,z)\, dx\, dy\, dz = \int_{-1}^{1}\int_{-1}^{1}\int_{-1}^{1} f(\xi,\eta,\zeta) \left|\mathbf{J}^{(e)}\right| d\xi\, d\eta\, d\zeta = \\ = \int_{-1}^{+1}\int_{-1}^{+1}\int_{-1}^{+1} g(\xi,\eta,\zeta)\, d\xi\, d\eta\, d\zeta \tag{8.81}$$

Gauss quadrature over the normalized cubic domain leads to

$$\int_{-1}^{+1}\int_{-1}^{+1}\int_{-1}^{+1} g(\xi,\eta,\zeta)\, d\xi\, d\eta\, d\zeta = \int_{-1}^{+1}\int_{-1}^{+1} \sum_{p=1}^{n_p} W_p\, g(\xi_p,\eta,\zeta)\, d\eta\, d\zeta = \\ = \int_{-1}^{+1} \sum_{q=1}^{n_q}\sum_{p=1}^{n_p} W_p W_q g(\xi_p,\eta_q,\zeta)\, d\zeta = \sum_{r=1}^{n_r}\sum_{q=1}^{n_q}\sum_{p=1}^{n_p} W_p W_q W_r g(\xi_p,\eta_q,\zeta_r) \tag{8.82}$$

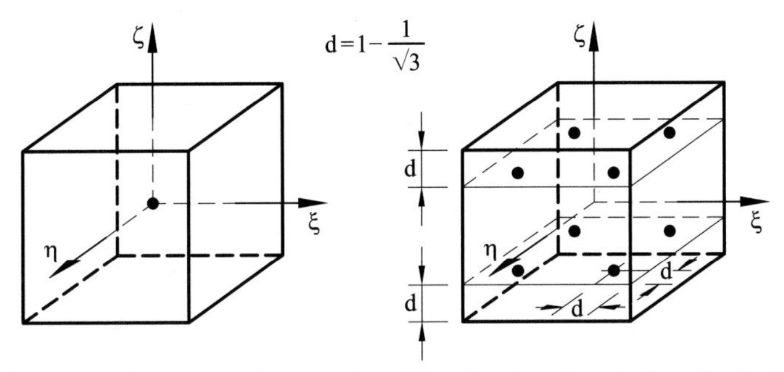

Fig. 8.20 Gauss quadratures of $1\times1\times1$ and $2\times2\times2$ points in hexahedral elements

where n_p, n_q and n_r are the integration points via the ξ, η, ζ directions, respectively, ξ_p, η_q, ζ_r are the coordinates of the integration point (p, q, r) and W_p, W_q, W_r are the weights for each natural direction.

The local coordinates and weights for each quadrature are deduced from Table 3.1 for the 1D case. We recall that a qth order quadrature integrates exactly a 1D polynomial of degree $2q - 1$. This rule helps us to identify the number of integration points in each natural direction. Figure 8.20 shows the sampling points for the $1 \times 1 \times 1$ and $2 \times 2 \times 2$ quadratures.

8.9.2 Tetrahedral elements

Gauss quadrature for tetrahedral elements formulated in terms of volume coordinates is written as

$$\int_0^1 \int_0^{1-L_1} \int_0^{1-L_1-L_2} f(L_1, L_2, L_3, L_4) dL_1 dL_2 dL_3 = \sum_{i=1}^{n_p} f(L_{1_i}, L_{2_i}, L_{3_i}, L_{4_i}) W_i \tag{8.83}$$

Figure 8.21 shows the position of the integration points and the corresponding weights for the linear, quadratic and cubic quadratures. Higher order quadratures can be found in [Be,GH,SC].

The weights in Figure 8.21 have been normalized so that their sum is 1/6. In this manner, the element volume is computed exactly. Thus

$$\begin{aligned} V^{(e)} &= \iiint_{V^{(e)}} dV = \int_{-1}^{+1} \int_{-1}^{+1} \int_{-1}^{+1} |\mathbf{J}^{(e)}| d\xi d\eta d\zeta = \\ &= |\mathbf{J}^{(e)}| \int_{-1}^{+1} \int_{-1}^{+1} \int_{-1}^{+1} d\xi d\eta d\zeta = \\ &= |\mathbf{J}^{(e)}| \sum_{i=1}^{n_p} W_i = 6V^{(e)} \sum_{i=1}^{n_p} W_i = \mathbf{V}^{(e)} \end{aligned} \tag{8.84}$$

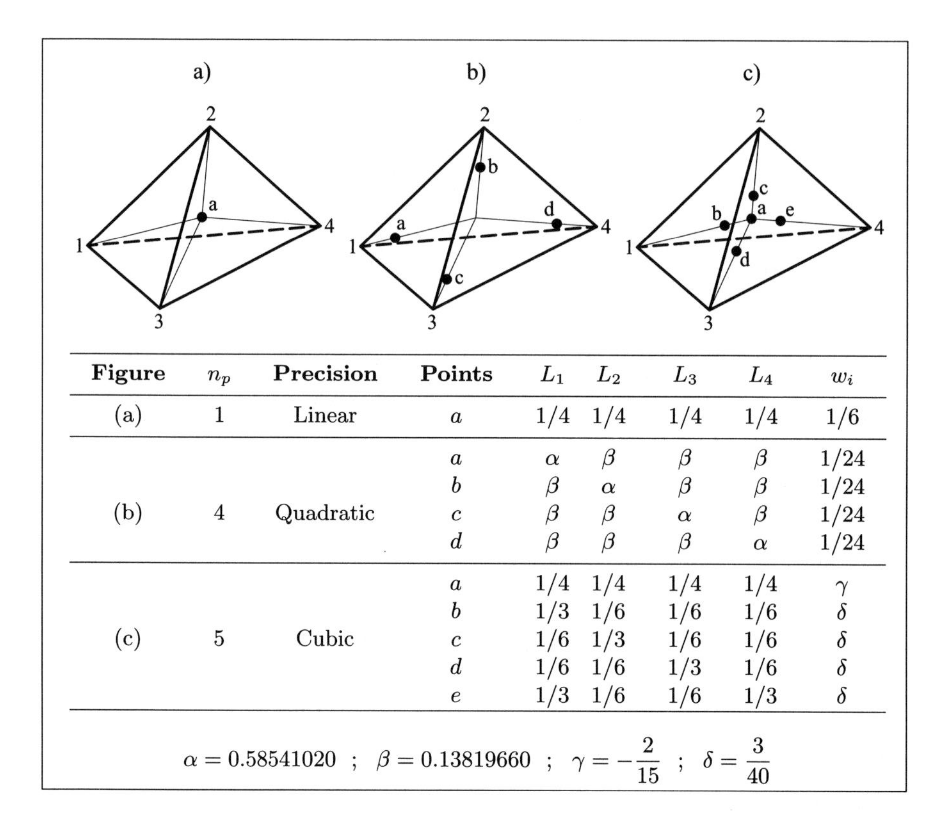

Figure	n_p	Precision	Points	L_1	L_2	L_3	L_4	w_i
(a)	1	Linear	a	1/4	1/4	1/4	1/4	1/6
(b)	4	Quadratic	a	α	β	β	β	1/24
			b	β	α	β	β	1/24
			c	β	β	α	β	1/24
			d	β	β	β	α	1/24
(c)	5	Cubic	a	1/4	1/4	1/4	1/4	γ
			b	1/3	1/6	1/6	1/6	δ
			c	1/6	1/3	1/6	1/6	δ
			d	1/6	1/6	1/3	1/6	δ
			e	1/3	1/6	1/6	1/3	δ

$$\alpha = 0.58541020 \ ; \ \beta = 0.13819660 \ ; \ \gamma = -\frac{2}{15} \ ; \ \delta = \frac{3}{40}$$

Fig. 8.21 Coordinates and weights for the Gauss quadrature for tetrahedra

8.10 NUMERICAL INTEGRATION OF ELEMENT MATRICES

8.10.1 Isoparametric hexahedral elements

Combining Eqs.(8.79a) and (8.82) yields the stiffness matrix for an isoparametric hexahedron as

$$\begin{aligned}\mathbf{K}_{ij}^{(e)} &= \iiint_{V^{(e)}} \mathbf{B}_i^T \mathbf{D} \mathbf{B}_j \, dx\, dy\, dz = \int_{-1}^{+1}\int_{-1}^{+1}\int_{-1}^{+1} \mathbf{B}_i^T \mathbf{D} \mathbf{B}_j \left|\mathbf{J}^{(e)}\right| d\xi\, d\eta\, d\zeta = \\ &= \sum_{p=1}^{n_p}\sum_{q=1}^{n_q}\sum_{r=1}^{n_r} \left[\mathbf{B}_i^T \ \mathbf{D} \ \mathbf{B}_j \left|\mathbf{J}^{(e)}\right|\right]_{p,q,r} W_p W_q W_r = \\ &= \sum_{p=1}^{n_p}\sum_{q=1}^{n_q}\sum_{r=1}^{n_r} [\mathbf{G}_{ij}]_{p,q,r} W_p W_q W_r \end{aligned} \tag{8.85}$$

where $\mathbf{G}_{ij}$ was given in Eq.(8.79b).

It is important to understand that, similarly as for 2D elements, the computation of matrix $\mathbf{G}_{ij}$ is not strictly necessary. Thus, the stiffness matrix can be simple obtained by sampling $\mathbf{B}_i$, $\mathbf{D}$ and the Jacobian determinant at each Gauss point and then performing the standard multiplications and summations shown in the second line of Eq.(8.85).

The computation of the equivalent nodal force vectors involving volume integrals follows an identical procedure. For the body force vector (Eq. (8.37b))

$$\begin{aligned}\mathbf{f}_b^{(e)} &= \iiint_{V^{(e)}} \mathbf{N}_i^T \mathbf{b}\, dx\, dy\, dz = \int_{-1}^{+1}\int_{-1}^{+1}\int_{-1}^{+1} \mathbf{N}_i^T \mathbf{b} \left|\mathbf{J}^{(e)}\right| d\xi\, d\eta\, d\zeta = \\ &= \sum_{p=1}^{n_p}\sum_{q=1}^{n_q}\sum_{r=1}^{n_r} \left[\mathbf{N}_i^T\ \mathbf{b} \left|\mathbf{J}^{(e)}\right|\right]_{p,q,r} W_p W_q W_r \end{aligned} \tag{8.86}$$

The flow chart of the subroutines for computing $\mathbf{K}_{ij}^{(e)}$ and $\mathbf{f}_b^{(e)}$ is basically identical to that shown in Figures 6.8 and 6.9 for the 2D case. The only modification is the introduction of a third integration loop ($r = 1, n_r$).

Computation of $f_t^{(e)}$ for surface tractions

The treatment of surface tractions is more complicated. Let us assume that a distributed force t_n acts *orthogonally* to the element face corresponding to $\zeta = +1$ defined by nodes 5 to 8 (Figure 8.22). The computation of $\mathbf{f}_t^{(e)}$ requires evaluating the term $\mathbf{t}\ dA$ (Eq.(8.38b)) where $\mathbf{t}$ contains the global components of the surface tractions acting on the element face and dA is the area differential. If $\mathbf{n}$ is the unit normal to the face we have

$$\mathbf{t} = t_n \mathbf{n} \qquad \text{with} \qquad \mathbf{n} = [n_x, n_y, n_z]^T \tag{8.87}$$

Vector $\mathbf{n}$ is obtained by the cross product of vectors $\mathbf{V}_1$ and $\mathbf{V}_2$ tangent to the lines η = constant and ξ = constant over the face, respectively (Figure 8.22). Thus

$$\begin{aligned}\mathbf{V}_1 &= \left(\frac{\partial x}{\partial \xi}\mathbf{i} + \frac{\partial y}{\partial \xi}\mathbf{j} + \frac{\partial z}{\partial \xi}\mathbf{k}\right)_{\zeta=+1} d\xi \\ \mathbf{V}_2 &= \left(\frac{\partial x}{\partial \eta}\mathbf{i} + \frac{\partial y}{\partial \eta}\mathbf{j} + \frac{\partial z}{\partial \eta}\mathbf{k}\right)_{\zeta=+1} d\eta\end{aligned} \tag{8.88}$$

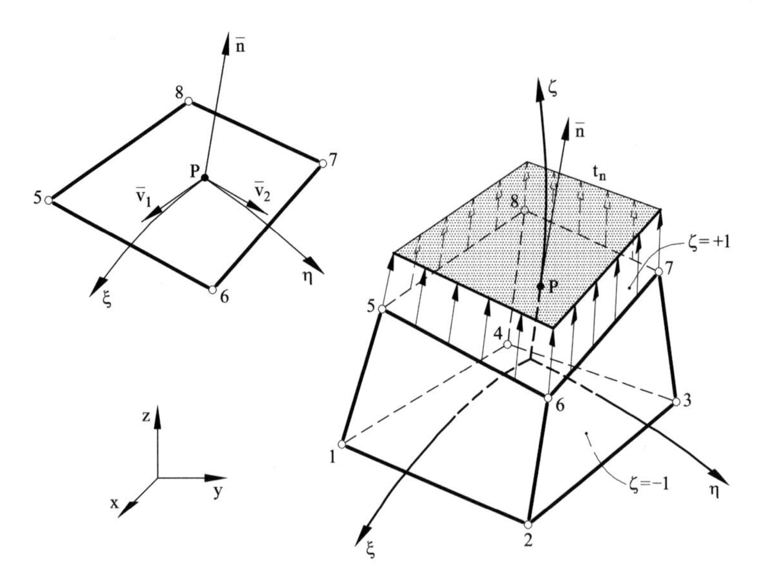

Fig. 8.22 Normal tractions acting on a face of a hexahedral element

Eq.(8.88) shows that the components of $\mathbf{V}_1$ and $\mathbf{V}_2$ coincide with the terms in the first and second row of the Jacobian matrix of Eq.(8.74) computed at the face $\zeta = +1$. The unit normal vector is

$$\mathbf{n} = \frac{\mathbf{V}_1 \times \mathbf{V}_2}{|\mathbf{V}_1 \times \mathbf{V}_2|} \tag{8.89}$$

Noting that $dA = |\mathbf{V}_1 \times \mathbf{V}_2|$ gives

$$\mathbf{n} = \frac{1}{dA} \begin{Bmatrix} J_{12}J_{23} - J_{22}J_{13} \\ J_{21}J_{13} - J_{11}J_{23} \\ J_{11}J_{32} - J_{21}J_{12} \end{Bmatrix}_{\zeta=+1}^{(e)} d\xi\, d\eta = \frac{1}{dA}\, \mathbf{j}^{(e)} d\xi\, d\eta \tag{8.90}$$

where the $J_{ij}^{(e)}$ terms are deduced from Eq.(8.75).

The final expression for the equivalent nodal force vector is

$$\begin{aligned} \mathbf{f}_{t_i}^{(e)} &= \iint_{A^{(e)}} \mathbf{N}_i^T\, t_n\, \mathbf{n}\, dA = \int_{-1}^{+1} \int_{-1}^{+1} \mathbf{N}_i^T\, t_n\, \mathbf{j}^{(e)}\, d\xi\, d\eta = \\ &= \sum_{p=1}^{n_p} \sum_{q=1}^{n_q} [\mathbf{N}_i^T\, \mathbf{j}^{(e)}\, t_n]_{p,q} W_p W_q \end{aligned} \tag{8.91}$$

where $\mathbf{j}^{(e)}$ is deduced from Eq.(8.90). Also note that in Eq.(8.91) $\mathbf{N}_i = \mathbf{N}_i(\xi, \eta, \zeta = +1)$.

8.10.2 Isoparametric tetrahedral elements

The element stiffness matrix and the equivalent nodal force vector for isoparametric tetrahedra are computed by

$$\mathbf{K}_{ij}^{(e)} = \sum_{p=1}^{n_p} \Big[\mathbf{G}_{ij}(\alpha,\beta,\gamma)\Big]_p W_p \tag{8.92a}$$

$$\mathbf{f}_{i}^{(e)} = \sum_{p=1}^{n_p} \Big[\mathbf{N}_i^T \mathbf{b} |\mathbf{J}^{(e)}|\Big]_p W_p \tag{8.92b}$$

where

$$\mathbf{G}_{ij} = \mathbf{B}_i^T(\alpha,\beta,\gamma)\mathbf{D}\mathbf{B}_j(\alpha,\beta,\gamma)|\mathbf{J}^{(e)}|$$

The expression of $\mathbf{G}_{ij}$ is deduced from Eq.(8.79b) simply substituting the coordinates ξ,η,ζ for α,β,γ, respectively.

For surface tractions we have

$$\mathbf{f}_{t_i}^{(e)} = \sum_{p=1}^{n_p} \Big[\mathbf{N}_i \mathbf{j}^{(e)} t_n\Big]_p W_p \tag{8.93}$$

where the different terms have the same meaning as for hexahedra.

8.10.3 Selection of the quadrature order

The selection of the quadrature order for hexahedral and tetrahedral elements follows the same rules given for 2D solid elements in Section 6.7. The full (exact) quadrature for linear and quadratic prisms is $2 \times 2 \times 2$ and $3 \times 3 \times 3$, respectively. These quadratures are also recommended for distorted shapes.

The reduced $1\times1\times1$ and $2\times2\times2$ quadratures induce spurious modes in the 8-noded and 27-noded hexahedral element, respectively. The performance of the 20-noded Serendipity elements generally improves with the $2\times2\times2$ reduced quadrature. However, this quadrature must be used with extreme care as it can lead to spurious mechanisms in certain problems, in a similar way as for 8-noded quadrilaterals (Section 6.7).

For linear and quadratic straight-sided tetrahedra the 1 and 4 points quadratures shown in Figure 8.21 yield exact integration. Cubic tetrahedra require a quadrature of quartic precission for the exact integration of all the stiffness matrix terms [Be,GH,SC].

8.11 PERFORMANCE OF 3D SOLID ELEMENTS

3D solid elements behave very similarly to their analogous 2D elements. Hexahedral elements are more accurate than tetrahedra of the same order. On the other hand, low order elements like the 8-noded hexahedron or 4-noded tetrahedron require fine meshes for bending dominated problems and higher order elements give more satisfactory results in these cases. Figure 8.23 shows a typical example of this situation where 8- and 20-noded prisms and 4- and 10-noded tetrahedra are compared for the analysis of a cantilever beam subjected to a pure bending state [Cl2,Ya]. The superiority of hexahedral elements is clear in this case. In particular, the 20-noded quadratic prism yields the exact solution of thick beam theory with 50% fewer variables than the equivalent quadratic tetrahedron.

The 27-noded Lagrange hexahedron performs better than the 20-noded Serendipity one for representing a quadratic function on trilinearly distorted shapes. The reasons are similar to those given in Section 6.2.2 to explain the good performance of the 9-noded quadrilateral when linearly distorted. Hence, despite its highest cost, the 27-noded Lagrange hexahedron is generally preferable for modelling smooth solutions on distorted geometries. The performance of the 20-noded Serendipity hexahedron in those cases can be improved by using $2 \times 2 \times 2$ reduced integration. However, as noted for 2D solid elements, great care should be taken when using a reduced quadrature for the quadratic Serendipity element (Section 6.8).

The application of the patch test to 3D solid elements follows precisely the concepts and rules given for 2D solid elements in Section 6.10.

Mesh generation is the crucial problem for practical 3D analysis. Here, tetrahedral elements are by far the more versatile option for the discretization of complex 3D geometries using unstructured meshes. Much research for the development of efficient mesh generators for tetrahedral and hexahedral elements has been carried out in recent years. This issue is of even greater importance if adaptive refinement strategies are used.

8.12 EXAMPLES

Some examples aiming to show the potential of 3D solid elements for solving complex structural problems in engineering are presented. Readers interested in the specific details of each example can contact the authors in the references given.

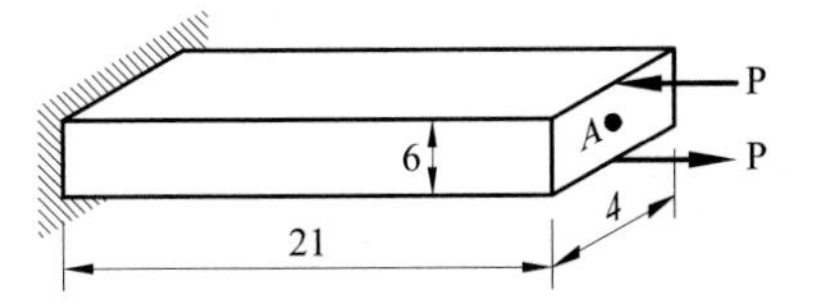

Element	Nodes	Mesh (a×b×c)	Degrees of freedom	Error in vertical displacement of point A
	4	7×3×2×5* = =210	84×3=252	−25.8%
	8	7×3×2=42	84×3=252	−10.2%
	10	7×3×2×5* = =210	448×3=1344	≃1%
	20	7×3×2=42	287×3=861	≃ Exact (Beam theory)

* Each hexahedron is divided into 5 tetrahedra

Fig. 8.23 Pure bending of a cantilever beam analyzed with different 3D elements [Cl2,Ya]

8.12.1 Analysis of a gravity dam

Figure 8.24 shows a perspective view of the geometry of the Mequinenza gravity dam in Spain and the mesh of 20-noded hexahedra used for the analysis [COG,COHO,OCOH]. The study of this dam as a 2D solid using 8-noded quadrilaterals was presented in Section 6.11.1. Figure 8.24 displays the contours of the displacement component in the downstream direction and several stress plots for a combination of self-weight and hydrostatic loading. The need for sophisticated graphic visualization techniques for practical 3D analysis such as this one is clear. This topic is treated in Chapter 10 and Annex D.

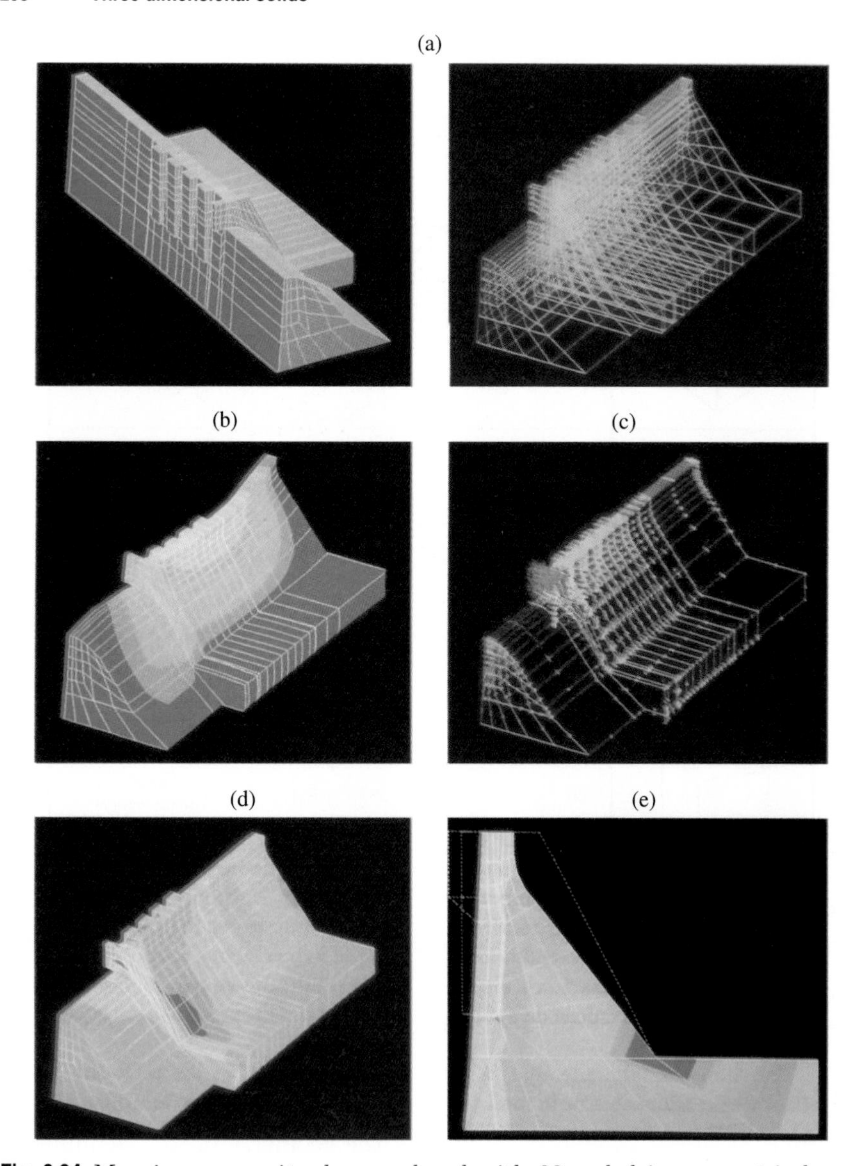

Fig. 8.24 Mequinenza gravity dam analyzed with 20-noded isoparametric hexahedra. (a) Geometry and mesh. (b) Contours of downstream displacement. (c) Principal stresses on dam surface. (d) Contours of transversal stress σ_y. (e) Contours of longitudinal stress σ_x in a cross section [CO,COHO,OCOH]

8.12.2 Analysis of a double curvature arch dam

Figure 8.25a displays the geometry of the Talvachia dam in Italy and the mesh of isoparametric 20-noded hexahedra used for the analysis. The dam

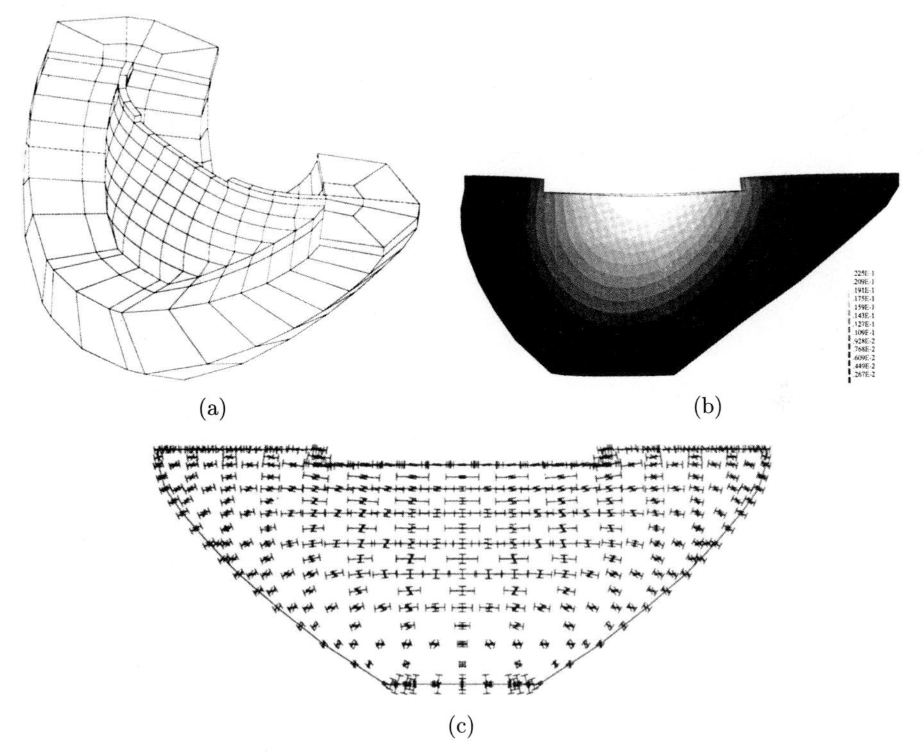

Fig. 8.25 Talvachia double curvature arch dam analyzed with 20-noded isoparametric hexahedra. (a) Geometry and mesh. (b) Contours of displacement modulus and distribution of compression stresses (N/m^2) for self-weight plus hydrostatic loading [MBBO]

foundation has also been discretized. Figure 8.25b shows the contours of the displacement vector modulus for self-weight plus hydrostatic loading. Also the compression principal stresses over the upstream face are displayed in Figure 8.25c [MBBO].

8.12.3 Analysis of arch dams

Figure 8.26 shows the analysis of the Melonares arch dam in Sevilla, Spain using again 20-noded Serendipity hexahedra. The interest was the study of the stress field in the spillway zone [SG2].

Figure 8.27 shows a snapshot of stress field contours for self-weight load in the surface of an arch dam in Jaen, Spain. The dam and the surrounding terrain were discretized with 4-noded linear tetrahedra in this case [SG3].

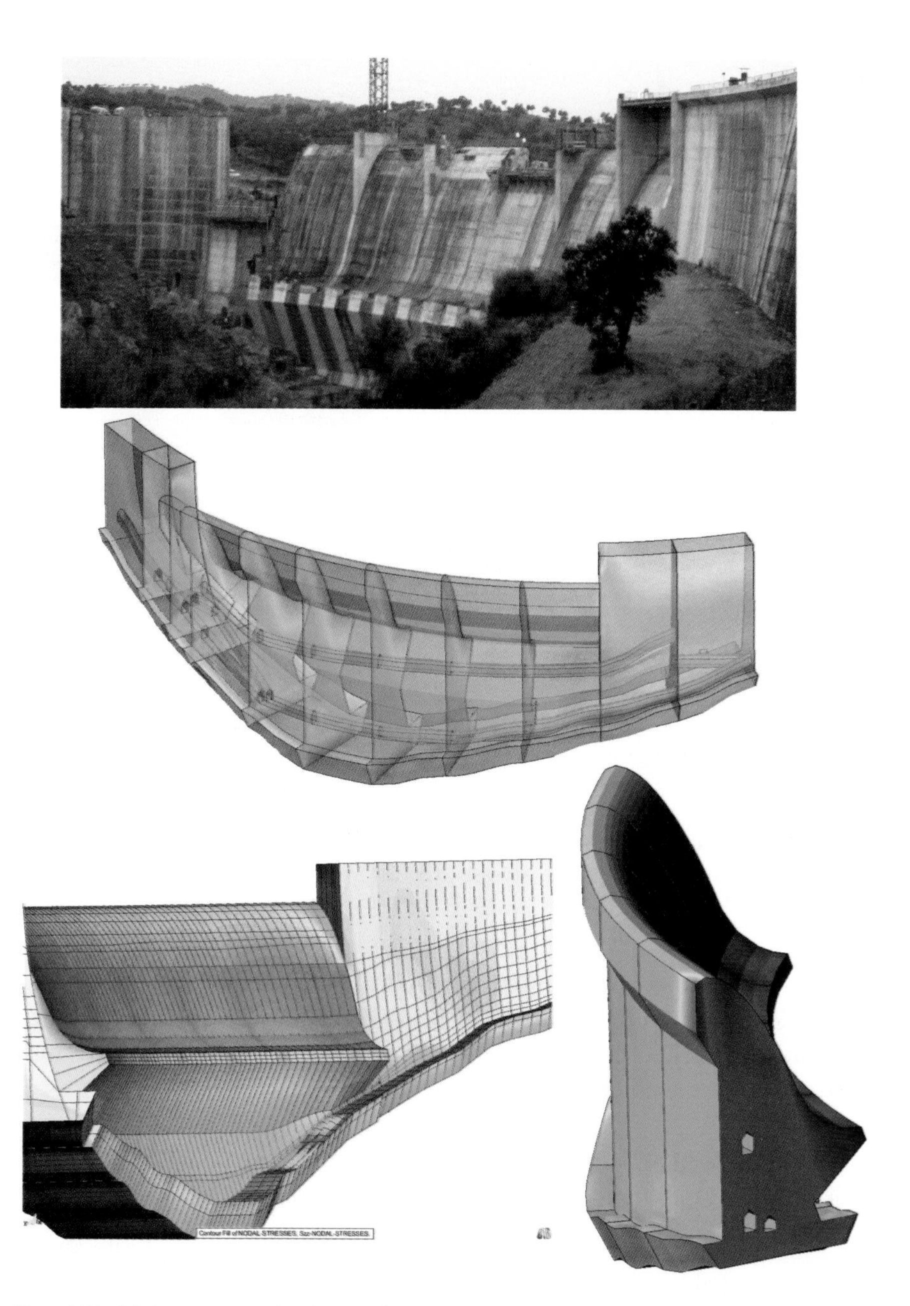

Fig. 8.26 Melonares arch dam. Geometry and discretization with 20-noded Serendipity hexahedra. Detail of stress field in one of the spillways for self-weight load [SG2]

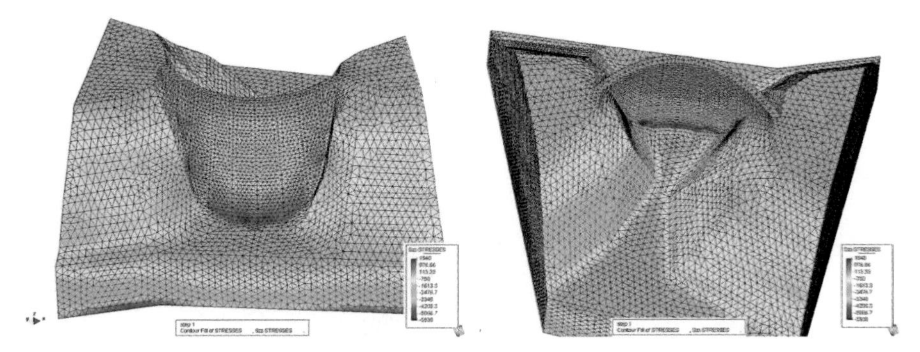

Fig. 8.27 Arch dam in Jaen, Spain. Discretization of dam and terrain with 4-noded tetrahedra. Snapshot of stress field under self-weight loading [SG3]

8.12.4 Analysis of a flat in a building

Figure 8.28 shows some results of the analysis of a whole flat in a building in the city of Barcelona, Spain. Figures 8.28a and 8.28b respectively show the discretization of the walls and upper and lower floors and the reinforced concrete beams using 20-noded hexahedra. Figures 8.28c show the contours of the displacement modulus plotted on the deformed shape for self-weight load (results have been amplified for visualization purposes). More information on this example can be found in [OH,On2].

8.12.5 Analysis of prismatic cellular caissons for harbour piers

Reinforced concrete caissons of prismatic cellular shape are commonly used for building piers in harbours. The caisson structure is typically built in the shore and is subsequently launched into the sea and transported as a floating body to the correct position in the pier. The caisson is then sunk into a stable position by filling the internal cells with concrete or rockfill material.

Figure 8.29a shows the discretization of a cellular caisson built in the Bilbao harbour in Spain (dimensions $\simeq 13 \times 7 \times 34$ mts) using a mesh of 20-noded hexahedra. Half the real geometry of the caisson has been discretized only due to symmetry. Figures 8.29b show results of the deformed shape of the caisson and the contours of the von Mises stress for a combination of lateral wave loading and internal pressure in some of the cells due to the filling. More information on the FEM analysis of this type of problems can be found in [PFGS,SFPC,SHM].

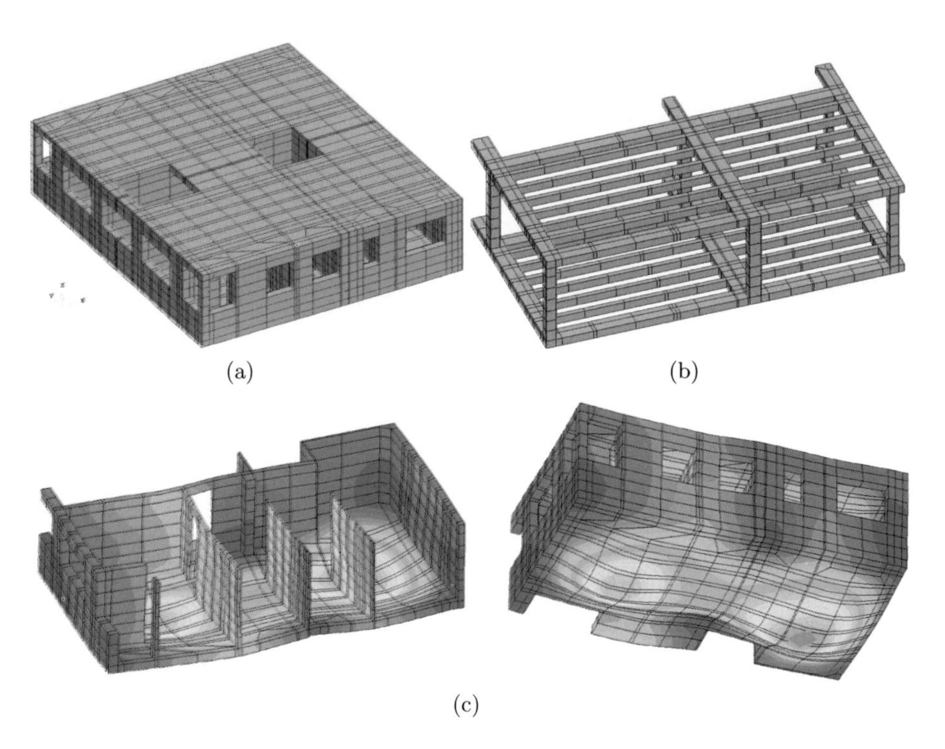

Fig. 8.28 Analysis of a flat in a building. Discretization of the walls and upper and lower floors (a) and the beams (b) using 20-noded hexahedra. Deformed shape and displacement modulus contours for self-weight loading (c). Results have been amplified for visualization purposes [OH,On2]

8.12.6 Analysis of a nuclear containment building

Figure 8.30a shows the discretization of the containment building in the Vandellós nuclear power plant in Spain using 20-noded hexahedra. Figure 8.30b shows the amplified deformed shape of the reinforced/prestressed concrete structure under a high internal pressure. Colours indicate the displacement modulus contours. Figures 8.30c,d show the damage (fractured) zones for two values of the increasing internal pressure. Details of this study can be found in [BCHC+].

8.12.7 Analysis of historical constructions

The next example presented is the structural analysis of Barcelona Cathedral. The numerical study was performed using four-noded tetrahedral

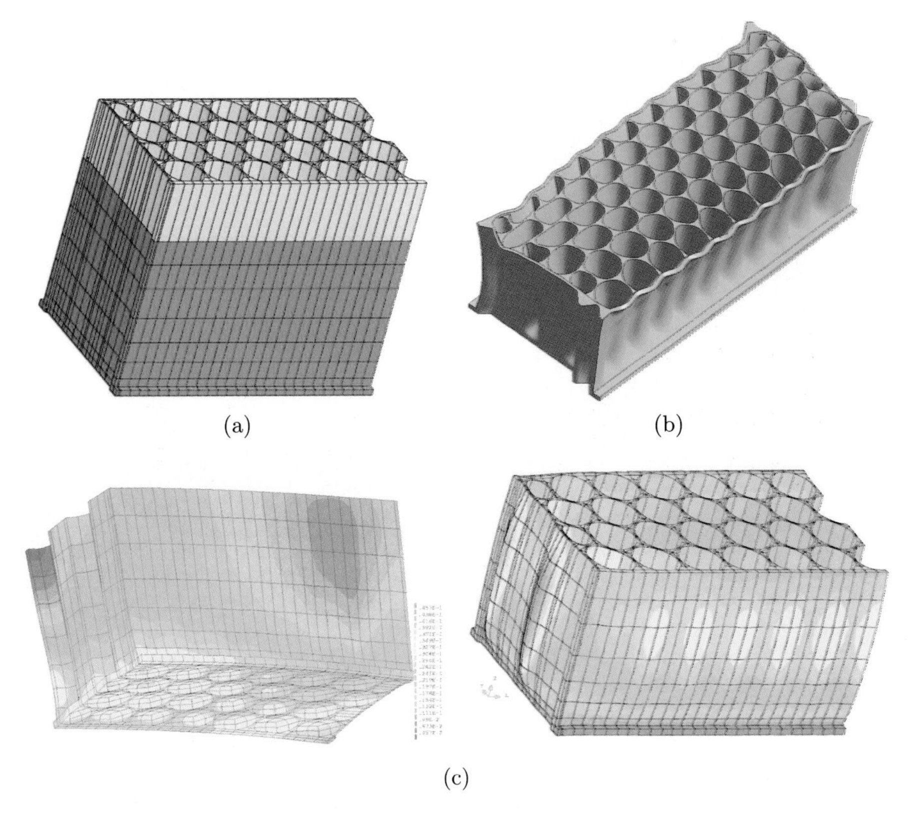

(a) (b) (c)

Fig. 8.29 Analysis of a prismatic cellular caisson in Bilbao harbour. (a) Discretization of half the caisson using 20-noded hexahedra. (b,c) Deformed shape of the caisson and contours of the von Mises stress under lateral wave loading and internal pressure in some cells [PFGS,SFPC,SHM]

elements. Figure 8.31 shows the mesh used for the analysis and contours of the displacement modulus under self-weight.

Figure 8.32 shows a similar example for the study of Saint Mark's Basilica in Venice. The figure shows the discretization of the five dome masonry and stone structure using 20-noded hexahedra, the contours of the displacement modulus under self-weight and an underneath view of damage zones in the structure for an increased value of the self-weight. For details see [OHBO+].

These two examples illustrate the possibilities of 3D solid elements for analysis of historical constructions. More details of these examples and similar ones can be found in [OHBO+,RCMR+,RGAA,RGMO,RGOL,RM, Ro2,RPOH].

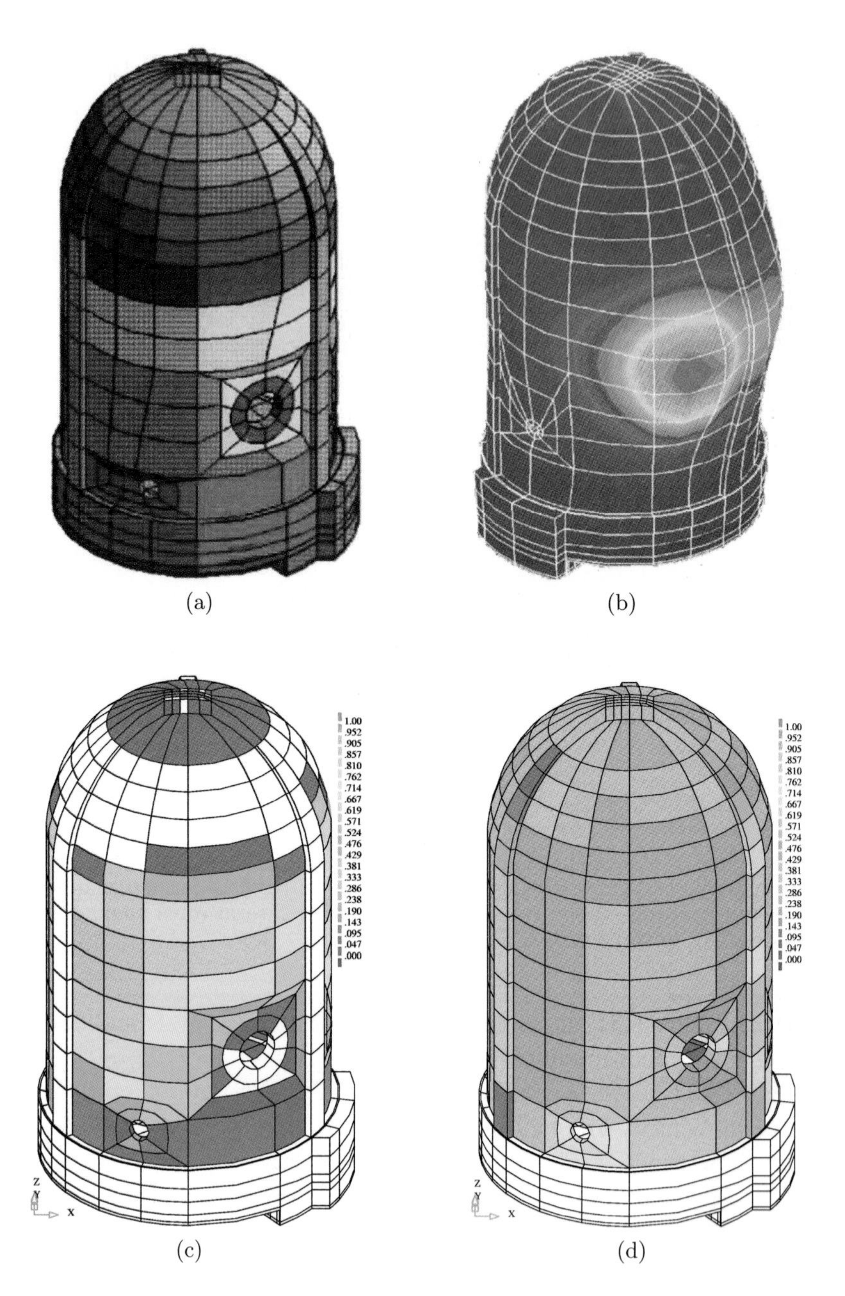

Fig. 8.30 Analysis of a nuclear containment building. (a) Discretization using 20-noded hexahedra. (b) Amplified deformed shape of the structure under internal pressure. (c,d) Damaged (fractured) zones for increasing internal pressure. Colours indicate the normalized value of damage parameter [BCHC+]

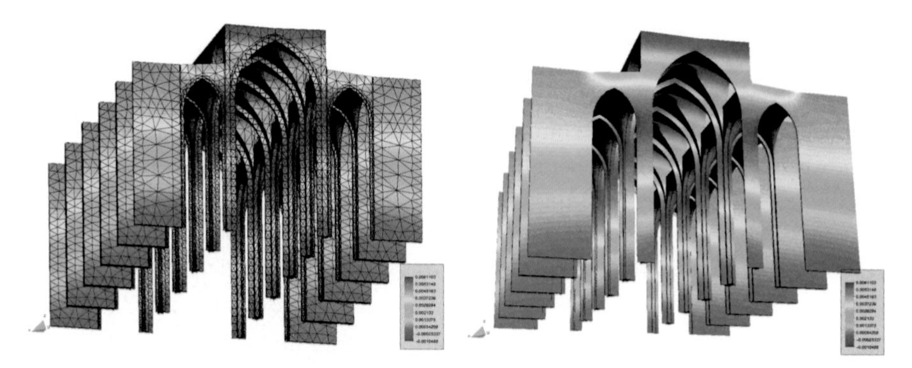

Fig. 8.31 Analysis of Barcelona Cathedral using 4-noded tetrahedra. Mesh and contours of displacement modulus under self-weight [RGAA,RM,Ro2,RPOH]

Fig. 8.32 Analysis of Saint Mark's Basilica in Venice using 20-noded hexahedra. (a) Mesh. (b) Contours of displacement modulus under self weight. (c) Underneath view of damaged zones for an increased value of the weight [OHBO+]

8.12.8 Analysis of mechanical parts

The next examples show the possibilities of 3D solid elements for the study of mechanical engineering components, such as that shown in Figure 8.33a. A mesh of 250 20-noded hexahedra has been used for this analysis. The loading considered is a combination of self-weight and uniform torsion. Figure 8.33b shows the contours of the displacement modulus. The distribution of tension and compression stresses are displayed in Figures 8.33c and d respectively [SM2].

The example shown in Figure 8.34 is the analysis of a mechanical component using 4-noded tetrahedra. The contours of the von Mises stress (Eq.(8.19)) under thermal loading are shown as an example of output for this type of problems.

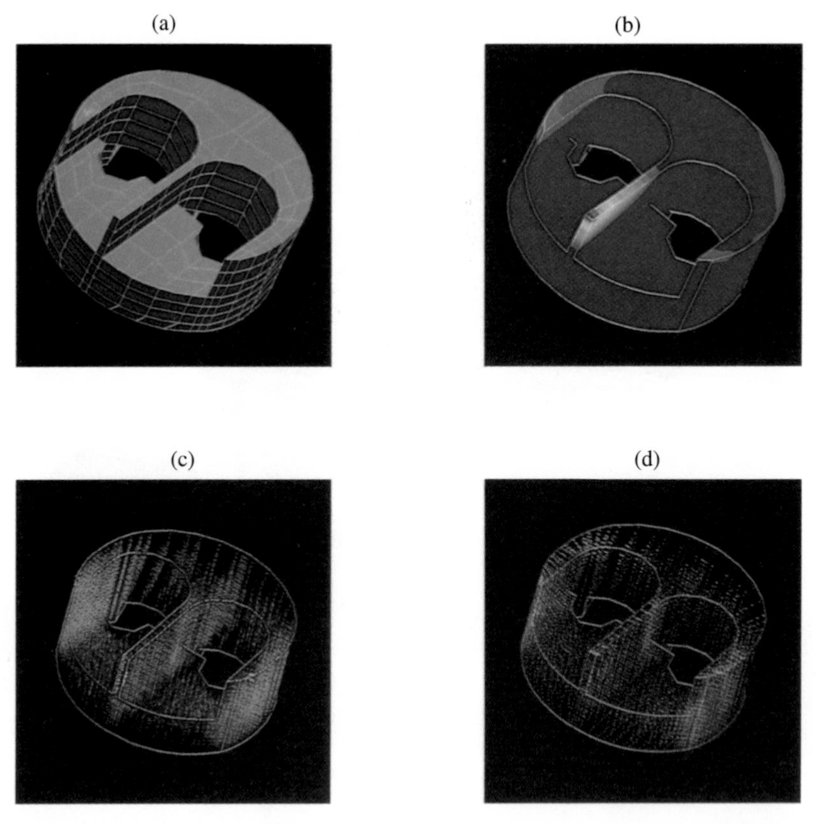

Fig. 8.33 Mechanical component analyzed with 20-noded hexahedra. (a) Geometry and mesh. (b) Contours of displacement modulus under self-weight and torsion loading. (c) Principal tension stresses. (d) Principal compression stresses

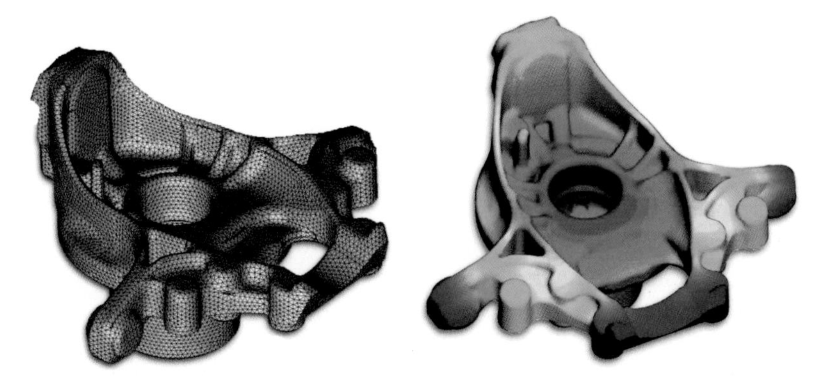

Fig. 8.34 Discretization of a mechanical component with 4-noded linear tetrahedra. Contour plot of von Mises stress under thermal loading

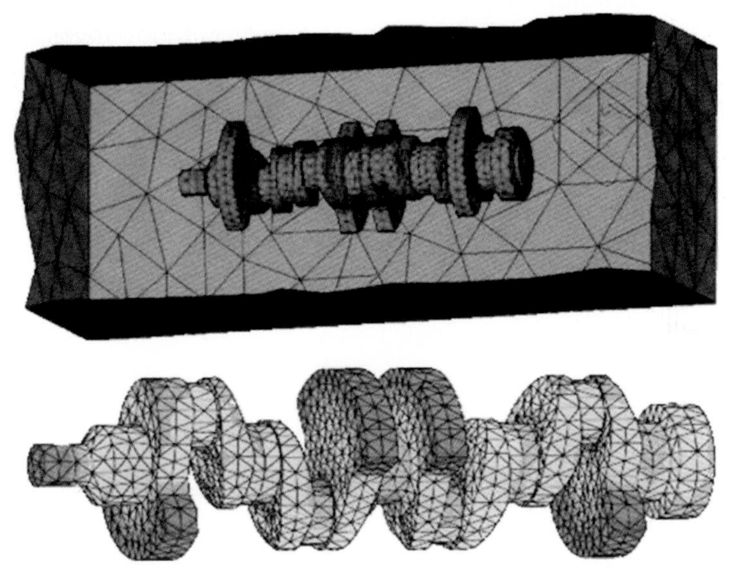

Fig. 8.35 Thermal-stress analysis of a crankshaft during the casting process using 4-noded tetrahedra to discretize the part and the mould. Colours indicate the von Mises stress field under thermal loading [CDOO,CVDO+,OCCO+]

Figure 8.35 finally presents a result of the thermal-mechanical analysis of a crankshaft for an automotive during the solidification and cooling phase of the casting process. Colours show the von Mises stress contours at a certain time of the analysis. A mesh of 4-noded tetrahedra was used to discretize the crankshaft and the surrounding mould. More information can be found in [CDOO,CVDO+,OCCO+].

8.13 FINAL REMARKS

The formulation of 3D solid elements is simple and most concepts are a straightforward extension of those explained for 2D solid elements. 3D solid elements also behave very similarly to the corresponding 2D elements. The 4-noded tetrahedron and the 8-noded hexahedron are the simplest of their respective families, although their accuracy is poorer than that of higher order solid elements. Among these, the 20- and 27-noded hexahedron and the 10-noded tetrahedron are good candidates for practical use.

The examples presented show the difficulties inherent in the discretization and the visualization of numerical results in 3D structural analysis using solid elements. Here the utilization of efficient mesh generators and advanced visualization techniques is essential. Mesh generation is the crucial issue and even more if adaptive mesh refinement procedures involving unstructured meshes of tetrahedra or hexahedra are to be used in an effective manner. This topic is treated in Chapter 10 and Appendix D.

9

MISCELLANEOUS: INCLINED SUPPORTS, DISPLACEMENT CONSTRAINS, ERROR ESTIMATION, MESH ADAPTIVITY ETC.

9.1 INTRODUCTION

This chapter deals with topics of general interest in finite element structural analysis not covered in previous chapters. Boundary conditions in inclined supports are presented first. Then, methods to link different element types and for prescribing general constraints in the nodal displacements are studied. The three following sections deal with condensation and recovery algorithms, mesh symmetries and elastic supports. The last part of the chapter is devoted to the computation of nodal stresses, the estimation of the solution error and its application to adaptive mesh refinement.

9.2 BOUNDARY CONDITIONS IN INCLINED SUPPORTS

Boundary nodes may have prescribed displacements in local directions different from those chosen for the global stiffness assembly process. Let us consider, for instance, a boundary node k in the mesh of Figure 9.1. The slope of the support at k enforces the node to move along the local tangent direction x', whereas the normal displacement v' is zero. The problem is solved by assembling the stiffness equation at node k in the *boundary directions* x', y'.

The global displacements of node k are transformed to the local boundary system x', y' by

$$\mathbf{a}_k^{(e)} = [u_k, v_k]^T = \mathbf{L}_k [u'_k, v'_k]^T = \mathbf{L}_k \mathbf{a}_k^{\prime(e)} \tag{9.1a}$$

where

$$\mathbf{L}_k = \begin{bmatrix} \cos\phi_k & -\sin\phi_k \\ \sin\phi_k & \cos\phi_k \end{bmatrix} \tag{9.1b}$$

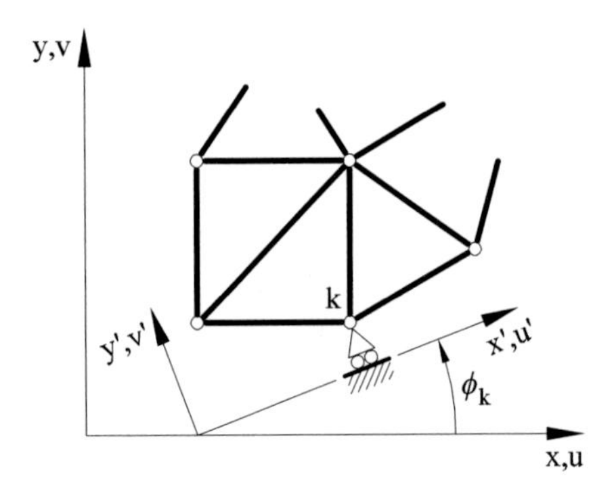

Fig. 9.1 Boundary condition in inclined supports. Mesh of linear triangles

and ϕ_k is the angle formed by axes x' and x (Figure 9.1). The equilibrium equations at node k are transformed as

$$\mathbf{q}_i'^{(e)} = \mathbf{L}_i^T \sum_{j=1}^{n} \mathbf{K}_{ij}^{(e)} \mathbf{a}_j^{(e)} = \mathbf{L}_i^T \sum_{j=1}^{n} \mathbf{K}_{ij}^{(e)} \mathbf{L}_j \mathbf{a}_j' = \sum_{j=1}^{n} \overline{\mathbf{K}}_{ij} \mathbf{a}_j'^{(e)} \tag{9.2}$$

The stiffness matrix of the elements sharing node k is modified as

$$\overline{\mathbf{K}}_{ij}^{(e)} = \hat{\mathbf{L}}_i^T \mathbf{K}_{ij}^{(e)} \hat{\mathbf{L}}_j \tag{9.3}$$

with

$$\hat{\mathbf{L}}_i = \begin{cases} \mathbf{L}_k & \text{if} \quad i = k \\ \mathbf{I}_2 & \text{if} \quad i \neq k \end{cases} \tag{9.4}$$

where $\mathbf{I}_2$ is the 2×2 unit matrix. The reactions at the inclined boundary nodes are directly computed in the local boundary system. Transformation to the global axes can be performed using $\mathbf{L}_i$. An alternative and more economical procedure is to compute the stiffness matrix of the elements sharing node k directly using a modified nodal strain matrix given by $\overline{\mathbf{B}}_k = \mathbf{B}_k \hat{\mathbf{L}}_k$. More details on the specification of skew conditions in the FEM are given in [MDR].

Example 9.1: Obtain the stiffness matrix of the plane stress element mnp shown in Figure 9.2 so that the boundary conditions $v_m' = v_n' = 0$ can be prescribed.

- ***Solution***

Nodes m and n can slide along the x' axis, while the displacements of the interior node p are kept in the global coordinate system x, y. The displacements

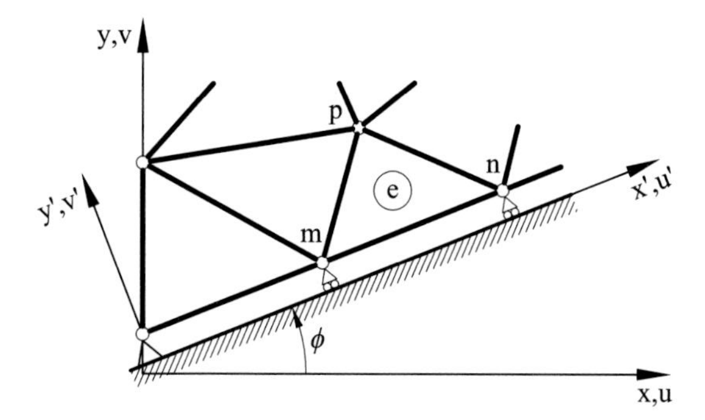

Fig. 9.2 Plane body sliding on inclined support

of nodes m and n are transformed to the boundary system x', y' by

$$\mathbf{a}'_m = \mathbf{L}^T \mathbf{a}_m \quad \text{and} \quad \mathbf{a}'_n = \mathbf{L}^T \mathbf{a}_n$$

where $\mathbf{a}'_i = [u'_i, v'_i]^T, \quad \mathbf{a}_i = [u_i, v_i]^T$ and $\mathbf{L} = \begin{bmatrix} \cos\phi & -\sin\phi \\ \sin\phi & \cos\phi \end{bmatrix}$.

The sought element stiffness matrix is found after application of the transformations of Eq.(9.3), with $\mathbf{L}_m = \mathbf{L}_n = \mathbf{L}$ as

$$\bar{\mathbf{K}}^{(e)} = \begin{array}{c} \begin{matrix} m & \quad n & \quad p \end{matrix} \\ \begin{bmatrix} \mathbf{L}^T\mathbf{K}_{11}^{(e)}\mathbf{L} & \mathbf{L}^T\mathbf{K}_{12}^{(e)}\mathbf{L} & \mathbf{L}^T\mathbf{K}_{13}^{(e)} \\ \mathbf{L}^T\mathbf{K}_{21}^{(e)}\mathbf{L} & \mathbf{L}^T\mathbf{K}_{22}^{(e)}\mathbf{L} & \mathbf{L}^T\mathbf{K}_{23}^{(e)} \\ \mathbf{K}_{31}^{(e)}\mathbf{L} & \mathbf{K}_{32}^{(e)}\mathbf{L} & \mathbf{K}_{33}^{(e)} \end{bmatrix} \begin{matrix} m \\ n \\ p \end{matrix} \end{array}$$

where $\mathbf{K}_{ij}^{(e)}$ is given by Eq.(4.63*b*).

9.3 JOINING DISSIMILAR ELEMENTS

Many practical situations require matching elements of different types. The key in the connection process is to express the displacement field in one of the two elements in terms of the nodal displacements of the other one.

Let us consider a 2-noded axial bar element connected to a 4-noded solid quadrilateral (Figure 9.3). This procedure is typical for introducing

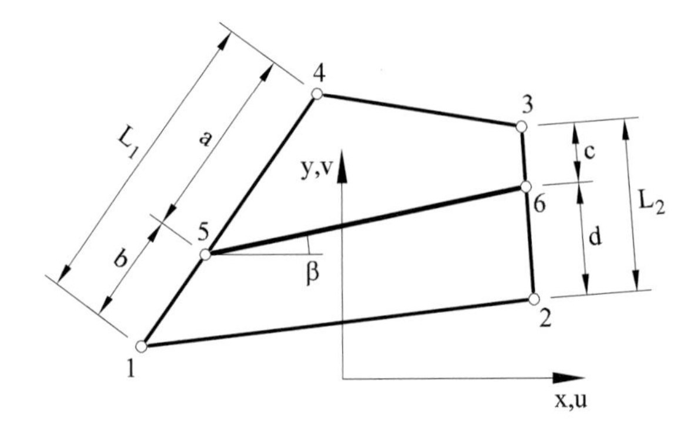

Fig. 9.3 Joining different elements. 4-noded solid quadrilateral connected to a 2-noded axial bar element [CMPW]

the effect of reinforcement steel bars in concrete solid elements. The local displacements of the bar nodes are expressed in terms of the global displacements of the quadrilateral. For node 5 we have

$$u_5' = u_5 \cos\beta + v_5 \sin\beta \tag{9.5}$$

where the prime denotes the local axial displacement along the bar direction. The global displacements u_5 and v_5 are now linearly interpolated in terms of the displacements of nodes 1 and 4 as

$$\begin{Bmatrix} u_5 \\ v_5 \end{Bmatrix} = \frac{1}{L_1}\begin{bmatrix} a & 0 & b & 0 \\ 0 & a & 0 & b \end{bmatrix}\begin{Bmatrix} u_1 \\ v_1 \\ u_4 \\ v_4 \end{Bmatrix} \tag{9.6}$$

Combining the last two equations and repeating the same process for node 6 gives

$$[u_5', u_6']^T = \mathbf{T}[u_1, v_1.u_2, v_2, u_3, v_3, u_4, v_4]^T \tag{9.7}$$

where

$$\mathbf{T} = \begin{bmatrix} \frac{\cos\beta}{L_1} & \frac{\sin\beta}{L_1} & 0 & 0 \\ 0 & 0 & \frac{\cos\beta}{L_2} & \frac{\sin\beta}{L_2} \end{bmatrix}\begin{bmatrix} a\,0\,0\,0\,0\,0\,b\,0 \\ 0\,a\,0\,0\,0\,0\,0\,b \\ 0\,0\,c\,0\,d\,0\,0\,0 \\ 0\,0\,0\,c\,0\,d\,0\,0 \end{bmatrix} \tag{9.8}$$

The stiffness matrix and the equivalent nodal force vector for the bar element are transformed as

$$\underset{8\times 8}{\mathbf{K}} = \mathbf{T}^T \underset{2\times 2}{\mathbf{K}_b} \mathbf{T} \quad \text{and} \quad \underset{8\times 1}{\mathbf{f}} = \mathbf{T}^T \underset{2\times 1}{\mathbf{f}_b}\,. \tag{9.9}$$

where $\mathbf{K}_b$ and $\mathbf{f}_b$ can be found in Chapter 2.

The expressions for $\mathbf{K}$ and $\mathbf{f}$ can now be directly assembled with those for the 4-noded quadrilateral element.

A particular case is when the bar is connected to the corner nodes. The nodal displacements in the bar coincide now with those in the solid element and distances a, b, c, d are either zero or the side length.

Rigid elements

Rigid elements may be necessary to model indeformable parts within a structure that undergoes rigid body motions. An alternative is to assume that the elastic modulus of the rigid elements is much higher than that of the surrounding elements. This can however introduce rounding off errors in the solution process of the type discussed in Section 3.13.4. Consequently, it is better to make the element perfectly rigid rather than stiffer.

Let us assume that the shaded triangle of Figure 9.4 is to be modelled as a perfectly rigid solid [CMPW]. The rigid body motion can be described by three DOFs, say u_1, v_1 and u_2. These DOFs are related to the original six DOFs by the transformation

$$\mathbf{a}'^{(e)} = \mathbf{T}\mathbf{a}^{(e)} \quad \text{or} \quad \begin{Bmatrix} u_1 \\ v_1 \\ u_2 \\ v_2 \\ u_3 \\ v_3 \end{Bmatrix} = \begin{bmatrix} 1 & 0 & 0 \\ 0 & 1 & 0 \\ 0 & 0 & 1 \\ -a/b & 1 & a/b \\ 1 & 0 & 0 \\ -a/b & 1 & a/b \end{bmatrix} \begin{Bmatrix} u_1 \\ v_1 \\ u_2 \end{Bmatrix} \tag{9.10}$$

with $u_3 = u_1$ and $v_2 = v_3 = v_1 - a\left[\dfrac{u_1 - u_2}{b}\right]$, where $\frac{u_1-u_2}{b}$ is a small rigid body rotation. Above transformation is applied to all the elements containing u_3, v_2 and v_3. Hence, these DOFs no longer appear in the final assembled system. The final result is independent of the values of the constitutive parameters for the rigid (undeformable) triangle. Note however that the choice of the rigid body displacements u_1, v_1 and u_2 is not unique. Care must be taken that $b = y_2 - y_1 \neq 0$ to avoid division by zero in Eq.(9.10). Also if 1-2 is a straight line, the use of u_1 and u_2 as independent DOFs contradicts the rigid body assumption as the deformation of side 1–2 is possible in this case [CMPW].

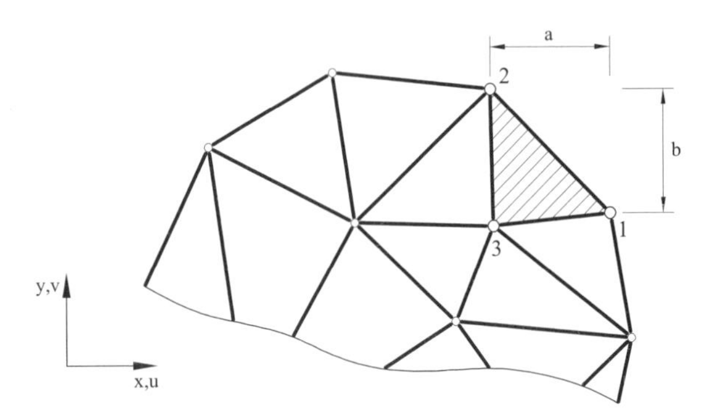

Fig. 9.4 Rigid element

9.4 DISPLACEMENT CONSTRAINTS

The transformations of the previous section can be generalized to prescribe any relationship (constraints) on or among nodal DOFs. Problems such as that shown in Figure 9.3 are, in fact, examples of constraints at the end nodes of 1D elements connected to quadrilateral elements.

A typical example of displacement constraint is the prescription of DOFs in a mesh. Each prescribed displacement can be eliminated from the assembled stiffness equations, thus reducing the total number of DOFs unknowns [Li]. Figure 9.5 shows examples of more complex displacement constraints. Nodes B and C in Figure 9.5a are considered to be separated but having the same local displacement $v'_B = v'_C$. Nodes A and B in Figure 9.5b are pin-jointed, thus enforcing the same displacements for both nodes. The displacements of the end nodes for the beam AB in the frame of Figure 9.5c can be constrained so as to ignore the axial deformation in the analysis. This implies enforcing $v_A = v_B = 0$ and $u_A = u_B$. The final system involves only 3 DOFs for each floor (u_A or u_B, θ_A and θ_B), thus allowing for a pure bending solution.

9.4.1 General procedure to eliminate constrained DOFs

Each constraint allows us to eliminate one DOF in the final assembled system as follows. Let us write the c independent constraints on the nodal displacements $\mathbf{a}$ as

$$\underset{c\times n}{\mathbf{C}}\;\underset{n\times 1}{\mathbf{a}} = \underset{c\times 1}{\mathbf{g}} \tag{9.11}$$

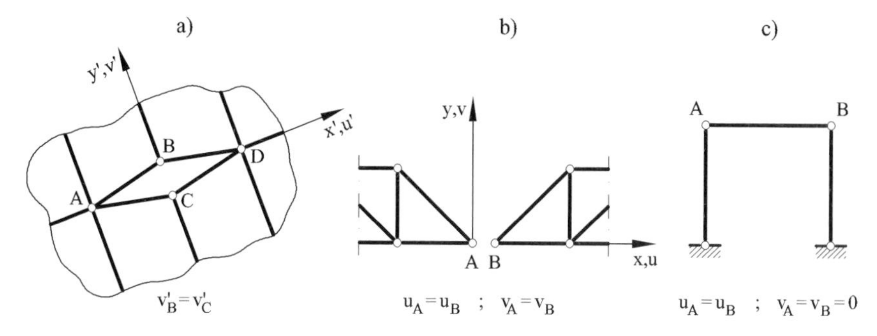

Fig. 9.5 Examples of displacement constraints [CMPW]

where $\mathbf{C}$ and $\mathbf{g}$ contain constraints. $\mathbf{C}$ has more columns than rows as there are more DOFs in $\mathbf{a}$ than constraints. Let us consider the following partition

$$\mathbf{C} = \underset{c\times(n-c)}{[\mathbf{C}_r} \quad \underset{c\times c}{\mathbf{C}_e]} \qquad \text{and} \qquad \mathbf{a} = \begin{Bmatrix} \mathbf{a}_r \\ \mathbf{a}_e \end{Bmatrix} \begin{matrix} (n-c) \\ c \end{matrix} \tag{9.12}$$

where $\mathbf{C}_e$ is a square non-singular matrix and indexes r and e denote the DOFs to be retained and those to be eliminated, respectively. Let us consider first the simplest case with $\mathbf{g} = 0$. From Eqs.(9.11) and (9.12) we have

$$\mathbf{a}_e = \mathbf{H}\mathbf{a}_r \quad \text{with} \quad \mathbf{H} = -[\mathbf{C}_e]^{-1}\mathbf{C}_r \tag{9.13}$$

Therefore we can write

$$\mathbf{a} = \mathbf{T}\mathbf{a}_r \quad \text{with} \quad \mathbf{T} = \begin{Bmatrix} \mathbf{I} \\ \mathbf{H} \end{Bmatrix} \tag{9.14}$$

Matrix $\mathbf{T}$ transforms the element stiffness matrix and the equivalent nodal force vector as $\mathbf{T}^T\mathbf{K}\mathbf{T}$ and $\mathbf{T}^T\mathbf{f}$, respectively and the final system involves the retained DOFs $\mathbf{a}_r$ only. Once these are computed, vector $\mathbf{a}_e$ can be obtained from Eq.(9.13).

The selection of $\mathbf{a}_e$ and $\mathbf{C}_e$ is not unique. An alternative is to choose the last c columns of $\mathbf{C}$ which are linearly independent. This guarantees that $\mathbf{C}_e$ is invertible. Note that an excessive number of constraints can lock the solution giving $\mathbf{a} = 0$.

If $\mathbf{g} \neq 0$ the value of $\mathbf{a}_e$ can be computed from Eqs.(9.11)-(9.13) as

$$\mathbf{a}_e = \bar{\mathbf{g}} + \mathbf{H}\mathbf{a}_r \tag{9.15a}$$

where

$$\bar{\mathbf{g}} = [\mathbf{C}_e]^{-1}\mathbf{g} \tag{9.15b}$$

The assembled (symmetric) stiffness matrix can be written before the constraints are enforced as

$$\mathbf{Ka} = \begin{bmatrix} \mathbf{K}_{rr} & \mathbf{K}_{re} \\ \mathbf{K}_{re}^T & \mathbf{K}_{ee} \end{bmatrix} \begin{Bmatrix} \mathbf{a}_r \\ \mathbf{a}_e \end{Bmatrix} = \begin{Bmatrix} \mathbf{f}_r \\ \mathbf{f}_e \end{Bmatrix} = \mathbf{f} \tag{9.16}$$

This equation can be transformed using Eq.(9.14) as

$$\begin{bmatrix} (\mathbf{K}_{rr} + \mathbf{K}_{re}\mathbf{H}) & \mathbf{0} \\ -\mathbf{H} & \mathbf{I} \end{bmatrix} \begin{Bmatrix} \mathbf{a}_r \\ \mathbf{a}_e \end{Bmatrix} = \begin{Bmatrix} \mathbf{f}_r - \mathbf{K}_{re}\bar{\mathbf{g}} \\ \bar{\mathbf{g}} \end{Bmatrix} \tag{9.17}$$

The first row of Eq.(9.17) gives $\mathbf{a}_r$. Then $\mathbf{a}_e$ can be obtained using the second row. Note that the system (9.17) is not symmetric. Symmetry can be recovered as follows: multiply first the second row of Eq.(9.16) by $\mathbf{H}^T$ and add the result to the first row of Eq.(9.17). Then multiply the second row of Eq.(9.17) by $-\mathbf{K}_{ee}$. This yields

$$\begin{bmatrix} (\mathbf{K}_{rr} + \mathbf{K}_{re}\mathbf{H} + \mathbf{H}^T\mathbf{K}_{re}^T) & \mathbf{H}^T\mathbf{K}_{ee} \\ \mathbf{K}_{ee}\mathbf{H} & -\mathbf{K}_{ee} \end{bmatrix} \begin{Bmatrix} \mathbf{a}_r \\ \mathbf{a}_e \end{Bmatrix} = \begin{Bmatrix} \mathbf{H}^T\mathbf{f}_e + \mathbf{f}_r - \mathbf{K}_{re}\bar{\mathbf{g}} \\ -\mathbf{K}_{ee}\bar{\mathbf{g}} \end{Bmatrix} \tag{9.18}$$

Eq.(9.18) can now be solved taking advantage of the symmetry of the system [Pr,Ral]. The negative sign in the lower diagonal terms can introduce problems in the solution process.

If $\mathbf{g} = \mathbf{0}$, then $\bar{\mathbf{g}} = \mathbf{0}$ and in Eq.(9.18) allows us to eliminate $\mathbf{a}_e$ from the second row. Substituting this value into the first row gives

$$(\mathbf{K}_{rr} + \mathbf{K}_{re}\mathbf{H} + \mathbf{H}^T\mathbf{K}_{re}^T + \mathbf{H}^T\mathbf{K}_{ee}\mathbf{H})\mathbf{a}_r = \mathbf{H}^T\mathbf{f}_e + \mathbf{f}_r \tag{9.19}$$

or

$$\mathbf{K}_r\mathbf{a}_r = \mathbf{f}_r \tag{9.20}$$

It is easy to verify that $\mathbf{K}_r$ and $\mathbf{f}_r$ can be obtained for $\mathbf{g} = 0$ by

$$\mathbf{K}_r = \mathbf{T}^T\mathbf{K}\mathbf{T} \quad \text{and} \quad \mathbf{f}_r = \mathbf{T}^T\mathbf{f} \tag{9.21}$$

where $\mathbf{K}$ and $\mathbf{f}$ are given by Eq.(9.16) and $\mathbf{T}$ is the transformation matrix of Eq.(9.14).

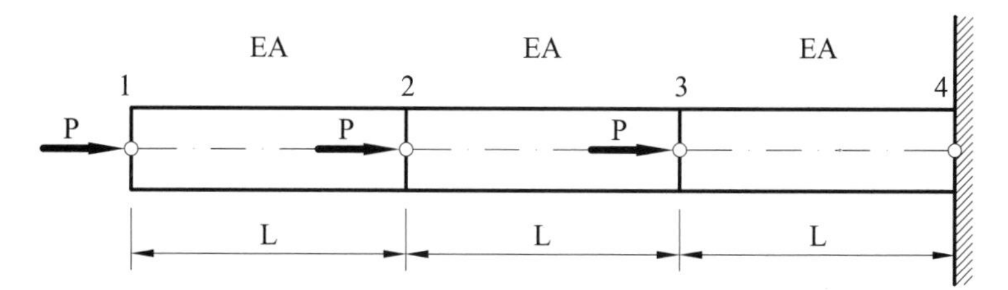

Fig. 9.6 Bar under point loads analyzed with three linear elements

---

Example 9.2: Compute the displacements in the mesh of three bar elements of Figure 9.6 imposing $u_1 = u_2$.

- ***Solution***

The assembled stiffness equations after eliminating the displacement at the fixed node $u_4 = 0$ are

$$\left(\frac{EA}{l}\right)\begin{bmatrix} 1 & -1 & 0 \\ -1 & 2 & -1 \\ 0 & -1 & 2 \end{bmatrix}\begin{Bmatrix} u_1 \\ u_2 \\ u_3 \end{Bmatrix} = \begin{Bmatrix} P \\ P \\ P \end{Bmatrix}$$

The constraint $u_1 = u_2$ is written as

$$\underset{\mathbf{C}_e\ \vdots\ \ \mathbf{C}_r}{[1,\ \vdots -1,\ \ 0]}\begin{Bmatrix} u_1 \\ \cdots\cdots \\ u_2 \\ u_3 \end{Bmatrix}\begin{matrix} \mathbf{a}_e \\ \cdots\cdots \\ \mathbf{a}_r \end{matrix} = 0$$

and

$$\mathbf{u}_1 = \underset{\mathbf{H}}{[1, 0]}\begin{Bmatrix} u_2 \\ u_3 \end{Bmatrix} \quad \text{and} \quad \mathbf{T} = \begin{bmatrix} 1 & 0 \\ 0 & 1 \\ 1 & 0 \end{bmatrix}$$

Therefore

$$\mathbf{K}_r = \mathbf{T}^T\mathbf{K}\mathbf{T} = \frac{EA}{l}\begin{bmatrix} 1 & -1 \\ -1 & 2 \end{bmatrix}$$

$$\mathbf{f}_r = \mathbf{T}^T\mathbf{f} = \begin{Bmatrix} 2P \\ P \end{Bmatrix}$$

The final system is

$$\frac{EA}{l}\begin{bmatrix} 1 & -1 \\ -1 & 2 \end{bmatrix}\begin{Bmatrix} u_2 \\ u_3 \end{Bmatrix} = \begin{Bmatrix} 2P \\ P \end{Bmatrix}$$

giving

$$u_2 = \frac{5Pl}{EA} \quad \text{and} \quad u_3 = \frac{3Pl}{EA}$$

9.4.2 Use of Lagrange multipliers

The Lagrange multipliers method is a technique to find the maximum or minimum of a function whose variables are related by constraints [Ral,ZTZ]. In structural mechanics the function is typically the total potential energy Π, the variables are the nodal DOFs and the constraints are those expressed by Eq.(9.11). An augmented function is written as

$$\overline{\Pi} = \Pi + \boldsymbol{\lambda}^T(\mathbf{Ca} - \mathbf{g}) \tag{9.22a}$$

where

$$\Pi = \frac{1}{2}\mathbf{a}^T\mathbf{Ka} - \mathbf{a}^T\mathbf{f} \tag{9.22b}$$

and $\boldsymbol{\lambda} = [\lambda_1, \lambda_2, \ldots, \lambda_r]^T$ is the vector containing the Lagrange multipliers equaling the number of constraints.

The problem is solved by imposing the stationarity of $\overline{\Pi}$ with respect to $\boldsymbol{\lambda}$ and $\mathbf{a}$ leading to the following system of equations

$$\begin{bmatrix} \mathbf{K} & \mathbf{C}^T \\ \mathbf{C} & \mathbf{0} \end{bmatrix} \begin{Bmatrix} \mathbf{a} \\ \boldsymbol{\lambda} \end{Bmatrix} = \begin{Bmatrix} \mathbf{f} \\ \mathbf{g} \end{Bmatrix} \tag{9.23}$$

from which $\mathbf{a}$ and $\boldsymbol{\lambda}$ can be obtained. The Lagrange multipliers can be interpreted as the linking forces necessary to enforce the constraints on the DOFs [ZTZ].

Example 9.3: Solve the problem of Figure 9.6 using the Lagrange multipliers method.

- ***Solution***

Eq.(9.23) is written, noting that $\mathbf{C} = [1, -1]$, as

$$\begin{bmatrix} k & -k & 0 & 1 \\ -k & 2k & -k & -1 \\ 0 & -k & 2k & 0 \\ 1 & -1 & 0 & 0 \end{bmatrix} \begin{Bmatrix} u_1 \\ u_2 \\ u_3 \\ \lambda \end{Bmatrix} = \begin{Bmatrix} P \\ P \\ P \\ 0 \end{Bmatrix} \quad \text{with} \quad k = \frac{EA}{l}$$

The solution is

$$[u_1, u_2, u_3, \lambda] = \left[\frac{5P}{k}, \frac{5P}{k}, \frac{3P}{k}, P\right]$$

The multiplier $\lambda = P$ coincides with the force applied along the resulting rigid bar 1-2.

9.4.3 Penalty method

The constraints (9.11) can be written as

$$\mathbf{p} = \mathbf{Ca} - \mathbf{g} \quad ; \quad \mathbf{p} = 0 \tag{9.24}$$

The potential energy of the structure Π is now augmented by the penalty function $\frac{1}{2}\mathbf{p}[\boldsymbol{\alpha}]\mathbf{p}$, where $[\boldsymbol{\alpha}]$ is a diagonal matrix of penalty numbers α_i.

Thus

$$\overline{\Pi} = \Pi + \frac{1}{2}\mathbf{p}^T[\boldsymbol{\alpha}]\mathbf{p} \tag{9.25}$$

Obviously, if $\mathbf{p} = 0$ then the constraints (9.24) are satisfied and there is no need to add anything to Π. Stationarity of $\overline{\Pi}$ yields

$$\delta\overline{\Pi} = \delta\Pi + \delta\mathbf{p}^T[\boldsymbol{\alpha}]\mathbf{p} = 0 \tag{9.26}$$

For $\alpha_i \to \infty$ Eq.(9.26) tends to $\delta\overline{\Pi} \simeq \delta\mathbf{p}^T[\boldsymbol{\alpha}]\mathbf{p} = 0$. This requires $\mathbf{p} = 0$ and the constraints are satisfied. As the α_i grow the solution of Eq.(9.26) evolves so that the constraint equation (9.24) is progressively better satisfied. In practice it suffices to choose a sufficiently large value for α_i. The equilibrium equation for finite values of the α_i's is deduced from Eqs.(9.22*b*), (9.24) and (9.26) as

$$[\mathbf{K} + \mathbf{C}^T[\boldsymbol{\alpha}]\mathbf{C}]\mathbf{a} = \mathbf{f} + \mathbf{C}^T[\boldsymbol{\alpha}]\mathbf{g} \tag{9.27}$$

from which $\mathbf{a}$ can be obtained.

Example 9.4: Solve the problem of Figure 9.6 using the penalty method.

- ***Solution***

Vector $\mathbf{p}$ of Eq.(9.24) is

$$\mathbf{p} = [1, -1, 0][u_1, u_2, u_3]^T = \mathbf{C}\,\mathbf{a}$$

A single value of α is needed as there is only one constraint. The matrix $\mathbf{C}^T[\boldsymbol{\alpha}]\mathbf{C}$ is

$$\mathbf{C}^T\alpha\mathbf{C} = [1, -1, 0]\alpha \begin{bmatrix} 1 \\ -1 \\ 0 \end{bmatrix} = \alpha \begin{bmatrix} 1 & -1 & 0 \\ -1 & 1 & 0 \\ 0 & 0 & 0 \end{bmatrix}$$

$\mathbf{C}^T\alpha\mathbf{C}$ can be interpreted as the stiffness matrix of a bar element (with $\alpha = \frac{EA}{l}$) linking nodes 1 and 2. Obviously $\alpha \to \infty$ implies increasing the rigidity of this element and the constraint is approximated. The new system of equations is

$$\begin{bmatrix} (k+\alpha) & -(k+\alpha) & 0 \\ -(k+\alpha) & -2k+\alpha & -k \\ 0 & -k & 2k \end{bmatrix} \begin{Bmatrix} u_1 \\ u_2 \\ u_3 \end{Bmatrix} = \begin{Bmatrix} P \\ P \\ P \end{Bmatrix}$$

which yields

$$u_1 = \frac{5P}{k} + \frac{P}{k+\alpha}, \quad u_2 = \frac{5P}{k}, \quad u_3 = \frac{3P}{k}$$

Note that as α grows then $\frac{P}{k+\alpha} \to 0$ and the value of u_1 tends to u_2.

The penalty method, despite its simplicity, can pose numerical problems. Consider, for instance, Eq.(9.27) with $\mathbf{g} = 0$ and $[\boldsymbol{\alpha}] = \alpha$. The system of equilibrium equations degenerates for large values of α to

$$\mathbf{C}^T\mathbf{C}\mathbf{a} = \frac{1}{\alpha}\mathbf{f} \simeq \mathbf{0} \tag{9.28}$$

Clearly the solution "locks" for $\alpha \to \infty$ giving $\mathbf{a} = 0$, unless matrix $\mathbf{C}^T\mathbf{C}$ is singular. This singularity can be anticipated by observing the number of rows and columns in this matrix. This problem is similar to the singularity requirement for the shear stiffness matrix in order to prevent shear locking in Timoshenko beams and Reissner-Mindlin plate and shell elements [On,ZT,ZTZ].

9.5 NODAL CONDENSATION AND SUBSTRUCTURES

In some occasions we may want to eliminate the DOFs which are internal to an element or a group of elements and work only with the DOFs at the boundary. The internal DOFs, as well as any other necessary information within the internal domain, are computed "a posteriori".

An application of the so called nodal condensation is the analysis by substructures treated in a next section. For the sake of clarity let us consider first the techniques for condensation and recovery of nodal DOFs.

9.5.1 Nodal condensation

Let $\mathbf{a} = [\mathbf{a}_r, \mathbf{a}_e]^T$ be the DOFs in a mesh where $\mathbf{a}_r$ are the boundary DOFs to be retained and $\mathbf{a}_e$ are the internal DOFs to be eliminated (Figure 9.7).

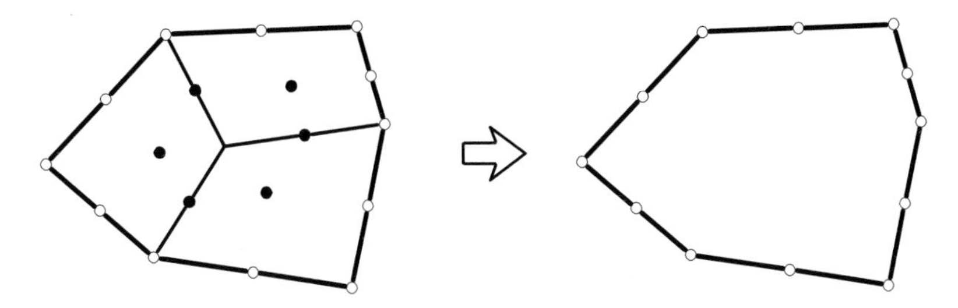

• Nodes to be eliminated
○ Nodes to be retained

Fig. 9.7 Condensation of internal nodes in a quadrilateral mesh

The matrix equilibrium equation is written as

$$\begin{bmatrix}\mathbf{K}_{rr} & \mathbf{K}_{re}\\ \mathbf{K}_{er} & \mathbf{K}_{ee}\end{bmatrix}\begin{Bmatrix}\mathbf{a}_r\\ \mathbf{a}_e\end{Bmatrix} = \begin{Bmatrix}\mathbf{f}_r\\ \mathbf{f}_e\end{Bmatrix} \tag{9.29}$$

Vector $\mathbf{a}_e$ is obtained using the second row of Eq.(9.29) as

$$\mathbf{a}_e = -[\mathbf{K}_{ee}]^{-1}[\mathbf{K}_{er}\mathbf{a}_r - \mathbf{f}_e] \tag{9.30}$$

Substituting this into the first row of Eq.(9.29) gives

$$\mathbf{K}_r\mathbf{a}_r = \bar{\mathbf{f}}_r \tag{9.31}$$

with

$$\begin{aligned}\mathbf{K}_r &= \mathbf{K}_{rr} - \mathbf{K}_{re}[\mathbf{K}_{ee}]^{-1}\mathbf{K}_{er}\\ \bar{\mathbf{f}}_r &= \mathbf{f}_r - \mathbf{K}_{re}[\mathbf{K}_{ee}]^{-1}\mathbf{f}_e\end{aligned} \tag{9.32}$$

If the loads act on the boundary of the element patch considered, then $\mathbf{f}_e = \mathbf{0}$ and Eqs.(9.32) can be simplified.

The condensed element is treated as an ordinary single element and $\mathbf{K}_r$ and $\mathbf{f}_r$ are assembled into the stiffness matrix and the equivalent nodal force vector of the remaining elements in the standard manner. The internal DOFs in the mesh $\mathbf{a}_e$ are "recovered" using Eq.(9.30). The stresses at any point are computed from the retained displacements $\mathbf{a}_r$ as follows. The stress vector is given by

$$\boldsymbol{\sigma} = [\mathbf{S}_r, \mathbf{S}_e][\mathbf{a}_r, \mathbf{a}_e]^T - \mathbf{D}\boldsymbol{\varepsilon}_o + \boldsymbol{\sigma}_o \tag{9.33}$$

where

$$\mathbf{S}_r = \mathbf{D}\mathbf{B}_r, \quad \mathbf{S}_e = \mathbf{D}\mathbf{B}_e, \quad \text{with} \quad \mathbf{B} = [\mathbf{B}_r, \mathbf{B}_e] \tag{9.34}$$

Indexes r and e in above denote the matrices associated to the retained and eliminated DOFs, respectively. Substituting $\mathbf{a}_e$ from Eq.(9.30) gives

$$\boldsymbol{\sigma} = \mathbf{S}^* \mathbf{a}_r + \boldsymbol{\tau}_o \tag{9.35}$$

where

$$\begin{aligned} \mathbf{S}^* &= \mathbf{S}_r - \mathbf{S}_e [\mathbf{K}_{ee}]^{-1} \mathbf{K}_{er} \\ \boldsymbol{\tau}_o &= \mathbf{S}_e [\mathbf{K}_{ee}]^{-1} \mathbf{f}_e - \mathbf{D} \boldsymbol{\varepsilon}_o + \boldsymbol{\sigma}_o \end{aligned} \tag{9.36}$$

It is interesting that if $\mathbf{a}_e$ contains a single DOF, the condensation process of Eqs.(9.29)-(9.32) coincides with the standard Gauss elimination technique for solving systems of algebraic equations. The condensation can, therefore, be interpreted as the first step in the solution of the original assembled system $\mathbf{K}\ \mathbf{a} = \mathbf{f}$.

9.5.2 Substructuring

Substructuring is the splitting of a structure into a number of parts (called substructures) whose assembly forms the original structure. Each substructure can be treated using the condensation technique explained in the previous section. This considerably reduces the computational cost of the solution.

The substructuring process includes the following steps: 1) Condensating the DOFs and the loads of each substructure to the boundary nodes. Each substructure is then transformed to a "super-element" connected to the other elements (or substructures) via the boundary nodes. 2) Assembly of the stiffness and nodal force contributions from the different substructures into a global stiffness matrix $\mathbf{K}$ and a force vector $\mathbf{f}$. 3) Solution of the system $\mathbf{Ka} = \mathbf{f}$. 4) Computation (recovery) of the nodal displacements $\mathbf{a}_e$, the strains and the stresses for each substructure.

In some occasions the structure can be modelled by repetition of the same substructure. This simplifies considerably the solution, since, with an adequate nodal numbering, the condensation process needs to be performed only once.

Substructuring is advantageous when a structure is formed by clearly identifiable and repeated parts. An example is the box girder bridge of Figure 9.8 where the substructures 1, 2 and 3 are repeated.

Substructuring has become popular as a parallel computing technique to solve large scale structural problems. Each substructure is solved in parallel in a different processor and hence the computational effort and

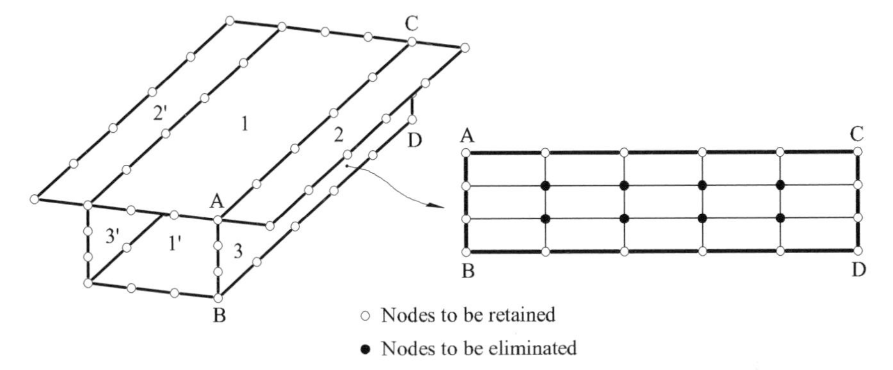

Fig. 9.8 Substructuring analysis of a box girder bridge. Retained and eliminated nodes in a substructure

memory requirements can be uniformly split among the processors. This makes substructuring attractive for distributed computing in networks of workstations and PCs working in parallel, as well as in shared memory parallel computers.

9.6 STRUCTURAL SYMMETRY

9.6.1 Symmetric solution

A structure is symmetric if the material and geometric properties and the boundary conditions have one or more axes or planes of symmetry. Structural symmetry can be reflective or rotational. A structure is symmetric with respect to an axis or a plane if one or more reflections and/or rotations bring the structure to a configuration indistinguishable from the original one. For the plate of Figure 9.9a each dotted line is a reflective symmetry axis. Also, a line normal to the plate through the central point C is a rotational symmetry axis, as successive $90°$ rotations bring the structure into the same position. Other examples of rotational symmetry include axisymmetric structures and the cyclic structures described in a next section.

A symmetric structure may carry symmetric or antisymmetric loads. A system of loads is antisymmetric if a single reflection of the structure with its loads followed by sign reversal of all the loads result in self coincidence (Figure 9.9c). For both symmetric and antisymmetric loading cases it *suffices to analyze the symmetric half of the structure* with the following boundary conditions:

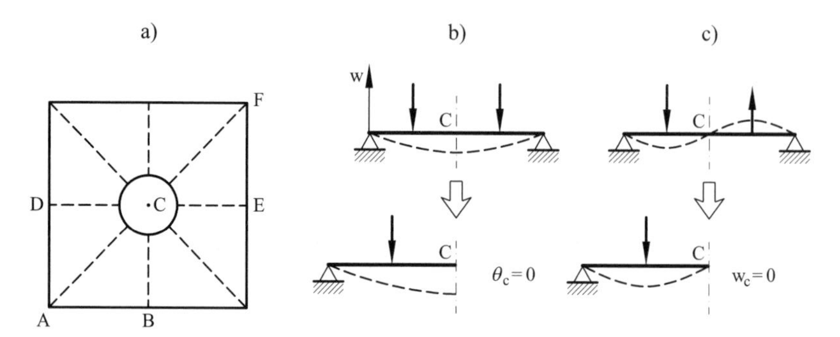

Fig. 9.9 Examples of symmetries: (a) Reflective and rotational symmetries; (b) symmetric loading and (c) antisymmetric loading in a beam

Symmetric loading

- Displacements are zero in a direction perpendicular to a symmetry plane.
- Rotation vectors have no component on a symmetry plane (Figure 9.9b).

Antisymmetric loading

- Displacements are zero in points contained on a plane of geometrical symmetry (Figure 9.9c).
- Rotation vectors have no component perpendicular to a plane of geometric symmetry.

Loads are *skew symmetric* if they require a rotation or more than one deflection to reproduce the original loading pattern. The plate of Figure 9.9a has skew symmetric loading if sectors ACD and FCE carry the same loads and the rest of the plate is unloaded. The deflection satisfies $w(r) = w(-r)$, where r is the radial coordinate measured from C. For the *skew antisymmetric* case the loads acting on ACD and FCE have the same module but opposite sign and the deflection satisfies $w(r) = -w(-r)$.

Symmetric structures under arbitrary loading can be analyzed by expressing the loads as the sum of two sets of symmetric and antisymmetric loads. One half of the structure is solved under each set of loads only and the final result is obtained by superposition. For symmetric and antisymmetric loads, the resulting displacements and stresses are respectively symmetric and antisymmetric. An example are the axisymmetric structures under arbitrary loading studied in Chapter 11 of Volume 2 [On].

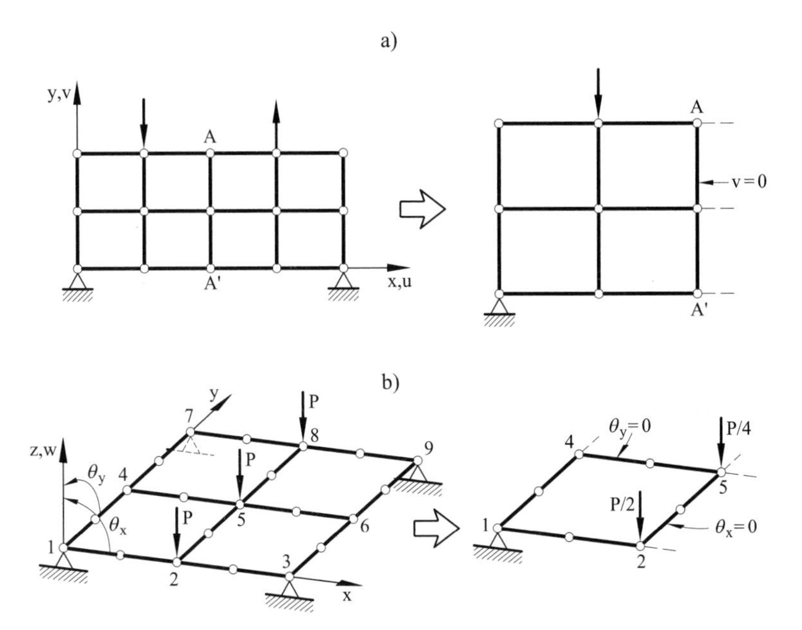

Fig. 9.10 Structures with antisymmetric and symmetric loads

Example 9.5: Impose the boundary conditions in the structures of Figure 9.10.

- ***Solution***

Structure 1 : Plane stress wall under antisymmetric point loads (Figure 9.10a). Half the structure can be analyzed prescribing zero vertical displacement in the nodes along the symmetry axis AA'.

Structure 2 : Square plate under symmetric point loads (Figure 9.10b). The double symmetry allows to analyze a quarter of plate with the following conditions on the rotations: $\theta_x = 0$ on side 2-5 and $\theta_y = 0$ on side 4-5.

9.6.2 Cyclic symmetry

In some structures a repetition of geometry and loading patterns can be identified although a symmetry axis can not be recognized as such. This property is called *cyclic symmetry* and it is usual in rotational parts such as fans, pump impellers, etc. Figure 9.11 shows a hypothetic 2D structure of this kind [CMPW,ZS]. Each triangle is identified with a repeatable structure. The displacements, strains and stresses have cyclic symmetry and this suffices to analyze a single substructure as follows.

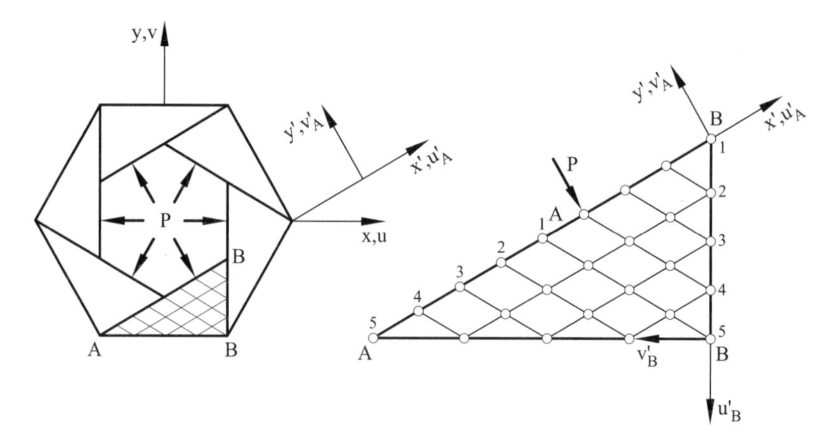

Fig. 9.11 Plane structure with cyclic geometry [CMPW,ZS]

Let us consider the triangular substructure shown in Figure 9.11. Nodes in the cyclic boundaries AA and BB have been equally numbered for convenience. This numbering affects only to the assembly process and the actual coordinates must be used for computing the element matrices.

It is also convenient to compute the stiffness matrices of the nodes on the boundaries AA and BB in the local boundary axes $x'y'$ shown in the figure. This can be done following the transformation of Section 9.2. Observation of the displacement pattern gives

$$\mathbf{a}'_A = \mathbf{a}'_B \tag{9.37}$$

where $\mathbf{a}'_A$ and $\mathbf{a}'_B$ are the local displacements on the boundaries AA and BB, respectively. Denoting by $\mathbf{a}_R$ the displacements of the rest of nodes it is deduced from Eq.(9.37).

$$\begin{Bmatrix} \mathbf{a}_R \\ \mathbf{a}'_A \\ \mathbf{a}'_B \end{Bmatrix} = \begin{bmatrix} \mathbf{I} & \mathbf{0} \\ \mathbf{0} & \mathbf{I} \\ \mathbf{0} & \mathbf{I} \end{bmatrix} \begin{Bmatrix} \mathbf{a}_R \\ \mathbf{a}'_A \end{Bmatrix} = \mathbf{T} \begin{Bmatrix} \mathbf{a}_R \\ \mathbf{a}'_A \end{Bmatrix} \tag{9.38}$$

The stiffness equations of the substructure are written as

$$\begin{bmatrix} \mathbf{K}_{RR} & \mathbf{K}_{RA} & \mathbf{K}_{RB} \\ \mathbf{K}_{RA}^T & \mathbf{K}_{AA} & \mathbf{K}_{AB} \\ \mathbf{K}_{RB}^T & \mathbf{K}_{AB}^T & \mathbf{K}_{BB} \end{bmatrix} \begin{Bmatrix} \mathbf{a}_R \\ \mathbf{a}'_A \\ \mathbf{a}'_B \end{Bmatrix} = \begin{Bmatrix} \mathbf{f}_R \\ \mathbf{f}'_A + \mathbf{r}'_A \\ \mathbf{r}'_B \end{Bmatrix} \tag{9.39}$$

where $\mathbf{f}'_A$ and $\mathbf{f}_R$ are the equivalent nodal forces due to external loads (rotation, pressure, etc.) acting on AA and the rest of the structure, respectively. Loads $\mathbf{r}'_A$ and $\mathbf{r}'_B$ are the reactions applied along AA and BB by the neighboring substructures. Note that loads $\mathbf{f}'_B$ need not be considered as the external loads on the interface boundaries must be assigned to only one boundary. From the condition of cyclic symmetry

$$\mathbf{r}'_A = -\mathbf{r}'_B \tag{9.40}$$

Using Eqs.(9.38)–(9.40) and the techniques to enforce constraints described in Section 9.4.1 gives

$$\mathbf{K}_g = \mathbf{T}^T\mathbf{K}\mathbf{T} = \begin{bmatrix} \mathbf{K}_{RR} & \mathbf{K}_{RA} + \mathbf{K}_{RB} \\ \mathbf{K}_{RA}^T + \mathbf{K}_{RB}^T & \mathbf{K}_{AA} + \mathbf{K}_{AB} + \mathbf{K}_{AB}^T + \mathbf{K}_{BB} \end{bmatrix} \tag{9.41a}$$

$$\mathbf{f}_g = \mathbf{T}^T\mathbf{f} = \begin{Bmatrix} \mathbf{f}_R \\ \mathbf{f}'_A \end{Bmatrix} \tag{9.41b}$$

The resulting equation is

$$\mathbf{K}_g \begin{Bmatrix} \mathbf{a}_R \\ \mathbf{a}'_A \end{Bmatrix} = \begin{Bmatrix} \mathbf{f}_R \\ \mathbf{f}'_A \end{Bmatrix} \tag{9.42}$$

The DOFs and forces on AA can now be transformed to the global axes. This is not strictly necessary and Eq.(9.42) can be solved directly for the values of $\mathbf{a}'_A$ and $\mathbf{f}'_A$ [CMPW,ZS].

9.7 STRUCTURES ON ELASTIC FOUNDATION

The assumption of undeformable supports does not apply in many practical cases. Typical examples are a rail on a roadbed, a pavement slab on soil, a dam on rock etc. The interest of the analysis is primarily concerned with the study of the rail, the pavement or the dam. However, the deformability of the foundation can not be disregarded as it can affect the response of the structure. The supporting effect of the foundation can be accurately modelled by treating the foundation domain as an additional structure which is discretized using standard finite elements. This allows us to account for variable material properties in different parts of the foundation, etc. An example of this kind is the analysis of the soil-dam interaction using 2D or 3D solid finite elements to discretize both the dam and the soil. The problem is more difficult if plate or shell elements are used to model the structure, as these must be adequately coupled with the 3D solid elements needed to discretize the foundation. A simpler alternative is to use an *elastic solid foundation model* characterized by an elastic modulus k. This yields a first approximation of the supporting effect of the foundation in an easy manner as follows.

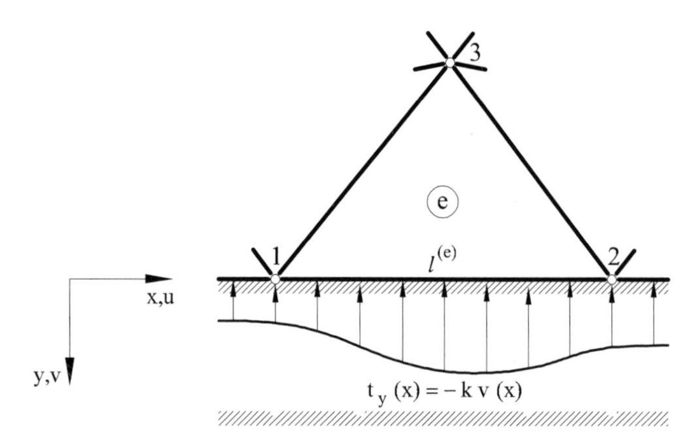

Fig. 9.12 2D solid triangle on elastic foundation

Let us consider a 2D solid laying on an elastic foundation characterized by an elastic modulus k assumed here to be constant, for simplicity (Figure 9.12). Proportionality between the vertical reaction pressure of the foundation t_y and the vertical deflection v is assumed (Winkler foundation), i.e.

$$t_y(x) = -kv(x) \tag{9.43}$$

Note that we have assumed here $v(x)$ to be positive in the downward direction for clarity (Figure 9.12).

The foundation pressure generates an additional virtual external work of amount $t_y(x)\delta v(x)$. The PVW for the triangle is

$$\iint_A \delta\boldsymbol{\varepsilon}^T \boldsymbol{\sigma} dA = \iint_A \delta\mathbf{u}^T \mathbf{b} dA + \int_\Gamma \delta v \; t_y(x) \; dx \tag{9.44}$$

where Γ is the boundary of the domain in contact with the foundation.

In Eq.(9.44) we have assumed a unit thickness for simplicity.

Substituting Eq.(9.43) into (9.44) gives

$$\iint_A \delta\boldsymbol{\varepsilon}^T \boldsymbol{\sigma} dA + \underline{\int_\Gamma \delta v \; k \; v \; dx} = \iint_A \delta\mathbf{u}^T \mathbf{b} dA \tag{9.45}$$

Accounting for the elastic deformation of the foundation therefore increases the virtual strain energy in the term underlined in Eq.(9.45).

We introduce now a standard finite element discretization. This leads to the following equilibrium equation for the element

$$[\mathbf{K}^{(e)} + \mathbf{H}^{(e)}]\mathbf{a}^{(e)} - \mathbf{f}^{(e)} = \mathbf{q}^{(e)} \tag{9.46}$$

where the additional stiffness matrix $\mathbf{H}^{(e)}$ is given by

$$\mathbf{H}^{(e)} = \begin{bmatrix} \mathbf{H}_{11}^{(e)} & \mathbf{H}_{12}^{(e)} & \mathbf{0} \\ & \mathbf{H}_{22}^{(e)} & \mathbf{0} \\ Symm. & & \mathbf{0} \end{bmatrix} \quad \text{with} \quad \mathbf{H}_{ij}^{(e)} = \int_{l^{(e)}} k \begin{bmatrix} 0 & 0 \\ 0 & N_i N_j \end{bmatrix} dx \tag{9.47}$$

where $l^{(e)}$ is the length of the element side in contact with the foundation (Figure 9.12).

The non-zero terms in matrix $\mathbf{H}^{(e)}$ involve the vertical DOFs of nodes 1, 2 in contact with the foundation.

The effect of the elastic foundation therefore increases the element stiffness. This leads to smaller nodal deflections values, as expected. For linear elements

$$\mathbf{H}_{ij}^{(e)} = \left[\frac{kl}{3}\right]^{(e)} \begin{bmatrix} 0 & 0 \\ 0 & \beta \end{bmatrix} \tag{9.48}$$

with $\beta = 1$ for $i + j = 2, 4$ and $\beta = 1/2$ for $i + j = 3$.

Note that $\mathbf{H}^{(e)}$ is a full matrix and this introduces a coupling between the deflection of nodes 1 and 2. A simplification is to diagonalize $\mathbf{H}^{(e)}$ by adding up the coefficients of each row. This is equivalent to assume a spring of elastic modulus $\frac{(kl)^{(e)}}{2}$ acting at each node. If equal length elements are used along the foundation, this simply implies adding the coefficient $(kl)^{(e)}$ to the diagonal term corresponding to the vertical deflection of the nodes laying on the foundation in the original element stiffness matrix.

Matrix $\mathbf{H}_{ij}^{(e)}$ has a similar form for 3D solid elements and is given by

$$\mathbf{H}_{ij}^{(e)} = \iint_{A^{(e)}} k^{(e)} \begin{bmatrix} 0 & 0 & 0 \\ 0 & 0 & 0 \\ 0 & 0 & N_i N_j \end{bmatrix} dA \tag{9.49}$$

where $A^{(e)}$ is the area of the element side in contact with the foundation. The case of beam and plate elements laying on an elastic foundation is treated in Volume 2 [On].

9.8 COMPUTATION OF NODAL STRESSES

The stresses at a node can be directly computed from the nodal displacements as

$$\boldsymbol{\sigma}_i = \mathbf{D}\mathbf{B}(\xi_i, \eta_i, \zeta_i)\mathbf{a}^{(e)} \tag{9.50}$$

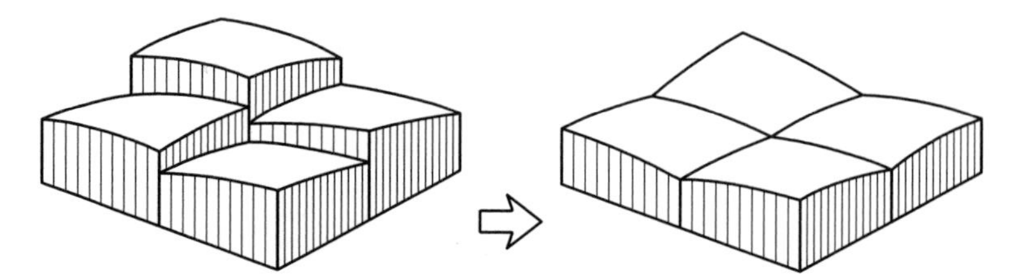

Fig. 9.13 Discontinuous and continuous (smoothed) stress field

where $\mathbf{B}(\xi_i, \eta_i, \zeta_i)$ denotes the strain matrix evaluated at node i.

Eq.(9.50) gives nodal stresses which are *discontinuous* between the adjacent elements (Figure 9.13). This is a consequence of the finite element formulation where continuity is only required to the displacements. Stress discontinuity is reduced as the mesh is refined. Single nodal stress values can be obtained via nodal averaging, or by using the smoothing techniques described in the next section.

The nodal stresses obtained via Eq.(9.50) are not as accurate as those obtained by extrapolating to the nodes the stresses computed at the Gauss points, which are the optimal stress sampling points within an element (Section 6.7). Different extrapolation techniques are presented next.

9.8.1 Global smoothing of stresses

The stresses at the Gauss points can be used to define a global extrapolation procedure giving directly a continuous nodal stress field. Let $\boldsymbol{\sigma}$ be the stresses at the Gauss points and $\boldsymbol{\sigma}_s$ the sought smoothed stress field (Figure 9.14) defined within each element as

$$\boldsymbol{\sigma}_s = \sum_{i=1}^{n} \mathbf{N}_i \hat{\boldsymbol{\sigma}}_{s_i}^{(e)} = \mathbf{N} \hat{\boldsymbol{\sigma}}_s^{(e)} \tag{9.51}$$

where

$$\mathbf{N}_i = N_i \mathbf{I}_{n_\sigma} \quad \text{and} \quad \hat{\boldsymbol{\sigma}}_s^{(e)} = \left\{ \begin{matrix} \hat{\boldsymbol{\sigma}}_{s_1}^{(e)} \\ \vdots \\ \hat{\boldsymbol{\sigma}}_{s_n}^{(e)} \end{matrix} \right\} \tag{9.52}$$

where $\hat{(\cdot)}$ denotes nodal values of the smoothed stress field, n_σ is the number of stress components (i.e. 3 for 2D solids, 6 for 3D solids, etc.), $\mathbf{I}$ is a unit matrix, N_i is the standard shape function of node i and n is

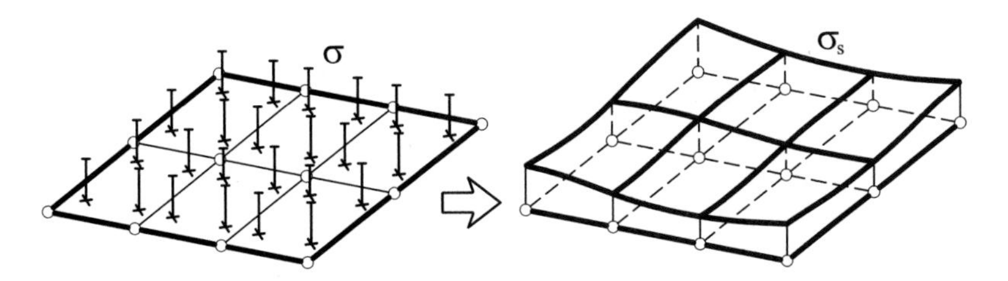

Fig. 9.14 Global smoothing of the nodal stresses from the Gauss point values

the number of element nodes. Higher of lower order interpolations can be used to define the stress field, if necessary [HC].

The error between the original and smoothed stress fields at each point is

$$\mathbf{e} = \boldsymbol{\sigma}_s - \boldsymbol{\sigma} = \mathbf{N}\hat{\boldsymbol{\sigma}}_s^{(e)} - \mathbf{DBa}^{(e)} \tag{9.53}$$

The values of $\hat{\boldsymbol{\sigma}}_{s_i}$ are obtained by minimizing the following functional

$$F = \iint_A \mathbf{e}^T \mathbf{e}\, dA = \iint_A (\boldsymbol{\sigma}_s - \boldsymbol{\sigma})^T (\boldsymbol{\sigma}_s - \boldsymbol{\sigma}) dA \tag{9.54}$$

i.e.

$$\frac{\partial F}{\partial \boldsymbol{\sigma}_i{}^{(e)}} = 0 \Longrightarrow \mathbf{M}\hat{\boldsymbol{\sigma}}_s = \mathbf{g} \tag{9.55}$$

where $\hat{\boldsymbol{\sigma}}_s$ contains the stresses at all nodes and

$$\mathbf{M}_{ij}^{(e)} = \iint_{A^{(e)}} \mathbf{N}_i^T \mathbf{N}_j dA \tag{9.56}$$

$$\mathbf{g}_i^{(e)} = \iint_{A^{(e)}} \mathbf{N}_i^T \mathbf{DBa}^{(e)} dA \quad , \quad i,j = 1,n \tag{9.57}$$

The smoothing matrix $\mathbf{M}^{(e)}$ and the "force" vector $\mathbf{g}^{(e)}$ for each element can be assembled in the usual manner to form the global expression of $\mathbf{M}$ and $\mathbf{g}$. Solution of the system (9.55) yields the stresses at all nodes. The stress field has the same continuity than the interpolation function N_i (typically C° continuity is chosen). The smoothing process can be applied independently to each stress component, although a different system of $n \times n$ equations must be solved for each component. The solution is simplified by using a Jacobi iteration scheme [Ral] for solving Eq.(9.55) as

$$\hat{\boldsymbol{\sigma}}_s^{n+1} = \mathbf{M}_D^{-1}[\mathbf{g} - (\mathbf{M} - \mathbf{M}_D)\hat{\boldsymbol{\sigma}}_s^n] \tag{9.58}$$

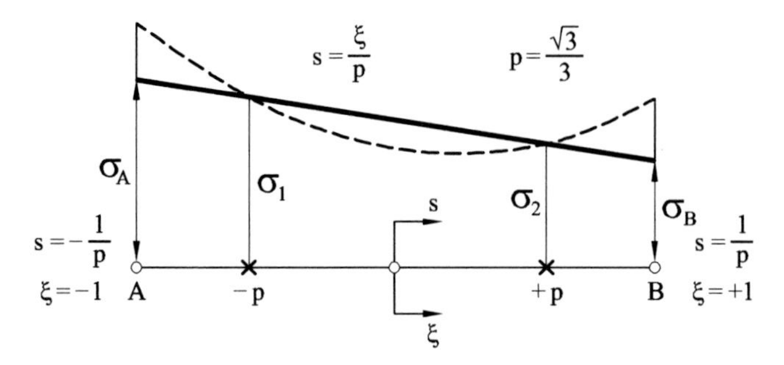

Fig. 9.15 Direct linear extrapolation of the two Gauss point stresses to the end nodes of 1D quadratic elements

where $\mathbf{M}_D = \text{diag. } \mathbf{M}$. Typically 4 or 5 iterations are needed to obtain a converged solution [ZTZ].

This technique (also called L_2 projection of stresses) can be applied to each individual element. A simpler stress extrapolation procedure can however be implemented at the element level as explained next.

9.8.2 Direct local extrapolation

Let us consider the 1D quadratic element of Figure 9.15. The Gauss point stresses σ_1 and σ_2 at $\xi = \pm p$ ($p = \frac{\sqrt{3}}{3}$) are linearly interpolated within the interval $-p < \xi < p$ as

$$\sigma = \left[\frac{1-s}{2}, \frac{1+s}{2}\right]\begin{Bmatrix}\sigma_1\\ \sigma_2\end{Bmatrix} \tag{9.59}$$

where $s = \frac{\xi}{p}$ satisfies $s = -1$ for $\xi = -p$ and $s = 1$ for $\xi = p$. The stresses at the end nodes A and B are obtained making $s = \pm 1/p$ in (9.59) giving

$$\begin{Bmatrix}\sigma_A\\ \sigma_B\end{Bmatrix} = \frac{1}{2}\begin{bmatrix}a & b\\ b & a\end{bmatrix}\begin{Bmatrix}\sigma_1\\ \sigma_2\end{Bmatrix} \tag{9.60}$$

where $a = 1 + \frac{1}{p}$ and $b = 1 - \frac{1}{p}$.

This technique is easily extended to two and three dimensions. For instance, using the values at the 2×2 Gauss points of a quadrilateral element (Figure 9.16) the following bilinear stress interpolation can be written

$$\sigma(s,t) = \sum_{i=I}^{IV} N_i(s,t)\sigma_i \tag{9.61}$$

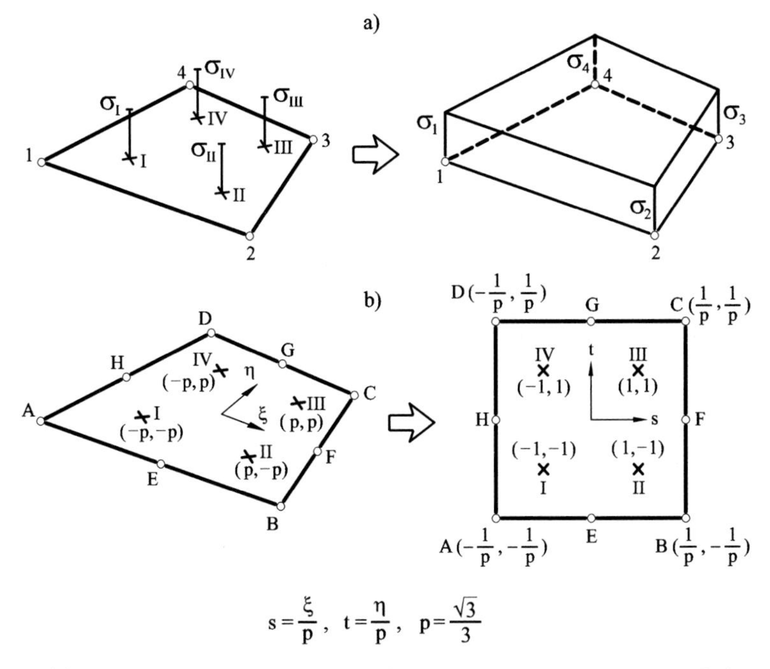

$$s=\frac{\xi}{p},\quad t=\frac{\eta}{p},\quad p=\frac{\sqrt{3}}{3}$$

Fig. 9.16 (a) Local extrapolation of 2×2 Gauss point stresses in a 4-noded quadrilateral using a bilinear interpolation. (b) Local coordinates ξ,η and s,t

where i denotes the Gauss point and $N_i = \frac{1}{4}(1+ss_i)(1+tt_i)$ are the bilinear interpolation functions with $s=\frac{\xi}{p}$ and $t=\frac{\eta}{p}$, $(p=\frac{\sqrt{3}}{3})$. Note that N_i takes a unit value at the i-th Gauss point and zero at the other three points. The values for each stress component $(\sigma_x,\sigma_y,\tau_{xy})$ at any point within the element are obtained by substituting the adequate values of the coordinates s and t in Eq.(9.61). Thus, $s=t=-\frac{1}{p}$ at the corner node A of the element in Figure 9.16b and

$$\sigma_A = \frac{1}{4}[a^2, ab, b^2, ab][\sigma_I, \sigma_{II}, \sigma_{III}, \sigma_{IV}]^T$$

where $a = 1+\frac{1}{p}$ and $b = 1-\frac{1}{p}$.

The procedure also applies for 8 and 9-noded quadrilaterals. Thus, for node E of Figure 9.16b with $s=0$, $t=-\frac{1}{p}$ we find

$$\sigma_E = \frac{1}{4}[a, a, b, b][\sigma_I, \sigma_{II}, \sigma_{III}, \sigma_{IV}]^T \ ; \ \text{etc.}$$

This technique can be used to extrapolate to the nodes the stresses from the $n\times n$ quadrature points using an adequate C° interpolation

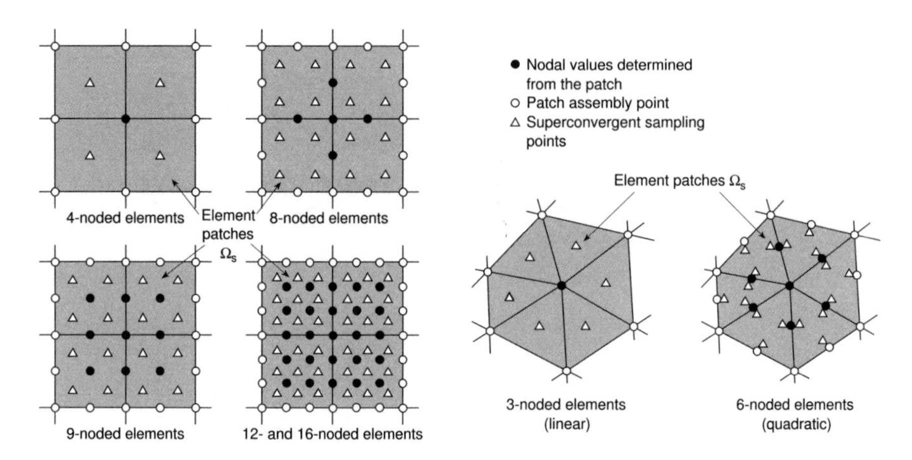

Fig. 9.17 Interior superconvergent patches for quadrilateral elements (linear, quadratic, and cubic) and triangles (linear and quadratic) [ZTZ]

[HC]. Thus, for a 3×3 quadrature the quadratic shape functions of the 9-noded Lagrange quadrilateral should be used, etc. This method is also applicable to 3D elements.

The nodal stresses obtained using this procedure are discontinuous between adjacent elements. A single nodal value can be obtained by nodal averaging.

9.8.3 Superconvergent patch recovery techniques

The stress extrapolation procedure of Section 9.8.1 can be applied to a patch of elements Ω_s surrounding a node. The method, called *superconvergent patch recovery* (SPR), yields enhanced values of the stresses at the nodes which are internal to the patch. The name of the method refers to the so called "superconvergence" property of the Gauss points (i.e., the stress values sampled at these points show an error which decreases more rapidly than elsewhere) (Section 6.7 and [ZTZ]). Figure 9.17 shows some patches for the nodal computation of the stresses for linear and quadratic quadrilateral and triangular elements.

For each element patch and for each stress component we minimize the following least square functional with n sampling points (the number of Gauss points)

$$\Pi = \frac{1}{2}\sum_{k=1}^{n} [\sigma_{s_i}(x_k, y_k) - \sigma_i(x_k, y_k)]^2 \tag{9.62}$$

where σ_{s_i} and σ_i are respectively the ith component of the sought smoothed stress (also called "recovered" stress in the SPR literature [ZTZ,ZZ,ZZ2]) and the stress directly computed from the finite element solution at the kth sampling point.

The smoothed stress σ_{s_i} is interpolated within the patch as

$$\sigma_{s_i} = \mathbf{p}^T(\bar{x}, \bar{y})\mathbf{a}_i \tag{9.63a}$$

where

$$\mathbf{p}(\bar{x}, \bar{y}) = [1, \bar{x}, \bar{y}, \cdots \bar{x}^p, \bar{y}^p]^T \quad \text{and} \quad \mathbf{a}_i = [a_1, a_2, \cdots, a_p]^T \tag{9.63b}$$

with $\bar{x} = x - x_c$ and $\bar{y} = y - y_c$ where x_c, y_c are the coordinates of the interior vertex node describing the patch.

Substituting Eq.(9.63a) into (9.62) gives after minimization of Π

$$\mathbf{A}\mathbf{a}_i = \mathbf{b}_i \tag{9.64}$$

where

$$\mathbf{A} = \sum_{k=1}^{n} \mathbf{p}_k \mathbf{p}_k^T \quad \text{and} \quad \mathbf{b}_i = \sum_{k=1}^{n} \mathbf{p}_k \sigma_i(x_k, y_k) \tag{9.65}$$

Solution of Eq.(9.64) yields the values of $\mathbf{a}_i$ from which the stresses can be computed at any node of the patch using Eq.(9.63a).

It should be noted that at external boundaries or on interfaces where stresses are discontinuous the nodal values should be calculated from interior patches as shown in Figure 9.18. Also for the nodes belonging to more than one patch, an average of the computed smoothed stresses is typically performed.

The SPR procedure outlined above has proved to be a powerful tool leading to superconvergent results on regular meshes and much improved results on irregular meshes [ZL,ZTZ].

The SPR idea can be also used to compute an enhanced displacement field (and from that an improved stress field) using a higher order displacement interpolation over element patches [LW,WL,ZTZ]. Boroomand and Zienkiewicz [BZ,BZ2] developed an improved nodal stress recovery method by constraining the smoothed stress field of Eq.(9.63a) to satisfy the discrete equilibrium equations. The smoothed nodal stresses obtained by any of the procedures explained in this and the previous sections, are not only useful for the analysis of the finite element results, but they also provide an estimate of an enhanced stress field for error estimation and adaptive mesh refinement purposes (Section 9.4) [ZTZ,ZZ2,ZZ3].

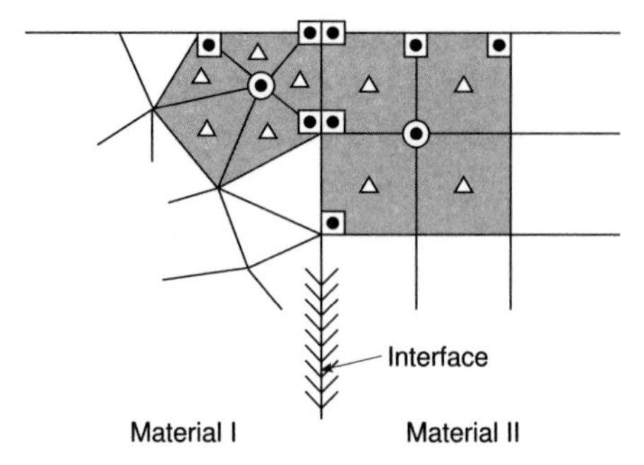

Fig. 9.18 Recovery of gradients at a boundary or an interface [ZTZ]

Figure 9.19 shows an example of the efficiency of the SPR technique taken from Chapter 4 of [ZTZ]. The problem is the analysis of the stress field around a hole in a plate under uniaxial loading. The recovered SPR stresses show much improved values compared with the original FE solution and also with the standard L_2 projection described in Section 9.8.1.

9.8.4 Iterative enhancement of the solution

Cantin *et al.* [CLT] proposed an iterative scheme to enhance the solution for the displacements and stresses. The starting point is the smoothed nodal stress field obtained by any of the procedures above explained. The stresses are interpolated within the element via Eq.(9.53).

The interpolated stress values are used to compute the unbalanced element nodal forces (typically called residual force vector) as

$$\mathbf{f}_r^{(e)} = \mathbf{f}^{(e)} - \int_{V^{(e)}} \mathbf{B}^T \boldsymbol{\sigma}_s dV \tag{9.66}$$

The element contributions are assembled to form the global residual force vector $\mathbf{f}_r$. A value of $\mathbf{f}_r = \mathbf{0}$ indicates that the smoothed stress field equilibrates the external loads $\mathbf{f}$. If $\mathbf{f}_r^n \neq \mathbf{0}$, where n is an iteration count, an enhanced solution for the displacements and stresses can be obtained as follows

$$\begin{aligned} &1) \text{ Solve } \mathbf{K}\Delta\mathbf{a}^n = \mathbf{f}_r^n \\ &2) \ \mathbf{a}^{n+1} = \mathbf{a}^n + \Delta\mathbf{a}^n \\ &\quad\ \boldsymbol{\sigma}^{n+1} = \mathbf{D}\ \mathbf{B}\ \mathbf{a}^{n+1} \end{aligned}$$

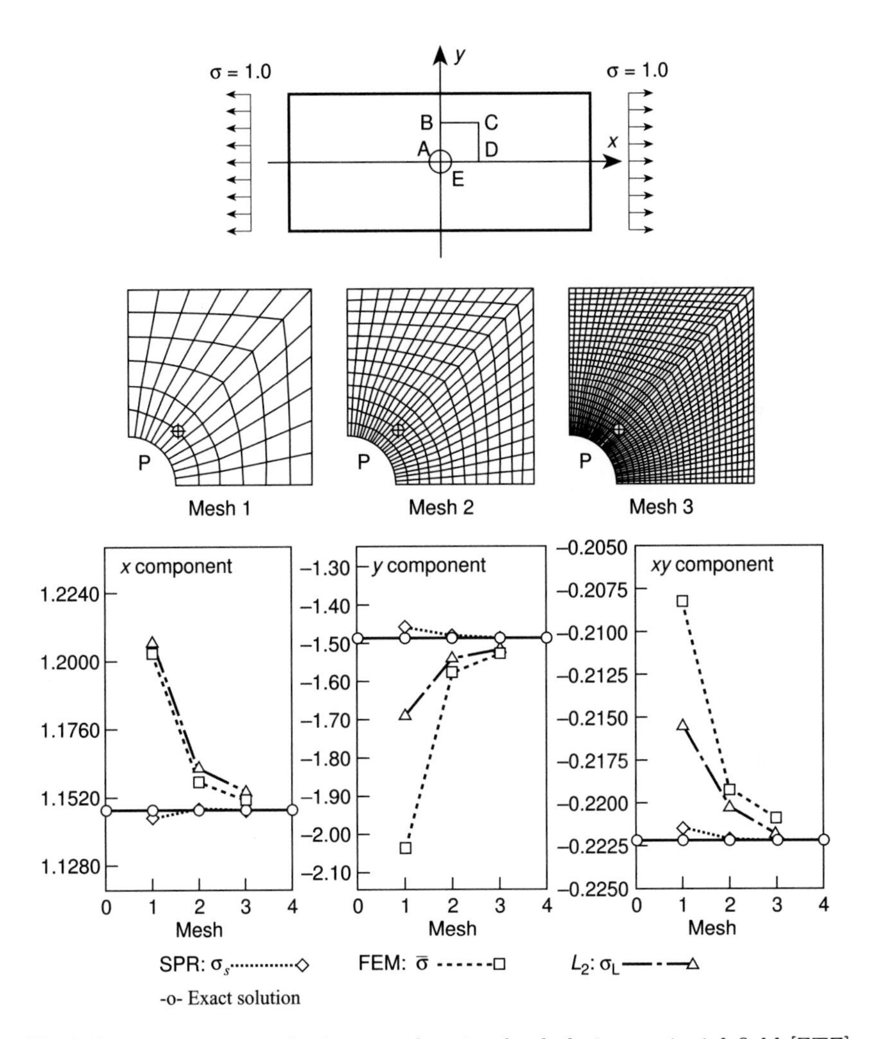

Fig. 9.19 Plane stress analysis around a circular hole in a uniaxial field [ZTZ]

3) Compute the smoothed nodal stresses $\hat{\boldsymbol{\sigma}}_s^{n+1}$ and the smoothed stress field $\boldsymbol{\sigma}_s^{n+1}$

4) Compute $\mathbf{f}_r^{n+1}$ from $\boldsymbol{\sigma}_s^{n+1}$

5) Error check

$$\frac{\|\mathbf{f}_r^{n+1}\|}{\|\mathbf{f}\|} \geq \varepsilon \text{ Yes. Stop the process} \tag{9.67}$$

No. Go back to 1

$\|\cdot\|$ in above denotes a quadratic norm, i.e. $\|\mathbf{a}\| = (\sum_i \mathbf{a}_i^2)^{1/2}$, and ε is a prescribed tolerance (typically $\varepsilon \simeq 10^{-2} - 10^{-3}$ is chosen). In practice only a few iterations are needed to improve the initial solution. Step 1 is inexpensive as the assembly and the reduction of matrix $\mathbf{K}$ is available from the initial solution.

9.9 ERROR ESTIMATION AND MESH ADAPTIVITY

9.9.1 Basic concepts of error estimation

The main error source in the finite element solution is the *discretization error*. This error combines the unaccuracies introduced by the interpolation and the mesh chosen. Other error sources are the limited storing capacity of computers (round-off errors), the bad approximation of the geometry, the unaccurate computation of the element integrals and the choice of a wrong constitutive equation (Section 3.13). These errors are more easily controlled and avoided *a priori*. We consider in this section the discretization error and the different strategies for reducing it to acceptable figures.

The prediction and reduction of the discretization error has been object of considerable research (see references on the subject listed in [GBFL+,LP2,ZTZ]). Much effort has been devoted to translate the mathematical concepts of error estimation into useful engineering expressions, as a basis for obtaining improved numerical solutions in an "intelligent" manner.

Enhanced accurate FEM solutions can be achieved by refining the mesh using a greater number of elements of smaller size (h method) [DH2,ZTZ]. An alternative approach is keeping the mesh topology constant and using higher order elements (p method) [ZTZ]. Obviously, a combination of h and p techniques is possible [ROD,Ve,ZTZ]. A review of adaptive finite element methods can be found in [LB3,SR,SRO].

The p method enjoyed some popularity among finite element practitioners in the early 1980's as it avoids the need for the redefinition of a mesh. Simple error estimators and p-adaptive procedures can be derived by using hierarchical procedures [KGZB,ZTZ]. However, p (and $h-p$) methods are difficult to implement in practice as the classical organization of a finite element program needs to be abandoned. The advances in mesh generation procedures has also favoured mesh adaptive h-methods which are increasingly popular. The main concepts behind the h-method are presented in the next sections.

9.9.2 Error measures

Let us consider the finite element solution of an elastic problem where the "exact" values of displacements, strains and stresses are approximated by

$$\mathbf{u} \simeq \overline{\mathbf{u}} = \mathbf{Na} \tag{9.68}$$

$$\boldsymbol{\varepsilon} \simeq \overline{\boldsymbol{\varepsilon}} = \mathbf{Ba} \quad \text{and} \quad \boldsymbol{\sigma} \simeq \overline{\boldsymbol{\sigma}} = \mathbf{DBa} \tag{9.69}$$

The approximate numerical solution $\overline{\mathbf{u}}$, $\overline{\boldsymbol{\varepsilon}}$, $\overline{\boldsymbol{\sigma}}$ differs from the exact one $\mathbf{u}$, $\boldsymbol{\varepsilon}$, $\boldsymbol{\sigma}$. The following error vectors for the displacements, strains and stresses at each point are defined as

$$\mathbf{e}_u = \mathbf{u} - \overline{\mathbf{u}}, \quad \mathbf{e}_\varepsilon = \boldsymbol{\varepsilon} - \overline{\boldsymbol{\varepsilon}}, \quad \mathbf{e}_\sigma = \boldsymbol{\sigma} - \overline{\boldsymbol{\sigma}} \tag{9.70}$$

A convenient global error measure is defined by the energy norm of the error as

$$\parallel \mathbf{e}_\varepsilon \parallel = \left[\int_\Omega \mathbf{e}_\varepsilon^T \mathbf{D} \mathbf{e}_\varepsilon d\Omega\right]^{1/2} \tag{9.71}$$

$$\parallel \mathbf{e}_\sigma \parallel = \left[\int_\Omega \mathbf{e}_\sigma^T \mathbf{D}^{-1} \mathbf{e}_\sigma d\Omega\right]^{1/2} \tag{9.72}$$

where Ω is the area or volume of the whole mesh domain. For linear elasticity $\parallel \mathbf{e}_\varepsilon \parallel \; = \; \parallel \mathbf{e}_\sigma \parallel$.

An alternative error measure is the L_2 norm. This is defined for the displacement error as

$$\parallel \mathbf{e}_u \parallel_{L_2} = \left[\int_\Omega \mathbf{e}_u^T \mathbf{e}_u d\Omega\right]^{1/2} \tag{9.73}$$

with similar expressions for the strain and stress errors.

The square value of the integral error norms can be computed by the sum of the element contributions. For instance

$$\parallel \mathbf{e}_\sigma \parallel^2 = \sum_{e=1}^{N} \left(\parallel \mathbf{e}_\sigma \parallel^{(e)}\right)^2 \tag{9.74}$$

where N is the number of elements in the mesh.

The key problem is that the exact solution is not known *a priori* (except for simple academic problems). It is therefore essential to find ways to estimate the error and this is treated in the next section.

9.9.3 Error estimation techniques

It is intuitive that the smoothed continuous stress field $\boldsymbol{\sigma}_s$ is a better approximation than the discontinuous distribution $\bar{\boldsymbol{\sigma}}$ directly provided by

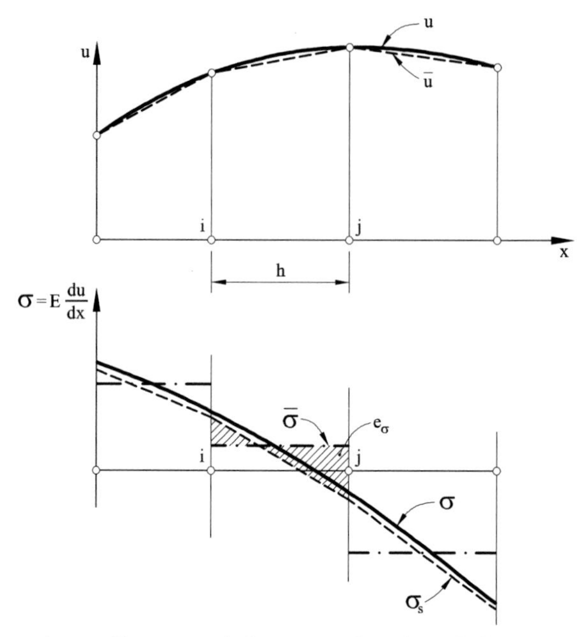

Fig. 9.20 Bar under uniform axial forces analyzed with linear elements. Exact stress distribution (σ), finite element solution ($\overline{\sigma}$), smoothed stresses (σ_s) and estimation of the error in stresses (e_σ) [ZTZ]

the finite element solution (Figure 9.20). A simple estimation of the error in the stresses at each point can be written as [ZTZ,ZZ]

$$\mathbf{e}_\sigma \simeq \boldsymbol{\sigma}_s - \overline{\boldsymbol{\sigma}} \tag{9.75}$$

The approximate expression of the energy norm of the error is

$$\|\mathbf{e}_\sigma\| \simeq \left[\int_\Omega [\boldsymbol{\sigma}_s - \overline{\boldsymbol{\sigma}}]^T \mathbf{D}^{-1} [\boldsymbol{\sigma}_s - \overline{\boldsymbol{\sigma}}] d\Omega\right]^{1/2} \tag{9.76}$$

The simplest procedure to obtain the smoothed stress field $\boldsymbol{\sigma}_s$ is to use a linear local extrapolation (Section 9.8.2) followed by a nodal averaging of the stresses. Alternatively any of the stress recovery techniques explained in Section 9.8 can be used.

The proof that $\mathbf{e}_\sigma$ is a good estimation of the error is immediate for the 1D case. Consider, for instance, the axial bar under uniformly distributed axial forces analyzed with 2-noded bar elements (Figure 9.20). As shown in Chapter 2, the displacement solution for 2-noded bar elements is exact at the nodes, whereas the stresses are constant within each element. A linear smoothed stress field gives also a linear distribution of $\mathbf{e}_\sigma$ within each element as the difference between the (linear) smoothed stresses and the constant FE values (the shadowed region in Figure 9.20). Clearly, $\mathbf{e}_\sigma$ will tend to zero as the element size reduces. This is also satisfied by the error norm $\|\mathbf{e}_\sigma\|$ if the material properties are constant. This also applies to higher order elements. In general $\|\mathbf{e}_\sigma\| \simeq O(h^p)$, where p is the order of the displacement approximation and h is the average element size in the mesh.

The superconvergent derivative recovery technique (Section 9.8.3) yields a higher approximation of the stress field which increases the convergence rate of the stresses. The alternative expression of the energy norm in this case is $\|\mathbf{e}_\sigma\| \simeq O(h^{p+\alpha})$, where $O \leq \alpha \leq 1$ [ZTZ,ZZ].

9.9.4 Mesh adaptation strategies

A finite element solution is "acceptable" if the two following conditions are satisfied:

a) Global error condition

The energy norm of the error is less than a percentage of the total strain energy norm, i.e.

$$\| \mathbf{e}_\sigma \| \leq \eta \| U \| \tag{9.77}$$

where η is the percentage of "admissible" relative error. This value is defined "a priori" before starting the analysis. The strain energy norm in Eq.(9.77) is obtained by

$$\| U \| = \left[\int_\Omega \boldsymbol{\sigma}_s^T \mathbf{D}^{-1} \boldsymbol{\sigma}_s d\Omega \right]^{1/2} \tag{9.78}$$

Eq.(9.77) allows to define a *global error parameter* as

$$\xi_g = \frac{\| \mathbf{e}_\sigma \|}{\eta \| U \|} \tag{9.79}$$

A value of $\xi_g = 1$ means that the global error condition (9.77) is exactly satisfied. Alternatively $\xi_g > 1$ and $\xi_g < 1$ indicate that the element sizes must be refined of derefined, respectively.

In the following we will assume $\| \mathbf{e}_\sigma \| = O(h^p)$, then $\xi_g = O(h^p)$ and the new size of an element $\bar{h}^{(e)}$ is obtained in terms of its actual size $h^{(e)}$ so that $\xi_g = 1$ in the refined mesh. This leads to

$$\xi_g = \left(\frac{h^{(e)}}{\bar{h}^{(e)}} \right)^p \quad \text{and hence} \quad \bar{h}^{(e)} = \frac{h^{(e)}}{\xi_g^{1/p}} \tag{9.80}$$

Computation of the new element size $\bar{h}^{(e)}$ from Eq.(9.80) implies that all elements refine (or derefine) in the same proportion, as ξ_g is a global parameter for the whole mesh. It is therefore important to allow for a selective change of the element sizes aiming for an "optimal mesh" using a local error criterion.

b) Optimum mesh condition

The element distribution in the mesh should satisfy a mesh optimality condition defined as

$$\| \mathbf{e}_\sigma \|^{(e)} = \| \mathbf{e}_\sigma \|_r^{(e)} \tag{9.81}$$

where $\| \mathbf{e}_\sigma \|^{(e)}$ is the energy norm of the error for each element and $\| \mathbf{e}_\sigma \|_r^{(e)}$ is the required value of this norm, defined according to a particular mesh optimality criterion.

A *local error parameter* is defined as

$$\bar{\xi}^{(e)} = \frac{\| \mathbf{e}_\sigma \|^{(e)}}{\| \mathbf{e}_\sigma \|_r^{(e)}} \tag{9.82}$$

A value of $\bar{\xi}^{(e)} = 1$ indicates that the element size is "optimal" (according to the mesh optimality criterion chosen). Values of $\bar{\xi}^{(e)} > 1$ and $\bar{\xi}^{(e)} < 1$ indicate that the size of element e must be reduced or enlarged, respectively.

A single error parameter for the element can be defined as

$$\xi^{(e)} = \xi_g \bar{\xi}^{(e)} = \frac{\|\mathbf{e}_\sigma\| \ \|\mathbf{e}_\sigma\|^{(e)}}{\eta \|U\| \ \|\mathbf{e}_\sigma\|_r^{(e)}} \tag{9.83}$$

General adaptive strategy

A mesh adaptive strategy can be designed so that the element sizes are modified with the following two objectives:

- To reach an optimum distribution of element sizes satisfying Eq.(9.81).
- To reduce the global error so that Eq.(9.77) is satisfied.

Conceptually, once a finite element solution is obtained two successive changes in the element sizes need to be made for achieving above two objectives. First the new element sizes $h_{\bar{\xi}}^{(e)}$ are defined in terms of the actual sizes $h^{(e)}$ and the local error parameter $\bar{\xi}^{(e)}$ as

$$h_{\bar{\xi}}^{(e)} = h^{(e)} [\bar{\xi}^{(e)}]^{-1/q} \tag{9.84}$$

where q is the convergence rate of $\bar{\xi}^{(e)}$ depending on the mesh optimality criterion chosen.

The second change aims to satisfying the global error condition. The element sizes are now uniformly changed using Eq.(9.80) to give

$$\bar{h}^{(e)} = h_{\bar{\xi}}^{(e)}[\xi_g]^{-1/p} \tag{9.85}$$

Eqs.(9.84) and (9.85) can be combined to give the final element sizes $\bar{h}^{(e)}$ in *a single* step as

$$\boxed{\bar{h}^{(e)} = \frac{h^{(e)}}{\beta^{(e)}}} \tag{9.86}$$

where the *element refinement parameter* $\beta^{(e)}$ is

$$\beta^{(e)} = [\bar{\xi}^{(e)}]^{1/q}\, \xi_g^{1/p} \tag{9.87a}$$

Clearly if $q = p$ then

$$\beta^{(e)} = [\xi^{(e)}]^{1/p} \tag{9.87b}$$

where $\xi^{(e)}$ is the single error parameter of Eq.(9.83).

Eq.(9.86) satisfies *both* the local and global error conditions. A key step is the definition of the required error for each element $\| \mathbf{e}_\sigma \|_r^{(e)}$, as this affects the local error parameter $\bar{\xi}^{(e)}$ and its convergence rate. Three typical choices for the required element error are presented next.

9.9.4.1 Mesh optimality criterion based on the equal distribution of the global energy error

This criterion states that a mesh is optimal when the global energy error is uniformly distributed between all elements [ZTZ,ZZ]. The required error norm for each element is defined as the ratio between the global error norm and the number of elements in the mesh N. Recalling that only the square norms are additive then

$$\| \mathbf{e}_\sigma \|_r^{(e)} = \frac{\| \mathbf{e}_\sigma \|}{\sqrt{N}} \tag{9.88}$$

Combining Eqs.(9.81) and (9.88) gives the local error parameter as

$$\bar{\xi}^{(e)} = \frac{\| \mathbf{e}_\sigma \|^{(e)}}{\| \mathbf{e}_\sigma \| \;\; N^{-1/2}} \tag{9.89}$$

From Eq.(9.76) it is deduced

$$\| \mathbf{e}_\sigma \|^{(e)} \simeq O(h^p)(\Omega^{(e)})^{1/2} \simeq O(h^{p+d/2}) \tag{9.90}$$

where h is a characteristic element size and d is the number of dimensions of the problem (i.e $d = 3$ for 3D problems). The convergence rate of $\| \mathbf{e}_\sigma \|^{(e)}$ dominates that of $\bar{\xi}^{(e)}$ and hence $q = \dfrac{2p+d}{2}$. The new element size is obtained from Eq.(9.87a) with

$$\beta^{(e)} = \left[\bar{\xi}^{(e)}\right]^{\frac{2}{2p+d}} \xi_g^{1/p} \tag{9.91}$$

Eq.(9.91) can be improved by taking into account that the number of elements in the refined mesh is not equal to the number of elements in the original mesh. The new expression for $\beta^{(e)}$ is (Appendix C)

$$\beta^{(e)} = [\bar{\xi}^{(e)}]^{\frac{1}{p+d}} \xi_g^{\frac{2}{2p+d}} N^{-\frac{1}{2p+d}} \left[\sum_{e=1}^{N} [\bar{\xi}^{(e)}]^{\frac{d}{p+d}}\right]^{\frac{1}{2p+d}} \tag{9.92}$$

where N is the number of elements in the original mesh. Eq.(9.91) can be simplified by accepting that the local error parameter is constant over the mesh (this is true for the optimal mesh). This gives (Appendix C)

$$\beta^{(e)} = [\xi^{(e)}]^{\frac{2}{2p+d}} \tag{9.93a}$$

where $\xi^{(e)}$ is the single error parameter for the element (Eq.(9.83)). Introducing Eqs.(9.79) and (9.89) into (9.83) gives

$$\xi^{(e)} = \xi_g \bar{\xi}^{(e)} = \frac{\| \mathbf{e}_\sigma \|^{(e)}}{\eta \ \| U \| \ N^{-1/2}} \tag{9.93b}$$

Eq.(9.93a) reassembles Eq.(9.91) and both lead to similar results in practice. The definition of $\beta^{(e)}$ of Eq.(9.91) has been used for solving the examples presented in a later section.

Some authors propose an alternative expression for $\beta^{(e)}$ as [ZTZ,ZZ]

$$\beta^{(e)} = [\xi^{(e)}]^{1/p} \tag{9.94}$$

Eq.(9.94) does not preserve the convergence rate of the local error norm and it can lead to oscillations during the mesh adaption process such as the alternate refinement/derefinement of the mesh [Bu,Bu2,OB]. These inconsistences can be overcome by introducing a relaxation parameter c $\xi^{(e)} = c\bar{\xi}^{(e)}\xi_g$, or by defining an "ad hoc" value of the power p in Eq.(9.94) [AH,AHS,LB2,LBBB,ZTZ,ZZ,ZZ2,ZZ3]. This problem disappears if $\beta^{(e)}$ is computed by either Eqs.(9.91), (9.92) or (9.93a) [OB].

9.9.4.2 Mesh optimality criterion based on the global distribution of the density of the energy error

A mesh can also be defined as optimal if the energy error per unit area (or volume) is uniformly distributed in the mesh. In this case

$$\frac{\| \mathbf{e}_\sigma \|^{(e)}}{[\Omega^{(e)}]^{1/2}} = \frac{\| \mathbf{e}_\sigma \|}{\Omega^{1/2}} \tag{9.95}$$

where Ω and $\Omega^{(e)}$ respectively denote the area (or volume) of the analysis domain and that of an element. Comparing Eqs. (9.81) and (9.95) gives the required element error as

$$\| \mathbf{e}_\sigma \|_r^{(e)} = \| \mathbf{e}_\sigma \| \left(\frac{\Omega^{(e)}}{\Omega}\right)^{1/2} \tag{9.96}$$

The local error parameter is deduced from Eqs. (9.82) and (9.96) as

$$\bar{\xi}^{(e)} = \frac{\| \mathbf{e}_\sigma \|^{(e)}}{\| \mathbf{e}_\sigma \|} \left(\frac{\Omega}{\Omega^{(e)}}\right)^{1/2} \tag{9.97}$$

and the single element error parameter is obtained from Eqs.(9.79) and (9.97) as

$$\xi^{(e)} = \bar{\xi}^{(e)} \xi_g = \frac{\| \mathbf{e}_\sigma \|^{(e)}}{\eta \| U \|} \left(\frac{\Omega}{\Omega^{(e)}}\right)^{1/p} \tag{9.98}$$

The definition of the required element error affects the convergence rate of $\bar{\xi}^{(e)}$. This can be obtained by noting that

$$\bar{\xi}^{(e)} = O \frac{\| \mathbf{e}_\sigma \|^{(e)}}{[\Omega^{(e)}]^{1/2}} = O(h^p) \tag{9.99}$$

from which it is deduced that $q = p$ in Eq.(9.84).

The new element size is obtained by Eq.(9.86) using the following expression for the element refinement parameter $\beta^{(e)}$

$$\beta^{(e)} = (\bar{\xi}^{(e)} \xi_g)^{1/p} = (\xi^{(e)})^{1/p} \tag{9.100}$$

The criterion of equal distribution of the specific error for mesh refinement was introduced by Oñate and Bugeda [BO2,BO3,OB] and has been successfully used by Bugeda and Oliver [BO] for structural optimization problems and by Oñate *et al.* [OC,OCK] for plate and shell analysis. A theoretical study of this criterion is reported in [LB2,LBBB]. Its advantage is that it concentrates more elements close to the higher stress gradients, such as in the vicinity of singular points or of stress discontinuities. Therefore it yields a better approximation for the stresses in those regions which is useful from a practical point of view. Its disadvantage is that it leads to optimal meshes with a larger number of elements than those generated with the criterion of equal distribution of the global error. These features are clearly shown in the examples presented in a next section.

We note again the importance of using an expression for $\beta^{(e)}$ consistent with the mesh optimality criterion chosen, in order to avoid oscillations in the mesh adaption process.

9.9.4.3 Mesh refinement strategy based on the point-wise error in stresses

Some refinement strategies are based on the control of a specific magnitude whose error at each point is limited to a maximum value everywhere. The advantage of this type of strategies is that they allow the control of magnitudes with a clear physical meaning, such as the stresses or the strains at each point. From an engineering perspective, the interpretation of this type of criteria can be easier than in the previous cases.

Assuming the necessary regularity conditions, the error in stresses at each point behaves as h^p [BS2]. For any point P, and for each component of the stress tensor σ_{ij} we can therefore write

$$|\sigma_{ij}(P) - \bar{\sigma}_{ij}(P)| \leq Ch^p \tag{9.101}$$

where C is a constant.

Expression (9.101) indicates that the size of all the elements in the mesh should be uniformly reduced in order to provide a specific value of the error in the component of σ_{ij} at point P. A uniform reduction of the element size h over the whole mesh ensures the reduction of the local error in stresses [BSGU,BSUG].

Expression (9.101) can be used for the definition of adaptive mesh refinement strategies based on the control of the stress error not only at a specific point, but over a global set of points using the stress recovery methods explained in the previous sections.

Expression (9.101) does not hold for zones around singularities due to the lack of regularity. At this zones, the behaviour of the error in stresses is governed by the intensity of the singularity λ instead of the degree of the shape function polynomials and the p power should be substituted by λ. In the next section we present an adaptive remeshing strategy based on Eq.(9.101) that does not account for singularities. Nevertheless, the substitution of p by λ at these zones provides the necessary alternative for regions around singular points.

Maximum error in stresses

An obvious possibility is to maintain the error in the stresses below a certain limit everywhere. This can be done by estimating the error in stresses $\mathbf{e}_\sigma$ (Eq.(9.70)). Due to the tensorial nature of the stresses, the error in stresses will also be a tensorial magnitude. Hence, we define

$$[e_\sigma] = \begin{bmatrix} e_{\sigma_x} & e_{\tau_{xy}} & e_{\tau_{xz}} \\ e_{\tau_{xy}} & e_{\sigma_y} & e_{\tau_{yz}} \\ e_{\tau_{xz}} & e_{\tau_{yz}} & e_{\sigma_z} \end{bmatrix} = \begin{bmatrix} \sigma_x - \bar{\sigma}_x & \tau_{xy} - \bar{\tau}_{xy} & \tau_{xz} - \bar{\tau}_{xz} \\ \tau_{xy} - \bar{\tau}_{xy} & \sigma_y - \bar{\sigma}_y & \tau_{yz} - \bar{\tau}_{yz} \\ \tau_{xz} - \bar{\tau}_{xz} & \tau_{yz} - \bar{\tau}_{yz} & \sigma_z - \bar{\sigma}_z \end{bmatrix} \tag{9.102}$$

Eq.(9.102) defines the error in stresses $[e_\sigma]$ as the difference between the exact and the approximated stress tensors. Clearly, the exact stresses in Eq.(9.102) are substituted by the smoothed continuous stress field $\boldsymbol{\sigma}_s$ (Eq.(9.75)). Tensor $[e_\sigma]$ can be written in terms of its eigenvalues:

$$\begin{bmatrix} e_{\sigma_I} & 0 & 0 \\ 0 & e_{\sigma_{II}} & 0 \\ 0 & 0 & e_{\sigma_{III}} \end{bmatrix} \tag{9.103}$$

A logical refinement strategy is based in limiting the maximum error in the stresses to a certain maximum value everywhere, i.e.

$$max(abs(e_{\sigma_I}), abs(e_{\sigma_{II}}), abs(e_{\sigma_{III}})) \leq e_{\sigma_{max}} \tag{9.104}$$

An error parameter ξ can then be defined at each point within the element as the ratio between the maximum error $max(abs(e_{\sigma_I}), abs(e_{\sigma_{II}}), abs(e_{\sigma_{III}}))$ and $e_{\sigma_{max}}$, thus

$$\xi = \frac{max(abs(e_{\sigma_I}), abs(e_{\sigma_{II}}), abs(e_{\sigma_{III}}))}{e_{\sigma_{max}}} \tag{9.105}$$

The element error parameter $\xi^{(e)}$ is taken as the maximum value of ξ for all the Gauss points in the element. The new element size is given by Eq.(9.86) using the value of $\beta^{(e)}$ of Eq.(9.100) [Bu2].

The error parameter ξ can be computed at any point of the domain. As the larger error in stresses typically occurs at the nodes, the safest option is to compute ξ there. Taking into account that different values of ξ will be obtained for each of the elements connected to a specific node, a natural strategy is to define the nodal error parameters as the maximum value of ξ for each node. The element error parameter is then computed as the average of the values of ξ for the element nodes.

Other point-wise error criteria based on the maximum error in the tensile and compressive stresses and in the von Mises stress are presented in [Bu2].

9.9.5 Construction of an adapted mesh

Once the error associated to a finite element solution has been computed, the existing mesh has to be modified accordingly to the new element sizes defined by Eq.(9.86) using the value of $\beta^{(e)}$ according to the mesh optimality criterion chosen. For the construction of an adapted mesh there are basically two procedures. The first one consists in *regenerating* a completely new mesh, so that the adapted mesh has the element sizes required by Eq.(9.86). This option changes the topology of the initial mesh and its success depends on the availability of a mesh generator able to generate non-structured meshes in a robust and efficient manner. The topic of mesh generation is studied in Chapter 10.

The second procedure is based in the simple *subdivision* of the elements in the existing mesh for which $\beta^{(e)}$ takes a value greater than one, while leaving the rest of the elements unaltered. The implementation of this method is extremely simple for triangular and tetrahedral elements. Every triangle and tetrahedra is split into three and four elements, respectively by introducing a new interior node, while leaving the adjacent elements unmodified. The disadvantage of this method is that it does not lead to adapted meshes satisfying accurately the element size distribution as defined by Eq.(9.86). It avoids however the regeneration of the whole mesh during the mesh adaption process.

A number of examples of mesh adaption are presented in the next section using the *mesh regeneration* procedure.

9.9.6 Examples of mesh adaptivity

Two examples have been chosen to compare the different adaptive mesh refinement (AMR) strategies presented: (1) thick cylinder under internal pressure and (2) hollow dam under water pressure and self-weight. The following notation is used for the different AMR strategies.

Strategy A. This is based on the criterion of equal distribution of the global energy error between all elements (Section 9.9.4.1) with $\beta^{(e)}$ given by Eq.(9.91).

Strategy B. It uses the same mesh optimality criterion as strategy A with the expression for $\beta^{(e)}$ as given by Eq.(9.94).

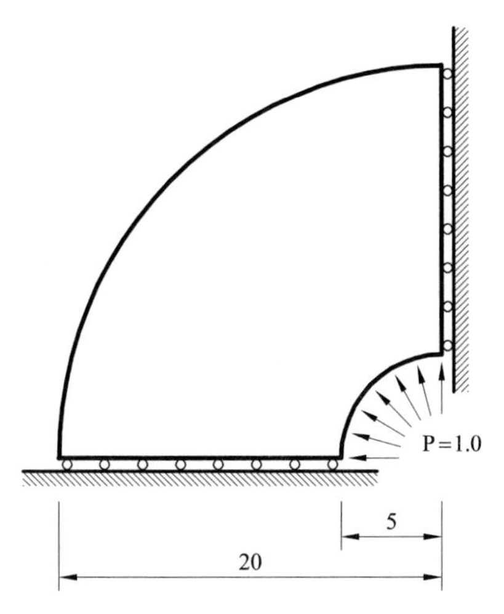

Fig. 9.21 Thick circular cylinder under internal pressure. Geometry and loads

Strategy C. It is based on the criterion of equal distribution of the error (Section 9.9.4.2) with $\beta^{(e)}$ as given by (9.100).

Strategy D. Control of the point-wise error in stresses everywhere (Section 9.9.4.3).

9.9.6.1 Thick circular cylinder under internal pressure

The first example is the analysis of the thick circular cylinder under internal pressure shown in Figure 9.21. Due to the symmetry of the problem only a quadrant has been studied under plane strain conditions. The material parameters are $E = 1.0 \times 10^5$ and $\nu = 0.3$ (units are in the International System). A value of the admisible relative error of the energy in the mesh of $\eta = 5\%$ has been chosen. Standard 3-noded linear triangles were used for the analyses. This example is typical of elliptic problems and it has many analogies in heat flow, ground water flow, etc. [ZTZ].

Table 9.1 shows some characteristic results obtained with the mesh refinement strategies A, B and C defined in the previous section like the number of elements, the global error parameter ξ_g, the average squared value of the local error parameter $[(\bar{\xi}^{(e)})^2]_a$ and its mean deviation $[(\bar{\xi}^{(e)})^2]_\sigma$ over each mesh for the three AMR strategies. From the numbers shown in the table we deduce:

- The three AMR strategies converge fast to the global permisible error chosen characterized by $\xi_g \leq 1$.

Strategy		NE	ξ_g	$[(\bar{\xi}^{(e)})^2]_a$	$[(\bar{\xi}^{(e)})^2]_\sigma$
	M0	200	4.133	1.000	8.401
A	M1	2180	0.839	1.000	0.604
	M2	1838	0.909	1.000	0.189
	M3	1797	0.925	1.000	0.193
	M0	200	4.133	1.000	8.401
B	M1	2820	0.919	1.000	8.469
	M2	2711	0.907	1.000	2.197
	M3	2758	0.839	1.000	4.484
	M0	200	4.133	1.190	15.37
C	M1	3028	0.884	2.224	2.071
	M2	6359	0.796	1.204	0.163
	M3	6026	0.835	1.204	0.151

Table 9.1 Thick circular under internal pressure. NE: number of elements

- Strategies A and C converge to an "optimal" mesh characterized by the values $\xi_g \leq 1$, $[(\bar{\xi}^{(e)})^2]_a \simeq 1.0$ and $[(\bar{\xi}^{(e)})^2]_\sigma \simeq 0$. However, the number of elements and its distribution is very different for these two AMR strategies.
- The mean deviation of the squared value of the local error parameter for strategy B oscillates and does not converge to zero.

More insight into the numbers of Table 9.2 can be gained by analyzing the sequence of refined meshes obtained with the three AMR strategies shown in Figure 9.22. First column of Figure 9.22 shows the results obtained with strategy A. The AMR process converges in a consistent manner.

Second column of Figure 9.22 depicts the results obtained with strategy B. The AMR process oscillates as clearly shown by the alternate re and derefinement of some mesh zones. This explains the lack of convergence of the mean deviation of the squared value of the local error parameter (Table 9.1) which precludes reaching an "optimal" mesh for this case.

Third column of Figure 9.22 shows the results for strategy C. It is seen that: (a) the AMR process evolves without oscillations, and (b) more and smaller elements concentrate in the vicinity of the internal edge where the error is greater due to the higher stress gradients. Larger elements than in Strategy A are however allowed in the rest of the mesh. The prize to be paid by Strategy C for the higher accuracy in capturing the high stress gradients is the larger total number of elements with respect to strategies A and B for the same global accuracy, as shown in Table 9.1. More details on this example can be found in [OB,BO3]. The same problem was studied under diametral loads in [BO2] with identical conclusions.

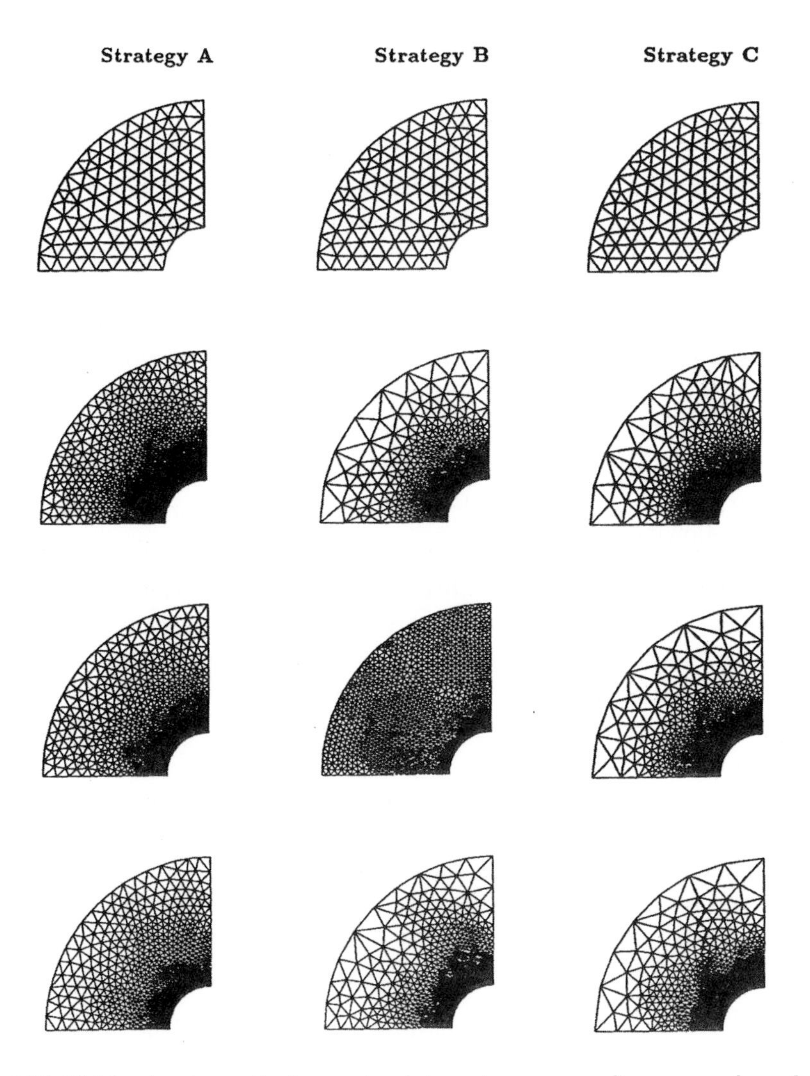

Fig. 9.22 Thick circular cylinder under internal pressure. Sequence of meshes M_0, M_1, M_2, M_3 obtained with AMR strategies A, B and C

9.9.6.2 2D hollow dam under water pressure and self-weight

The behavior of the different AMR strategies A, C and D are illustrated here through the analysis of a 2D section of a hollow gravity dam assuming a plane strain model. Strategy B has been disregarded as it produces oscillations in the refinement process (see previous example). Figure 9.23 shows the geometry of the dam section. All the possible corner points of the geometry have been conveniently rounded in order to eliminate singularities that could mask the behavior of each strategy. The main data used for the analysis are the following (units are in the International System): total height of the dam h=33.5, total height of water 32.5, Young modulus E=31.0 GPa, Poisson's ratio ν=0.25, density of concrete ρ=2.3.

The applied loads are the water pressure and the self-weight. The displacements of all the nodal points placed at the bottom line have been prescribed to zero in both the vertical and horizontal directions, whereas the displacements of the nodal points placed at the left and right sides have been prescribed to zero only in the horizontal direction. Figure 9.23 also shows the initial mesh used for all the adaptive strategies. Quadratic triangular elements have been used for the analyses.

For each AMR strategy A, C and D, the remeshing procedure has converged to the final mesh in a few remeshing steps. The cases studied are the following:

- The first set of results has been obtained by prescribing a 1.50% admisible error for the global energy norm (η=0.015) in strategy A, and prescribing the refinement parameters in strategies C and D in order to obtain a final mesh with a similar number of DOFs as for strategy A. Results for this set of meshes are displayed in Figures A1, C1 and D1 of Figure 9.23.
- The second set of results has been obtained by prescribing a 1.50% admisible error for the global energy norm in strategy C, and prescribing the refinement parameters in strategies A and D in order to obtain a final mesh with a similar number of DOFs as for strategy C. Results for this set of meshes are displayed in Figures A2, C2 and D2 of Figure 9.23.

Table 9.2 shows the quality parameters obtained in the final mesh for each strategy.

Strategy	Nr. of elements	Nr. of nodes	% error
A1	803	1715	1,40
C1	809	1693	2.75
D1	809	1713	3.36
A2	1923	4023	0.55
C2	1896	3968	1.32
D2	1906	3982	1.34

Table 9.2 Mesh parameters and error in the final mesh for each AMR strategy

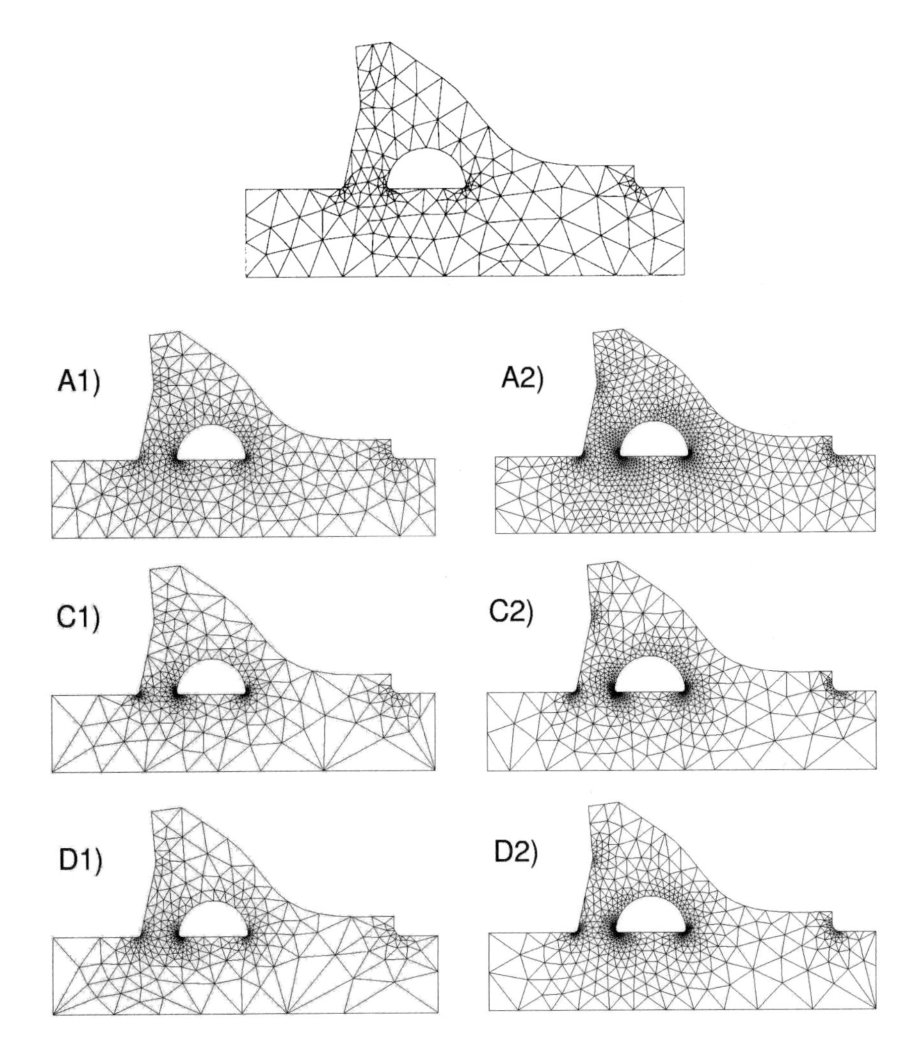

Fig. 9.23 Geometry of the section of a hollow gravity dam and initial mesh. Final meshes obtained in the first set of application of strategies A1, C1, D1 and final meshes obtained in the second set of application of strategies A2, C2, D2 [Bu2]

This example shows the coincidence of the results in the total number of elements and the percentage of error obtained with the AMR strategy based on prescribing the equal distribution of the density of the energy error (strategy C2) and the those obtained with the criteria based on the control of the point-wise error in stresses (strategy D2).

Strategy A1 based on the control of the equal distribution of the global energy error reaches the error target with less than half the number of elements needed for strategies C2 and D2, as expected. However, the point-wise error in stresses (as given by strategy D1) is larger in this case (3.36%). More details on this example can be found in [Bu2].

9.9.7 Conclusions from the examples

From the examples presented we can draw the following conclusions:

a) The AMR criterion based on the equidistribution of the energy error density captures better the zones with high stress gradients. It also leads to meshes with many small elements in the high gradient zones, while it generates larger elements in regions where the stress distribution is uniform. It is therefore recommended to limit the maximum and minimum element sizes when using this AMR criterion. This limiter is essential in problems with stress singularities, as the optimal mesh size in these regions tends to zero and hence the mesh adaption would progress indefinitely.
b) The AMR criterion based on the equal distribution of the global error leads to meshes with a smaller number of elements than those obtained with the error density criterion. It also leads to a more uniform distribution of element sizes. This criterion tends to smooth the effect of stress singularities over the whole mesh and hence the element size limiters are not needed in this case. The disadvantage is its reduced capacity for accurately capturing sharp stress (or strain) gradient zones.
c) The mesh adaption criterion based on the point-wise error in stresses leads to very similar meshes and results as with the criteria based on the equidistribution of the energy error density.

The choice of one or other AMR criterion should be decided by the needs of each specific analysis and the interest of the user. If a strict control of the stresses is required, then the strategy based on the equal distribution of the energy error density (Strategy C) or any of the equivalent strategies based on the control of the point-wise error in stresses are preferible.

We finally note that maintaining a prescribed accuracy in the results can lead to a very large terms of number of elements, higher memory size or excessive computing time. Computational constraints can be introduced in the AMR strategy in order to make it affordable for practical applications [BCF,CB].

The topic of error estimation and mesh adaption is continuously evolving. Interested readers are recommended to consult the work of the references in Chapters 13 and 14 of [ZTZ] and also in [LP2].

10

GENERATION OF ANALYSIS DATA AND VISUALIZATION OF NUMERICAL RESULTS

10.1 INTRODUCTION

A significant task in the finite element analysis of a structure is the generation and the specification of all the data required for the computations. This work, usually called "preprocessing", includes the definition of the geometry of the structure in parametric form, either by hand or, what is more usual, by means of advanced computed-aided design (CAD) tools, the generation of a mesh and the assignment of the material properties, the boundary conditions and the loading. These tasks which are trivial for simple academic structural shapes, can be extremely complex for real structures. Here the use of advanced preprocessing tools is mandatory in practice.

A similar difficulty arises for the visualization of the results from the finite element computations. The so called "postprocessing" of the numerical outputs in the form of vector isolines or contours of displacements, strains and stresses is usually required. For practical problems these displays can only be performed with the help of modern graphical tools, specially for 3D problems.

The issues of analysis data input and the graphic visualization of numerical results cover many specialized topics which fall outside the scope of this book. The aim of this chapter is to present a panoramic view of the main aspects of the pre- and post-processing tasks with a particular emphasis in mesh generation. Interested readers are recommended to download from Internet and test the GiD pre and postprocessing system (www.gidhome.com) developed at CIMNE. GiD contains many utilities for data input and visualization of results for finite element structural analysis. GiD can be easily interfaced to any academic, research or commercial FEM package. A more detailed description of GiD is given in Appendix D.

10.2 THE IMPORTANCE OF A GOOD PRE AND POST PROCESSING SYSTEM

The practical use of computer simulation in engineering had its origin in the 1950's and 1960's with the advent of numerical methods such as the finite difference method (FDM) and the finite element method (FEM). Aerospace engineering was the focal point of activity during that time. By the late 1960's the first FEM based commercial computer programs (ASKA, NASTRAN, Stardyne, etc.) appeared. Subsequently, the FEM, the FDM and related numerical techniques (such as the finite volume method and the boundary element method) spread to other engineering and scientific disciplines, and now its use is widespread and many finite element-based commercial programs are available. Despite the fact that geometry modelling and mesh generation is the underpinning of computer analysis design, CAD as we know it today, had its origins later in the 1970's and 1980's. This perhaps explains why the geometric representation of objects in finite element analysis and CAD are so different. Many finite element programs were technically mature long before modern CAD was widely adopted. The typical situation in engineering practice is that designs are encapsulated in CAD systems and finite element meshes are generated from CAD data. This amounts to adopting a totally different geometric description for design and one that is only approximate for analysis. In some instances mesh generation can be done automatically but in most circumstances it can be done at best semi-automatically. There are still situations in major industries in which drawings are made of CAD designs and meshes are built from them. It is estimated that about 80% of the overall analysis time is devoted to mesh generation in practical computer-aided engineering work [HCB]. In the automotive industry, a mesh for an entire vehicle typically takes nowadays about *four months* to create. Design changes are made on a daily basis, limiting the utility of analysis in design if new meshes cannot be generated within that time frame. Once a mesh is constructed, refinement requires communication with the CAD system during each refinement iteration. This link is not straightforward and it is often unavailable, which perhaps justifies why, as today, adaptive mesh refinement is still not widely used at industrial level.

Above facts explain why generation and visualisation of analysis data for scientific and engineering computation has become the bottleneck in the practical application of the FEM. With the increasing trend of using networked PC's for analysis of real life problems, the need of efficient and versatile pre/postprocessing systems operating under Windows, Linux and Unix has become a universal problem. The possibility of performing computer simulations from home, either for teaching, research or engineering consultancy is becoming a habit for many graduate and undergraduate students and engineers. This invariably increases the need for powerful and economical procedures for preparation of analysis data for complex geometries, probably generated by CAD elsewhere, for handling the analysis results with support from visualisation tools and for producing high quality reports in written or electronic form.

In summary, there is a general *need for efficient and affordable pre-/post-processing systems* which incorporate all the requirements for user-friendly generation of analysis data, easy linking to any computer code and fast visualization of numerical results.

10.3 THE GEOMETRICAL REPRESENTATION OF STRUCTURES

A structure is typically defined by its boundary which consists of a closed loop of curved boundary segments (in 2D) or surfaces (in 3D). Curved boundary segments are in general represented by parametric spline curves. C^1 continuity is at least required in order to preserve a smooth stress field on the boundary curve and to satisfy the continuity conditions required by mesh generation algorithms. Hermite cubic splines can be used for this purpose.

Curved surfaces can be defined in a number of ways using polynomial functions represented by a variety of composite spline surfaces and curves (e.g. Bézier, B-splines, etc.). A common way to define arbitrary curved surfaces is by means of the so called Non-Uniform Rational B-Splines (NURBS). NURBS are typically used for defining complex surfaces in many engineering applications (Figure 10.1). A good description of the NURBS technique and other methods for the definition of curves and surfaces can be found in [BY,Fa,FP,Fa2,HCB,Or,PS,PT,Yam].

The efficient manipulation of geometries of objects by means of mathematical models has been a permanent objective of scientists and engineers since ancient times. The pioneer work of Descartes on analytical geometry meant a step forward in the graphical representation of complex bodies using mathematical expressions. Descartes work is considered by many the seed of the modern CAD methods widely used in industry. The big challenge nowadays is the effective linkage of CAD data with computer

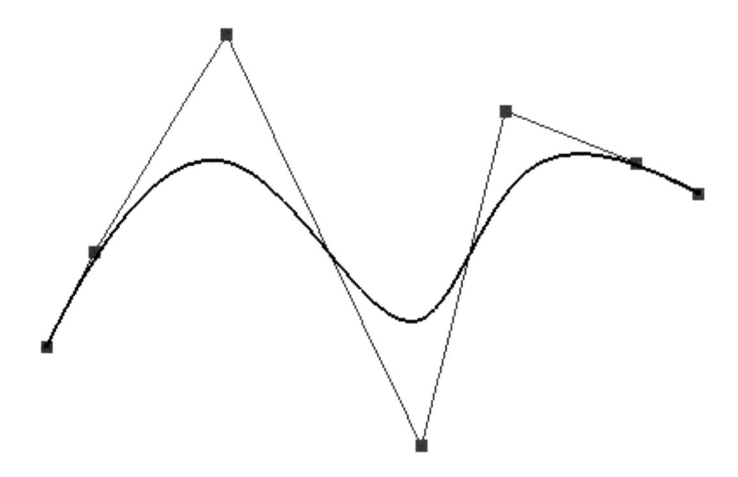

Fig. 10.1 NURBS line with 7 control points, degree=3 and knot vector={00000.30.580.81111}

simulation codes to solve relevant problems in science and engineering. Indeed the modern pre/post processing systems will help to make true the *dream of Descartes* of a rational knowledge of the work via geometry and mathematics, and nowadays also via computer simulations [DH].

A NURBS line is defined by its degree, a number of control points defining a polygon, weights associated to these points and a knot vector (Figure 10.1) [Fa,FP,HCB,PS,PT,Yam].

The NURBS curve smoothly approximates the control points polygon, without interpolating them. An exception are the end control points which are exactly interpolated. The knot vector is an increasing list of real numbers that divides the parametric space. The degree is the polynomial degree of each span.

The same concept can be extended to NURBS surfaces with a NxM net of control points (Figure 10.2).

A limitation is that NURBS surfaces are topologically an square, and can not be adapted to a general shape. To avoid this limitation usually surfaces are trimmed by a loop of curves on the surface that act as a mask (only the internal part is considered).

Figure 10.3 shows a trimming surface on the surface of Figure 10.2, i.e. cutting the surface with a pentagonal prism.

Every NURBS patch is perfectly defined by a parametric function. Typically the definition of the geometry is aimed to provide full details for

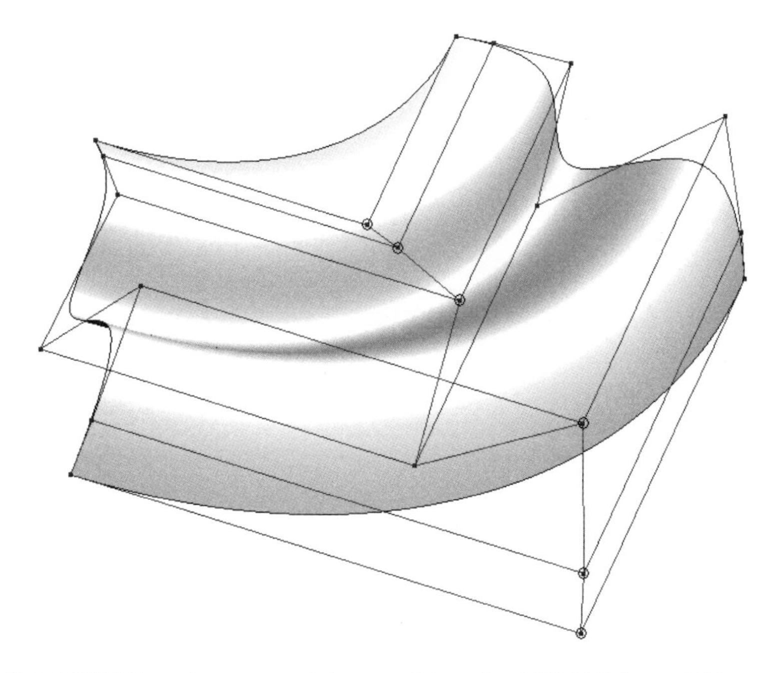

Fig. 10.2 NURBS surface created by revolving the NURBS line of Figure 10.1 around an axis. NURBS with 7×3 control points, degree $= 3 \times 2$. Two knot vectors $=\{00000.30.580.81111\} \times \{000111\}$. Weights $= 0.81$ for the middle row of control points (marked with circle) and 1.0 for the rest. Courtesy of the GiD Team at CIMNE

the manufacturing work and hence it aims to providing more information than that strictly required for structural analysis. Some simplifications are typically introduced in the NURBS data prior to mesh generation.

CAD systems allow to export the parametric definition of a line or a surface in different formats. Some popular CAD output formats are: IGES, DXF, VDA, STEP, ACIS, Parasolid, Solid Works, etc. [Fa,Fa2,FP,Yam]. Indeed, CAD data can contain errors or redundancies that must be "cleaned" prior to the mesh generation step and this can be a tedious and costly task. Preparing good CAD data suitable to the needs of the FEM is one of the challenges in the preprocessing step.

Figures 10.4 and 10.5 show the NURBS representation of a mechanical part and an aircraft surface and the finite element mesh of 3-noded shell triangles generated with the advancing front method (Section 10.4.1).

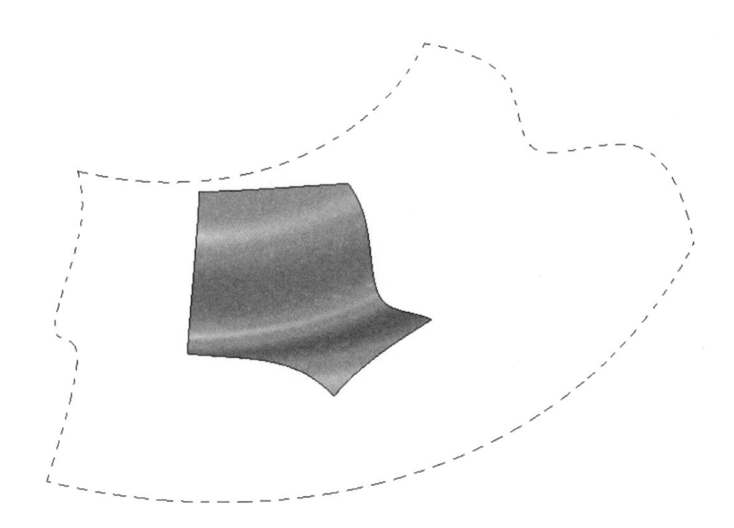

Fig. 10.3 Trimmed surface obtaining by cutting the NURBS surface of Figure 10.2 with a pentagonal prism. Courtesy of the GiD Team at CIMNE

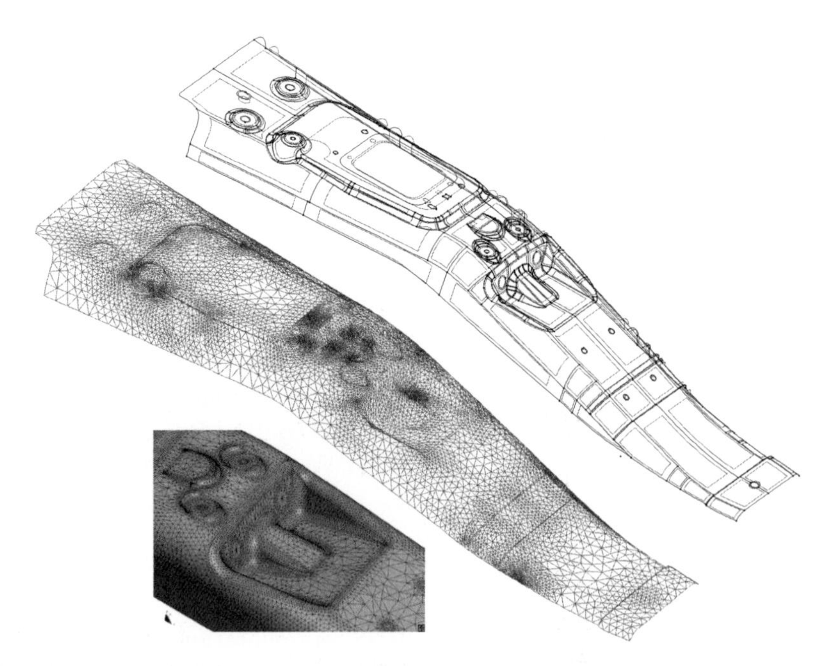

Fig. 10.4 NURBS definition of a mechanical part. Finite element mesh of 3-noded shell triangles generated with the advancing front method [Courtesy of Quantech ATZ, S.A., www.quantech.es]

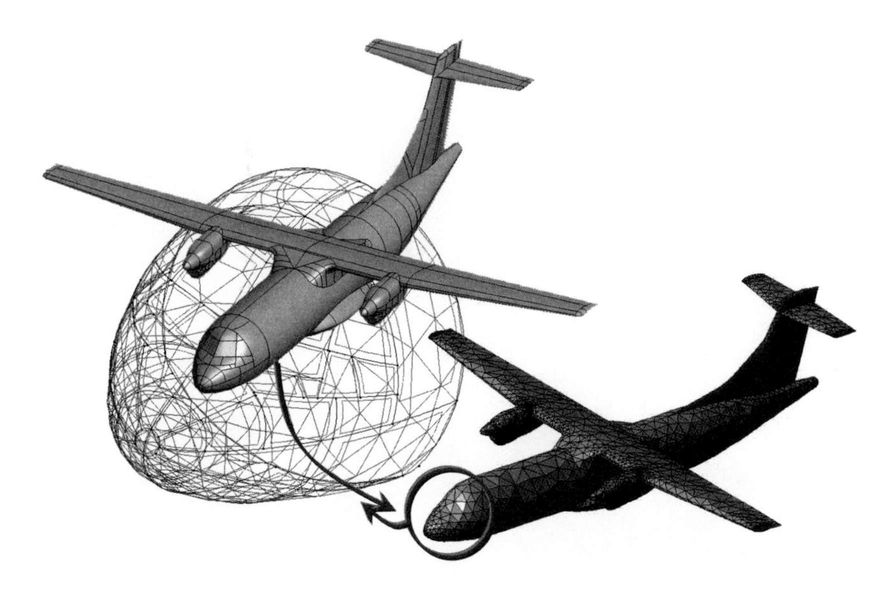

Fig. 10.5 Definition of an aircraft geometry with NURBS. Surface mesh of 3-noded triangles generated with the advancing front method and Gid [GiD]

10.4 MESH GENERATION

The generation of a finite element mesh is a key task prior to the analysis of a structure. While the generation of 1D and 2D meshes can still be attempted by hand, 3D mesh generation is an extremely difficult task which requires sophisticated algorithms and specialized software. Mesh generation has therefore become over the years a kind of science in itself. Only the basic concepts of mesh generation will be given here and the interested reader can found many details in papers and text books on the subject [Ca,FG,GB,GBBF+,Ge,KS,TSW,TWM].

Finite element meshes can be globally split into *structured* and *unstructured meshes* (Figure 10.6). A 2D structured mesh generally consists of two sets of lines. The lines in the same set do not interest each other and they intersect the lines in the other set only once. Structured meshes can be of uniform or non uniform element size. The same concept extends naturally to 3D structured meshes (Figure 10.4). Structured meshes can be generated semi-automatically by the so called *mapping methods.* These are based on isoparametric mapping techniques [ZP] by using blending functions [GH2] or by solving a partial differential equation on simple

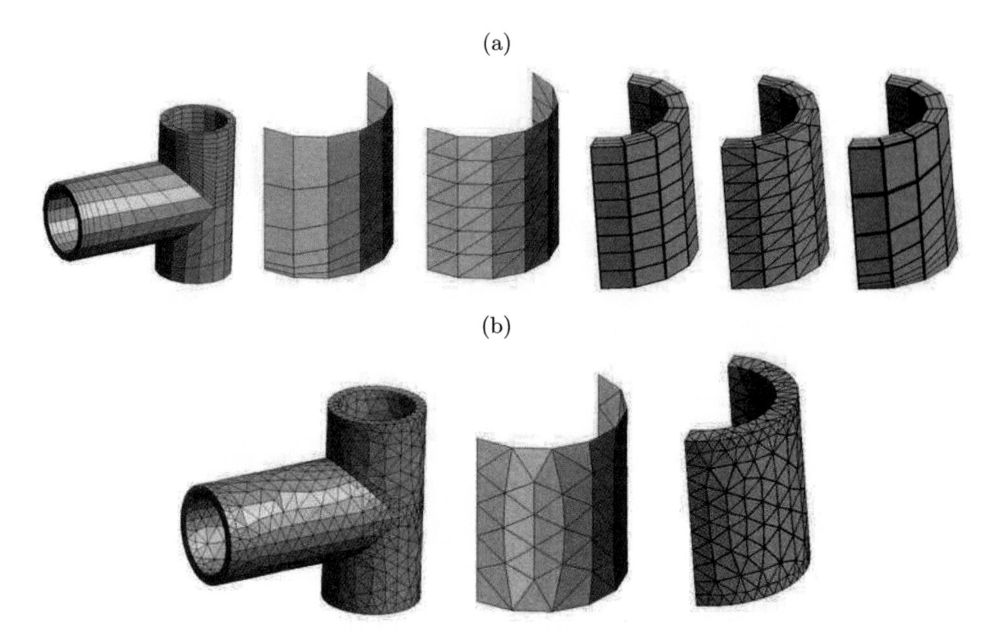

Fig. 10.6 Examples of (a) structured and (b) unstructured meshes

subregions of the analysis domain created manually [TTM,TW]. These subdomains are then mapped into regular grids to produce a mesh. This manual process is tedious and difficult for complex 3D geometries. Despite many efforts made to automate mapping methods no fully automatic mesh generator using a mapping method has been achieved.

The difficulties in obtaining a universal mesh generator via mapping methods has lead to the development of algorithms for generating unstructured meshes. These meshes are not the result of the intersection of sets of parallel lines and hence do not have a specific internal structure. Most of the unstructured mesh generation methods are designed for generating triangular elements (in 2D) and tetrahedral elements (in 3D) (Figure 10.6b). Unstructured meshes are suitable for discretizing objects with complex irregular geometry and they typically involve elements of different sizes. They are also ideal for adaptive mesh refinement. Abrupt changes in the geometry or the element size locally should however be avoided in practice.

Many automatic unstructured mesh generation algorithms have been proposed in the literature. The most popular are the advancing front method [GS,JT,Lo,LP,PVMZ], the Delaunay triangulation method [Ba,

FG,Ge,GBFL+,Re,SS,WH] and the tree methods (the finite quadtree method in 2D and the finite octree in 3D) [BWSG+,SG,YS]. All these methods can be extended to automatically generate unstructured quadrilateral meshes in 2D by collapsing two adjacent triangles which share a common edge into a quadrilateral [JSK,RSS]. Despite much research, the automatic generation of a hexahedral mesh is still a challenge [BM,CI,Dho, Joe,TBM].

Due to their wider applicability in the next sections we present the basis of the advancing front method (and its variant the paving method) and of the Delaunay method. The interested reader will find a description of the most popular generation procedures in [CBFL+,FG,Ge,TMIP+].

10.4.1 The advancing front method

The advancing front method is a simple, intuitive and powerful algorithm to generate meshes of triangles (2D) and tetrahedra (3D). A mesh is progressively created starting from an initial "front" formed by the elements covering the boundary of the domain (i.e. line segments in 2D). Elements in the interior of the domain are generated *one by one* by joining the nodes laying at the front with interior nodes placed at an appropriate distance, so as to define the specified element size. The generation of every new element changes the front profile which advances into the region to be discretized until the whole domain is completely covered by elements.

The advancing front method was proposed by Lo [Lo] who constructed a triangularization over a set of a priori generated points inside the domain. Jin and Wiberg [JW] presented an advancing from algorithm in 2D. The procedure was improved by Peraire *et al.* [PVMZ] who presented a method for generating points and elements at the same time with the assistance of a *background mesh* used to define the characteristics of the mesh. This allows to generate a non uniform distribution of element sizes, according to a particular specification, such as the need to use a highly refined mesh at a certain zone of the domain, as typically occurs in adaptive mesh refinement. Any directional orientation of the elements can be implemented by introducing stretches in certain directions. Mesh directionality is a significant feature of the advancing front method, which makes it superior to the Delaunay technique, in particular for adaptive mesh refinement applications.

Extensions of the advancing front method to 3D problems where reported by Moller and Hansbo [MH], Löhner and Parikh [LP] and Peraire *et al.* [PPFM+]. An adaptive mesh refinement algorithm for 3D Euler flows was presented by Peraire *et al.* [PPM]. The robustness of the advancing from method was studied by Jin and Tanner [JT] and by Moller and Hansbo [MH] who proposed various techniques for increasing its robustness. This improvement can also be achieved by incorporating principles of Delaunay triangulation into the algorithm [GB]. Considerable research has been invested with the objective of increasing the computational efficiency of the advancing front method. We note, for instance, the work of Bonet and Peraire [BP] and Kwok *et al* [KHK] who developed useful data structures to aid the efficiency of the method. Similar efforts were reported in [Ge]. Computational speed in generating a mesh obviously depends on the efficiency of the particular mesh generation code. Indeed, the speed of mesh generation can be considerably improved by using parallel computing techniques. The implementation of the advancing front method in a parallel computer was presented by Topping and Khan [TK2].

The main steps of the advancing front algorithm for the generation of a mesh of 3-noded triangles are described next.

Background mesh. The background mesh is employed to accurately control the distribution of the geometrical characteristics of the new mesh. Typically is formed by simple 3-noded triangular elements (in 2D) or 4-noded tetrahedra (in 3D) covering the domain to be meshed (Figures 10.7 and 10.8). Mesh parameters data, such as element sizes and stretching, are assigned to the nodes of the background mesh, and they are subsequently interpolated linearly to any point inside the domain or on its boundary. A sphere at each vertex shows the size that the element is required in the definitive mesh (Figure 10.8).

Generation front. The initial generation front is a collection of all the sides which form the discretized boundary of the domain (Figure 10.9). If the domain is composed of multiple connected regions, an initial generation front is formed for each of the regions. Once a new element is created the generation front is updated. The updating procedure ensures that the generation front always forms the boundary of the region to be meshed. The sides and the nodes in the generation front are referred to as *active sides* and *active nodes*, respectively.

Element generation. An active side with connecting nodes is selected from the generation front to generate a new element. For instance, take side 2-11 in the upper right frame of Figure 10.9. The position of point 12 to form a new triangle $2, 11, 12$ is placed over the direction normal to side 2-11 and located at a distance h^e from that side. This distance is found by interpolating the required element sizes from the background

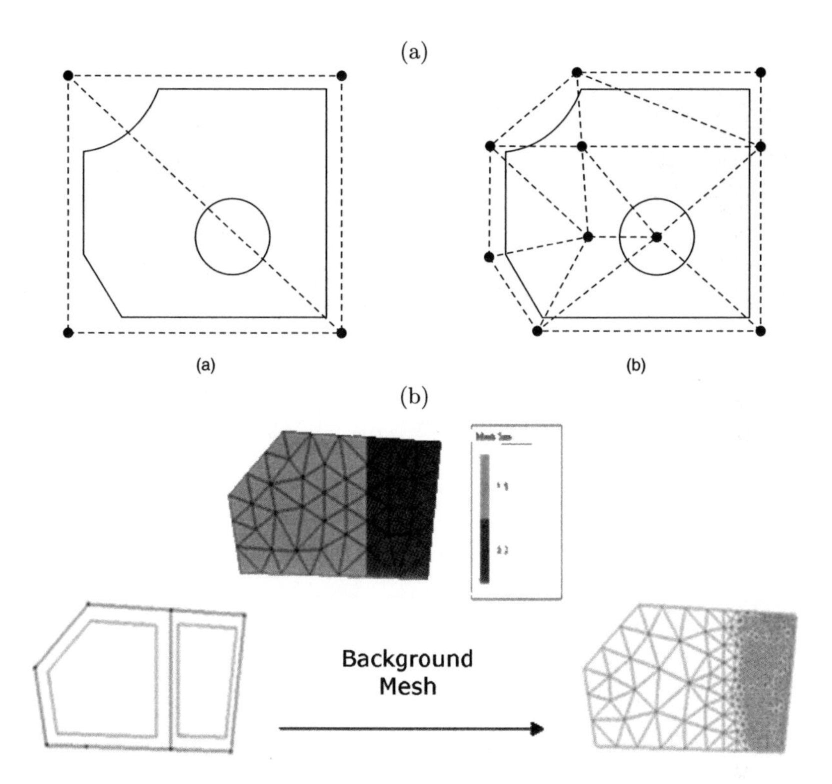

Fig. 10.7 (a) Background meshes for a typical 2D domain. (b) Mesh size assigned to a domain using a background mesh

mesh. A check is performed to verify that the sides of the new triangle $2, 11, 12$ do not intersect any of the existing sides in the generation front [ZTZ].

Mesh quality enhancement. Mesh quality enhancement is needed to improve the shape of the elements at the final stage of the mesh generation process. This is particularly important for strongly graded meshes with element sizes varying rapidly.

Figures 10.10–10.13 show examples of meshes generated with the advancing front method.

The quality of a mesh can be improved by *mesh smoothing*. The method is based on the so called *Laplacian smoothing* technique which repositions the internal node at the centroid of the polygon formed by its neighboring nodes [Ca]. The smoothing process is repeated over 3-5 iterations and typically leads to a mesh with better shaped elements. The algorithm can

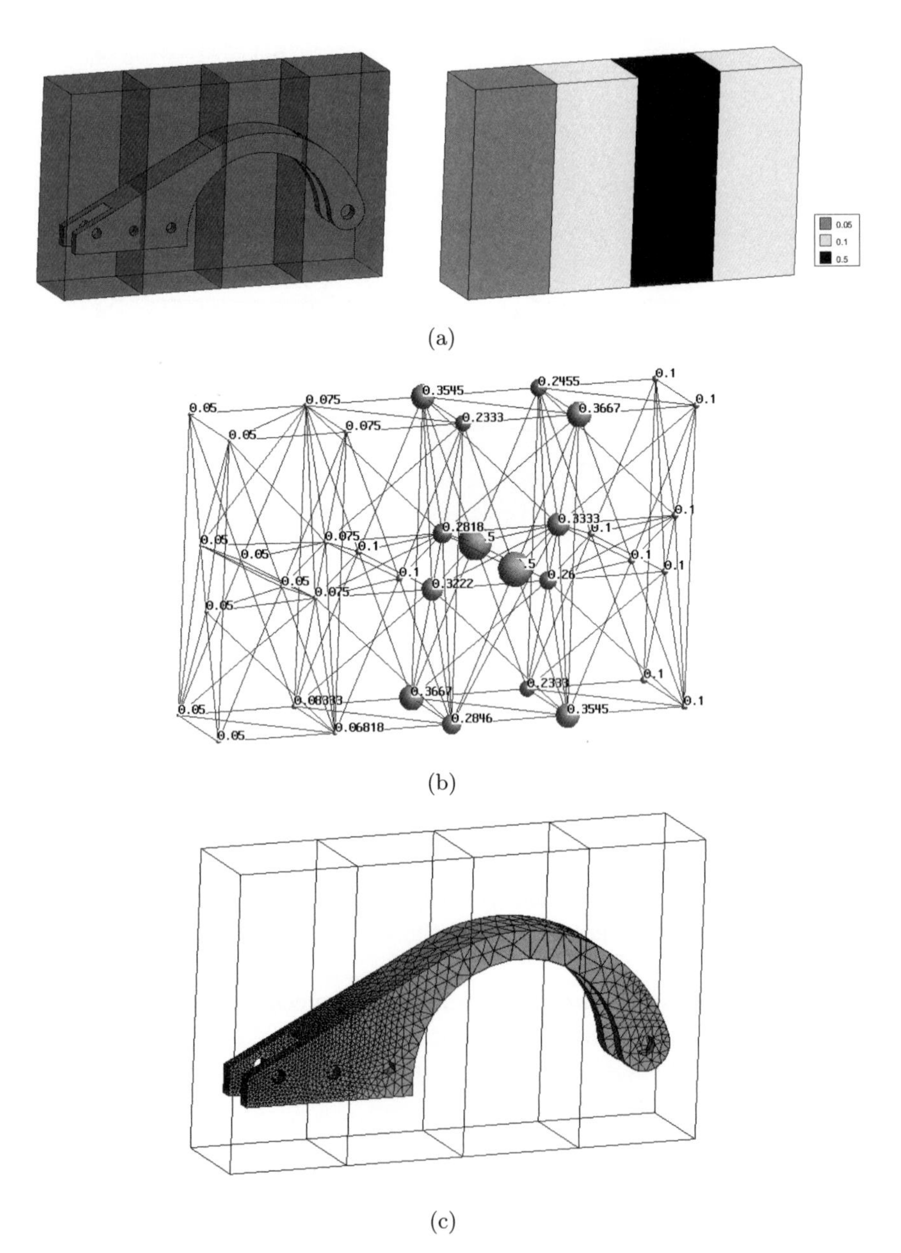

Fig. 10.8 3D background mesh. (a) A mechanical part immersed in four boxes where sizes are assigned and sizes assigned to each volume. (b) Nodes of a background mesh of 4-noded tetrahedra. Numbers denote element size. (c) Final mesh of linear tetrahedra for the mechanical part with the required element sizes

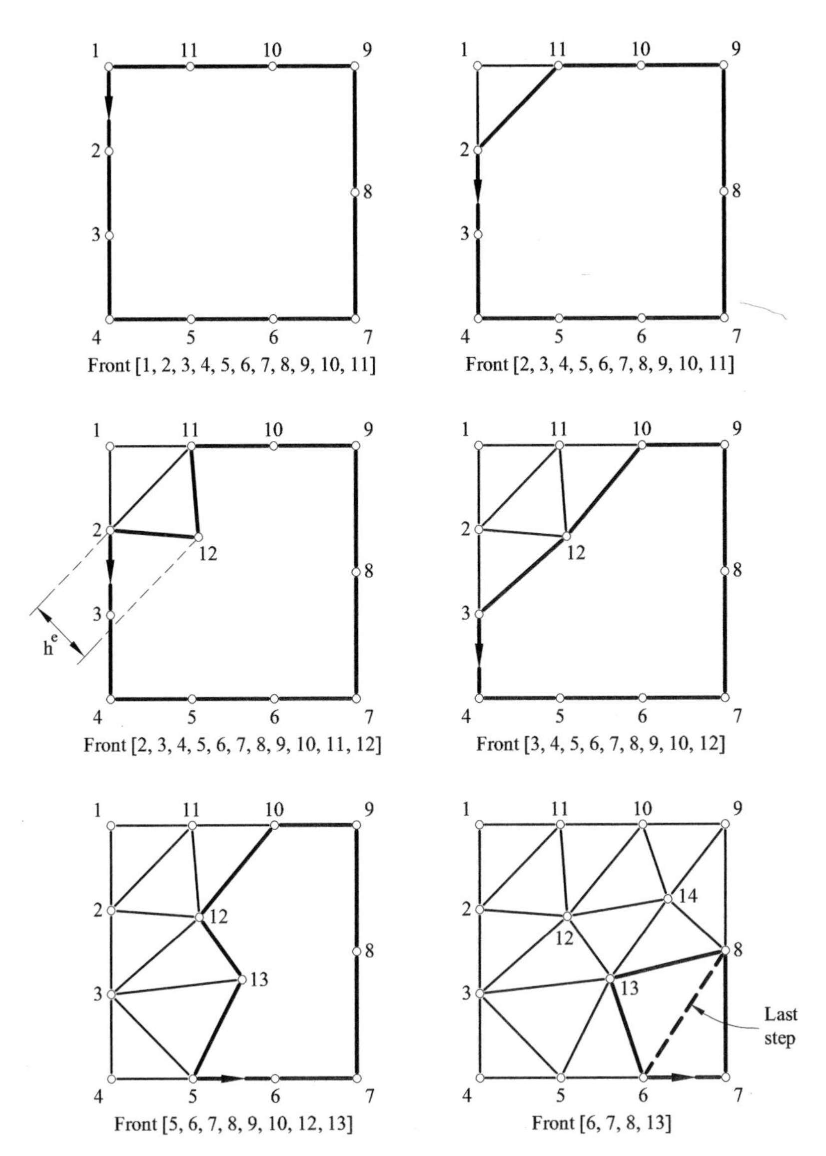

Fig. 10.9 Generation front and its updating during the element generation process. Active sides are identified by a thicker line. Active nodes at the front are listed within the bracket in each figure

however fail for nodes adjacent to concave boundaries and the smoothing procedure should be avoided for these nodes.

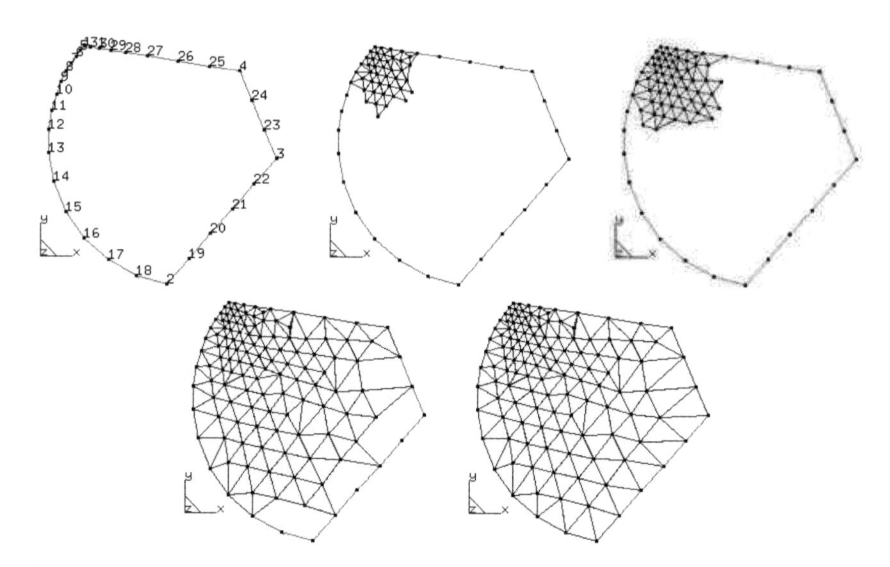

Fig. 10.10 Generation of a mesh of triangles with the advancing front method

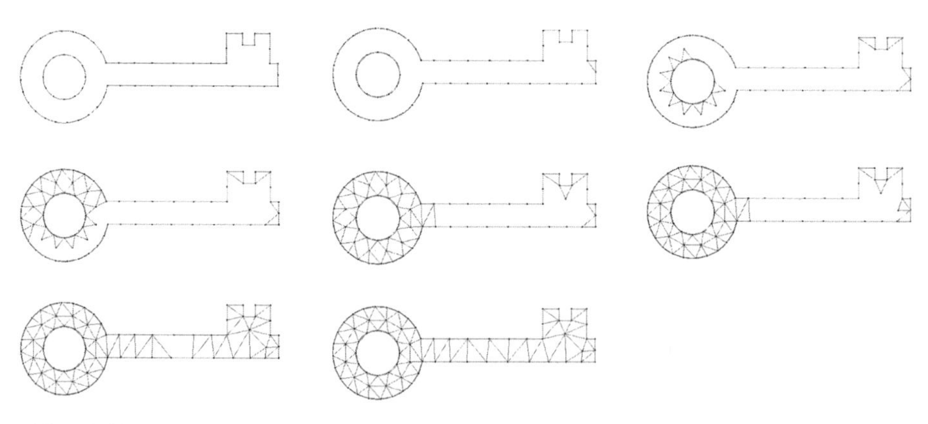

Fig. 10.11 Example of mesh generation with the advancing front method. Two starting fronts originating at an external and an internal boundary

An alternative procedure to reduce element distortion to a minimum in a generated mesh is the *elimination* of nodes creating elements with an undesirable aspect ratio, or the swapping of diagonals aiming to reducing the maximum angle of the existing elements in the mesh (Figure 10.14).

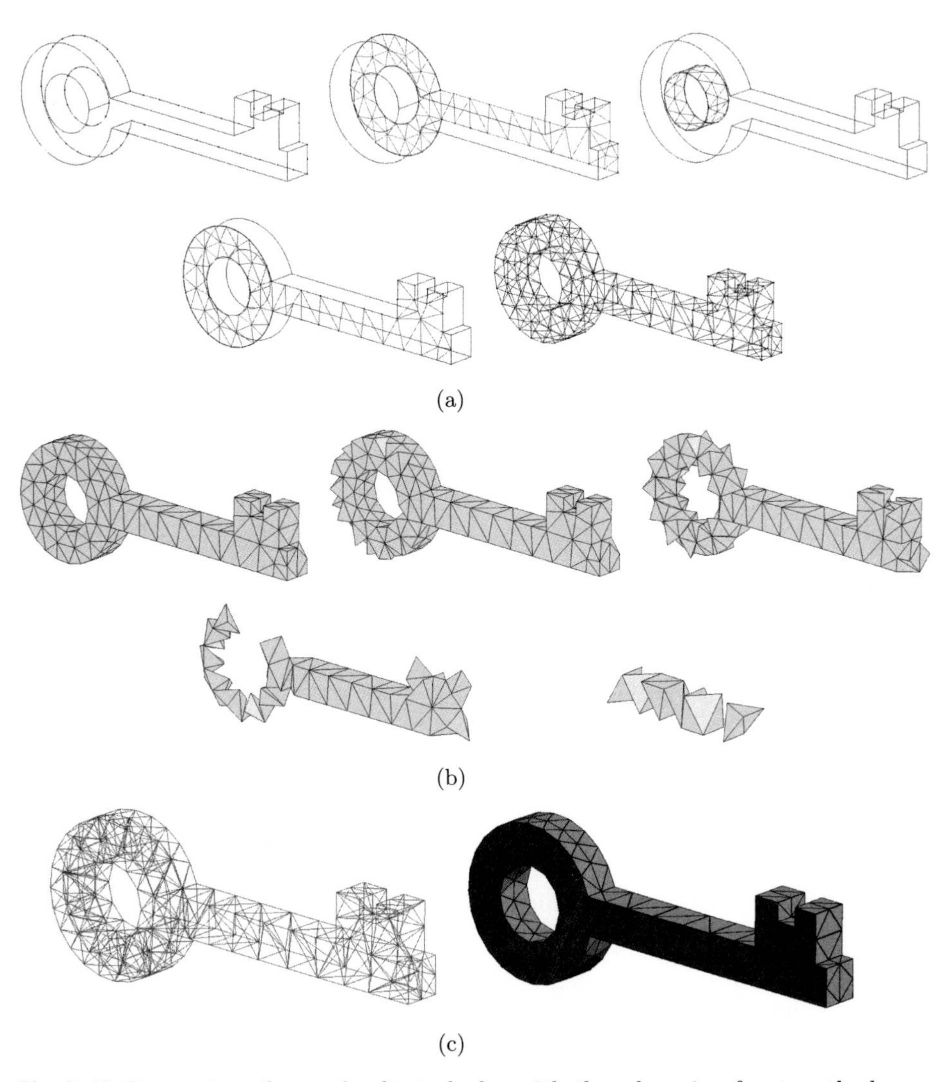

(a)

(b)

(c)

Fig. 10.12 Generation of a mesh of tetrahedra with the advancing front method. (a) Definition of external and internal surfaces. (b) Evolution of the front surface. (c) Final mesh

The advancing front method can also be applied for generating a surface mesh for an arbitrary 3D geometry. The basic idea is to perform the mesh generation, according to the prescribed element size distribution, in a 2D parametric plane using the advancing front method as described above, and then map the 2D mesh onto the 3D surface. The process re-

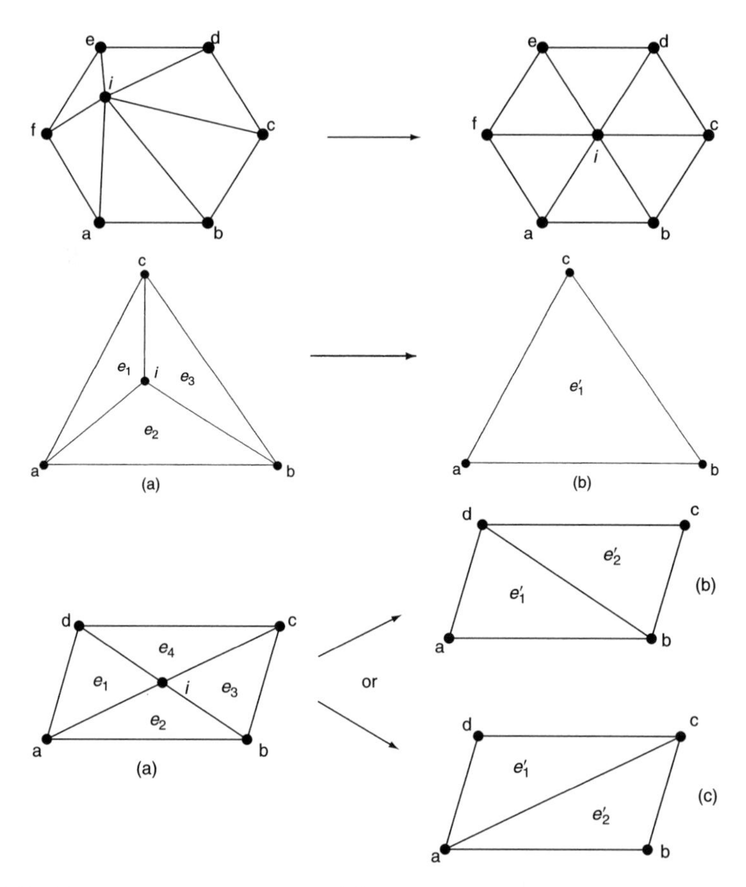

Fig. 10.13 (a) Laplacian smoothing; (b) Elimination of nodes and (c) Diagonal swapping

quires the transformation of the mesh parameters given for the 3D surface to the parametric plane. Details of an algorithm of this kind can be found in [Pe,Pe2,PPM].

As mentioned above the advancing front method is extremely powerful to generate triangular and tetrahedral meshes in complex geometries. Some examples are presented in Figure 10.14.

Generation of quadrilateral meshes via the advancing front method is possible and a procedure was reported by Zhu *et al.* [ZZHW]. A simple pragmatic approach is converting a triangular mesh form into mesh of quadrilaterals by collapsing two triangles into a quadrilateral [JSK,RSS]. In any case, the generation of a "good" hexahedral mesh is still nowadays a challenge.

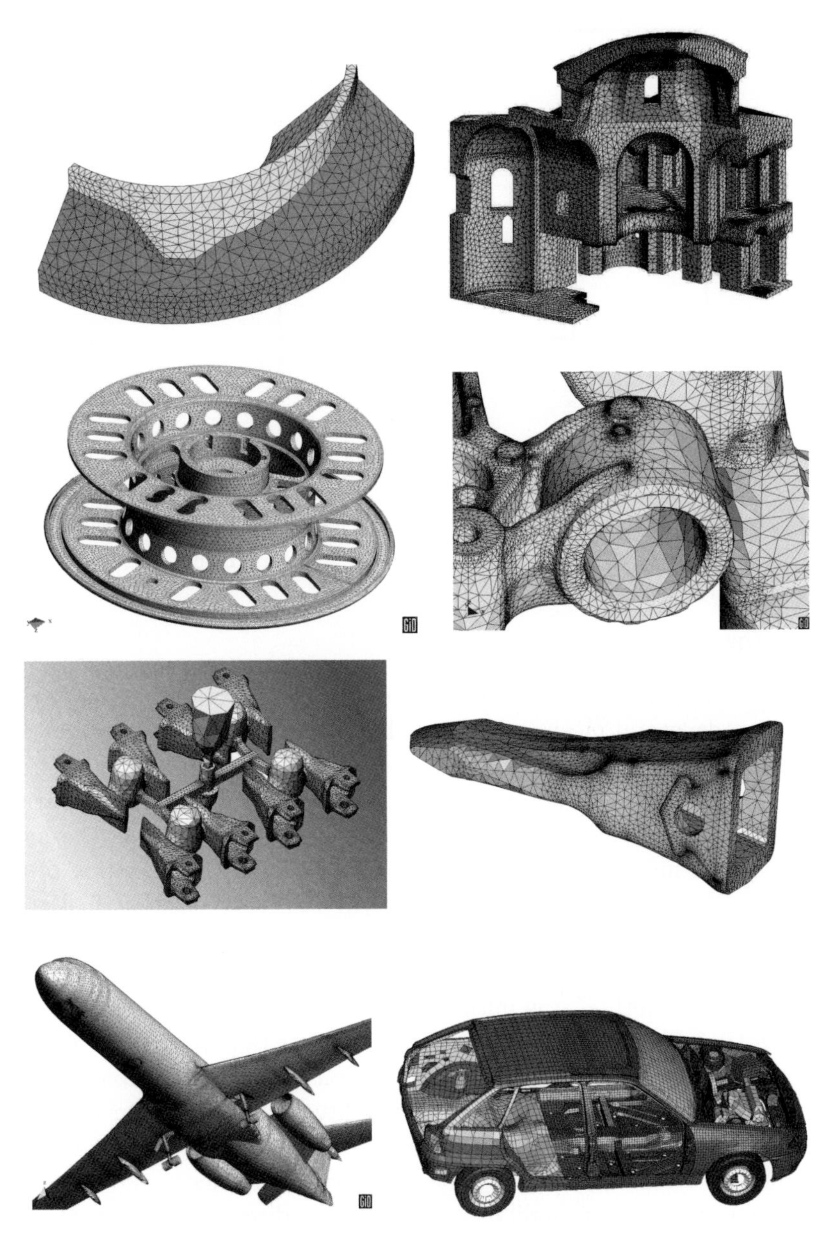

Fig. 10.14 Meshes generated with the advancing front method. Courtesy of the GiD team at CIMNE [GiD], Compass Ingeniería y Sistemas S.A. (www.compassis.com) and Quantech ATZ S.A. (www.quantech.es)

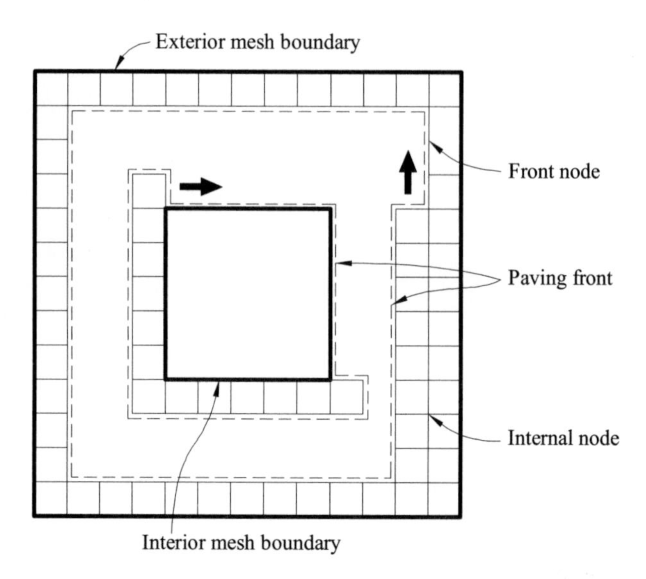

Fig. 10.15 The paving method. Mesh boundaries, paving fronts and node types

10.4.2 The paving method

Blacker *et al.* [Bl,BS,BSC] extended the advancing front technique to generate quadrilateral meshes. The so called *paving method* is based on the element by element addition of 4-noded quadrilaterals to an active front line so that the mesh generation propagates towards the internal parts of the mesh. An incremental smoothing technique is used during the mesh generation to improve the quality of the final mesh. A description of the paving method can be found in [CBMB,OSCS,TMIP+,vRBBB]. Extensions of the paving method for generating triangular meshes are reported in [vRBBB].

The paving method was initially implemented in the Sandia National Labs. in USA [Bl,LTV]. Despite its initial success (a patent of the method was registered in USA [Bl2]), its application to the generation of hexahedral meshes is quite difficult. A possibility to obtain an unstructured hexahedral mesh is by splitting each tetrahedral element into four hexahedra. This increases very rapidly the number of elements in the mesh and does not provide good quality hexahedra. The direct generation of good quality hexahedral meshes on 3D solids with arbitrary geometry is one of the current challenges of mesh generation.

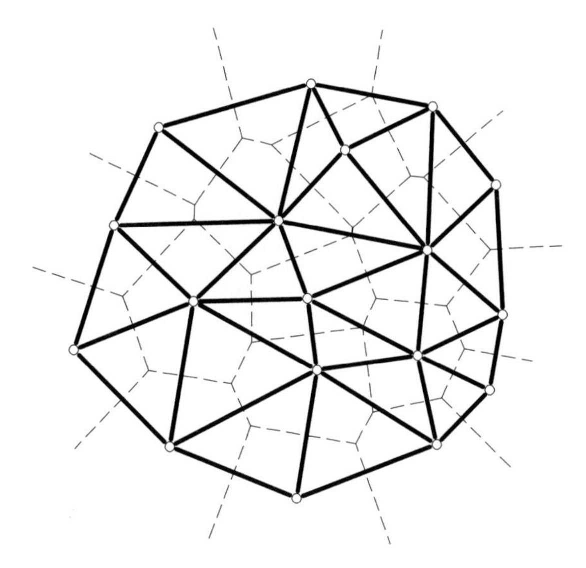

Fig. 10.16 The Voronoi diagram (in dotted lines) and the Delaunay triangulation (in solid lines) of a group of eighteen vertices (forming points)

Higher order elements can be created easily by adding intermediate notes to each element edge. The position of any interior node can be interpolated from the position of the element perimeter nodes.

10.4.3 The Delaunay method

The Delaunay method is used for the generation of triangular and tetrahedral meshes. The method yields unstructured meshes and can also be used for refining of existing meshes. The Delaunay technique has attracted much attention in research and software development due to its conceptual simplicity, mathematical rigor and algorithmic robustness.

A mesh of 3-noded triangles (the same property applies for 4-noded tetrahedra) can be generated using the Delaunay method as the dual of the so called Voronoi diagram [De,Vo]. The mesh generation process is explained next for the 2D case.

Let us have a collection of distinct points P_i, $i = 1, N$ in a 2D region. They are referred to as the *forming points* in the mesh generation literature. The *Voronoi region* $V(P_i)$ is defined as the set of points that are at least as close to P_i as to any other forming point. As an example Figure 10.17 shows the Voronoi regions defined by seventeen forming points. The Voronoi region $V(P_i)$ represents a convex polygonal region,

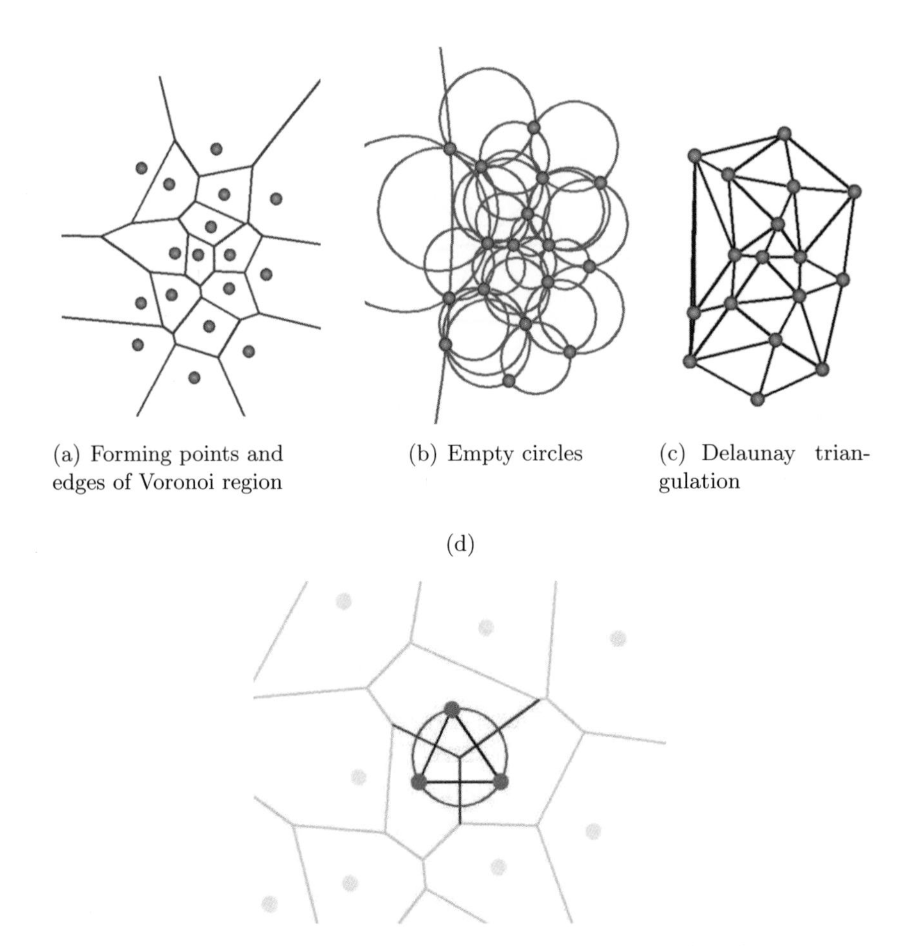

Fig. 10.17 Generation of a mesh with the Delaunay method. (a,b,c) Voronoi diagram, empty circles and Delaunay triangulation. (d) Duality between the Delaunay triangulation and the Voronoi diagram

possibly unbounded. The points that belong to more than one region form the edges of the Voronoi regions. The points at which the edges intersect are called the Voronoi vertexes. The union of the Voronoi regions is the Voronoi diagram of the forming point set P.

The dual graph of the Voronoi diagram is formed by connecting by straight lines joining the forming points of the neighbouring Voronoi regions sharing a common edge. This graph forms the Delaunay triangulation of the Voronoi forming points, i.e. the 3-noded finite element mesh sought. There is a perfect duality between the Voronoi diagram and the Delaunay triangulation (Figure 10.17d).

The more relevant properties of the Delaunay triangulation for mesh generation are the following:

1. The Delaunay method fails to generate triangles only if four of the forming points P lay on the same circle (then a quadrilateral is formed). This situation should be avoided if an all triangular mesh is sought.
2. Each Delaunay triangle corresponds to a Voronoi vertex, which is the centre of circumcircle of the triangle. This property is used to construct data structures.
3. The interior of the circumcircle contains no forming points P. This is the well known *empty-circle-criterion* which should be satisfied by any new point introduced in the Delaunay triangulation (Figure 10.17b).
4. The boundary of the Delaunay triangulation is the convex hull of the forming points. This is the basis for using a convex hull which contains all the mesh points in 3D mesh generation.

Each Voronoi diagram corresponds to a set of forming points which constitutes the Delaunay triangulation. Adding a new forming point will inevitably result in a modification of the Voronoi diagram and the Delaunay triangulation. The process of constructing a new Voronoi diagram and Delaunay triangulation after the insertion of a new node is frequently used in automatic mesh generation and is illustrated here in the two dimensional setting shown in Figure 10.18 [ZTZ].

Let the new forming point n be inserted in the Delaunay triangulation shown in Figure 10.18a. It falls into the circumcircles of Delaunay triangles afg, abf and bef, therefore violating property 3. This causes the removal of the three Voronoi vertices which are the centers of the circumcircles and their corresponding Delaunay triangles, as illustrated in Figure 10.18b. The new Delaunay triangulation is constructed by linking the new forming point n and its contiguous forming points that form a face of the neighboring triangle followed by the construction of the new Voronoi diagram as shown in Figure 10.18c.

As we have indicated previously, the process used in the last example is applicable to three dimensions.

10.4.4 3D mesh generation with the Delaunay method

The crucial issues in the practical application of the Delaunay method for mesh generation are how Delaunay triangulation can be formed effec-

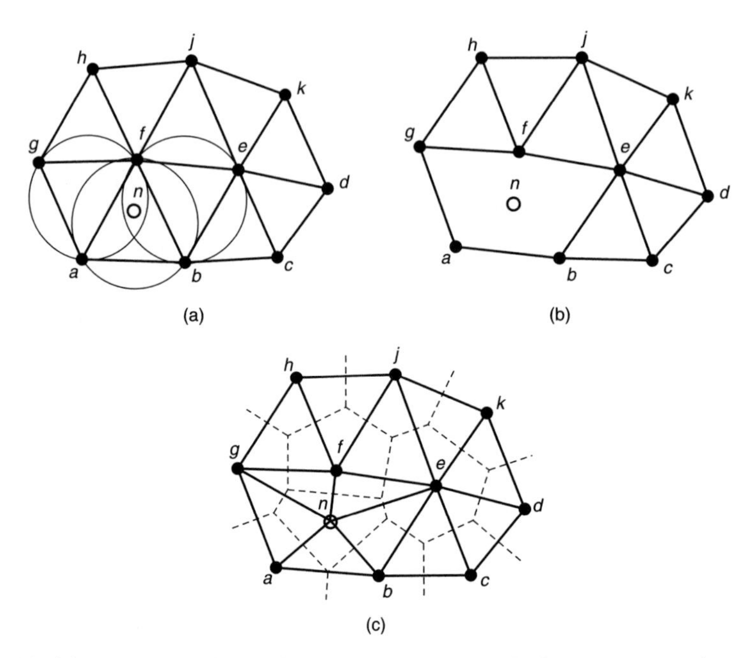

Fig. 10.18 (a) Insertion of new forming point n into Delaunay triangulation. (b) Removal of Delaunay triangles, deleted Voronoi vertices are not shown. (c) New Delaunay triangulation and Voronoi diagram

tively [Ba,Bo,CFF,GHS,SS,Wat,We], how to generate the points that will be inserted in the triangulation [MW,Re,WH] and how to preserve the boundary of a region when the forming points are from the boundary of a concave region [GHS,We2,WH].

Indeed these problems are more relevant for 3D mesh generation. In the following lines we summarize a procedure for 3D mesh generation using the Voronoi method:

a) Input the triangular surface mesh and derive the topological data of the surface mesh such as edges of surface elements and node-element connections. Figure 10.19a shows the surface mesh of a simple 3D object.
b) Build on convex hull that contains all the mesh points. An eight node convex hull is shown in Figure 10.19a.
c) Perform a Delaunay triangulation using nodes of the surface mesh to form tetrahedra. Figure 10.18b shows a Delaunay triangulation of the surface nodes.

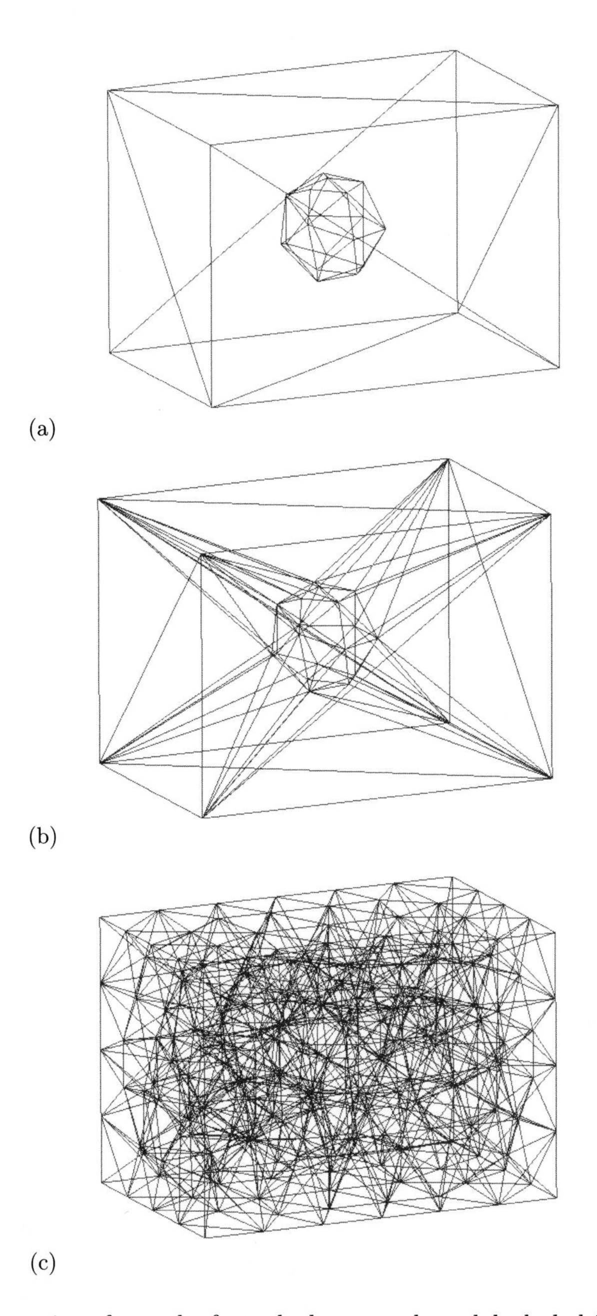

Fig. 10.19 Generation of a mesh of tetrahedra around a polyhedral object with the Delaunay method. (a) Convex hull and surface mesh. (b) Delaunay triangulation of surface nodes. (c) Final Delaunay mesh

d) Create interior points following a specified element size distribution, then perform Delaunay triangulation to form tetrahedra. Details of an automatic node generation algorithm and a 3D Delaunay triangulation can be found in [Bo,Wat,ZTZ].
e) Recover any missing edges and triangular faces of the surface mesh to ensure that the input surface triangulation is preserved after the volume triangulation. A surface mesh recovery method is described in Chapter 8 of [ZTZ].
f) Identify and remove all the tetrahedra outside the domain of interest to give the final 3D mesh (Figure 10.19c).

Figure 10.20 shows another example of the generation of a 3D tetrahedral mesh around a mechanical part with the Delaunay method. Details of this 3D mesh generation algorithm can be found in [WH,ZTZ].

Mesh quality enhancement is crucial in 3D mesh generation using either the Delaunay method or the advancing front method. Poorly shaped elements (typically called *slivers*) can be obtained using the Delaunay technique, therefore rendering the 3D mesh unusable for practical applications. Mesh quality can be improved by node addition or elimination, or by mesh smoothing similarly as for 2D problems. We note that the standard Laplacian smoothing can not be applied directly to a tetrahedral mesh as it reduces the quality of the mesh. The smoothing procedure here is typically based in moving a node incrementally and iteratively towards each of its connecting nodes until it reaches a position that increases the quality of the worst adjacent element. A description of 3D mesh quality enhancement procedures can be found in [PGH, LJ, LZG].

The Delaunay method and the advancing front method produce very similar meshes in many cases (Figure 10.21). Indeed the application of one or other method very much depends on the robustness and efficiency of the computer implementation. Parallel processing may be key in the future development of these methods for generation and adaptive refinement of large meshes involving millions of 3D elements in affordable times.

10.5 VISUALIZATION OF NUMERICAL RESULTS

The graphical representation of the numerical results resulting from the finite element computation is mandatory for practical purposes. Typical results to be displayed are the mesh deformation, the displacement vector of the nodes, the isolines or the contour fill plots of individual components of the nodal displacement vector, the strain and the stress vectors and the

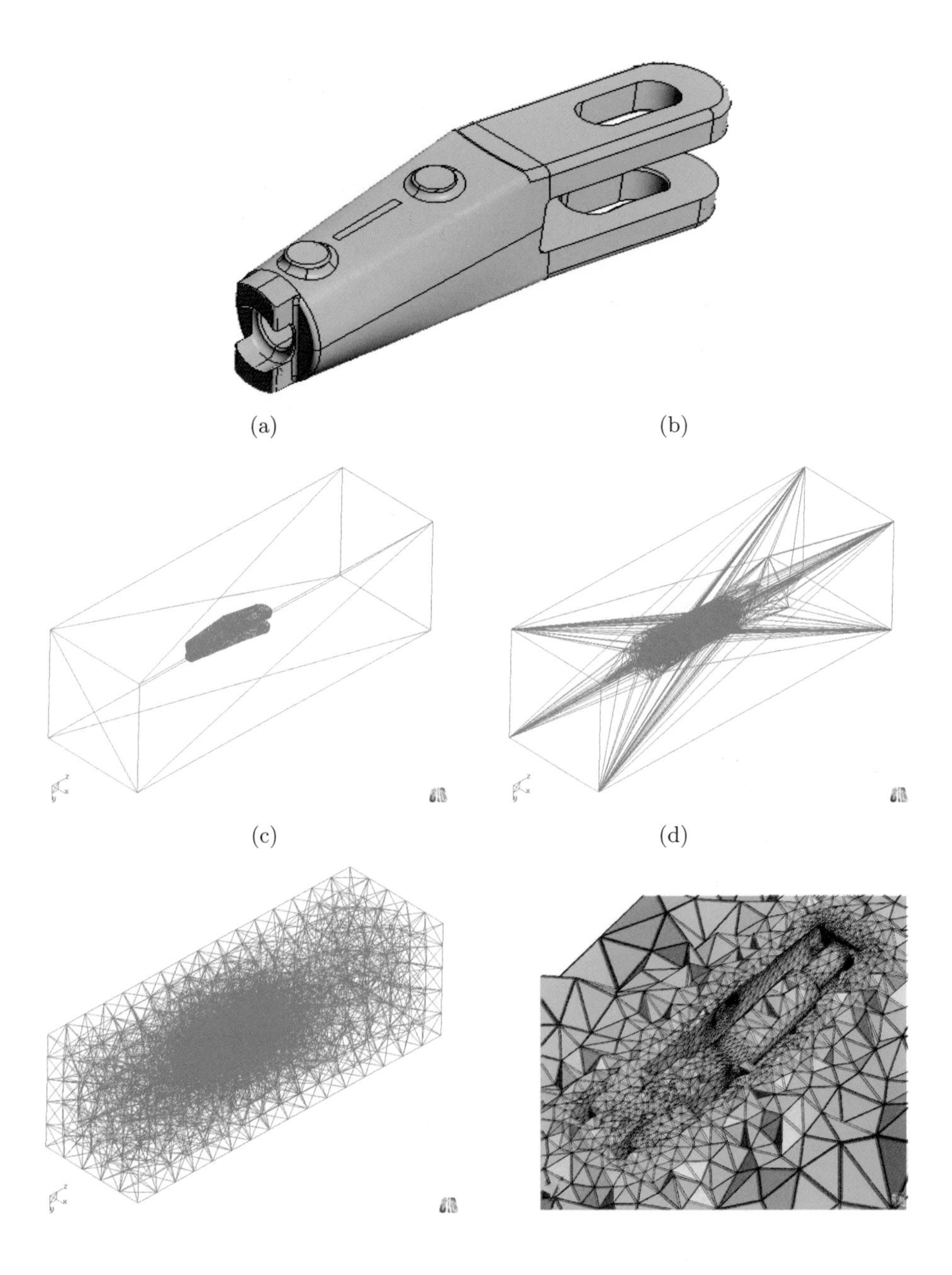

Fig. 10.20 Generation of a tetrahedral mesh around a mechanical part with the Delaunay method. (a) Convex hull and surface mesh. (b) Delaunay triangulation of surface nodes. (c) Final Delaunay mesh. (d) Detail of 3D mesh near the part

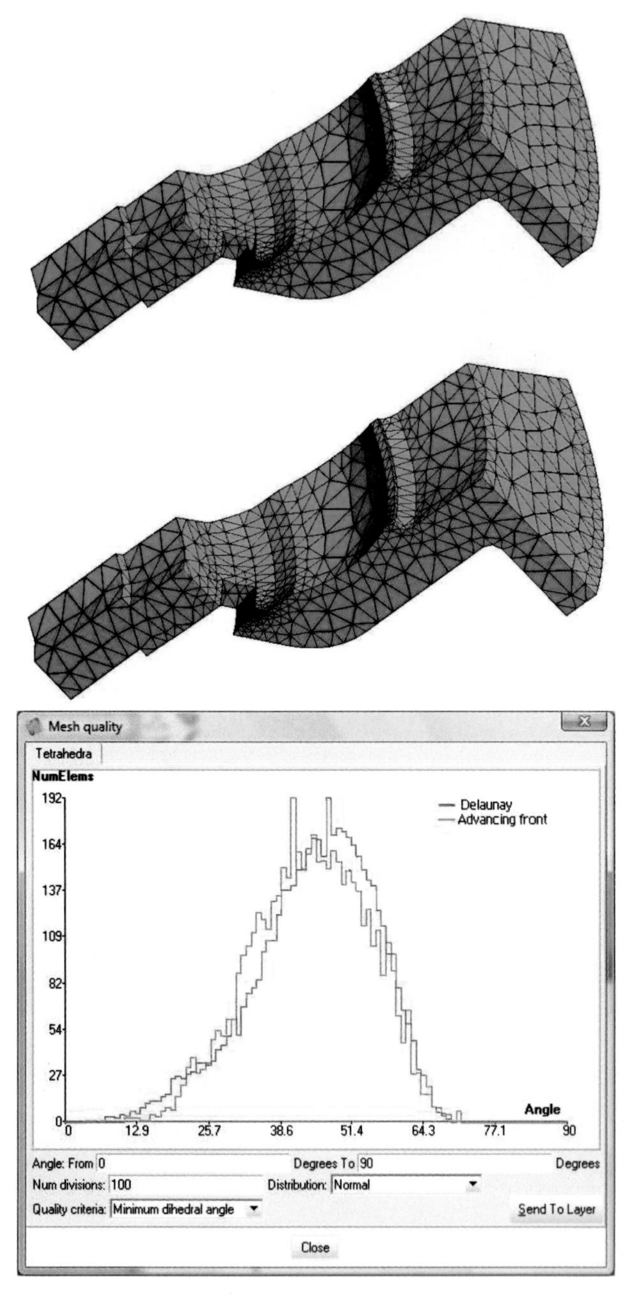

Fig. 10.21 Distribution of tetrahedral elements on the basis of the minimum dihedral angle on two meshes generated with the advancing front (a) and the Delaunay (b) method. Note the similarity of the two meshes

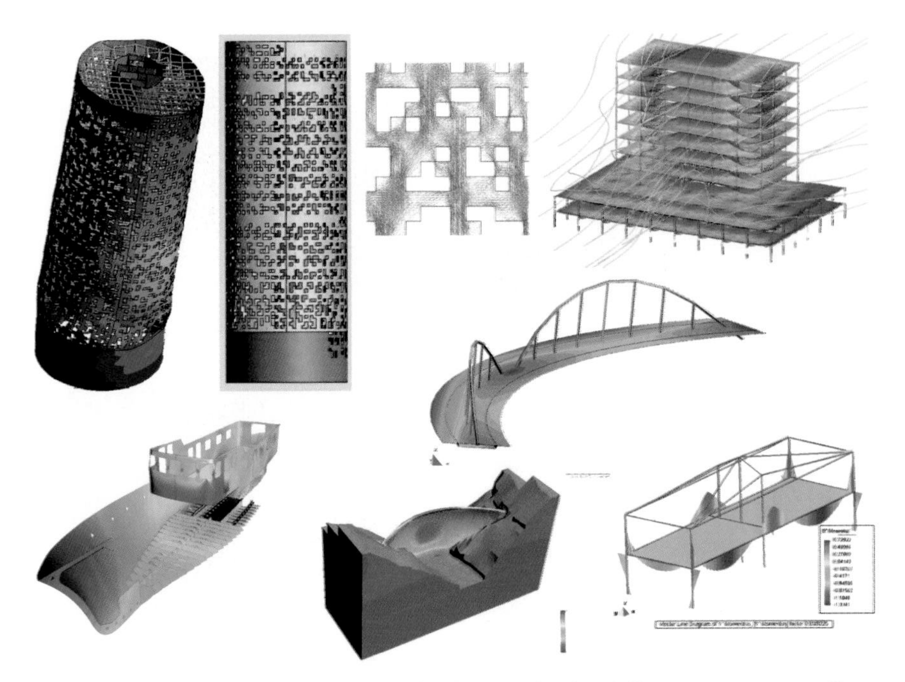

Fig. 10.22 Examples of visualization of FEM results for different structures. Courtesy of Compass Ingeniería y Sistemas S.A. (www.compassis.com)

principal stresses. Figures 10.22 and 10.23 show examples of visualization of results of structural analysis with the FEM.

Stresses and strains can be plotted using directly the values computed at the element Gauss points. Alternatively the smoothed nodal values of stresses and strains can be used (Section 9.8).

The graphical visualization of the finite element results is a complex task which requires deep knowledge of computer graphics. A review of the most popular techniques for the visualization of FEM results can be found in [SS2].

FEM developers and users can make use of the visualization facilities implemented in the GiD pre/postprocessing software available from Internet [GiD] (see also Appendix D).

10.6 CONCLUDING REMARKS

Some hundred years after Descartes times, the need for controlling the geometrical description of objects in order to solve practical problems is

Fig. 10.23 Results of the FEM analysis of Tarazona Cathedral (Spain). Colours indicate displacement and stress contours under self-weight [RCMR+,RGAA,Ro2]

as important as in ancient times. The size of the problems has changed, our knowledge of geometry, mathematics and computation has increased, the available computer resources grow every day, but the need to link physical objects with geometrical models represented by mathematical equations has remained the same over the years. The transformation of these models into data for the finite element analysis and the graphic representation of the numerical results are nowadays the key challenges for solving many problems of interest using the FEM. The attempt to solve these problems has motivated much work for the development of new mathematical methods, innovative and efficient algorithms and, of course, pre and postprocessing codes facilitating the activity of academics, scientists and engineers interested in computations, as well as in teaching the new analysis procedures to the next generation of students. The motivation which inspired the work of Descartes: the rational knowledge of the world with the help of geometry and mathematics, has guided much of this work and will continue to do so in the future.

11

LEARNING TO PROGRAM THE FEM WITH MATLAB AND GID

written by Francisco Zárate[1]

11.1 INTRODUCTION

As for any other numerical method, the application of the FEM is linked to the programming language and software tools chosen. Historically the first programming language for practical use of the FEM was FORTRAN. Since then many routines, algorithms and programs associated to the method have been programmed in this language. With the development of computers new languages have appeared, each one with capabilities and specific tools for diverse fields of application. The common objective is to simplify the coding of the algorithms and to optimize the computer resources.

Although FORTRAN continues being a language of reference for the FEM, the new languages and programming tools allow simplifications in the coding work. At the same time specific libraries can be used that optimize the memory and computer resources. This is a key feature of MATLAB that besides being a research tool, it allows us to write codes that it can be interpreted at the time of execution. From an optimal programming point of view, interpretive languages are quite slow. However, MATLAB allows us to make use of all the implemented matrix routines for optimizing the calculations up to the point to compete efficiently with other compiled languages.

MATLAB is a software code designed to work with matrices, facilitating the matrix algebra operations from the numerical and storage points of view, while providing also a simple and easy way to handle complex routines.

[1] Dr. F. Zárate can be contacted at zarate@cimne.upc.edu

Having an efficient analysis code is not the only requirement to work with the FEM. It is necessary to rely on a suitable interface to prepare the analysis data, to generate meshes adapted to the kind of problem to be solved and to display the results so that their interpretation is clear and simple. MATLAB is very efficient in the treatment of matrices but quite poor in graphical capacities. An ideal complement to MATLAB is the pre/postprocessor program GiD (www.gidhome.com and Appendix D).

GiD is a tool designed to treat any geometry via CAD and to easily assign to it the data needed for FE computations, i.e. material properties, boundary conditions, loads, etc. Different efforts, such as the discretization and data writing levels in a pre-defined format become a transparent task for the user with GiD.

Data processing by means of GiD is also a simple task. The easy visualization of the analysis data and the numerical results allows one concentrating in their interpretation.

MAT-fem has been written thinking on the close interaction of GiD with MATLAB for FEM analysis. GiD allows manipulating geometries and discretizations, writing the input data files required by MATLAB. The calculation program is executed in MATLAB without losing any of the MATLAB advantages. Finally GiD gathers the output data files for graphical visualization and interpretation.

This scheme allows us understanding the development and application of a FEM program in detail, following step by step each one of the code lines if desired, and making it possible to solve examples that by their dimensions would fall outside the capabilities of any program with educational aims.

In the following sections the MAT-fem program is described in some detail. The description starts with the input data file instructions, automatically generated by GiD, and it follows with the information to understand the operations within MAT-fem.

Finally, the user interface implemented in GiD is described by means of an example of application.

11.2 MAT-fem

MAT-fem is a top-down execution program. The program flow chart is shown in Figure 11.1. The input data module is implemented in the same file were the data is defined, as described in the next section.

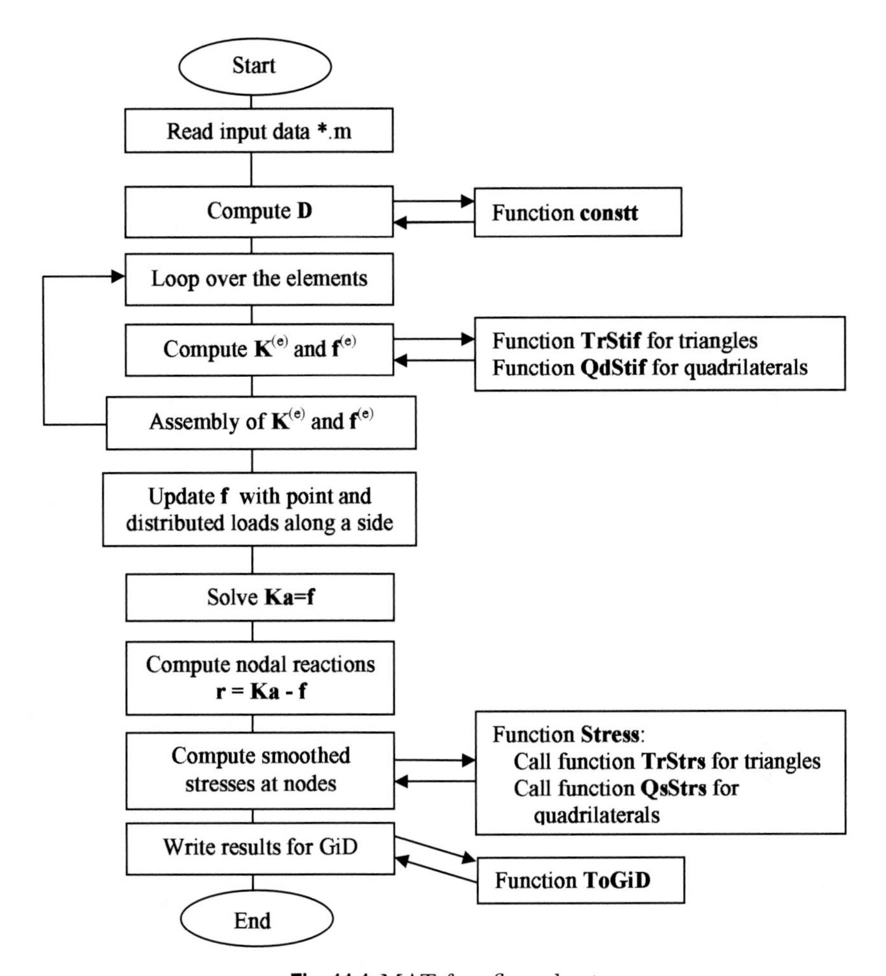

Fig. 11.1 MAT-fem flow chart

We consider that all elements have the same material properties so the constitutive matrix is evaluated outside the loop over the elements within which the element stiffness matrix and the body force vector are evaluated.

To save memory the element stiffness matrix and the equivalent nodal force vector are assembled as they are evaluated for each element.

Outside the element loop the equivalent nodal force vector is updated with the nodal point forces and the distributed loads acting along a side.

Once the unknown DOFs are found, the program evaluates the nodal reactions at the prescribed nodes and the smoothed stresses at the nodes.

The final step is the writing of the numerical results to visualize them in GiD.

Because the aim of MAT-fem is to show how a FEM code works just two elements are implemented. The 3-noded triangular element for which the stiffness matrix is programmed in an explicit form, and the 4-noded quadrilateral element for which the stiffness matrix is computed by means of numerical integration.

11.3 DATA FILES

Before executing MAT-fem it is necessary to feed it with information on the nodal coordinates, the element discretization, the boundary conditions, the material properties and the loading. In the following, the input data file is described in order to become familiar with the programming style and the variables used. As previously mentioned, MATLAB is an interpreter code and we will use this property to define the input data. This means that the input data file is in fact a subroutine of the program in where the values corresponding to the problem are assigned directly to the variables.

This avoids having to define a special reading syntax for the program and the need to implement an I/O interface.

The input data file uses MATLAB syntax. The program variables are defined directly in that file. The name of the file will take the MATLAB extension `.m`.

Inside the data file we distinguish three groups of variables: those associated to the material properties, those defining the topology of the problem and those defining the boundary conditions. With the intention of simplifying the code, an isotropic linear elastic material for the whole domain is used. Hence the material data appears only once in the data file.

Figure 11.2 shows the variables associated to the material data: `pstrs` indicates a plane stress (`pstrs = 1`) or plane strain (`pstrs =0`) problem. `young` contains the Young modulus and `poiss` the Poisson's ratio. `thick` and `denss` define the thickness of the domain and the density of the material, respectively. For a plane strain problem the thickness value is the unity, as usual.

It is important to note that the program is free of data validation mechanisms. Hence we will not check up aspects such the Poisson's ratio rank ($0 <=$ `poiss` < 0.5) and others. The reason is that these kind of

```
%
% Material Properties
%
 pstrs  = 1;
 young  = 1000.0;
 poiss  = 0.2;
 thick  = 0.1;
 denss  = 1.0;
```

Fig. 11.2 Input data file. Definition of material properties

```
%
% Coordinates
%
global coordinates
coordinates = [
  0.00 , 0.00;
  0.50 , 0.00;

  2.00 , 1.00;
  2.50 , 1.00  ];
%
% Elements
%
global elements
elements = [
    1,   2,   7 ;
    2,   3,   8 ;

   17,  16,  11 ;
   18,  17,  12 ];
```

Fig. 11.3 Input data file: topology definition

details, although they are important in practice, would hide the core of the FEM algorithm.

The variable group that describes the problem topology is defined with the attribute of a global variable to be accessible within the code by any subroutine. Figure 11.3 shows the definition of the coordinates and the nodal connectivities for each element by means of the variables `coordinates` and `elements`.

`coordinates` is a matrix with as many rows as nodes in the mesh and columns as the number of dimensions of the problem (i.e. 2 for a 2D problem). This variable lists the coordinates for all the nodes in the mesh. The number of any node corresponds to the position that its coordinates have in the `coordinates` matrix, i.e. node number 25 has the position 25 in `coordinates`.

Matrix `elements` defines the number of elements and their nodal connectivities. Each element has as many rows as number of elements in the

```
%
% Fixed Nodes
%
fixnodes = [
    1, 1, 0.0 ;
    1, 2, 0.0 ;

   13, 1, 0.0 ;
   13, 2, 0.0 ];
%
% Point loads
%
pointload = [
             6, 2, -1.0 ;

            18, 2, -1.0 ];
%
% Side loads
%
sideload = [
            11,12, 2.0, 3.0;

            14,15, 2.0, 3.0 ];
```

Fig. 11.4 Input data file: Boundary conditions definition

mesh and columns as number of nodes (`nelem` $\times$ `nnode`). Three-noded triangular elements have three columns in `elements` while 4-noded quadrilateral elements have four. The number of an element corresponds with the row number where its nodes are stored in `elements`.

The last group of variables defines the boundary conditions of the problem, as shown in Figure 11.4.

The `fixnodes` matrix defines the DOFs prescribed for the particular problem to be solved. `fixnodes` is a matrix where the number of rows corresponds to the number of prescribed DOFs and the number of columns describes in the following order: the prescribed node number, the fixed DOF code (1 if the node is fixed in the x direction and 2 if it is fixed in the y direction) and the prescribed DOF value. In this way if a node is prescribed in both directions two lines are necessary to define this condition.

The `pointload` matrix is used to define nodal point loads. As for the previous variables, this is a matrix where the number of rows is the number of point loads acting on the structure and each of the three columns describes the number of the loaded node, the direction in which the load acts and the magnitude of the load. Point loads are defined in the global system of coordinates. If there are no point loads, `pointload` is defined as an empty matrix by means of the command `pointload = []`;

```
%% MAT-fem
%
% Clear memory and variables.
  clear

  file_name = input('Enter the file name :','s');

  tic;                      % Start clock
  ttim = 0;                 % Initialize time counter
  eval (file_name);         % Read input file

% Find basic dimensions
  npnod  = size(coordinates,1);        % Number of nodes
  nndof  = 2*npnod;                    % Number of total DOF
  nelem  = size(elements,1);           % Number of elements
  nnode  = size(elements,2);           % Number of nodes per element
  neleq  = nnode*2;                    % Number of DOFs per element

  ttim = timing('Time needed to read the input file',ttim);
```

Fig. 11.5 Program initialization and data reading

Finally, `sideload` contains the information for uniformly distributed loads acting on the element sides. `sideload` is a matrix with as many rows as the number of loaded element sides. The first two columns define the nodes on the loaded side and columns three and four list the values of the distributed load by unit length in directions x and y, respectively. If no uniform loads act `sideload` is defined as an empty matrix by means of the command `sideload = [];`

The name of the data file is up to the user; nevertheless, the extension must be `.m` so that MATLAB can recognize it.

11.4 START

MAT-fem begins making all variables equal to zero with the `clear` command. Next it asks the user the name of the input data file that he/she will use (the `.m` extension in not included in the filename). Figure 11.5 shows the first lines of the code corresponding to the variables boot as well as the clock set up, which stores the total time of execution in `ttim`.

Data reading, as previously said, is a direct variable allocation task in the program. From the data matrices it is possible to extract the basic dimensions of the problem, such as the number of nodal points, `npnod`, which corresponds to the number of lines in the `coordinates` matrix. The number of total DOFs of the problem, `nndof`, will be twice the number of nodes (`2*npnod`). `nelem` is the number of elements and is equal to the number of lines in the `elements` matrix, whereas the number of nodes for

```
% Dimension the global matrices.
  StifMat = sparse ( nndof , nndof );  % The global stiffness matrix
  force   = sparse ( nndof , 1 );      % The global equivalent nodal force vector

%  Material properties (Constant over the domain).
  dmat = constt(young,poiss,pstrs);
```

Fig. 11.6 Initialization of the global stiffness matrix and the equivalent nodal force vector

each element `nnode` is the number of columns in `elements`. In this way triangular elements are identified if the number of columns in `elements` is three, whereas the number of columns will be four for quadrilateral elements.

The total number of equations per element, `neleq`, is the number of element nodes `nnode` multiplied by the number of DOFs for each node, i.e. two for 2D problems.

Note that these variables are defined in the data structure, which simplifies the code interpretation.

Throughout the program the `timing` routine is used to calculate the run time between two statements in the code. In this way the user can check the program sections that require higher computational effort. Inside `timing` the `tic` and `toc` MATLAB commands are used.

11.5 STIFFNESS MATRIX AND EQUIVALENT NODAL FORCE VECTOR FOR SELF-WEIGHT

11.5.1 Generalities

The code lines shown in Figure 11.6 define the global stiffness matrix and the equivalent nodal force vector as a `sparse` matrix and vector, respectively. MAT-fem uses sparse matrices to optimize the memory using MATLAB tools. In this manner and without additional effort, MAT-fem makes use of very powerful algorithms without losing its simplicity.

As the program's main purpose is to demonstrating the implementation of the FEM, some simplifications are made like using a single material for the whole domain. Consequently, the constitutive matrix does not vary between adjacent elements and it is evaluated before initiating the computation of the element stiffness matrix.

The subroutine `constt` makes use of the Young modulus, the Poisson's ratio and a flag that allows us to distinguish between a plane stress and

```
function D = constt (young,poiss,pstrs)

%  Plane Stress
   if (pstrs==1)
       aux1 = young/(1-poiss^2);
       aux2 = poiss*aux1;
       aux3 = young/2/(1+poiss);
%  Plane Strain
   else
       aux1 = young*(1-poiss)/(1+poiss)/(1-2*poiss);
       aux2 = aux1*poiss/(1-poiss);
       aux3 = young/2/(1+poiss);
       thick= 1.0;
  end

   D = [aux1,aux2,0;aux2,aux1,0;0,0,aux3];
```

Fig. 11.7 Constitutive matrix definition

a plane strain problem to form the constitutive matrix which is stored in `dmat` as shown in Figure 11.6. In Figure 11.7 the subroutine `constt` shows the explicit form of `dmat` for an isotropic linear elastic material.

MAT-fem recalculates the values for each variable instead of storing them. The recalculation process is performed in a fast manner and does not reduce significantly the program's efficiency. This leaves more memory for solving larger problems.

Figure 11.8 shows the element loop within which the stiffness matrix and the equivalent nodal vector for self-weight are calculated and assembled for each element. The loop begins recovering the geometrical properties for each element. Vector `lnods` stores the nodal connectivities for the element and the `coord` matrix stores the coordinates for these nodes.

In the next step the element stiffness matrix is calculated. The subroutines `TrStif` and `QdStif` are called for triangular and quadrilateral elements, respectively. The same subroutine evaluates the element stiffness matrix and the equivalent nodal force vector for the element. The use of the same integration quadrature allows this simplification. The calculation of the element stiffness matrix for each one of the two elements considered is detailed in the following section.

Before the assembly of the equations vector `eqnum` is defined. It contains the global equations number for each one of the equations in the element stiffness matrix. The number conversion is simple because two equations correspond to each node (one for each DOF).

The equations assembly process is implemented by means of two loops from `1` to `neleq` (number of equations for each element). In the first loop the equivalent nodal force vector is assembled and in the second one the

```
%  Element loop.
  for ielem = 1 : nelem

% Recover element properties
    lnods = elements(ielem,:);                           % connectivity
    coord(1:nnode,:) = coordinates(lnods(1:nnode),:); % coordinates

% Evaluate the element stiffness matrix and the equivalent nodal force vector.
    if (nnode == 3)
      [ElemMat,ElemFor] = TrStif(coord,dmat ,thick,denss); % Triangle
    else
      [ElemMat,ElemFor] = QdStif(coord,dmat ,thick,denss); % Quadrilateral
    end

% Find the equation number list for the i-th element
    eqnum = [];                                     % Clear the list
    for i =1 : nnode                                % Node loop
      eqnum = [eqnum,lnods(i)*2-1,lnods(i)*2];   % Build the equation
    end                                             % Number list

% Assemble the equivalent nodal force vector and the stiffness matrix for each
element
    for i = 1 : neleq
      force(eqnum(i)) = force(eqnum(i)) + ElemFor(i);
      for j = 1 : neleq
         StifMat(eqnum(i),eqnum(j)) = StifMat(eqnum(i),eqnum(j)) + ...
                                      ElemMat(i,j);
      end
    end

  end  % End element loop
```

Fig. 11.8 Evaluation and assembly of the stiffness matrix and the equivalent nodal force vector

element stiffness matrix is assembled term by term. This scheme avoids storing the element matrices temporarily.

11.5.2 Computation and assembly of $\mathbf{K}^{(e)}$ and $\mathbf{f}^{(e)}$ (self-weight) for 3-noded triangles and 4-noded quadrilaterals

The stiffness matrix for the 3-noded triangle is calculated in explicit form while that for the 4-noded quadrilateral is calculated by numerical integration. Both routines require exactly the same input data and also return the same variables: `ElemMat` for the stiffness matrix and `ElemFor` for the equivalent nodal force vector.

The `TrStif` subroutine (called as a function) is defined for triangular elements as shown in Figure 11.9. Note that the Cartesian derivatives of the linear shape functions are calculated directly. These derivatives are constant over the element. In this way the strain matrix `bmat` is built simply by placing each of the Cartesian derivatives (`b(i)/area2` and `c(i)/area2`) in the adequate position of the strain matrix (Eq.(4.39)).

```
3-noded triangle: stiffness matrix and equivalent node force vector
(self weight)

function [M,F] = TrStif (nodes,dmat,thick,denss)

  b(1) = nodes(2,2) - nodes(3,2);     % bi = yj - yk
  b(2) = nodes(3,2) - nodes(1,2);
  b(3) = nodes(1,2) - nodes(2,2);

  c(1) = nodes(3,1) - nodes(2,1);     % ci = xk - xj
  c(2) = nodes(1,1) - nodes(3,1);
  c(3) = nodes(2,1) - nodes(1,1);

  area2 = abs(b(1)*c(2) - b(2)*c(1));
  area = area2 / 2;

  bmat = [b(1),  0 ,b(2),  0 ,b(3),  0 ;        % Matrix B
           0 ,c(1),  0 ,c(2),  0 ,c(3);
          c(1),b(1),c(2),b(2),c(3),b(3)];

  bmat = bmat / area2;

  M = (transpose(bmat)*dmat*bmat)*area*thick;  % Element stiffness matrix

  force = area*denss*thick/3;
  F = [0,-force,0,-force,0,-force]; % Element equivalent nodal force vector
```

Fig. 11.9 Stiffness matrix and equivalent nodal force vector for self-weight loading for the 3-noded triangular element

The element stiffness matrix is calculated by the classic expression `B`^{T}`DB dA` and stored in matrix `M`. Here one of the fundamental advantages of MATLAB is observed as the multiplication of matrices is performed by means of a single instruction, eliminating the need for writing troublesome loops.

The equivalent nodal force vector for self-weight loading is computed and stored in vector `F` (Figure 11.9). As the gravity is defined in the opposite direction of the y axis, this yields negative nodal forces in the y direction.

Figure 11.10 shows the evaluation of the stiffness matrix for the quadrilateral element which is performed using numerical integration. Initially the element shape functions and their natural derivatives are defined (`fform` and `deriv`). This definition is made by means of an intrinsic function. This facility of MATLAB avoids the use of additional subroutines.

At this level also the Gauss point coordinate values (`pospg`) and their weights (`pespg`) corresponding to the 2×2 integration rule are defined. The loop over the Gauss points for computing the stiffness matrix and the equivalent nodal load vector for the element are also shown in Figure 11.10.

```
function [M,F] = QdStif ( nodes,dmat,thick,denss)

   fform = @(s,t)[(1-s-t+s*t)/4,(1+s-t-s*t)/4,(1+s+t+s*t)/4,(1-s+t-s*t)/4];
   deriv = @(s,t)[(-1+t)/4,( 1-t)/4,( 1+t)/4,(-1-t)/4 ;
                  (-1+s)/4,(-1-s)/4,( 1+s)/4,( 1-s)/4 ];

   pospg = [ -0.577350269189626E+00 , 0.577350269189626E+00 ];
   pespg = [  1.0E+00 ,  1.0E+00];
   M = zeros(8,8);
   fy = zeros(1,4);

   for i=1 : 2
      for j=1 : 2
         lcffm = fform(pospg(i),pospg(j)) ;     % SF at gauss point
         lcder = deriv(pospg(i),pospg(j)) ;     % SF Local derivatives
         xjacm = lcder*nodes ;                  % Jacobian matrix
         ctder = xjacm\lcder ;                  % SF Cartesian derivates
         darea = det(xjacm)*pespg(i)*pespg(j)*thick;

         bmat = [];
         for inode = 1 : 4
           bmat = [ bmat , [ctder(1,inode),            0 ;
                                         0 ,ctder(2,inode);
                           ctder(2,inode),ctder(1,inode) ] ] ;
         end

         M = M + (transpose(bmat)*dmat*bmat)*darea;

         fy = fy + lcffm*denss*darea;

      end
   end

   F = [ 0, -fy(1), 0, -fy(2), 0, -fy(3), 0, -fy(4)];
```

Fig. 11.10 Stiffness matrix and equivalent nodal force vector for the 4-noded quadrilateral element

Once the variables are initialized two loops define the Gauss integration process. The `lcffm` vector contains the values of the shape functions evaluated at the integration point `i,j` and the `lcder` matrix stores the values for the natural derivatives of the shape functions at each integration point. The Jacobian matrix (`xcjacm`) is evaluated by multiplication of the `lcder` matrix by the coordinates of the element nodes. The values of the Cartesian derivatives of the shape functions (`ctder`) at each integration point are obtained by multiplying the inverse of the Jacobian matrix by the natural derivatives of the shape functions computed at the integration point. The area differential (`darea`) is computed as the determinant of the Jacobian matrix multiplied by the two weighting functions at the integration point and the element thickness.

The strain matrix is obtained by placing the Cartesian derivates of the shape functions in the matrix array `bmat`. The element stiffness matrix

```
% Add side forces to the force vector
  for i = 1 : size(sideload,1)
     x=coordinates(sideload(i,1),:)-coordinates(sideload(i,2),:);
     l = sqrt(x*transpose(x));        % Finds the length of the side
     ieqn = sideload(i,1)*2;          % Finds eq. number for the first node
     force(ieqn-1) = force(ieqn-1) + l*sideload(i,3)/2;    % add x force
     force(ieqn  ) = force(ieqn  ) + l*sideload(i,4)/2;    % add y force

     ieqn = sideload(i,2)*2;          % Finds eq. number for the second node
     force(ieqn-1) = force(ieqn-1) + l*sideload(i,3)/2;    % add x force
     force(ieqn  ) = force(ieqn  ) + l*sideload(i,4)/2;    % add y force
  end
```

Fig. 11.11 Equivalent nodal force vector for a uniform distributed load acting on the element sides

is obtained by integrating the standard `B^TDB` expression. Numerical integration requires evaluating the sum of the product of `B^TDB` and `darea` calculated at all the Gauss points. This sum is stored in matrix `M` (Figure 11.10).

The equivalent nodal force vector for the self-weight case requires integrating over the element area the product of the shape functions and the specific weight for the element. The vector is stored in the variable `fy`.

Finally the equivalent nodal force components are placed in the `F` vector, recalling that only the negative force component along the y axis exists.

The routines shown in Figures 11.9 and 11.10 clearly demonstrate the two alternatives for computing the element stiffness matrix using the explicit form (3-noded triangle) or by means of numerical integration (4-noded quadrilateral).

11.6 EXTERNAL LOADS

Besides the self-weight load we consider uniformly distributed loads acting on the element sides and nodal point loads.

As both the elements considered have linear shape functions, the calculation of the nodal contribution for uniformly distributed side loads is exactly the same for both cases. The evaluation is made in the main routine of the MAT-fem program after the assembly of the stiffness matrix. The code is shown in Figure 11.11 where the loop over the number of loads defined by `sideload` can be seen. The nodal contributions are the same for each node (due the linear shape functions) and they are stored in the `force` variable.

```
% Add point loads to the global equivalent nodal force vector
  for i = 1 : size(pointload,1)
    ieqn = (pointload(i,1)-1)*2 + pointload(i,2);        % Finds eq. number
    force(ieqn) = force(ieqn) + pointload(i,3);          % add the force
  end
```

Fig. 11.12 Equivalent nodal force vector for point loads

```
% Apply prescribed displacement conditions and adjust the right hand side.
  u = sparse ( nndof, 1 );
  for i = 1 : size(fixnodes,1)
    ieqn = (fixnodes(i,1)-1)*2 + fixnodes(i,2);  %Finds the equation number
    u(ieqn) = fixnodes(i,3);                     %and store the solution in u
    fix(i) = ieqn;                             % and mark the eq as a fix value
  end
  force = force - StifMat * u;  % adjust the rhs with the prescribed values
```

Fig. 11.13 Updating of the equivalent nodal force vector due to the prescribed DOFs

We recall that the loads are defined in the global coordinate system.

The calculation steps for nodal point loads are as simple as adding the value of the point load acting at the node to the global equivalent nodal force vector in the position corresponding to the adequate DOF of the loaded node. A loop over the number of nodal point loads is implemented, finding for each node the equation number associated to it and adding the value of the point load to the `force` vector (Figure 11.12).

11.7 PRESCRIBED DISPLACEMENTS

Figure 11.13 shows the loop over the prescribed displacement DOFs and how the values defined by the `fixnodes` matrix are assigned to the nodal displacement vector `u`. Also the `fix` vector is defined to store the equation numbers for the prescribed DOFs.

Finally the `force` vector is updated with the product of the `StifMat` matrix and the `u` vector following the procedure described in Section 1.10. Vector `u` at this moment contains the values of the prescribed DOFs only.

11.8 SOLUTION OF THE EQUATIONS SYSTEM

The strategy used in MAT-fem basically consists in solving the global equation system without considering those DOFs whose values are known

```
% Compute the solution by solving StifMat * u = force for the
% remaining unknown values of u.
  FreeNodes = setdiff ( 1:nndof, fix ); % Finds the free nodes list and
                                        % solve for it.
  u(FreeNodes) = StifMat(FreeNodes,FreeNodes) \ force(FreeNodes);
```

Fig. 11.14 Solution of the equations system

```
% Compute the reactions at the fixed nodes as a R = StifMat * u - F
  reaction = sparse(nndof,1);
  reaction(fix) = StifMat(fix,1:nndof) * u(1:nndof) - force(fix);
```

Fig. 11.15 Computation of nodal reactions

(i.e. prescribed). The `FreeNodes` vector contains the list of the equations to be solved (Figure 11.14).

The `FreeNodes` vector is used as a DOF index and allows us to write the solution of the equations system in a simple way. MATLAB takes care of choosing the most suitable algorithm to solve the system. The solution step is totally transparent for the user. The routines implemented in the MATLAB kernel nowadays compete in speed and memory optimization with the best existing algorithms.

11.9 NODAL REACTIONS

The solution to the equations system is stored in the `u` vector containing the nodal displacements (Figure 11.14). Nodal reactions are computed by means of the expression: `reaction = StifMat*u - force`. Obviously the value of the reactions at the prescribed nodes is not zero. In order to avoid unnecessary calculations we use vector `fix` which contains the list of the equations associated to the prescribed DOFs as shown in Figure 11.15.

11.10 STRESSES

11.10.1 Generalities

Once the nodal displacements have been found it is possible to evaluate the stresses in the elements by means of the `DBu` expression. Since the strain matrix `B` was previously computed at the integration points, the stresses are also computed at these points which are also optimal for evaluation of stresses (Section 6.7). The next step is to transfer the values of the stresses from the integration points to the element nodes (Section 9.8).

```
% Compute the stresses
  Strnod = Stress(dmat,poiss,thick,pstrs,u);
```

Fig. 11.16 Call for evaluating the nodal stresses

```
function S = Stress (dmat,poiss,thick,pstrs,u)

%% Evaluates the stresses at the gauss points and smooth the values
%          to the nodes.
%
  global coordinates;
  global elements;

  nelem  = size(elements,1);          % Number of elements
  nnode  = size(elements,2);          % Number of nodes per element
  npnod  = size(coordinates,1);       % Number of nodes

  if (pstrs==1)
      nstrs= 3;                       % Number of Strs. Sx Sy Txy
  else
      nstrs= 4;                       % Number of Strs. Sx Sy Sz Txy
  end
  nodstr = zeros(npnod,nstrs+1);
```

Fig. 11.17 Variables boot for computation of stresses

Figure 11.16 shows the call for the subroutine for computing the nodal stresses which are stored in the `Strnod` matrix.

11.10.2 Computation of the stresses at the nodes

The stress calculation in the elements requires the use of a specific subroutine, not only for the stress computation itself but also to project the stresses from the integration points to the nodes.

The stress subroutine controls the program flow towards the element routines. For the 3-noded triangular element the stresses are constant and nodal extrapolation is trivial. This is not the case for the 4-noded quadrilateral element where the stresses have a bilinear variation and the stress extrapolation is performed using the element shape functions.

Figure 11.17 presents the initial part of the stress subroutine were the input data are: the material constitutive matrix `dmat`, the Poisson's ratio `poiss`, the thickness `thick`, the flag for the problem type `pstrs` and the nodal displacements `u`. Additionally the nodal coordinates and the element connectivities will be used (defined as global variables). In order to simplify the reading of the routine some variables are extracted like `nelem` indicating the number of elements, `nnode` indicating the number of

```
   for ielem = 1 : nelem

% Recover element properties
     lnods = elements(ielem,:);
     coord(1:nnode,:) = coordinates(lnods(1:nnode),:);
     eqnum = [];
     for i =1 : nnode
       eqnum = [eqnum,lnods(i)*2-1,lnods(i)*2];
     end
     displ = u(eqnum);
```

Fig. 11.18 Recovering the element coordinates and the nodal displacements

nodes per element and `npnod` defining the total number of nodes in the mesh.

We recall the number of stresses to be computed for plain stress problems (σ_x, σ_y and τ_{xy}) and plane strain problems (σ_x, σ_y, σ_z and τ_{xy}) were σ_z is a function of σ_x and σ_y.

The `nodstr` matrix is initialized to a null matrix to store the nodal stresses. In the last column the number of elements that share a node are stored. This is necessary for the nodal averaging of the stresses.

Similarly as for computing the stiffness matrix, the stress evaluation requires a loop over the elements, recovering the element connectivities (`lnods`), the coordinates for these nodes (`coord`) and the nodal displacements `displ`, as shown in Figure 11.18.

The computation of the nodal stresses for the triangular element is shown in Figure 11.19. Note that the `B` matrix is recalculated. The stresses at the center of the element (a single Gauss point is used) are directly computed by the `DBu` product.

Depending on the problem selected, the computed stresses are three for plane stress and four for plane strain (with $\sigma_z = -\nu(\sigma_x + \sigma_y)$) as shown in the same figure.

The computation of the nodal stresses for the 4-noded quadrilateral follows the direct nodal extrapolation procedure explained in Section 9.8.2. The nodal value for each stress component σ is obtained by Eq.(9.61) as

$$\sigma_j = \sum_{i=I}^{IV} N_i(s_j, t_j)\sigma_i \qquad j = 1, 4 \tag{11.1}$$

where σ_j is the value of the stress at the jth node (j is the local number of the node), σ_i is the value of the stress component at each Gauss point and the coordinates s and t range from $1/p$ to $-1/p$ for the four element nodes as shown in Figure 11.20 (see also Figure 9.16).

```
function S = TrStrs (nodes,dmat,displ,poiss,thick,pstrs)

  b(1)  = nodes(2,2)  - nodes(3,2);
  b(2)  = nodes(3,2)  - nodes(1,2);
  b(3)  = nodes(1,2)  - nodes(2,2);

  c(1)  = nodes(3,1)  - nodes(2,1);
  c(2)  = nodes(1,1)  - nodes(3,1);
  c(3)  = nodes(2,1)  - nodes(1,1);

  area2 = abs(b(1)*c(2) - b(2)*c(1));
  area = area2 / 2;

  bmat = [b(1),  0 ,b(2),  0 ,b(3),  0 ;
            0 ,c(1),  0 ,c(2),  0 ,c(3);
          c(1),b(1),c(2),b(2),c(3),b(3)];

  se = (dmat*bmat*displ)/area2;
%  Plane Stress
  if (pstrs==1)
       S = se ;
%  Plane Strain
   else
       S = [se(1),se(2),-poiss*(se(1)+se(2)),se(3)];
 end
```

Fig. 11.19 Computation of stresses at the center of the element for the 3-noded triangle

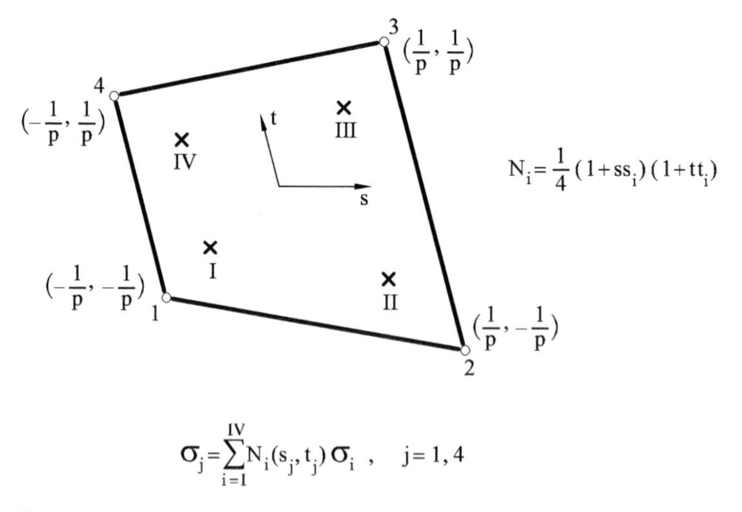

$$\sigma_j = \sum_{i=I}^{IV} N_i(s_j, t_j)\,\sigma_i \quad , \quad j = 1,4$$

Fig. 11.20 Extrapolation of the Gauss point stresses to the nodes for a 4-noded quadrilateral

Figure 11.21 shows the code for computing the nodal stresses for the 4-noded quadrilateral. The stresses are computed first at the Gauss points

```
function S = QdStif (nodes,dmat,displ,poiss,thick,pstrs)

    fform = @(s,t) [(1-s-t+s*t)/4,(1+s-t-s*t)/4,(1+s+t+s*t)/4,(1-s+t-
s*t)/4];
    deriv = @(s,t) [(-1+t)/4,( 1-t)/4,( 1+t)/4,(-1-t)/4 ;
                    (-1+s)/4,(-1-s)/4,( 1+s)/4,( 1-s)/4 ];

    pospg = [ -0.577350269189626E+00 , 0.577350269189626E+00 ];
    pespg = [  1.0E+00 , 1.0E+00];

    strsg = [];
    extrap = [];
    order = [ 1 , 4 ; 2 , 3 ]; % Align the Gauss points with the element
corners

    for i=1 : 2
       for j=1 : 2
          lcder = deriv(pospg(i),pospg(j)) ;     % SF Local derivatives
          xjacm = lcder*nodes ;                  % Jacobian matrix
          ctder = xjacm\lcder ;                  % SF Cartesian derivates

          bmat = [];
          for inode = 1 : 4
            bmat = [ bmat , [ctder(1,inode),              0 ;
                                       0 ,ctder(2,inode);
                         ctder(2,inode),ctder(1,inode) ] ] ;
          end

          strsg(:,order(i,j)) = (dmat*bmat*displ) ;

          a = 1/pospg(i);
          b = 1/pospg(j);

         extrap(order(i,j),:) = fform(a,b) ;
       end
    end

    se = transpose(extrap*transpose(strsg));
%   Plane Sress
  if (pstrs==1)
       S = se ;
%   Plane Strain
   else
       S = [se(1,:) ; se(2,:) ; -poiss*(se(1,:)+se(2,:)) ; se(3,:)];
   end
```

Fig. 11.21 Computation of nodal stresses for the 4-noded quadrilateral

using the **B** matrix and then they are extrapolated to the nodes using Eq.(11.1).

Like for the triangular element case, the stresses are calculated for plain stress or plane strain problems.

Figure 11.22 shows the general steps for computing the stresses at the nodes in a mesh of 3-noded triangles and 4-noded quadrilaterals by nodal averaging. For triangular elements the routine builds the `ElemStr`

```
% Computation of stresses at the nodes.
    if (nnode == 3)

% Triangular element
      ElemStr = TrStrs(coord,dmat,displ,poiss,thick,pstrs);

      for j=1 : nstrs
        nodstr(lnods,j) = nodstr(lnods,j) + ElemStr(j);
      end
      nodstr(lnods,nstrs+1) = nodstr(lnods,nstrs+1) + 1;

    else
% Quadrilateral element
      ElemStr = QdStrs(coord,dmat,displ,poiss,thick,pstrs);

      for j=1 : 4
        for i = 1 : nstrs
          nodstr(lnods(j),i) = nodstr(lnods(j),i) + ElemStr(i,j);
        end
      end
      nodstr(lnods,nstrs+1) = nodstr(lnods,nstrs+1) + 1;
    end
  end
% Find the mean stress value at the nodes

  S = [];
  for i = 1 : npnod
   S = [S ; nodstr(i,1:nstrs)/nodstr(i,nstrs+1)];
  end
```

Fig. 11.22 Computation of nodal stresses by nodal averaging

vector that contains the nodal stresses for the element. These values are accumulated in `nodstr` were the last column is the number of elements that share the node. This is needed in order to evaluate the nodal average of the stresses in a final stage as shown in the last lines of Figure 11.22.

For quadrilateral elements `ElemStr` is a matrix that contains the stresses at each one of the element nodes (computed as shown in Figure 11.21). Like in the previous case. The last column of `ElemStr` contains the number of elements that share the node. Once the stresses at all the element nodes have been calculated and accumulated in `nodstr`, a nodal averaging is performed to compute a smoothed stress field at the nodes.

11.11 POSTPROCESSING STEP

Once the nodal displacements, the reactions and the stresses have been calculated their values are transferred to the postprocessing files from where GiD will be able to display them in graphical form. This is performed in the subroutine `ToGiD` shown in Figure 11.23.

```
% Graphic representation.
  ToGiD (file_name,u,reaction,Strnod);
```

Fig. 11.23 Call for the postprocessing step

During the program execution the total time used by the program will appear in the MATLAB console as well as the time consumed in each subroutine. The largest time consumption in the academic problems solved with MAT-fem is invested in the calculation and assembly of the global stiffness matrix, whereas the solution of the equations system represents a small percentage of the consumed time. This is not typically the case for the solution of larger problems for real size structures.

Once the program execution is finished, the variables are still recorded inside MATLAB in order to experiment with the collection of internal functions available.

11.12 GRAPHICAL USER INTERFACE

11.12.1 Preprocessing

In this section the Graphical User Interface (GUI) implemented in GiD is described. In order to access the GUI is necessary to select from the GiD's DATA menu the module corresponding to MAT-fem in the option Problem Type. When selected, the image shown in Figure 11.24 appears.

All the GiD capabilities are part of the MAT-fem module: geometry generation, import and handling, as well as the GiD discretización techniques provide MAT-fem with capacities difficult to surpass for an educational code.

There is plenty of information on GiD available in Internet. We recommend visiting the GiD web site at www.gidhome.com. However, solving a problem with MAT-fem is very simple once the geometry has been defined. Just follow the icons of the MAT-fem graphical menu.

Figure 11.25 shows the graphical menu that appears when MAT-fem is activated. The first button works to identify the geometrical elements (point or lines) that have nodes with prescribed displacements. When pressing on, an emergent window will appear to select the points or lines where the displacements are prescribed (Figure 11.26) The check boxes identify the prescribed directions. Also it is possible to assign a non-zero value to the constraint.

Fig. 11.24 MAT-fem GUI start up page

Fig. 11.25 MAT-fem graphical menu

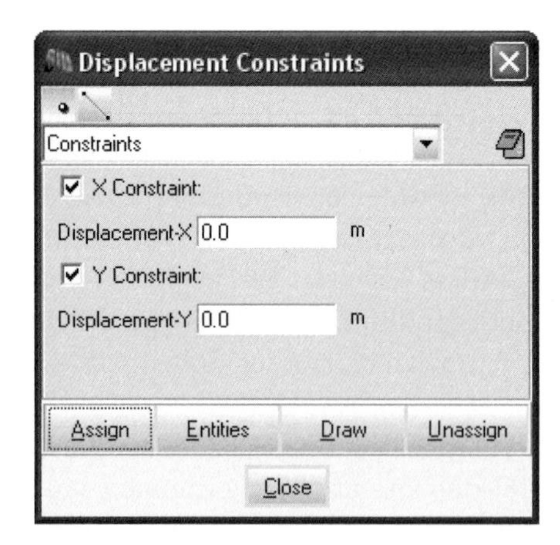

Fig. 11.26 Fixed displacement conditions

The second button shown in Figure 11.25 is used for point loads allocation. When selected, an emergent window (Figure 11.27a) allows introducing the point load value in the global coordinate system. Then it is necessary to select the nodes were the load is applied.

The third button is associated to uniformly distributed loads along the element sides and permits to assign this condition on geometry lines. The

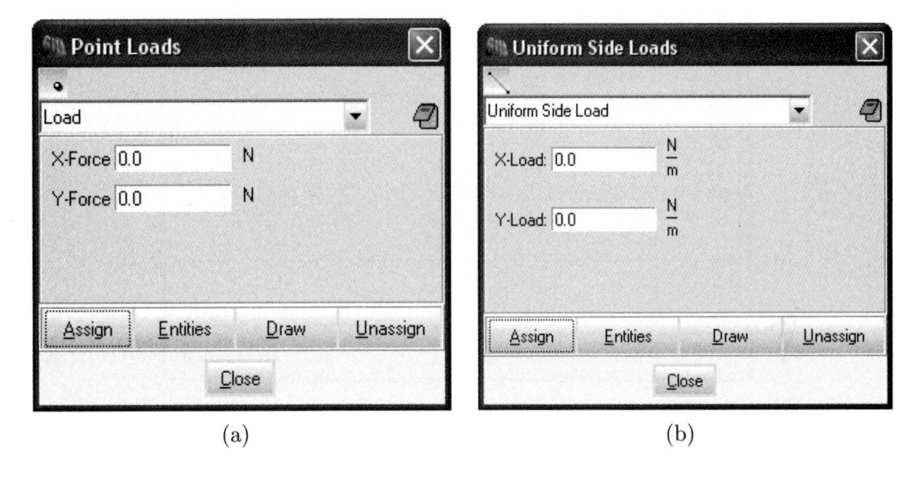

(a) (b)

Fig. 11.27 Point load condition

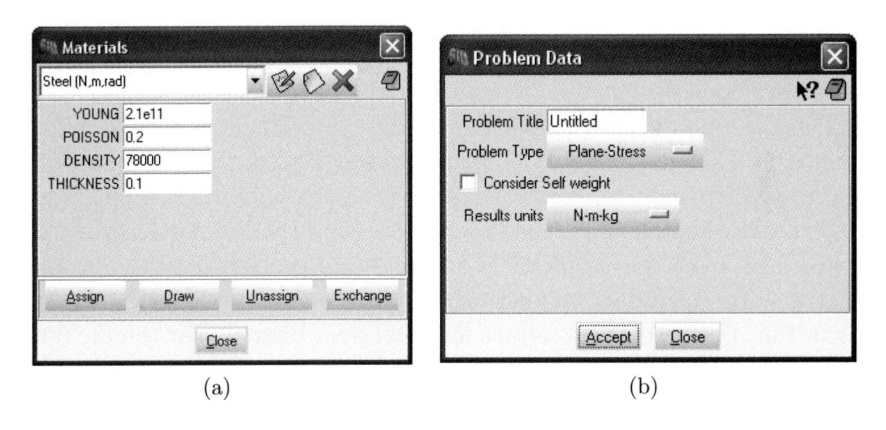

(a) (b)

Fig. 11.28 Material properties definition

emergent window (Figure 11.27b) allows introducing the value of the side load per unit length referred to the global coordinate system.

The material properties are defined with the fourth button in Figure 11.25 which leads to the emergent window shown in Figure 11.28a. This allows defining the material parameters like the Young modulus, the Poisson's ratio, the density and the thickness. It is necessary to assign these properties over the surfaces that define the domain. As mention earlier, only one type of material is allowed in MAT-fem for the sake of simplicity.

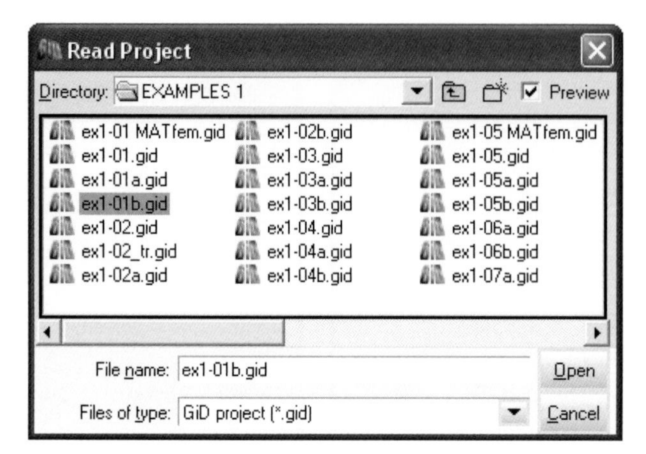

Fig. 11.29 Input data file definition

The general properties button (the fifth button in Figure 11.25) allows to access the window shown in Figure 11.28b were the title of the problem is defined as well as the problem type (plane stress or plane strain), the self-weight load option and the units for the results.

Once the boundary conditions and the material properties have been defined it is necessary to generate the mesh. The sixth button shown in Figure 11.25 is used to create the mesh with the GiD toolbox facilities.

The data file writing is made when pressing the last button shown in Figure 11.25. All the geometrical and material properties of the problem as well as the boundary conditions and the loads are written on the data file in the specific reading format for MAT-fem. Recall that the file name needs the `.m` extension as shown in Figure 11.29.

11.12.2 Program execution

The problem calculation is performed with MATLAB. The execution does not have other complications than knowing the directory where the output file will be written. A good practice is to set this directory as the working directory were the postprocessing file will be also written.

11.12.3 Postprocessing

Once the problem execution in MATLAB is concluded it is necessary to return to GiD for the file postprocessing step in order to analyze the

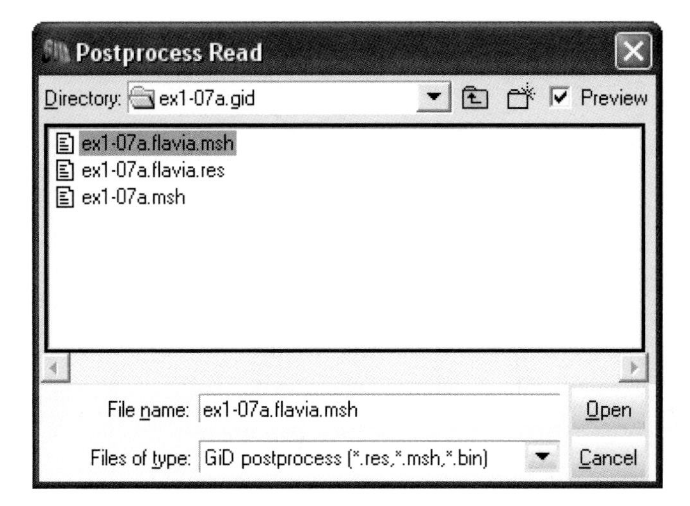

Fig. 11.30 Postprocessing file reading

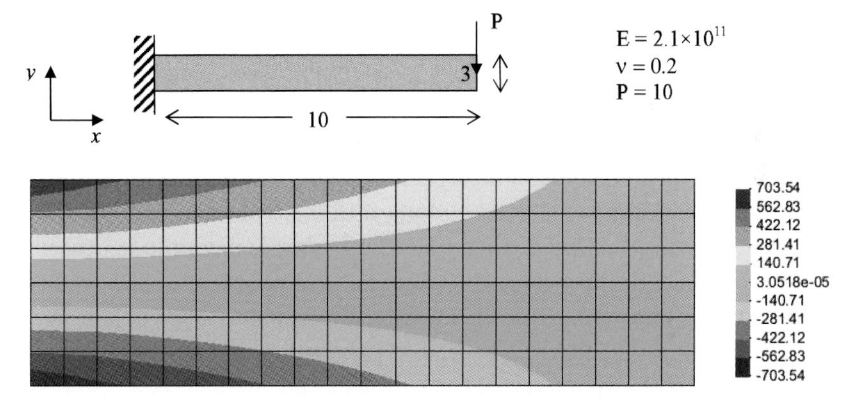

Fig. 11.31 Iso values of the stress σ_x for a plane beam

results. The next step is therefore to open any of the generated files that contain the extension *.flavia.msh or *.flavia.res.

The results visualization step is performed using the GiD graphical possibilities which permit to visualize the results by means of iso-lines, cuts and graphs. This facilitates the interpretation of the MAT-fem results.

Figure 11.31 shows an example showing the contours of the stress σ_x in a deep beam clamped at its left end and loaded by a vertical point load acting on its right end.

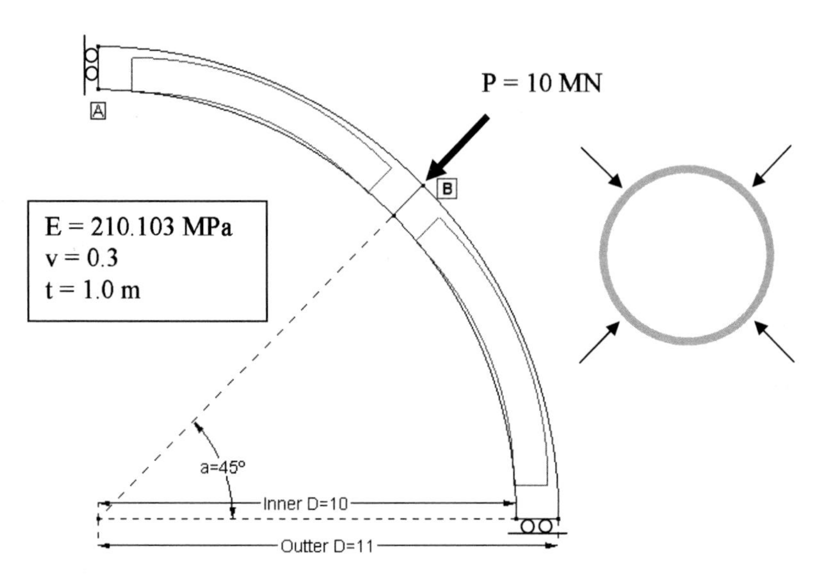

Fig. 11.32 Circular ring under point load. NAFEMS IC6 test [HKD]

11.13 EXAMPLE

An example is presented to show the simplicity of using MAT-fem for computing the displacements and stresses in a circular thick ring subjected to two opposite point loads acting along a diametral line.

In the following the input data file of this simple example is described in detail for quadrilateral elements. Also the convergence of the results is presented using triangular and quadrilateral elements.

The example is the NAFEMS IC6 benchmark proposed in [HKD]. Figure 11.32 shows the geometry of the problem and the material properties. The problem is solved under plain strain conditions. Only one quarter of the ring is analyzed due to symmetry.

The ring is fixed at both ends (the normal displacement is zero). The point load is applied at 45° in the center of the outer middle side. The goal is to compute σ_x at point A in the lower fibre of the upper clamped end (Figure 11.32). The exact value is $\sigma_x = -53.2$ MPa.

The problem definition for MAT-fem is done using the menu described in the GUI section. The prescribed displacements, point loads, material properties and boundary conditions are shown in Figures 11.33 and 11.34.

The finite element mesh and the input data files are created with the last two buttons of the MAT-fem menu.

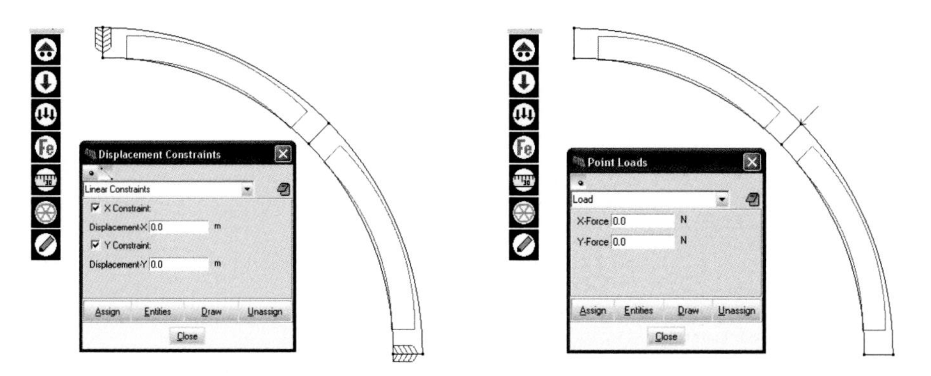

Fig. 11.33 Boundary conditions and loads over the ring

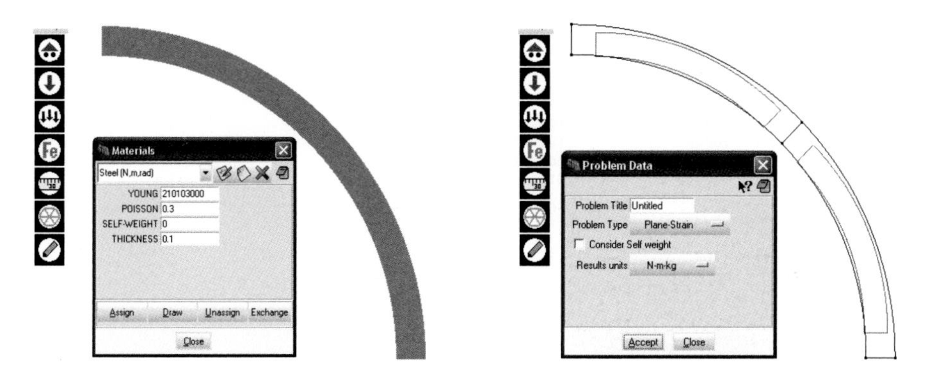

Fig. 11.34 Definition of material parameters and problem type

The problem has been solved first with different structured meshes of 4-noded quadrilateral elements. Figure 11.35 shows the numbering of the nodes and the elements for a mesh of 8 elements and 15 nodes. In the same figure the input data file is presented.

The file contains all the information necessary for the analysis: the nodal coordinates, the material properties, the coordinates, the element connectivity, the boundary conditions, the point load and the uniform side loads that are defined here by an empty matrix.

The program execution is performed with the MAT-fem command. Figure 11.36 shows that the larger time consumption is in the stiffness matrix assembly. This is due to the storage system where the internal indices of the sparse matrix must be updated. The total running time for this problem is about 0,15 seconds.

```
%=======================================================================
% MAT-fem 1.0  - MAT-fem is a learning tool for understanding
%                the Finite Element Method with MATLAB and GiD
%=======================================================================
% PROBLEM TITLE = NAFEMS IC6
%  Material Properties
%
  young =      210103000.00000 ;
  poiss =               0.30000 ;
  denss = 0.00 ;
  pstrs =  0 ;
  thick =  1 ;
%
% Coordinates
%
global coordinates
coordinates = [
        11.00000   ,        0.00000  ;
        10.50000   ,        0.00000  ;
        10.00000   ,        0.00000  ;
         9.23880   ,        3.82683  ;
         9.70074   ,        4.01818  ;
        10.16267   ,        4.20952  ;
         7.07107   ,        7.07107  ;
         7.42462   ,        7.42462  ;
         7.77817   ,        7.77817  ;
         3.82683   ,        9.23880  ;
         4.01818   ,        9.70074  ;
         4.20952   ,       10.16267  ;
         0.00000   ,       10.00000  ;
         0.00000   ,       10.50000  ;
         0.00000   ,       11.00000  ]                                  ;
%
% Elements
%
global elements
elements = [
      2   ,     5   ,     4   ,     3   ;
      1   ,     6   ,     5   ,     2   ;
      5   ,     8   ,     7   ,     4   ;
      6   ,     9   ,     8   ,     5   ;
      8   ,    11   ,    10   ,     7   ;
      9   ,    12   ,    11   ,     8   ;
     11   ,    14   ,    13   ,    10   ;
     12   ,    15   ,    14   ,    11   ] ;
%
% Fixed Nodes
%
fixnodes = [
      1  , 2 ,    0.00000  ;
      2  , 2 ,    0.00000  ;
      3  , 2 ,    0.00000  ;
     13  , 1 ,    0.00000  ;
     14  , 1 ,    0.00000  ;
     15  , 1 ,    0.00000  ] ;
%
% Point loads
%
pointload = [
      9  , 1 , -7071067.81200  ;
      9  , 2 , -7071067.81200  ] ;
%
% Side loads
%
sideload = [ ];
```

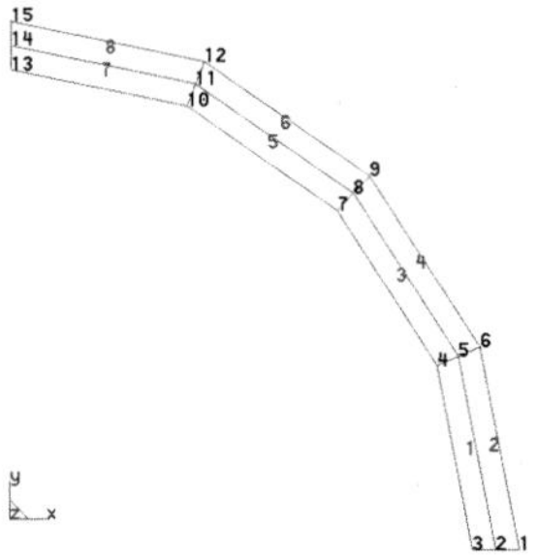

Fig. 11.35 Input data file for the circular thick ring problem

```
>> MATfem
Enter the file name :cilQ2
Time needed to read the input file        0.003399
Time needed to set initial values         0.000912
Time to assemble the global system        0.032041
Time for applying side and point loads    0.000196
Time for solving the stiffness equations  0.026293
Time to compute the nodal stresses        0.024767
Time used to write the solution           0.066823

 Total running time      0.154431
>>
```

Fig. 11.36 MATLAB console for running the circular ring problem

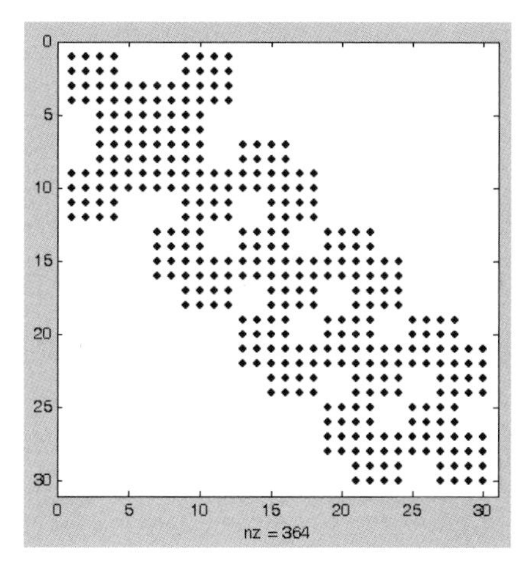

Fig. 11.37 Stiffness matrix profile

Some of the facilities given by MATLAB are shown in Figure 11.37 where the spy(StifMat) command allows to visualize the profile of the global stiffness matrix. With the aid of other commands it is possible to know the properties of this matrix, such as its rank, eigenvectors, determinant, etc.

Figure 11.38 shows some of the meshes of 3-noded triangles and 4-noded quadrilaterals used for the analysis. In Figure 11.39 the deformed mesh is presented as well as the distribution of the σ_x stress inside the ring for the simple mesh of 8 quadrilateral element.

For the coarse eight element mesh the σ_x stress at point A has a value of -14.186 MPa, far from the reference solution. The solution however converges to the target value as the number of elements (and DOFs) is increased, as shown in Table 11.1 where the convergence of the vertical dis-

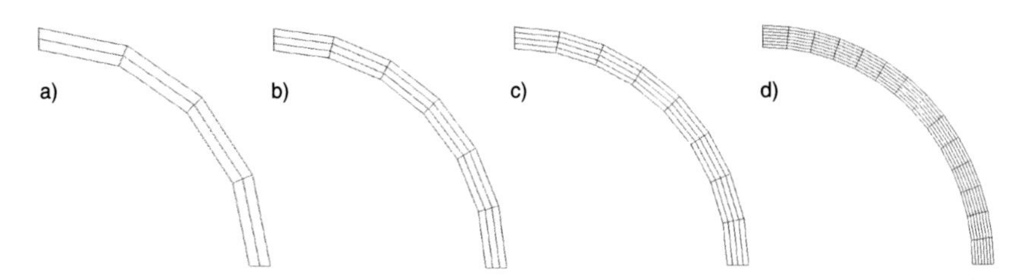

Quadrilateral meshes for a) 8 elements, b) 18 elements, c) 32 elements and d) 98 elements

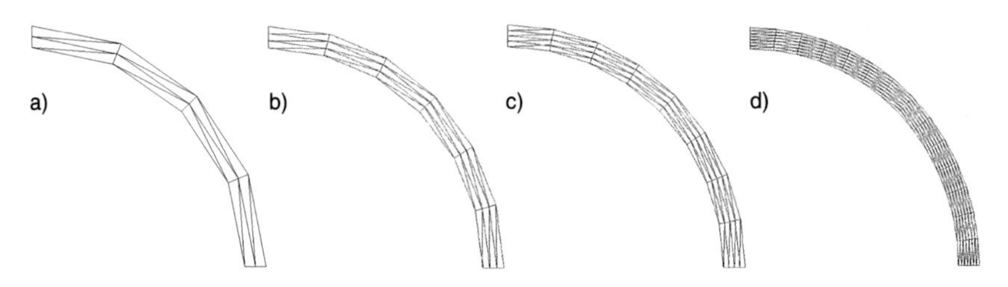

Triangular meshes for a) 16 elements, b) 32 elements, c) 64 elements and d) 196 elements

Fig. 11.38 Different meshes for quadrilateral and triangular elements

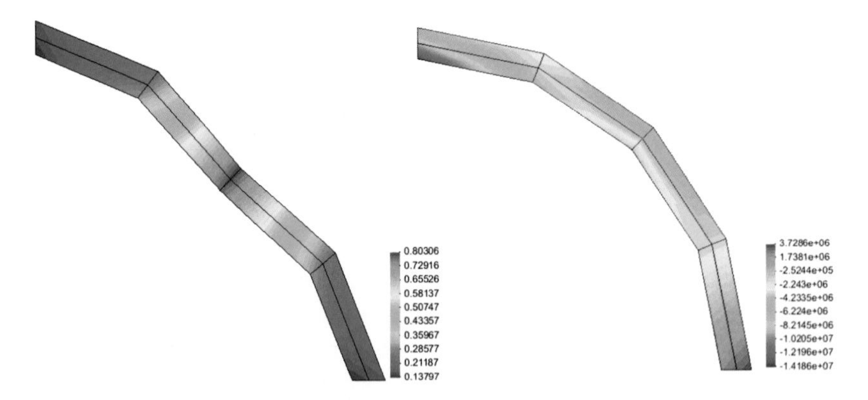

Fig. 11.39 Deformed mesh and σ_x stress distribution

placement at point B and the σ_x stress at point A is presented for meshes of 3-noded triangles and 4-noded quadrilaterals. A plot of the convergence of $(\sigma_x)_A$ with the number of DOFs for the two elements considered is shown in Figure 11.40.

Note that the convergence is slower for the triangular element, as expected (see Sections 4.5 and 5.7).

TRIANGLES					QUADRILATERALS				
Elem.	Nodes	DOF	v_A	$(\sigma_x)_A$	Elem.	Nodes	DOF	v_A	$(\sigma_x)_A$
16	15	30	0.586	-7.88E+06	8	15	30	0.803	-1.41E+07
36	28	56	0.808	-9.84E+06	18	28	56	1.308	-2.14E+07
64	45	90	1.069	-1.25E+07	32	45	90	1.799	-2.75E+07
100	66	132	1.348	-1.56E+07	50	66	132	2.222	-3.27E+07
196	120	240	1.888	-2.21E+07	98	120	240	2.842	-4.00E+07
400	231	462	2.540	-3.05E+07	200	231	462	3.368	-4.58E+07
2500	1326	2652	3.733	-4.68E+07	800	861	1722	3.911	-5.15E+07
3600	1891	3782	3.848	-4.84E+07	1800	1891	3782	4.039	-5.26E+07
6400	3321	6642	3.972	-5.03E+07	3200	3321	6642	4.090	-5.30E+07

Table 11.1 Convergence of the vertical displacement and σ_x (Pa) at point A

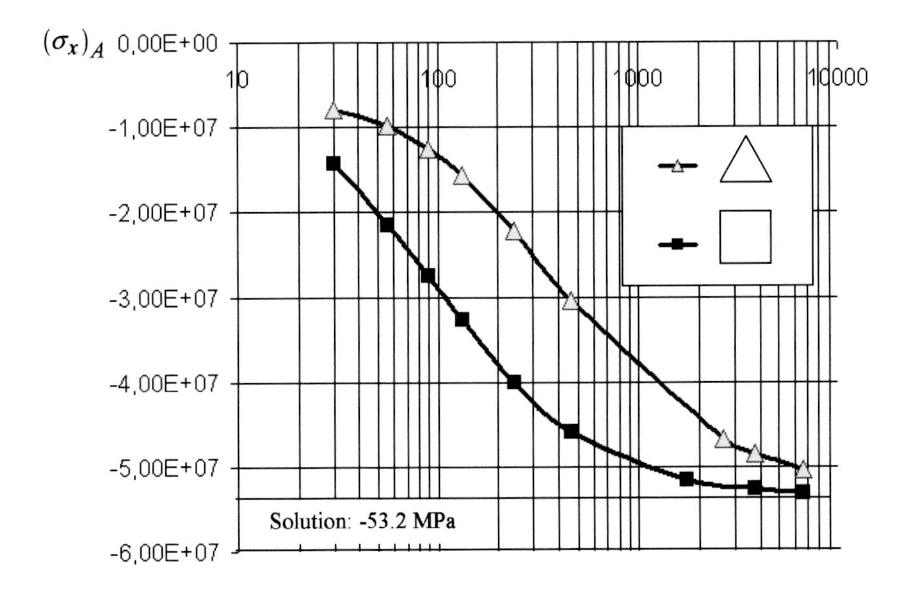

Fig. 11.40 Convergence of σ_x at point A with the number of DOFs

APPENDIX A. MATRIX ALGEBRA

We present the basic concepts of matrix algebra needed for the study of the book.

A.1 DEFINITION OF MATRIX

The linear relationship between a system of variables x_i and b_i

$$\begin{aligned} a_{11}x_1 + a_{12}x_2 + a_{13}x_3 + a_{14}x_4 &= b_1 \\ a_{21}x_1 + a_{22}x_2 + a_{23}x_3 + a_{24}x_4 &= b_2 \\ a_{31}x_1 + a_{32}x_2 + a_{33}x_3 + a_{34}x_4 &= b_3 \end{aligned} \tag{A.1}$$

can be written in abbreviated form as

$$\mathbf{A}\mathbf{x} = \mathbf{b} \tag{A.2}$$

where

$$\mathbf{A} = \begin{bmatrix} a_{11} & a_{12} & a_{13} & a_{14} \\ a_{21} & a_{22} & a_{23} & a_{24} \\ a_{31} & a_{32} & a_{33} & a_{34} \end{bmatrix} \quad , \quad \mathbf{x} = \begin{Bmatrix} x_1 \\ x_2 \\ x_3 \\ x_4 \end{Bmatrix} \quad , \quad \mathbf{b} = \begin{Bmatrix} b_1 \\ b_2 \\ b_3 \end{Bmatrix} \tag{A.3}$$

Eq.(A.1) contains the definition of matrix and the multiplication process. Matrices are defined as sets of numbers organized in rows and columns as in Eq.(A.3). Thus, matrix **A** has twelve elements organized in three rows and four columns. A matrix of order $m \times n$ has m rows and n columns.

A *vector*, or a *column matrix*, is a particular case of a matrix where all numbers are grouped in a column (i.e. the matrix is of order $m \times 1$). Hence **x** and **b** in Eq.(A.2) are vectors containing 4 and 3 elements, respectively.

In the book we denote matrices by bold capital letters and vectors by bold lower case letters.

A row matrix of order n is a matrix containing a single row and n columns. Hence

$$\mathbf{c}^T = [c_{11}, c_{12}, \cdots, c_{in}] \tag{A.4a}$$

where $\mathbf{c}^T$ is the "transpose" of vector $\mathbf{c}$

$$\mathbf{c} = \begin{Bmatrix} c_{11} \\ c_{12} \\ \vdots \\ c_{in} \end{Bmatrix} \tag{A.4b}$$

A.1.1 Transpose of a matrix

The transpose of a matrix $\mathbf{A}$ of order $(m \times n)$ (denoted by $\mathbf{A}^T$) has the rows equal to the columns of the original matrix. Thus

$$\mathbf{A}^T = \begin{bmatrix} a_{11} & a_{12} & \cdots & a_{1n} \\ a_{21} & a_{22} & \cdots & a_{2n} \\ \vdots & \vdots & \cdots & \vdots \\ a_{m1} & a_{m2} & \cdots & a_{mn} \end{bmatrix}^T = \begin{bmatrix} a_{11} & a_{21} & \cdots & a_{m1} \\ a_{12} & a_{22} & \cdots & a_{m2} \\ \vdots & \vdots & \cdots & \vdots \\ a_{1n} & a_{2n} & \cdots & a_{mn} \end{bmatrix} \tag{A.5}$$

A.1.2 Square matrix

A matrix is square if it has as many rows as columns. The square matrix of order 3 is

$$\mathbf{A} = \begin{bmatrix} a_{11} & a_{12} & a_{13} \\ a_{21} & a_{22} & a_{23} \\ a_{31} & a_{32} & a_{33} \end{bmatrix} \tag{A.6}$$

As an example, the stiffness matrix of an element or of the whole structure is always a square matrix.

A.1.3 Symmetric and antisymmetric matrix

A matrix is symmetric if is square and

$$a_{ij} = a_{ji} \tag{A.7}$$

Example

$$\mathbf{A} = \begin{bmatrix} a_{11} & a_{12} & a_{13} \\ a_{12} & a_{22} & a_{23} \\ a_{13} & a_{23} & a_{33} \end{bmatrix}$$

The transpose of a symmetric matrix coincides with itself. Hence

$$\mathbf{A}^T = \mathbf{A} \tag{A.8}$$

As an example, the stiffness matrix of an element (or of the structure) is always square and symmetric.

A matrix is antisymmetric if it is square and its elements satisfy the conditions

$$a_{ii} = 0 \qquad \text{and} \qquad a_{ij} = -a_{ji} \tag{A.9}$$

A.1.4 Null matrix

A null matrix has all its elements equal to zero, i.e. $a_{ij} = 0$ for $i = 1, 2, \cdots, m$ and $j = 1, 2, \cdots, n$.

A.1.5 Diagonal matrix

It is a square matrix which elements satisfy

$$\begin{aligned} a_{ij} &= 0 \quad \text{for} \quad i \neq j \\ a_{ij} &\neq 0 \quad \text{for} \quad i = j \end{aligned} \tag{A.10}$$

A.1.6 Identity matrix

It is a diagonal matrix with $a_{ij} = 1$ for $i = 1, 2, \cdots, n$. In the book it is represented as $\mathbf{I}$. The unit matrix of order 3×3 is

$$\mathbf{I} = \begin{bmatrix} 1 & 0 & 0 \\ 0 & 1 & 0 \\ 0 & 0 & 1 \end{bmatrix} \tag{A.11}$$

A.1.7 Triangular matrix

It is a square matrix which elements satisfy the conditions

$$a_{ij} = 0 \qquad \text{for} \qquad i > j \tag{A.12a}$$

or

$$a_{ji} \neq 0 \qquad \text{for} \quad j > i \tag{A.12b}$$

A.2 OPERATION WITH MATRICES

A.2.1 Multiplication of matrices

Let us assume that a relationship similar to (A.1) exists for a set of variables x'_i and b'_i as

$$\begin{aligned} a_{11}x'_1 + a_{12}x'_2 + a_{13}x'_3 + a_{14}x'_4 &= b'_1 \\ a_{21}x'_1 + a_{22}x'_2 + a_{23}x'_3 + a_{24}x'_4 &= b'_2 \\ a_{31}x'_1 + a_{32}x'_2 + a_{33}x'_3 + a_{34}x'_4 &= b'_3 \end{aligned} \tag{A.13}$$

We can therefore write

$$\mathbf{AX} = \mathbf{B} \tag{A.14}$$

where

$$\mathbf{X} = \begin{bmatrix} x_1 & x'_1 \\ x_2 & x'_2 \\ x_3 & x'_3 \\ x_4 & x'_4 \end{bmatrix} , \quad \mathbf{B} = \begin{bmatrix} b_1 & b'_1 \\ b_2 & b'_2 \\ b_3 & b'_3 \\ b_4 & b'_4 \end{bmatrix} \tag{A.15}$$

which implies grouping expressions (A.1) and (A.13) as

$$\begin{bmatrix} a_{11}x_1 + \cdots + a_{14}x_4 \ , \ a_{11}x'_1 + \cdots + a_{11}x'_4 \\ a_{21}x_1 + \cdots + a_{24}x_4 \ , \ a_{21}x'_1 + \cdots + a_{24}x'_4 \\ a_{31}x_1 + \cdots + a_{34}x_4 \ , \ a_{31}x'_1 + \cdots + a_{34}x'_4 \end{bmatrix} = \begin{bmatrix} b_1 & b'_1 \\ b_2 & b'_2 \\ b_4 & b'_3 \end{bmatrix} \tag{A.16}$$

Expression (A.14) defines the multiplication of matrices and it obviously has a meaning only if the number of columns of **A** is equal to the number of rows of **X**.

More generally we can define the multiplication of a matrix **A** of order $n \times m$ by a matrix **X** of order $m \times r$, as a new matrix **B** of order $n \times r$ which elements are obtained as

$$b_{ij} = \sum_{k=1}^{m} a_{ik}x_{kj} \qquad \begin{aligned} i &= 1, 2, \cdots, n \\ j &= 1, 2, \cdots, r \end{aligned} \tag{A.17}$$

The multiplication of matrices is not commutative, this means that

$$\mathbf{AB} \neq \mathbf{BA} \tag{A.18}$$

The following rules are however satisfied.

A.2.2 Associative rule

$$(\mathbf{AB})\mathbf{C} = \mathbf{A}(\mathbf{BC}) = \mathbf{ABC} \tag{A.19}$$

A.2.3 Distributive rule

$$\mathbf{A}(\mathbf{B} + \mathbf{C}) = \mathbf{AB} + \mathbf{AC} \tag{A.20}$$

The product of the identity matrix by a matrix $\mathbf{A}$ gives the same matrix $\mathbf{A}$, i.e.

$$\mathbf{IA} = \mathbf{AI} = \mathbf{A} \tag{A.21}$$

The product of the transpose of two matrices $\mathbf{A}^T$ and $\mathbf{B}^T$ is

$$\mathbf{A}^T\mathbf{B}^T = (\mathbf{BA})^T \tag{A.22a}$$

Obviously it is also satisfied

$$(\mathbf{AB})^T = \mathbf{B}^T\mathbf{A}^T \tag{A.22b}$$

A.2.4 Product of a matrix by a vector and a scalar

As shown in Eq.(A.2), the product of matrix $\mathbf{A}$ of order $n \times m$ and a vector $\mathbf{x}$ of order $m \times 1$ is a vector $\mathbf{b}$ of order $n \times 1$ defined by the expression

$$\mathbf{b} = \mathbf{Ax} \qquad \text{con} \quad b_i = \sum_{k=1}^{m} a_{ik}x_k \quad , \qquad i = 1, 2, \cdots, m \tag{A.23}$$

The product of a matrix $\mathbf{A}$ by a scalar α is another matrix $\mathbf{B}$ obtained by multiplying all the elements of $\mathbf{A}$ by α, i.e.

$$\mathbf{B} = \alpha\mathbf{A} \qquad \text{con} \quad b_{ij} = \alpha a_{ij} \tag{A.24}$$

We finally note that the product of a file vector and a column vector of equal order is a scalar number. Thus

$$\mathbf{a}^T\mathbf{b} = [a_{11}, a_{12}, \cdots, a_{1n}] \begin{bmatrix} b_{11} \\ b_{12} \\ \vdots \\ b_{1n} \end{bmatrix} = a_{11}b_{11} + a_{12}b_{12} + \cdots + a_{1n}b_{1n} = c \tag{A.25}$$

This type of product appears frequently in the book. An example is the expression of the virtual work principle.

A.2.5 Sum and subtraction of matrices

The sum or subtraction of two matrices $\mathbf{A}$ and $\mathbf{B}$ is a new matrix $\mathbf{C}$ which terms are the sum or the subtraction of matrices $\mathbf{A}$ and $\mathbf{B}$. Thus

$$\begin{aligned} \mathbf{A}+\mathbf{B}=\mathbf{C} \qquad & \text{with} \quad c_{ij}=a_{ij}+b_{ij} \\ \mathbf{A}-\mathbf{B}=\mathbf{C} \qquad & \text{with} \quad c_{ij}=a_{ij}-b_{ij} \end{aligned} \tag{A.26}$$

Obviously, these operations are only possible if $\mathbf{A}$ and $\mathbf{B}$ are of equal order.

A.2.6 Partition of a matrix

A matrix can be partitioned into submatrices. For example we can write

$$\mathbf{A}=\left[\begin{array}{ccc|ccc} a_{11} & a_{12} & a_{13} & a_{14} & \cdots & a_{15} \\ a_{21} & a_{22} & a_{23} & a_{24} & \cdots & a_{25} \\ \hline a_{31} & a_{32} & a_{33} & a_{34} & \cdots & a_{35} \end{array}\right] = \left[\begin{array}{c|c} \mathbf{A}_{11} & \mathbf{A}_{12} \\ \hline \mathbf{A}_{21} & \mathbf{A}_{22} \end{array}\right] \tag{A.27}$$

where matrices $\mathbf{A}_{11}, \mathbf{A}_{12}, \mathbf{A}_{21}$ and $\mathbf{A}_{22}$ have the elements contained within each of the four partitions marked in Eq.(A.27). Thus

$$\mathbf{A}_{11}=\begin{bmatrix} a_{11} & a_{12} & a_{13} \\ a_{21} & a_{22} & a_{23} \end{bmatrix} \qquad \text{etc.} \tag{A.28}$$

The partition of a matrix is useful in order to simplify to product of two matrices. Thus the product of matrix $\mathbf{A}$ of Eq.(A.27) and matrix $\mathbf{B}$ defined as

$$\mathbf{B}=\left[\begin{array}{cc} b_{11} & b_{12} \\ b_{21} & b_{22} \\ b_{31} & b_{32} \\ \hline b_{41} & b_{42} \\ b_{51} & b_{52} \end{array}\right] = \begin{bmatrix} \mathbf{B}_1 \\ \mathbf{B}_2 \end{bmatrix} \tag{A.29}$$

can be obtained by

$$\mathbf{AB}=\begin{bmatrix} \mathbf{A}_{11} & \mathbf{A}_{12} \\ \mathbf{A}_{21} & \mathbf{A}_{22} \end{bmatrix} \begin{bmatrix} \mathbf{B}_1 \\ \mathbf{B}_2 \end{bmatrix} = \begin{bmatrix} \mathbf{A}_{11}\mathbf{B}_1+\mathbf{A}_{12}\mathbf{B}_2 \\ \mathbf{A}_{21}\mathbf{B}_1+\mathbf{A}_{22}\mathbf{B}_2 \end{bmatrix} \tag{A.30}$$

The identity in expression (A.30) requires that the products $\mathbf{A}_{ij}\mathbf{B}_j$ are computable, which means that the number of columns of $\mathbf{A}_{ij}$ is equal to the number of rows of $\mathbf{B}_j$. If this condition holds then the product of $\mathbf{A}$ and $\mathbf{B}$ can be computed in terms of the submatrices $\mathbf{A}_{ij}\mathbf{B}_i$, treating each one of them as an scalar number.

A.2.7 Determinant of a matrix

The determinant of a square matrix $\mathbf{A}$ of order n is denoted as $|\mathbf{A}|$ and can be obtained by [Ral]

$$|\mathbf{A}| = \sum_{i=1}^{n} (-1)^{i+1} a_{1i} |\mathbf{A}_{1i}| \tag{A.31}$$

where $\mathbf{A}_{1i}$ is the matrix of order $(n-1)(n-1)$ resulting from the elimination of the first row and the ith column of $\mathbf{A}$. The computation of the determinants $|\mathbf{A}_{1i}|$ can be performed by the recusive application of Eq.(A.31).

A.2.8 Inverse of a matrix

If $\mathbf{A}$ is a square matrix in Eq.(A.2), i.e. the number of unknowns x_i is equal to that of the number of simultaneous equations, then it is possible to compute the elements of the unknown vector $\mathbf{x}$ in terms of those of vector $\mathbf{b}$. This operation can be written as

$$\mathbf{x} = \mathbf{A}^{-1}\mathbf{b} \tag{A.32}$$

where $\mathbf{A}^{-1}$ is the *inverse* of the square matrix $\mathbf{A}$. Clearly $\mathbf{A}^{-1}$ is also a square matrix and has the same order as $\mathbf{A}$.

The obvious property of the inverse matrix is

$$\mathbf{A}\mathbf{A}^{-1} = \mathbf{A}^{-1}\mathbf{A} = \mathbf{I} \tag{A.33}$$

where $\mathbf{I}$ is the identity matrix.

The necessary conditions for the existence of the inverse of a square matrix of $\mathbf{A}$ is that its determinant has a non zero value.

Eq.(A.32) can be obtained by multiplying both sides of Eq.(A.2) by $\mathbf{A}^{-1}$. This gives

$$\mathbf{A}^{-1}\mathbf{A}\mathbf{x} = \mathbf{A}^{-1} b \tag{A.34}$$

and since $\mathbf{A}^{-1}\mathbf{A} = \mathbf{I}$, then $\mathbf{x} = \mathbf{A}^{-1}\mathbf{b}$ (as $\mathbf{I}\mathbf{x} = \mathbf{x}$).

$$\mathbf{A} = \begin{bmatrix} a_{11} & a_{12} \\ a_{21} & a_{22} \end{bmatrix}$$

$$\mathbf{A}^{-1} = \frac{1}{|\mathbf{A}|} \begin{bmatrix} a_{22} & -a_{12} \\ -a_{21} & a_{11} \end{bmatrix}$$

$$\text{con } |\mathbf{A}| = a_{11}a_{22} - a_{12}a_{21}$$

$$\mathbf{A} = \begin{bmatrix} a_{11} & a_{12} & a_{13} \\ a_{21} & a_{22} & a_{23} \\ a_{31} & a_{32} & a_{33} \end{bmatrix}$$

$$\mathbf{A}^{-1} = \frac{1}{|\mathbf{A}|} \begin{bmatrix} A_{11} & A_{12} & A_{13} \\ A_{21} & A_{22} & A_{23} \\ A_{31} & A_{32} & A_{33} \end{bmatrix}$$

$$\begin{array}{ll} A_{11} = a_{22}a_{33} - a_{32}a_{23} & A_{23} = -(a_{11}a_{23} - a_{21}a_{13}) \\ A_{12} = -(a_{12}a_{33} - a_{13}a_{32}) & A_{31} = a_{21}a_{32} - a_{31}a_{22}) \\ A_{13} = a_{12}a_{23} - a_{22}a_{13} & A_{32} = -(a_{11}a_{32} - a_{31}a_{12}) \\ A_{21} = -(a_{21}a_{33} - a_{31}a_{23}) & A_{33} = a_{11}a_{22} - a_{12}a_{21} \\ A_{22} = a_{11}a_{33} - a_{13}a_{31} & \end{array}$$

$$|\mathbf{A}| = a_{11}a_{22}a_{33} + a_{13}a_{21}a_{32} + a_{31}a_{12}a_{23} - a_{31}a_{22}a_{13} - a_{33}a_{12}a_{21} - a_{11}a_{23}a_{32}$$

Box A.1 Inverse of 2×2 and 3×3 matrices

Other properties of the inverse matrix are

- the inverse of a product of two matrices is

$$(\mathbf{AB})^{-1} = \mathbf{B}^{-1}\mathbf{A}^{-1} \tag{A.35}$$

- the inverse of a symmetric matrix is also symmetric
- the inverse of the transpose of a matrix is the transpose of the inverse of the original matrix, i.e.

$$[\mathbf{A}^T]^{-1} = [\mathbf{A}^{-1}]^T \tag{A.36}$$

The general expression of the inverse of a matrix is quite elaborate and it is not given here. On the other hand the solution of systems of equations such as that of Eq.(A.2) is performed in practice by numerical techniques such as the Gauss elimination method, which does not require the knowledge of the inverse matrix (Appendix B). Box A.1 shows the expression for the inverse of the Jacobian matrix for 2D and 3D solid elements. The readers interested in the general expression for the inverse of a matrix are addressed to references [Ral,PFTW].

A.3 EIGENVALUES AND EIGENVECTORS OF A MATRIX

An *eigenvalue* of a symmetric matrix $\mathbf{A}$ of size $n \times n$ is a scalar λ_i which allows the solution of

$$(\mathbf{A} - \lambda_i \, \mathbf{I}) \, \boldsymbol{\phi}_i = \mathbf{0} \quad \text{and} \quad det \mid \mathbf{A} - \lambda_i \, \mathbf{I} \mid = 0 \tag{A.37}$$

where $\boldsymbol{\phi}_i$ is called the *eigenvector*.

There are, of course, n such eigenvalues λ_i to each of which corresponds an eigenvector $\boldsymbol{\phi}_i$. Such vectors can be shown to be orthonormal and we write

$$\boldsymbol{\phi}_i^T \boldsymbol{\phi}_j = \delta_{ij} = \begin{cases} 1 \text{ for } i = j \\ 0 \text{ for } i \neq j \end{cases} \quad .$$

The full set of eigenvalues and eigenvectors can be written as

$$\boldsymbol{\Lambda} = \begin{bmatrix} \lambda_1 & & 0 \\ & \ddots & \\ 0 & & \lambda_n \end{bmatrix} \qquad \boldsymbol{\Phi} = \begin{bmatrix} \phi_1, \dots \phi_n \end{bmatrix} \quad .$$

Using these the matrix $\mathbf{A}$ may be written in its *spectral form* by noting from the orthonormality conditions on the eigenvectors that

$$\boldsymbol{\Phi}^{-1} = \boldsymbol{\Phi}^T$$

then from

$$\mathbf{A} \, \boldsymbol{\Phi} = \boldsymbol{\Phi} \, \boldsymbol{\Lambda}$$

it follows immediately that

$$\mathbf{A} = \boldsymbol{\Phi} \, \boldsymbol{\Lambda} \, \boldsymbol{\Phi}^T \quad . \tag{A.38}$$

The condition number of $\mathbf{A}$ which is related to the equation solution roundoff error (Section 3.13.4) is defined as

$$\text{cond}(\mathbf{A}) = \frac{\mid \lambda_{max} \mid}{\mid \lambda_{min} \mid} \quad . \tag{A.39}$$

APPENDIX B. SOLUTION OF SIMULTANEOUS LINEAR ALGEBRAIC EQUATIONS

A finite element problem leads to a large set of simultaneous linear algebraic equations whose solution provides the nodal and element parameters in the formulation. In this section methods to solve the simultaneous algebraic equations are summarized. We consider both *direct* methods where an *a priori* calculation of the number of numerical operations can be made, and *indirect or iterative* methods where no such estimate can be made.

B.1 DIRECT SOLUTION

Consider first the general problem of direct solution of a set of algebraic equations given by

$$\mathbf{K}\mathbf{a} = \mathbf{f} \tag{B.1}$$

where $\mathbf{K}$ is a square coefficient matrix, $\mathbf{a}$ is a vector of unknown parameters and $\mathbf{f}$ is a vector of known values. The reader can associate these with the quantities described previously: namely, the stiffness matrix, the nodal displacement unknowns and the equivalent nodal forces.

In the discussion to follow it is assumed that the coefficient matrix $\mathbf{K}$ has properties such that row and/or column interchanges are unnecessary to achieve an accurate solution. This is true in cases where $\mathbf{K}$ is symmetric positive (or negative) definite. Pivoting may or may not be required with unsymmetric, or indefinite, conditions which can occur when the finite element formulation is based on some weighted residual methods [ZTZ]. In these cases some checks or modifications may be necessary to ensure that the equations can be solved accurately.

Let us assume that the coefficient matrix can be written as the product of a lower triangular matrix with unit diagonals and an upper triangular matrix. Accordingly,

$$\mathbf{K} = \mathbf{L}\mathbf{U} \tag{B.2}$$

where

$$\mathbf{L} = \begin{bmatrix} 1 & 0 & \cdots & 0 \\ L_{21} & 1 & \cdots & 0 \\ \vdots & & \ddots & \vdots \\ L_{n1} & L_{n2} & \cdots & 1 \end{bmatrix} \tag{B.3}$$

and

$$\mathbf{U} = \begin{bmatrix} U_{11} & U_{12} & \cdots & U_{1n} \\ 0 & U_{22} & \cdots & U_{2n} \\ \vdots & & \ddots & \vdots \\ 0 & 0 & \cdots & U_{nn} \end{bmatrix} \tag{B.4}$$

This form is called a *triangular decomposition* of $\mathbf{K}$. The solution to the equations can now be obtained by solving the pair of equations

$$\mathbf{L}\mathbf{y} = \mathbf{f} \tag{B.5}$$

and

$$\mathbf{U}\mathbf{a} = \mathbf{y} \tag{B.6}$$

where $\mathbf{y}$ is introduced to facilitate the separation [Ral,ZTZ].

The reader can easily observe that the solution to these equations is trivial. In terms of the individual equations the solution is given by

$$\begin{aligned} y_1 &= f_1 \\ y_i &= f_i - \sum_{j=1}^{i-1} L_{ij} y_j \qquad i = 2, 3, \ldots, n \end{aligned} \tag{B.7}$$

and

$$\begin{aligned} a_n &= \frac{y_n}{U_{nn}} \\ a_i &= \frac{1}{U_{ii}} \left(y_i - \sum_{j=i+1}^{n} U_{ij} a_j \right) \qquad i = n-1, n-2, \cdots, 1 \end{aligned} \tag{B.8}$$

Equation (B.7) is commonly called *forward elimination*, while Eq. (B.8) is called *back substitution.*

The problem remains to construct the triangular decomposition of the coefficient matrix. This step is accomplished using variations on the Gaussian elimination method. In practice, the operations necessary for the triangular decomposition are performed directly in the coefficient array. Details on this step may be found in [ZTZ].

Once the triangular decomposition of the coefficient matrix is computed, several solutions for different right-hand sides $\mathbf{f}$ can be obtained using Eqs (B.7) and (B.8). This process is often called a *resolution* since it is not necessary to recompute the $\mathbf{L}$ and $\mathbf{U}$ arrays. For large size coefficient matrices the triangular decomposition step is very costly while a resolution is relatively cheap; consequently, a resolution capability is necessary in any finite element solution system using a direct method [Dem,Str,Tay,WR].

The above discussion considered the general case of equation solving (without row or column interchanges). In coefficient matrices resulting from a finite element formulation some special properties are usually present. Typically the coefficient matrix is symmetric ($K_{ij} = K_{ji}$) and it is easy to verify in this case that

$$U_{ij} = L_{ji} U_{ii} \tag{B.9}$$

For this problem class it is not necessary to store the entire coefficient matrix. It is sufficient to store only the coefficients above (or below) the principal diagonal and the diagonal coefficients. This reduces by almost half the required storage for the coefficient array as well as the computational effort to compute the triangular decomposition.

The required storage can be further reduced by storing only those rows and columns which lie within the region of non-zero entries of the coefficient array. Structural problems formulated by the finite element method normally have a symmetric profile which further simplifies the storage form [Tay].

In 2D formulations, problems with many thousand DOFs can be solved on today's personal computers. In 3D however problems are restricted to several hundred thousand equations. To solve larger size problems there are several options. The first is to retain only part of the coefficient matrix in the main array with the rest saved on backing store (e.g., hard disk). This can be quite easily achieved but the size of problem is not greatly increased due to the very large solve times required and the rapid growth in the size of the profile-stored coefficient matrix in 3D problems.

A second option is to use sparse solution schemes. These lead to significant program complexity over the procedure discussed above but can lead to significant savings in storage demands and computing time – specially for 3D problems [Dem,Ral,WR]. Nevertheless, capacity limitations in terms of storage and computing time are again rapidly encountered and alternatives are needed.

B.2 ITERATIVE SOLUTION

One of the main problems in direct solution methods is that terms within the stiffness matrix which are zero from a finite element formulation become non-zero during the triangular decomposition step. This leads to a very large increase in the number of non-zero terms in the factored coefficient matrix. To be more specific consider the case of a 3D linear elastic problem solved using 8-noded isoparametric hexahedron elements. In a regular mesh each interior node is associated with 26 other nodes, thus, the equation of such a node has 81 non-zero coefficients – three for each of the 27 associated nodes. On the other hand, for a rectangular block of elements with n nodes on each of the sides the typical column height in the stiffness matrix is approximately proportional to n^2 and the number of equations to n^3.

Typically, the demands for a direct solution grow very rapidly (storage is approximately proportional to n^5) while at the same time the demands for storing the non-zero terms in the stiffness matrix grows proportional to the number of equations (i.e., proportional to n^3 for the block).

Iterative solution methods use the terms in the stiffness matrix directly and thus for large problems have the potential to be very efficient for large 3D problems. On the other hand, iterative methods require the resolution of a set of equations until the residual of the linear equations, given by

$$\boldsymbol{\Psi}^{(i)} = \mathbf{f} - \mathbf{K}\mathbf{a}^{(i)} \tag{B.10}$$

becomes less than a specified tolerance [Dem,Ral,WR]. Index i in Eq.(B.10) denotes the number of iterations in the iterative solution process.

In order to be effective the number of iterations i to achieve a solution must be quite small – generally no larger than a few hundred. Otherwise, excessive solution costs will result. The subject of iterative solution for general finite element problems remains a topic of intense research. There are some impressive results available for the case where $\mathbf{K}$ is symmetric positive (or negative) definite; however, those for other classes (e.g., unsymmetric or indefinite forms) are generally not efficient enough for reliable use of iterative methods in the solution of general problems.

For the symmetric positive definite case methods based on a preconditioned conjugate gradient method are particularly effective [Dem,Ral,ZTZ]. The convergence of the method depends on the condition number of the matrix $\mathbf{K}$ (Eq.(A.39)) – the larger the condition number, the slower the convergence [Ral].

Usually, the condition number for an elasticity problem modelled by the finite element method is too large to achieve rapid convergence and a *preconditioned conjugate gradient* method (PCG) is used. A symmetric form of preconditioned system for Eq.(B.1) is written as

$$\mathbf{K_p z} = \mathbf{PKP}^{\mathrm{T}}\mathbf{z} = \mathbf{Pf} \tag{B.11}$$

where $\mathbf{P}$ is the preconditioner matrix, and

$$\mathbf{P}^{\mathrm{T}}\mathbf{z} = \mathbf{a} \tag{B.12}$$

The convergence of the PCG algorithm depends on the condition number of $\mathbf{K_p}$. The problem remains to construct a preconditioner matrix which adequately reduces the condition number of $\mathbf{K_p}$. The simplest option is to use the diagonal of $\mathbf{K}$. More efficient schemes are discussed in [Dem,Fe2,Ral,Str].

APPENDIX C. COMPUTATION OF THE ELEMENT REFINEMENT PARAMETER FOR AN EQUAL DISTRIBUTION OF THE ERROR AND ACCOUNTING FOR THE CHANGE IN THE NUMBER OF ELEMENTS

The discretization error in energy norm is (Section 9.9.1)

$$\|\mathbf{e}_\sigma\| = \left[\sum_{e=1}^{N} \left(\|\mathbf{e}_\sigma\|^{(e)}\right)^2\right]^{1/2} \tag{C.1}$$

where N is the actual number of elements in the mesh where the error is computed and

$$\|\mathbf{e}_\sigma\|^{(e)} = \int_{\Omega^{(e)}} [\mathbf{e}_\sigma^{(e)}]^T \mathbf{D}^{-1} \mathbf{e}_\sigma^{(e)} d\Omega \tag{C.2}$$

with

$$\mathbf{e}_\sigma^{(e)} = \boldsymbol{\sigma}_s - \bar{\boldsymbol{\sigma}} \tag{C.3}$$

where $\boldsymbol{\sigma}_s$ are the smoothed stresses (Section (9.8)).

We will assume that the optimal mesh satisfies the criterion of equal distribution of the global error.

Let us call N_{opt} the number of elements in the optimal refined mesh, $\|\mathbf{e}_\sigma^p\|$ the prescribed (global) discretization error for the optimal mesh and $\|\bar{\mathbf{e}}_\sigma\|^{\text{uni}}$ the uniform value of the discretization error for each element of the optimal mesh. Then

$$\|\mathbf{e}_\sigma^p\| = N_{\text{opt}}^{1/2} \|\mathbf{e}_\sigma\|^{\text{uni}} \tag{C.4}$$

Typically $\|\mathbf{e}_\sigma^p\| \equiv \eta \|U\|$ where is user defined global error parameter and $\|U\|$ is the strain energy norm (see Eq.(9.78)).

The optimal number of elements can be estimated as

$$N_{\text{opt}} = \sum_{e=1}^{n} \left(\frac{h^{(e)}}{\bar{h}^{(e)}} \right)^{d} \tag{C.5}$$

where $h^{(e)}$ and $\bar{h}^{(e)}$ are the actual and optimal element sizes, respectively and d is the number of space dimensions of the problem (i.e. $d = 2$ for 2D problems, etc.).

Clearly for a single element

$$\|\mathbf{e}_\sigma^{\text{opt}}\|^{(e)} = \left(\frac{h^{(e)}}{\bar{h}^{(e)}} \right)^{d/2} \|\mathbf{e}_\sigma\|^{\text{uni}} \tag{C.6}$$

where $\|\mathbf{e}_\sigma^{\text{opt}}\|^{(e)}$ is the error on the element e (belonging to the initial mesh) after refinement. From (C.2) and (C.4)

$$\|\mathbf{e}_\sigma^p\| = \|\mathbf{e}_\sigma\|^{\text{uni}} \left[\sum_{e=1}^{N} \left(\frac{h^{(e)}}{\bar{h}^{(e)}} \right)^{d} \right]^{1/2} = \left[\sum_{e=1}^{N} \left(\|\mathbf{e}_\sigma^{\text{opt}}\|^{(e)} \right)^2 \right]^{1/2} \tag{C.7}$$

The convergence ratio of the element error norm is (Eq.(9.90)) $\|\mathbf{e}_\sigma\|^{(e)} = O(h^p)(\Omega^{(e)})^{1/2} \simeq O(h^{p+d/2})$. This gives

$$\frac{\|\mathbf{e}_\sigma^{\text{opt}}\|^{(e)}}{\|\mathbf{e}_\sigma\|^{(e)}} = \left(\frac{\bar{h}^{(e)}}{h^{(e)}} \right)^{p+d/2} \tag{C.8}$$

Combining Eqs.(C.6) and (C.8) gives

$$\frac{\|\mathbf{e}_\sigma\|^{\text{uni}}}{\|\mathbf{e}_\sigma\|^{(e)}} = \left(\frac{\bar{h}^{(e)}}{h^{(e)}} \right)^{p+d} \tag{C.9}$$

and hence

$$\bar{h}^{(e)} = h^{(e)} \left(\frac{\|\mathbf{e}_\sigma\|^{\text{uni}}}{\|\mathbf{e}_\sigma\|^{(e)}} \right)^{\frac{1}{p+d}} \tag{C.10}$$

From Eq.(C.7)

$$\|\mathbf{e}_\sigma\|^{\text{uni}} = \|\mathbf{e}_\sigma^p\| \left[\sum_{e=1}^{N} \left(\frac{h^{(e)}}{\bar{h}^{(e)}} \right)^{d} \right]^{-1/2} = \|\mathbf{e}_\sigma^p\| \left[\sum_{e=1}^{N} \left(\frac{\|\mathbf{e}_\sigma\|^{(e)}}{\|\mathbf{e}_\sigma\|^{\text{uni}}} \right)^{\frac{d}{p+d}} \right]^{-1/2} \tag{C.11}$$

Hence

$$\|\mathbf{e}_\sigma\|^{\text{uni}} = \|\mathbf{e}_\sigma^p\|^{\frac{2(p+d)}{2p+d}} \left[\sum_{e=1}^{N} \left(\|\mathbf{e}_\sigma\|^{(e)}\right)^{\frac{d}{p+d}}\right]^{-\frac{p+d}{2p+d}} \tag{C.12}$$

Substituting (C.12) into (C.10) gives

$$\bar{h}^{(e)} = \frac{h^{(e)}}{\beta^{(e)}} \tag{C.13}$$

with

$$\beta^{(e)} = \|\mathbf{e}_\sigma^p\|^{-\frac{2}{2p+d}} \left(\|\mathbf{e}_\sigma\|^{(e)}\right)^{\frac{1}{p+d}} \left[\sum_{e=1}^{N} \left(\|\mathbf{e}_\sigma\|^{(e)}\right)^{\frac{d}{p+d}}\right]^{\frac{1}{2p+d}} \tag{C.14}$$

Expression (C.14) can be written in terms of the global and local error parameters. From Eqs.(9.77) and (9.89)

$$\begin{aligned} \|\mathbf{e}_\sigma^p\| &= \eta\|U\| = \xi_g^{-1}\|\mathbf{e}_\sigma\| \\ \|\mathbf{e}_\sigma\|^{(e)} &= \bar{\xi}^{(e)}\|\mathbf{e}_\sigma\|N^{-1/2} \end{aligned} \tag{C.15}$$

Substituting (C.13) and (C.15) into (C.12) gives

$$\begin{aligned} \beta^{(e)} &= \xi_g^{\frac{2}{2p+d}}\|\mathbf{e}_\sigma\|^{-\frac{2}{2p+d}}[\bar{\xi}^{(e)}]^{\frac{1}{p+d}}N^{-\frac{1}{2(p+d)}}\left[\sum_{e=1}^{N}[\bar{\xi}^{(e)}]^{\frac{d}{p+d}}\|\mathbf{e}_\sigma\|^{\frac{d}{p+d}}N^{-\frac{d}{2(p+d)}}\right]^{\frac{1}{2p+d}} \\ &= \xi_g^{\frac{2}{2p+d}}\|\mathbf{e}_\sigma\|^{-\frac{d}{(2p+d)(p+d)}}[\bar{\xi}^{(e)}]^{\frac{1}{p+d}}N^{-\frac{1}{2p+d}}\|\mathbf{e}_\sigma\|^{\frac{d}{(2p+d)(p+d)}}N^{-\frac{d}{2(2p+d)(p+d)}} \\ &\left[\sum_{e=1}^{N}[\bar{\xi}^{(e)}]^{\frac{d}{p+d}}\right]^{\frac{1}{2p+d}} = \xi_g^{\frac{2}{2p+d}}[\bar{\xi}^{(e)}]^{\frac{1}{p+d}}N^{-\frac{1}{2p+d}}\left[\sum_{e=1}^{N}[\bar{\xi}^{(e)}]^{\frac{d}{p+d}}\right]^{\frac{1}{2p+d}} \end{aligned} \tag{C.16}$$

which coincides with Eq.(9.92). Accepting that $\bar{\xi}^{(e)}$ is constant gives

$$\beta^{(e)} = \xi_g^{\frac{2}{2p+d}}[\bar{\xi}^{(e)}]^{\frac{1}{p+d}}N^{-\frac{1}{2p+d}}N^{\frac{1}{2p+d}}[\bar{\xi}^{(e)}]^{\frac{d}{(2p+d)(p+d)}} = \xi_g^{\frac{2}{2p+d}}[\bar{\xi}^{(e)}]^{\frac{2}{2p+d}} = [\xi^{(e)}]^{\frac{2}{2p+d}} \tag{C.17}$$

with

$$\xi^{(e)} = \xi_g\bar{\xi}^{(e)} \tag{C.18}$$

Eq.(C.17) coincides with Eq.(9.93a).

Above derivation is based on the ideas presented in [Fu].

APPENDIX D

D.1 THE GID PRE/POSTPROCESSING SYSTEM

Extensive research has been carried out in the last twenty years at the International Center for Numerical Methods in Engineering (CIMNE, www.cimne.com) in the development of advanced mathematical and numerical methods and software tools for: (a) the adaptation of geometrical models to the need of computer simulation codes, (b) the user-friendly definition of analysis data for the numerical solution of problems in science and engineering, (c) the interfacing of the analysis data with any academic or commercial computer simulation code, and (d) the graphic visualization of the results from the computation.

The outcome of this research has lead to software tools which were progressively integrated over the years into an innovative computer system named GiD (for **G**eometry **I**ntegration with **D**ata) [GiD,OPSE]. GiD is downloadable from *www.gidhome.com.*

GiD is a pre/postprocessing system which incorporates all the functionalities for the user-friendly generation of finite element analysis data, for the simple linking to any FEM code and for the fast visualization of numerical results from finite element computations (Figure D.1).

A good pre/post processing system should invariably provide the analyst with a fast overview of the simulation data: geometry, material properties, boundary and loading conditions and numerical results at a glimpse, thus allowing its integration in critical decision loops and systems.

Until GiD was created, research teams around the world were grouped in the following categories: the ones that developed its own pre and post-processing tools because the programs available on the market were expensive and hard, if ever possible, to personalize, and the ones who developed interfaces between its analysis programs and the existing professional pre

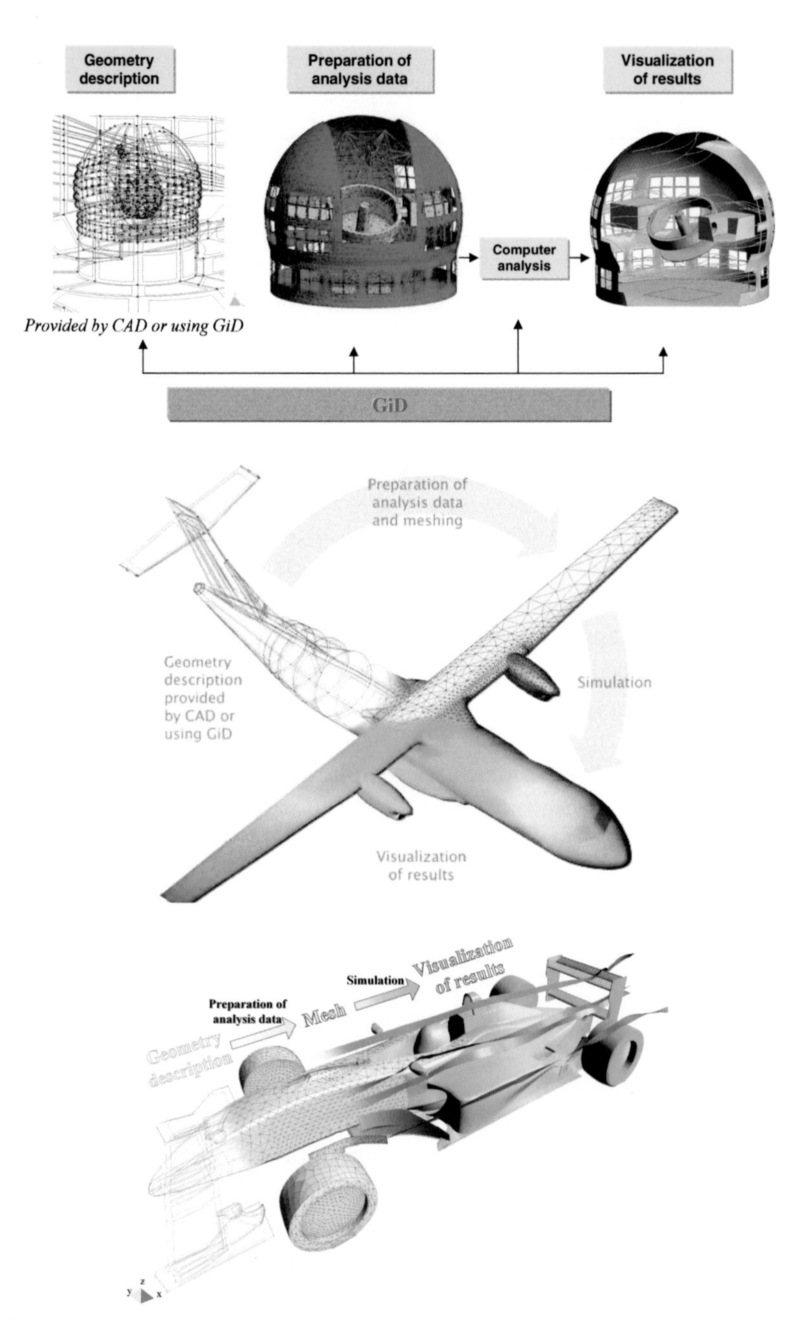

Fig. D.1 From geometry to results with GiD. From above: Aerodynamic analysis of a large telescope building [CMOS]. Aerodynamic and structural analysis of an airplane [FO,FO2,REMFI]. Aerodynamic analysis of a racing car

and postprocessors. And there is yet a third group: the undergraduate and graduate students who want to learn about computer simulation methods (such as FEM) but can not afford costly pre and postprocessing products.

For a research center such as CIMNE, employing some 180 scientists working in the development of new numerical methods and software, there was no such altruist philosophy, but the urgency to solve a practical need.

In any research centre involved in computer simulation, the development of state of the art simulation techniques requires innovative data management and visualization tools that generally are not available in commercial codes until several years later. These techniques are easily incorporated into GiD, as a software product in constant evolution, which can be customized by the user. The development of GiD is therefore feed by the comments and suggestions from its users.

GiD has become nowadays an indispensable tool for the widespread development and use of computer simulation codes in many research organizations and universities worldwide, as well as in industry. GiD has also proven to be the ideal partner for software developers in academic and industrial environments. A unique advantage of GiD is its capability for providing the indispensable pre and postprocessing modules to existing and emerging software codes in order to create integrated computer simulation packages in a simple and user-friendly manner. GiD has been successfully applied to a wide range of problems in science and engineering including civil, mechanical, aerospace, naval, telecommunications, bio-medical and food processing engineering, architecture, computational physics and chemistry, computational biology and others.

D.1.1 General features of GiD

The key features of GiD can be summarized as being: *universal, adaptive, user friendly, product generator* and a *tool for engineering practice, teaching and research.*

Universal. GiD is ideal for generating all the information (structured and unstructured meshes, boundary and loading conditions, material parameters, visualisation of results, computational parameters etc.) required for the analysis of any problem in science and engineering using numerical methods. Typical problems that can be successfully tackled with GiD include most situations in solid and structural mechanics, fluid dynamics, electromagnetics, heat transfer, geomechanics, etc. using finite element, fi-

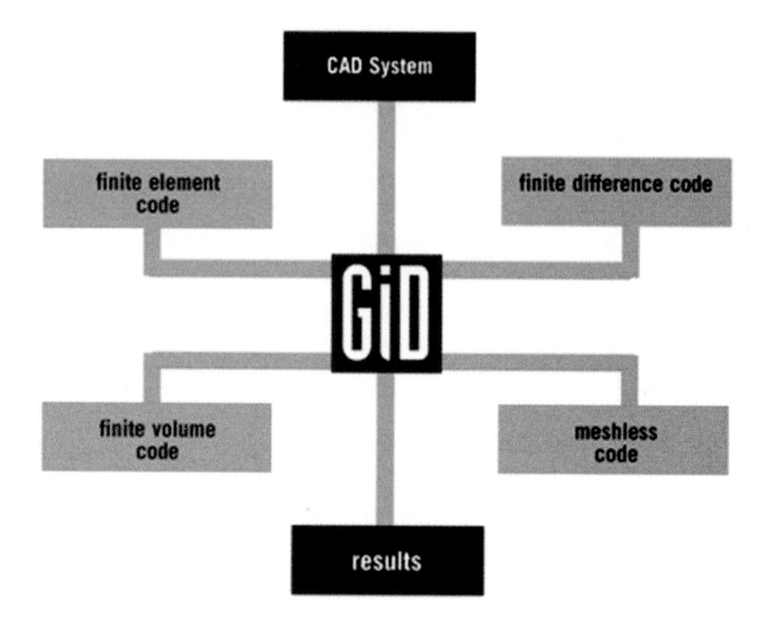

Fig. D.2 GiD can be linked to any numerical simulation code

nite volume, boundary element, finite difference or point-based (meshless) numerical procedures (Figure D.2).

Adaptive. GiD is extremely easy to adapt to any numerical simulation code. In fact, GiD can be customized by the user to read and write data in an unlimited number of formats. GiD's input and output formats can be made compatible with any existing in-house software. The different menus for data input and results visualisation can be tailored to the specific needs of the user.

User friendly. The development of GiD has been focused on the needs of the user and on the simplicity, speed, effectiveness and accuracy he or she demands at input data preparation and results visualisation levels. GiD can read and write data accordingly to the specific needs of each individual code. The GiD environment can therefore support an unlimited number of numerical simulation codes.

Product generator. GiD is the ideal complement to any software code in order to create a new integrated computer simulation package with full pre and postprocessing facilities. An example of this facility is the different simulation packages developed by industrial partners containing "GiD inside" (see www.gidhome.com).

A tool for engineering practice. GiD is the ideal tool for interfacing all analysis codes typically used in a company. This enables the unification of the pre and postprocessing tasks in a common environment (GiD) while the analysis can be performed in different software codes launched from GiD (Figure D.2).

A unique tool for teaching and research. GiD allows research and undergraduate students to introduce themselves in the solution of practical problems of any complexity level. This indeed simplifies enormously the teaching process in computational science and engineering in a simple manner (Figure D.3). The capability of using GiD via internet facilitates the work for multi-site organizations, as well as the academic and research tasks (Figure D.4).

The many features of GiD help students to enter into the fascinating world of computer simulation beyond the simple academic examples.

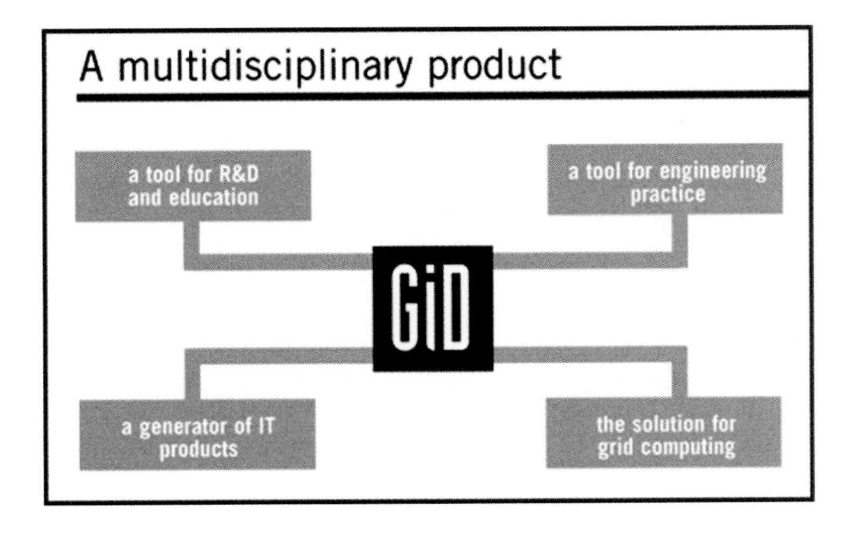

Fig. D.3 GiD is a multidisciplinary tool

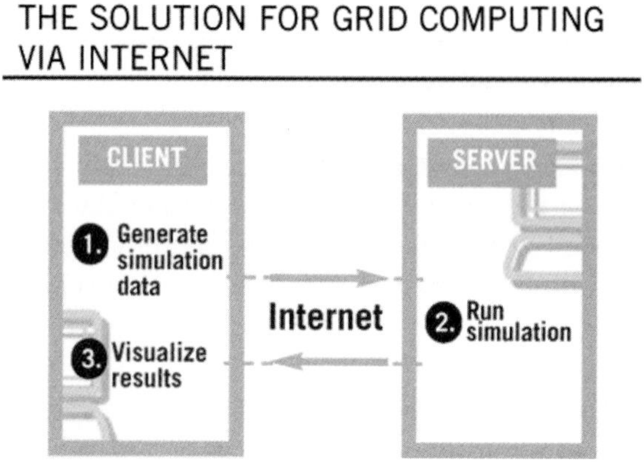

Fig. D.4 GiD allows computing via internet

D.1.2 More features of GiD

GiD is easy to adapt to any calculation code, including those developed by students, research centers and indeed commercial codes like NE NASTRAN, BEASY, TDYN, RAM-SERIES, STAMPACK, VULCAN, etc [see www.gidhome.com for details]. The adaptation reaches so far that the user can run the code from within GiD. This code can be executed locally or remotely in a manner fully transparent to the users. The transferred data are protected by an encryption system. This remote data export feature allows running multiple problems on several machines in a distributed mode using grid computing facilities via intranet or internet [Figure D.4]. This utility is essential for optimisation processes involving massive large scale computations. The user can also group several frequent used actions into macros, which can be executed just by clicking a button.

GiD implements the idea of a C++/OpenGL kernel together with a Tcl/Tk GUI interface which has become a standard for the customization of FEM interfaces.

One of the advantages of GiD is that it runs on a simple PC. Therefore it does not need a sophisticated computer and can exploit the continuous advances in computer technology, such as GPU's (Graphical Processing Units) which incorporate the latest technology in parallelism into graphical cards.

Special effort has been invested in making of GiD a multi-platform tool. As such it runs on PCs equipped with Linux, Microsoft Windows,

SGI, Sun Hp workstations, each machine running UNIX and OpenGL. GiD is also available on a PDA, so that, for instance, a user can verify the computational results remotely.

The use of innovative tools and macros allows the translation of GiD in several languages (english, spanish, japanese, french and russian are already available). The advanced customization tools have made of GiD a 'product generator'. Moreover, the 'look and feel' of GiD can be changed to create, when linked to an analysis code, a completely new software package.

This approach for developing GiD via OpenGL, a C++ kernel and a Tcl/Tk GUI interface has allowed GiD to be cross-platform. GiD now runs on every UNIX/Linux + OpenGL machines like Sun, SGI, Macintosh, Hewlett Packard, PC's and on Microsoft Windows.

D.1.3 The handling of geometrical data with GiD

One of the main problems in industry is to prepare/adapt the geometrical models from the design room to the simulation stage. This process needs specific tools to sew (join) surfaces and simplify the model.

Among other tools, GiD incorporates several facilities for geometrical model discretizations using NURBS technology (Chapter 10). Its own NURBS modelling library is the basis of the internal powerful geometrical modelling engine which enables GiD to communicate with the most widespread CAD and CAE programs on the market. This allows GiD to be used in state of the art research fields of numerical simulation dealing with standard mesh-based method (finite element, finite volume, finite differences, etc.), as well as with meshless methods, particle methods, discrete element methods and others.

Within its own NURBS library, GiD uses a hierarchical model of entities, so that a volume is a closed group of surfaces, and each surface is a closed group of lines. This hierarchy is automatically reflected in the mesh used for the numerical simulation.

The internal library also allows the intersection, trimming and joining of surfaces, and all other necessary geometrical management operations needed for the preparation of analysis data. User-friendly connectivity with geometrical data emanating from existing CAD system is an essential feature of GiD. GiD can import and export in the most common used geometry interchange formats (IGES, ACIS, VDA, Parasolid, DXF,STL, VRML, Shapefile, etc.).

Generally the designers of new products, like cars, airplanes or ships do not work thinking on the subsequent simulation step. Therefore the geometrical models created in CAD systems are too detailed and too focused on visualization rather than in a correct spatial specification. Several hours (days or months in some cases) must be spent to prepare the geometrical models for the simulation to allow a proper generation of meshes. GiD incorporates several innovative tools that simplify and automate the geometry preparation process and, if the process fails, the problematic geometrical areas can be easily and graphically localized and corrected.

GiD also incorporates the latest technology in mesh generation based on state of the art advancing front and Delaunay meshing algorithms (Chapter 10). Specially focused on automatic mesh generation, GiD includes several element types, detailed size control and innovative methods to create adaptive structured and unstructured meshes, also incorporating possibilities for element re-generation over a previous mesh (Figures D.5 and D.6).

The research on mesh edition tools includes the control of coherent orientations, mesh smoothing to optimise an specific objective function, edge collapse, split, mapping over a geometry, element check and correction determinant check and tools to display the mesh quality graphically.

D.1.4 Generation of analysis data and interfacing with computer simulation codes via GiD

The adaptation of GiD to a code in order to create the data for a computer simulation run, is as easy as just filling a couple of text files. Specific algorithms have been developed to create automatically the boundary and loading conditions, material properties and problem data windows, so that they can be directly applied to the geometry. When meshing, these properties are automatically transferred to the mesh used for the computations. A template file provides the format to write the data to be exported to the simulation code. With this approach the data inside GiD can be exported to an infinite number of data input formats, just by providing the appropriate template.

This adaptation is called *problem type*. This concept allows creating the input data for almost every existent simulation code and the future ones. Typically each research student involved in code development can generate a *problem type* for his/her own code in a matter of few hours (sometimes minutes). Although widely used by code developers and research organizations, the *problem type* concept is not restricted to research codes. It

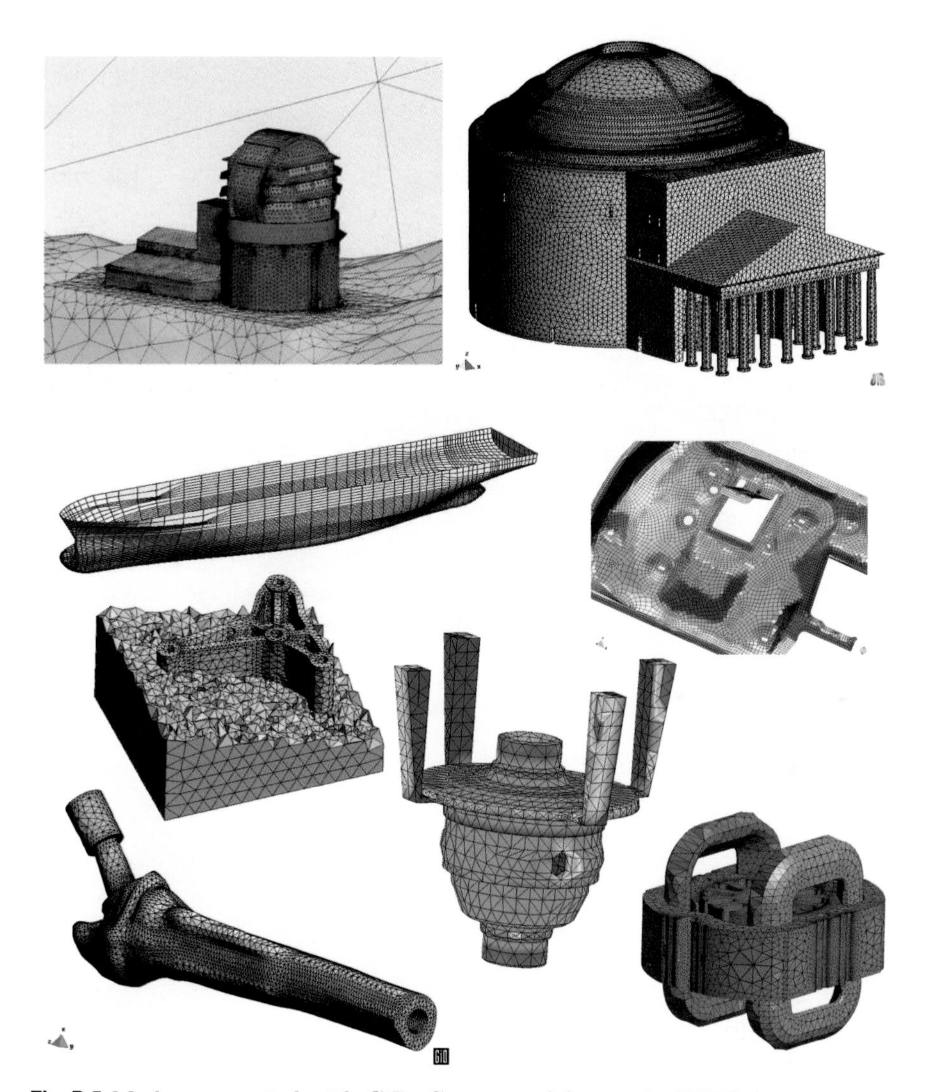

Fig. D.5 Meshes generated with GiD. Courtesy of Quantech ATZ S.A. (www.quantech.es), Compass Ingeniería y Sistemas S.A. (www.compassis.com) and the GiD team at CIMNE

can also be implemented into commercial codes, such as the NASTRAN problemtype which interfaces GiD with the commercial code NASTRAN. Some of the GiD "problemtypes" are available on-line and can be downloaded directly from www.gidhome.com.

Fig. D.5 Cont. Meshes generated with GiD for analysis of casting and sheet stamping processes and aerodynamic and structural analysis of helicopter and airplane. Courtesy of Quantech ATZ S.A. (www.quantech.es), Compass Ingeniería y Sistemas S.A. (www.compassis.com) and the GiD team at CIMNE

The customisation tools in GiD go even further by allowing the creation of new windows and macros. These options allow the user to automatically create the geometry in a parametric form.

The appearance (look and feel) of GiD can also be completely changed to create the image for a new computer simulation package.

The success of this customisation approach is demonstrated by the big acceptance of GiD around the world: from South America to Japan, many universities, research centres and small, medium and large size companies use GiD as their favourite pre/postprocessing system for a wide number of applications.

GiD is a code in constant evolution and is a bridge between the research and the industrial world. The GiD RTD team is happy to receive comments, suggestions and feedback from the users and tries to incorporate these into the future versions of GiD. The GiD team can be reached via www.gidhome.com.

D.1.5 Visualization of numerical results with GiD

Easy to use, easy to see. Special emphasis has been put in making GiD practical for educational purposes. Students interested in computer simulation using GiD do not need to read voluminous handbooks, nor have to click on several windows to visualize their results. The visualization options available in GiD like contour fill, animations, isosurfaces, stream lines, etc. are one or two clicks away. Much research work was invested so that the visualization facilities do not decrease the performance of GiD when handling huge amount of data (over 5 Gigabytes), like the ones which are typical in industrial applications such as aerospace or metal forming processes. Powerful options to customize the visualization step with specific tools suitable for each problem have been developed. GiD incorporates state of the art 3D techniques to visualize complex results, like gluing a texture to a surface so that a deformed metal sheet can be compared with experimental data obtained in laboratory.

Other specific visualization facilities developed within GiD include: deformed shapes, vector and contour plots, beam diagrams and isosurfaces from static and dynamic analyses; animated sequences; graphs of quantities along geometrical lines; particle line flow diagrams; user-definable interfaces; import of neutral files containing results produced by other codes; customisable menus, etc. (Figure D.6).

D.1.6 Who can benefit from using GiD?

GiD is an extremely useful tool for:

- Engineering companies wanting to unify their input data and results visualisation environment for a variety of numerical simulation codes.

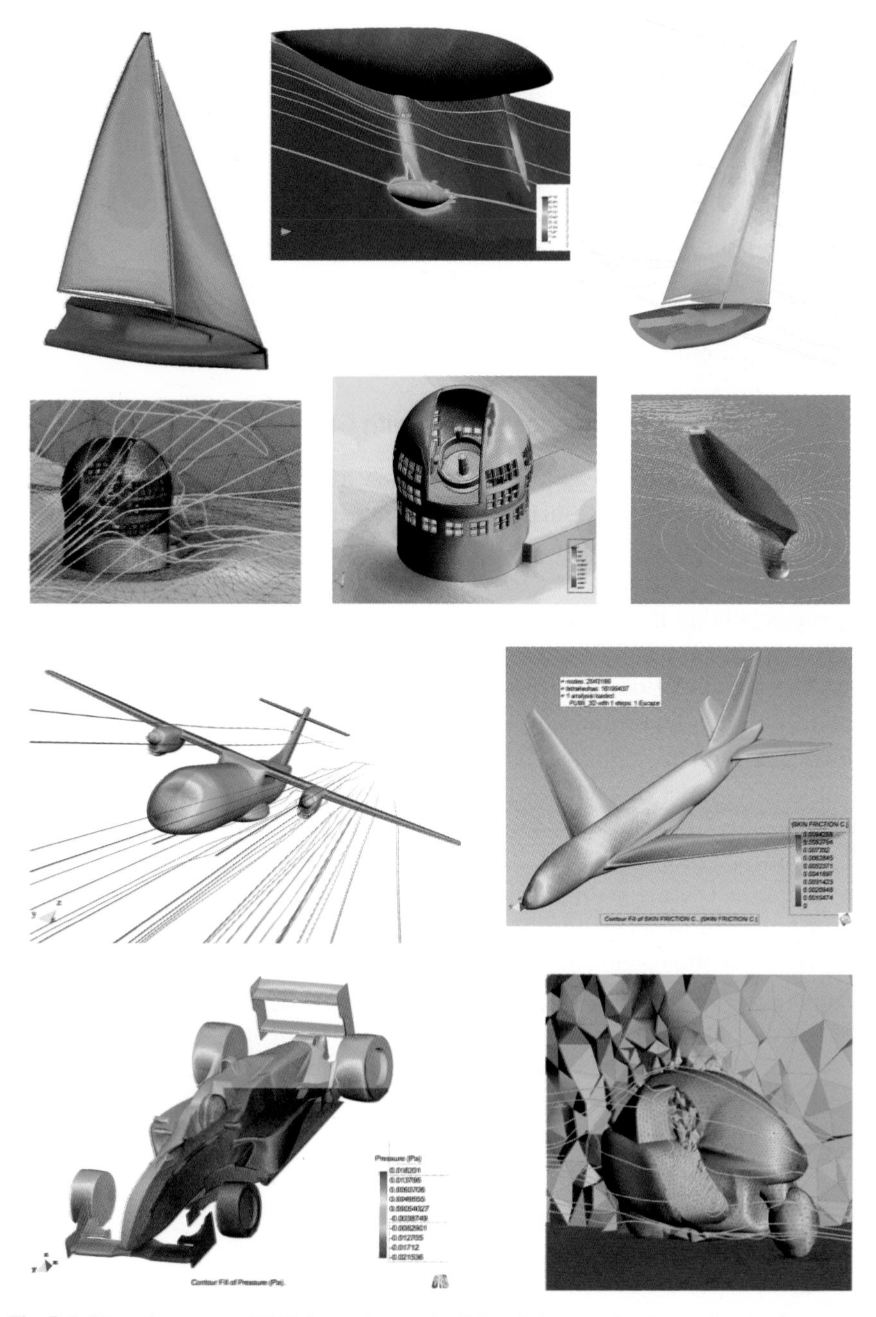

Fig. D.6 Visualization of FEM results with GiD of ship hydrodynamic analysis and aerodynamic analysis of sailing boat, telescope building, aircraft, racing car and motorbike. Images in first three rows taken from [OGI,CMOS,FO,FO2,REMFI]

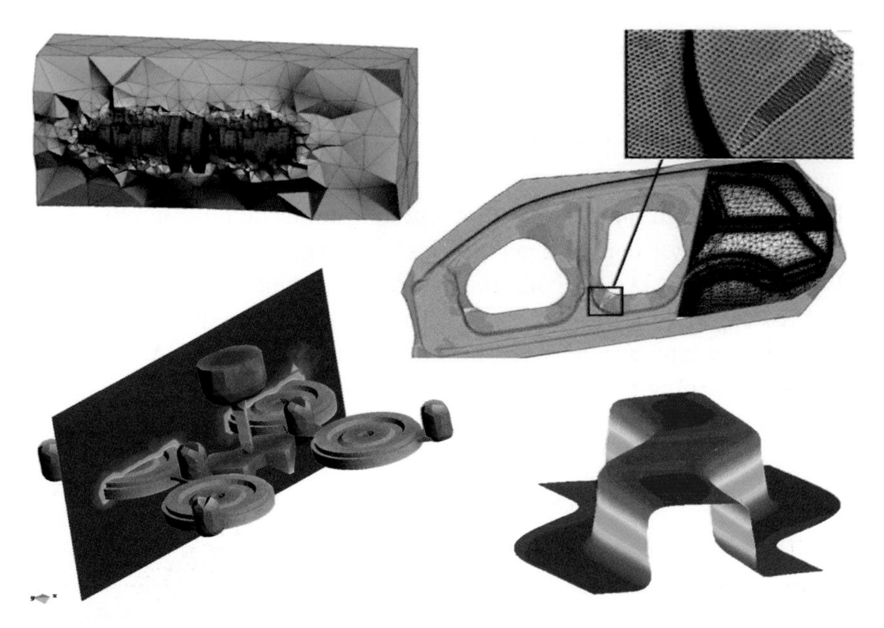

Fig. D.6 Cont. Visualization of FEM results with GiD for casting and sheet stamping process analyses. Courtesy of Quantech ATZ S.A. (www.quantech.es)

- Universities and RTD centers wishing to provide a multi-user environment for pre/post processing allowing the development and application of a diversity of software in computational science and engineering by students and RTD staff. The availability of such a "personal" pre/postprocessing system will invariantly help universities and research centres for the widespread use of numerical methods by undergraduate and graduate students wanting to solve practical problems using commercial or in-house software tools.
- Engineers and scientists wanting to have individual access from home or the office to a powerful pre and post processing system for research and applications of numerical simulation codes in either UNIX workstations or the simplest PC. The possibility of e-work is becoming for many engineers and scientists a habit. This invariably increases the need for new powerful and economical distributed computing procedures (generally via internet) for preparation of analysis data for complex geometries, probably generated via CAD elsewhere, and for handling the analysis results with support from visualisation tools for preparation of high quality reports in written or electronic form. GiD is the ideal tool for solving these problems.

D.1.7 Which are the advantages of using GiD?

- Reduces time and cost associated with numerical analysis through high performance data input and post processing in the simplest computer environment.
- High speed, high quality meshing and data input definition reduces the analysis time for complex geometries and large models.
- Direct use of CAD data combined with GiD geometric modelling facilities reduces redundancy and costs associated with model development.
- Allows easy interface with in-house software codes and CAD systems.
- Reduces learning time and improves efficiency with an intuitive graphical interface.

D.1.8 Why is it worth using GiD?

The GiD system incorporate a comprehensive collection of methods and software tools which solve the three main bottlenecks for the practical solution of complex problems using computer simulation: the user friendly preparation of analysis data, the easy link to any computer simulation code and the graphic visualisation of numerical results using either windows, Linux and Unix in a single computer, in an intranet or via internet.

This will help to change the order of magnitude of problems in science and engineering to be solved from office and/or home using computer simulation tools either in PC's, or supercomputers with the help fast communication networking tools and internet.

D.1.9 How can one learn to use GiD?

A number of tutorials with the aim of introducing newcomers to the use of GiD are available in the GiD web page (www.gidhome.com). Questions on the use of GiD can also be addressed to the GiD developing team via the GiD web page.

D.1.10 How can one access GiD?

GiD is freely downloadable from Internet (www.gidhome.com). An academic version of GiD suitable for research and university type projects can be freely used for an unlimited period.

The professional version of GiD requires a permanent password which can be purchased for the web page. It is the policy of GiD that passwords

Fig. D.7 Version 9.0 of GiD available from www.gidhome.com

have an unlimited duration and have an affordable prize for individuals and organizations. For details visit www.gidhome.com.

D.1.11 The GiD team

The GiD team is formed by the following group of researchers and engineers from CIMNE. In alphabetic order:

Abel Coll
Enrique Escolano
Sergio González
Adrià Melendo
Anna Monros
Miguel Pasenau
Jorge Suit

External adviser: Dr. Ramón Ribó (Compass Ingeniería y Sistemas, S.A., www.compassis.com).

You are welcomed to contact the GiD team at gid@cimne.upc.edu and www.gidhome.com.

References

[AFS] Argyris, J.H., Fried, I. and Scharpf, D.W., The TUBA family of plate elements for the matrix displacement method. *Aeronautical Journal of the Royal Aeron. Society*, **72**, 701–9, 1968.

[AH] Atamaz-Sibai, W. and Hinton, E., Adaptive mesh refinement with the Morley plate element. *Proc. of NUMETA 90 Conference*, Swansea, January 1990, **2**, 1049–57, Elsevier Appl. Science, London, 1990.

[AHS] Atamaz-Sibai, W., Hinton, E. and Selman, A., Adaptive mesh refinement with the Mindlin-Reissner elements. Proc. of *2nd Int. Conf. an Computer Aided Analysis and Design of Concrete Structures.*, Zell-an-see, Austria, Abril 1990, Pineridge Press, 1990.

[AK] Argyris, J.H. and Kelsey, S. *Energy theorems and structural analysis*. Butterworth, London, 1960 (originally published in a series of articles in *Aircraft Engineering*, 1954–55).

[ASME] ASME. Guide for Verification and Validation in Computational Solid Mechanics. ASME V& V 10-2006 American Society of Mechanical Engineers, 2006. Available at
http://catalog.asme.org/Codes/PrintBook/VV_10_2006_Guide_Verification.cfm

[Ba] Baker, T.J., Automatic mesh generation for complex three-dimensional regions using a constrained Delaunay triangulation. *Engineering Computations*, **5**, 161–175, 1989.

[Bat] Bathe, K.J., *Finite element procedures*. Prentice Hall, Inc., 1996.

[BB] Belytschko, T. and Bachrach, W.E., The efficient implementation of quadrilaterals with high coarse mesh accuracy. *Comput. Meth. in Appl. Mech. Engrg*, **54**, 276–301, 1986.

[BCHC+] Barbat A.H., Cervera M., Hanganu A., Cirauqui C. and Oñate E., Failure pressure evaluation of the containment building of a large dry nuclear power plant. *Nuclear Engineering and Design*, **180**, 251–270, 1998.

[BCF] Boussetta, R., Coupez, T. and Fourment, L., Adaptive remeshing based on a posteriori error estimation for forging simulation. *Comput. Meth. in Appl. Mech. Engrg*, **155**, 6626–6645, 2006.

[BD] Batoz, J.L. and Dhatt, G., *Modelisation des structures par élements finis. Vol. 1: Solides elastiques*. Hermes, Paris, 1990.

[Be] Bedrosian, G., Shape functions and integration formulas for three-dimensional finite element analysis. *Int. J. Num. Meth. Engrg.*, **35**(1), 95–108, 1992.

[Bey] Beyer, W.H., *CRC Standard Mathematical Tables.* CRC Press, 28th ed., 1988.

[Bl] Blacker, T.D., FASTQ Users Manual Version 1.2. Technical Report SAND88-1326, Sandia National Laboratories, Albuquerque, New Mexico, 1994.

[Bl2] Blacker, T.D., Automated quadrilateral surface discretization method and apparatus usable to generate mesh in a finite element analysis system. Unites States Patent, 5,315,537, May 1994.

[BM] Blacker, T.D. and Meyers, R.J., Seams and wedges in plastering: A 3D hexahedral mesh generation algorithm. *Engineering with Computers*, **2**(9), pp. 83–93, 1993.

[Bo] Bowyer, A., Computing Dirichlet tessellations. *The Computer Journal*, **24**(2), 162–166, 1981.

[BO] Bugeda, G. and Oliver, J., A general methodology for structural shape optimization problems using automatic adaptive remeshing. *Int. J. Num. Meth, Engrg.*, **36**(18), 3161–3185, 1993.

[BO2] Bugeda, G. and Oñate, E., New adaptative technique for structural problems. *First European Conference on Numerical Methods in Engineering*, Ch. Hirsch, O.C. Zienkiewicz and E. Oñate (Eds.), pp. 135–142, Elsevier, 7–11 September 1992.

[BO3] Bugeda, G. and Oñate, E., Adaptive techniques for elliptic problems. In *Advances in Post and Preprocessing for Finite Element Technology,* Papadrakakis M. and Topping B.H.V. (Eds.), Civil-Comp Press, 77–84, 1994.

[BP] Bonet, J. and Peraire, J., An alternate digital tree algorithm for geometric searching and intersection problems. *Int. J. Num. Meth, Engrg.*, **31**, 1–17, 1991.

[BS] Blacker, T.D. and Stephenson, M.B., Paving: A new approach to automated quadrilateral mesh generation. *Int. J. Num. Meth, Engrg.*, **32**, 811–847, 1991.

[BS2] Brenner S.C. and Scott L.R., *The mathematical theory of finite element methods.* Springer-Verlag, 1994.

[BSC] Blacker, T.D., Stephenson, M.B. and Canann, S.A., Analysis automation with paving: A new quadrilateral meshing technique. *Advances in Engineering Software*, **56**, 332–337, 1991.

[BSGU] Babuška, I., Stroubolis, T., Gangaraj, S.K. and Upadhyay, C.S., Pollution error in the h-version of the finite element method and the local quality of the recovered derivatives. *Comput. Meth. in Appl. Mech. Engrg*, **140**, pp. 1–37, 1997.

[BSUG] Babuška, I., Stroubolis, T., Upadhyay, C.S. and Gangaraj, S.K., A posteriori estimation and adaptive control of the pollution error in the h-version of the finite element method. *Int. J. Num. Meth, Engrg.*, **38**, pp. 4207–4235, 1995.

[Bu] Bugeda, G., *Error estimation technique and mesh generation procedure in structural shape optimization*, Ph.D. Thesis (in spanish), Civil Engineering School, Technical University of Catalonia (UPC), Barcelona, 1990.

[Bu2] Bugeda, G., A comparison between new adaptive remeshing strategies based on point wise stress error estimation and energy norm error estimation, *Communications in Numerical Methods in Engineering*, **18**, pp. 469–482, 2002.

[BWSG+] Baehmann, P.L., Wittchen, S.L., Shephard, M.S., Grice, K.R. and Yerry, M.A., Robust geometrically-based, automatic two-dimensional mesh generation. *Int. J. Num. Meth. Engrg.*, **24**, 1043–1078, 1987.

[BY] Boissonnat, J.D. and Yvinec, M., *Algorithmic Geometry*. Cambridge University Press, 1997.

[BZ] Boroomand, B. and Zienkiewicz, O.C., Recovery by equilibrium patches (REP). *Int. J. Num. Meth, Engrg.*, **40**, 137–154, 1997.

[BZ2] Boroomand, B. and Zienkiewicz, O.C., An improved REP recovery and the effectivity robustness test. *Int. J. Num. Meth, Engrg.*, **40**, 3247–3277, 1997.

[Ca] Carey, G.F., *Computational Grids: Generation, Adaptation and Solution Strategies*. Taylor and Francis, 1997.

[CB] Coorevits, P. and Bellenger, E., Alternative mesh optimality criteria for $h-$adaptive finite element method. *Finite Element in Analysis and Design*, **40**, 1195–1215, 2004.

[CBMB] Cass, R.J., Benzley, S.E., Meyers, R.J. and Blacker, T.D., Generalized 3D paving: An automated quadrilateral surface mesh generation algorithm. *Int. J. Num. Meth. Engrg.*, **39**, 1475–1489, 1996.

[CDOO] Celentano, D., Dabir, A., Oller, S. and Oñate, E., Thermal-mechanical analysis of a S.G. cast iron crankshaft during cooling and solidification in green molding sand. Technical Report CIMNE, IT 182, November 1995.

[CFF] Cavendish, J.C., Field, D.A. and Frey, W.H., An approach to automatic three dimensional finite element mesh generation. *Int. J. Num. Meth, Engrg.*, **21**, 329–347, 1985.

[CI] Calvo, N.A. and Idelsohn, S.R., All-hexahedral element meshing: Generation of the dual mesh by recurrent subdivision. *Comput. Meth. in Appl. Mech. Engrg*, **182**, pp. 371–378, 2000.

[Cl] Clough, R.W., The finite element method in plane stress analysis. *Proc. 2nd A.S.C.E. Conf. in Electronic Computation*, Pittsburgh, Pa., Sept. 1960.

[Cl2] Clough, R.W., Comparison of three dimensional finite elements. *Proceedings Symposium on Applications of Finite Element Methods in Civil Engineering*, Vanderbilt Univ., Nashville, Tenn., (published by ASCE), 1–26, 1969.

[CLT] Cantin, G., Loubignac, G. and Touzot, G., An iterative algorithm to build continous stress and displacement solutions. *Int. J. Num. Meth. Engrg.*, **12**, 1493–1506, 1978.

[CMOS] Codina, R., Morton, C. Oñate, E. and Soto, O., Numerical aerodynamic analysis of large buildings using a finite element model with application to a telescope building. *Int. Journal of Numerical Methods for Heat and Fluid Flow*, **10**(6), 616–633, 2000.

[CMPW] Cook, R.D., Malkus, D.S., Plesha, M.E. and Witt, R.J., *Concepts and application of finite element analysis*. 4º ed., John Wiley and Sons, 2002.

[Co] Courant, R., Variational methods for the solution of problems of equilibrium and vibration. *Bull. Am. Math. Soc.*, **49**, 1–23, 1943.

[Co2] Courtney, T.H., *Mechanical behaviour of materials*. McGraw-Hill, 1990.

[COG] Cervera, M., Oliver, J. and Galindo, M., Numerical simulation of damage in concrete dams (in Spanish), *Monograph* N. 4, CIMNE, Barcelona, 1991.

[COHO] Cervera, M., Oliver, J., Herrero, E. and Oñate E., A computational model for progressive cracking in large dams due to swelling of concrete. *Int. Journal of Fracture*, **35**(1,2,3), 573–85, 1990.

[COHR+] Casteleiro, M., Oñate, E., Huerta, A., Roig, J. and Alonso, E., Three dimensional analysis of no tension materials. *4th. Int. Conf. on Rock Mech*, Melbourne, Australia, 1983.

[Cow] Cowper, G.R., Gaussian quadrature formulae for triangles. *Int. J. Num. Meth. Engrg.*, **7**, 405–8, 1973.

[Cr] Crisfield, M.A., *Finite element and solution procedures for structural analysis, I: Linear analysis*. Pineridge Press, 1986.

[CVDO+] Celentano, D., Visconte, D., Dardati, P., Oller, S. and Oñate, E., A thermomechanical model for solidification problems. Experimental validation. In *Proceedings of Fourth International Conference on Computational Plasticity (COMPLAS IV)*, R. Owen, E. Oñate and E. Hinton (Eds.), Barcelona, Spain, 3–6 April 1995, Pineridge Press/CIMNE, pp. 2385–2396.

[Dao] de Arantes Oliveira, E.R., The patch test and the general convergence criteria of the finite element method. *Int. J. of Solids and Structures*, **13**, 159–178, 1977.

[De] Delaunay, B., Sur la sphère vide. *Izv. Akad. Nauk SSSR*, Otdelenie Matematicheskii i Estestvennyka Nauk, **7**, 793–800, 1934.

[Dem] Demmel, J., *Applied Numerical Linear Algebra*. Society for Industrial and Applied Mathematics, Philadelphia, PA, 1997.

[DH] Davis, P.J. and Hersch, R., *The Dream of Descartes*. Ed. Labor, 1989.

[DH2] Díez, P. and Huerta, A., A unified approach to remeshing strategies for finite element h-adaptivity. *Comput. Meth. in Appl. Mech. Engrg*, **176**(1-4), 215–229, 1999.

[Dho] Dhondt, G., A new automatic hexahedral mesher based on cutting. *Int. J. Num. Meth, Engrg.*, **50**, 2109–2126, 2001.

[Du] Dumavanet, D.A., High degree efficient symmetrical Gaussian quadrature rules for the triangle. *Int. J. Num. Meth. Engrg.*, **21**, **7**, 1985.

[DV] Dvorkin, E.N. and Vassolo, S.I. A quadrilateral 2D finite element based on mixed interpolation of tensorial components. *Engineering Computations*, **6**, 217–24, 1989.

[EIZ] Ergatoudis, J.G., Irons, B.M. and Zienkiewicz, O.C., Curved, isoparametric, quadrilateral elements for finite element analysis. *Int. J. of Solids and Structures*, **4**, 31–42, 1968.

[Fa] Farin, G., *Curves and Surfaces for Computer Aided Geometric Design.* Academic Press, 1990.

[Fa2] Farin, G., *Curves and Surfaces for CAGD. A Practical Guide.* 5th Edition, Morgan Kaufmann Publishers Inc., San Francisco, CA, USA, pp. 499, 2001.

[FdV] Fraeijs de Veubeke, B., Variational principles and the patch test. *Int. J. Num. Meth. Engrg.*, **8**, 783–801, 1974.

[Fe] Ferrers, N.M., *An elementary treatise on trilinear coordinates, the method of reciprocal polars and the theory of projections.* MacMillan , London, 1961.

[Fe2] Ferencz, R.M., *Element-by-element preconditioning techniques for large-scale, vectorized finite element analysis in nonlinear solid and structural mechanics.* Ph.D thesis, Department of Mechanical Engineering, Stanford University, Stanford, California, 1989.

[FG] Frey, P.J. and George, P.L., *Mesh Generation Application to Finite Elements.* Hermes Science Publishing, Oxford, Paris, 816 , 2000.

[FNS] Fröier, M., Nilsson, L. and Samuelsson, A., The rectangular plane stress element by Turner, Pian and Wilson. *Int. J. Num. Meth. Engrg.*, **8**, 2, 433–7, 1974.

[FP] Faux, I.D. and Pratt, M.J., *Computational Geometry for Design and Manufacture.* Ellis Horwood, 1985.

[FO] Flores, R. and Ortega, E., Pumi: An explicit 3D unstructured finite element solver for the Euler equations. Publication CIMNE PI 326, November 2008.

[FO2] Flores, R., Ortega, E. and Oñate, E., A numerical study of wing deformation effects in wind tunnel testing using twin-sting supports. Publication CIMNE, 2008. Submitted to *Engineering Computations.*

[Fu] Fourment, L., Notes on the mesh optimality criteria of Oñate and Bugeda [OB]. *Private Communication*, 1995.

[Ga] Gallagher, R.H., *Finite Element Analysis Fundamentals.* Prentice–Hall, Englewood Cliffs, N.J., 1975.

[GB] George, P.L. and Borouchaki, H., *Delaunay Triangulation and Meshing. Application to Finite Elements.* Editions Hermes, Paris, 1998.

[GBFL+] George, P.L., Borouchaki, H., Frey, P.J., Laug, P. and Saltel, E., Mesh Generation and Mesh Adaptivity. In *Encyclopedia of Computational Mechanics.* Chapter 17, 1 Fundamentals, E. Stein, R. De Borst, J.R. Thomas (Eds.), J. Wiley, 2004.

[Ge] George, P.L., *Automatic mesh generation. Applications to the finite element method.* J. Wiley, 1991.

[GH] Geller, M. and Harbord, R., Moderate degree cubature formulas for 3D tetrahedral finite element approximations. *Communications in Applied Numerical Methods*, **7**(6), 487–495, 1991.

[GH2] Gordon, W.J. and Hall, C.A., Construction of curvilinear co-ordinate systems and application to mesh generation. *Int. J. Num. Meth, Engrg.*, **3**, 461–477, 1973.

[GHS] George, P.L., Hecht, F. and Saltel, E., Automatic mesh generator with specified boundary. *Comput. Meth. in Appl. Mech. Engrg*, **92**, 269–288, 1991.

[GiD] GiD. The personal pre and postprocessor. CIMNE, Barcelona, 2008, www.gidhome.com.

[GS] George, P.L. and Seveno, E., The advancing-front mesh generation method revisited. *Int. J. Num. Meth, Engrg.*, **37**, 3605–3619, 1994.

[HC] Hinton, E. and Cambell, J.S., Local and global smoothing of discontinous element functions using a least square method. *Int. J. Num. Meth. Engrg.*, **8**, 461–80, 1979.

[HCA] Huerta, A., Casteleiro, M. and Alonso, E., Non linear numerical analysis of earth dam constructions, in *Numerical Methods for non Linear Problems*, C. Taylor, E. Hinton, R. Owen y E. Oñate (Eds.), Pineridge Press, 1984.

[HCB] Hughes, T.J.R., Cottrell, J.A. and Bazilevs, Y., Isogeometric analysis: CAD, finite elements, NURBS, exact geometry and mesh refinement. *Comput. Meth. in Appl. Mech. Engrg*, **194**, 4135–4195, 2005.

[He] Hearmon, R.F.S., *An introduction to applied anisotropic elasticity.* Oxford Univ. Press, 1961.

[HH] Hooley, R.F. and Hibbert, P.D., Bounding plane stress solutions by finite elements. *Journal Struct. Division*, ASCE, **92**, ST1, 39–48, 1966.

[HKD] Hitchings, D., Kamoulakos, A. and Davies, G.A.O., *Linear Static Benchmarks.* Vol 1, National Engineering Laboratory Report, January 1987 NAFEMS LSB1.

[HO] Hinton, E. and Owen, D.R.J., *Finite element programming*. Academic Press, 1979.

[HO2] Hinton, E. and Owen, D.R.J., *Introduction to finite element computations*. Pineridge Press, 1980.

[Hr] Hrenikoff, A., Solution of problems in elasticity by the framework method. *J. Appl. Mech.*, **18**, 169–75, 1991.

[Hu] Hughes, T.J.R., *The finite element method. Linear static and dynamic finite element analysis*. Prentice-Hall, 1987.

[IA] Irons, B.M. and Ahmad, S., *Techniques of finite elements*. Ellis Harwood, Chichester, 1980.

[Ir] Irons, B.M., Numerical integration applied to finite element method. *Conf. Use of Digital Computers in Struct. Eng.*, Univ. Newcastle, 1966.

[IR] Irons, B.M. and Razzaque, A., Experience with the patch test for convergence of finite element method. *The Mathematics of Finite Elements with Application to Partial Differential Equations*, 557–587, A.R. Aziz (ed.), Academic Press, 1972.

[Joe] Joe, B., Quadrilateral mesh generation in polygonal regions. *Computer Aided Design*, **27**, 209–222, 1995.

[Jor] Jordan, W.B., The plane isoparametric structural element. *General Electric C_o, Report KAPL–M–7112*, Schenectadly, New York, 1970.

[JSK] Johnston, B.P., Sullivan, J.M. and Kwasnik, A., Automatic conversion of triangular element meshes to quadrilateral elements. *Int. J. Num. Meth. Engrg.*, **31**, 67–84, 1991.

[JT] Jin, H. and Tanner, R.I., Generation of unstructured tetrahedral meshes by advancing front technique. *Int. J. Num. Meth. Engrg.*, **36**, 1805–1823, 1993.

[JW] Jin, H. and Wiberg, N.E., Two dimensional mesh generation, adaptive remeshing and refinement. *Int. J. Num. Meth. Engrg.*, **29**(7), 1501–1526, 1990.

[KF] Kosloff, D. and Frasier, G.A., Treatment of hour glass patterns in low order finite element codes. *Int. J. for Numerical Analysis Methods in Geomechanics*, **2**, 57–72, 1978.

[KGZB] Kelly, D.W., De S.R. Gago, J.P., Zienkiewicz, O.C. and Babuška, I., *A posteriori error* analysis and adaptive processes in the finite element method: Part I: Error analysis. *Int. J. Num. Meth, Engrg.*, **19**, 1593–1619, 1983.

[KHK] Kwok, W., Haghighi, K. and Kang, E., An efficient data structure for the advancing-front triangular mesh generation technique. *Communications in Numerical Methods in Engineering*, **11**(5), 465–473, 1995.

[KS] Knupp, P. and Steinberg, S., *The Fundamentals of Grid Generation.* CRC Press, 1993.

[LB] Lee, N.-S. and Bathe, K.-J., Effects of element distortion on the performance of isoparametric elements. *Int. J. Num. Meth. Engrg.*, **36**, 3553–3576, 1993.

[LB2] Li, L.Y. and Bettess, P., Notes on mesh optimal criteria in adaptive finite element computation. *Communications in Numerical Methods Engineering*, **11**(11), 911–915, 1995.

[LB3] Li, L.Y. and Bettess, P., Adaptive finite element methods: a review. *Appl. Mech. Reviews*, **50**(10), 581–591, 1997.

[LBBB] Li, L.Y., Bettess, P., Bull, J., Bond, T. and Ipplegarth, I., Theoretical formulations for adaptive finite element computations. *Communications in Numerical Methods Engineering*, **11**(10), 857–868, 1995.

[Le] Lekhnitskii, S.G., *Theory of elasticity of an anisotropic elastic body.* Holden Day, San Francisco, 1963.

[Li] Livesley, R.K., *Matrix methods in structural analysis.* 2nd ed., Pergamon Press, 1975.

[LJ] Liu, A. and Joe, B., Relationship between tetrahedron shape measures. *BIT*, **34**, 268–287, 1994.

[Lo] Lo, S.H., A new mesh generation scheme for arbitrary planar domains. *Int. J. Num. Meth, Engrg.*, **21**, 1403–1426, 1985.

[LP] Löhner, R. and Parikh, P., Three-dimensional grid generation by the advancing front method. *Int. J. Num. Meth. Engrg.*, **8**, 1135–1149, 1988.

[LP2] Ladeveze, P. and Pelle, J.P., *Mastering calculations in linear and non linear mechanics.* Springer, 2005.

[LTV] Lober, R.R., Tautges, T.J. and Vaugham, C.T., Parallel paving: An algorithm for generating distributed, adaptive, all-quadrilateral meshes on parallel computers. Technical Report SAND97-0545, Sandia National Laboratory, Alburquerque, New Mexico, USA, 1997.

[LW] Li, X.D. and Wiberg, N.-E., A posteriori error estimate by element patch postprocessing, adaptive analysis in energy and L2 norm. *Computers and Structures*, **53**, 907–919, 1994.

[LZG] Lewis, R.W., Zheng, Y. and Gethin, D.T., Three-dimensional unstructured mesh generation: Part 3. Volume meshes. *Comput. Meth. in Appl. Mech. Engrg*, **134**, 285–310, 1996.

[MBBO] Miquel, J., Botello, S., Buil, J. and Oñate, E., The Talvachia double curvature arch dam. Static and dynamic analysis (in spanish). *Monograph* No. 6, CIMNE, Barcelona, 1991.

[Mc] Mc. Henry, D., A lattice analogy for the solution of plane stress problems. *J. Inst. Civ. Engng.*, **21**, 59–82, 1943.

[MDR] Mangalgiri, P.D., Dattaguru, B. and Rammamurthy, T.S., Specification of skew conditions in finite element formulation. *Int. J. Num. Meth. Engrg.*, **12**(6), 1037–1041, 1978.

[MH] Moller, P. and Hansbo P., On advancing front mesh generation in three dimensions. *Int. J. Num. Meth. Engrg.*, **38**, 3551–3569 1995.

[MW] Marcum, D.L. and Weatherill, N.P., Unstructured grid generation using iterative point insertion and local reconnection. *AIAA Journal*, **33**(9), pp. 1619–1625, 1995.

[Na] Narayanaswani, R., Dependence of plate bending finite element deflections and eigenvalues on Poisson's ratio. *A.I.A.A.J.*, **12**. 1420–21, 1974.

[Ng] Nguyen, V.P., Automatic mesh generation with tetraedon elements. *Int. J. Num. Meth. Engrg.*, **18**, 273–89, 1982.

[No] Noor, A., Bibliography of books and monographs on finite element technology. *Appl. Mech. Rev.*, **44**(6), 307–17, June, 1991.

[OB] Oñate, E. and Bugeda, G., A study of mesh optimality criteria in adaptive finite element analysis. *Engineering Computations*, **10**(4), 307–321 1993.

[OC] Oñate, E. and Castro, J., Adaptive mesh refinement techniques for structural problems. In *The Finite Element Method in the 1990's, A book dedicated to O.C. Zienkiewicz* , E. Oñate, J. Periaux and A. Samuelsson (Eds.), Springer–Verlag/CIMNE, Barcelona, 1991.

[OCCO+] Oñate, E., Celentano, D., Codina, R., Oller, S. and Soto, O., Finite element model for analysis of mould filling and solidification in casting. In *Proceedings of the Conference on Modelling of Casting, Welding and Advanced Solidification Processes VII*, London, 10–15 September, 1995.

[OCK] Oñate, E., Castro, J. and Kreiner, R., Error estimation and mesh adaptivity techniques for plate and shell problems. In *The 3rd. International Conference on Quality Assurance and Standars in Finite Element Methods*, NAFEMS, Stratford-upon-Avon, England, 10–12 September, 1991.

[OCOH] Oliver, J., Cervera, M., Oñate, E. and Herrero, E., A case study of gravity dam subjected to severe internal actions: Description and numerical simulations. In *Sixth Congress on Large Dams. International Committee for Large Dams (ICOLD)*, San Francisco, C.29, 1387–409, 1988.

[OGI] Oñate, E., García, J. and Idelsohn, S.R., Ship hydrodynamics. *Encyclopedia of Comput. Mechanics*, John Wiley & Sons Ltd, Vol. **3**, Chapter 18, 579–607, 2004.

[OH] Oñate, E. and Hanganu, A., Advanced method for the analysis of the failure load of concrete structures (in Spanish). Publication CIMNE, PI 176, September 1999.

[OHBO+] Oñate, E., Hanganu, A., Barbat, A., Oller, S., Vitaliani, R., Saetta, A. and Scotta, R., Structural analysis and durability assessment of historical constructions using a finite element damage model. Publication CIMNE, Pl73, pp. 26, 1995.

[OHBO+2] Oñate, E., Hanganu, A., Barbat, A., Oller, S., Vitaliani, R., Saetta, A. and Scotta, R., Structural Analysis and durability assessment of historical constructions using a finite element damage model. Publication CIMNE, Pl73, 1995. Also in *Structural Analysis of Historical Constructions. Possibilities of Numerical and Experimental Techniques*, Roca *et al.* (Eds.), CIMNE, Barcelona, 1997.

[On] Oñate, E., Reliability analysis of concrete structures. Numerical and experimental studies. Publication CIMNE, PI 107, January 1997.

[On2] Oñate, E., *Structural Analysis with the Finite Element Method. Linear Statics. Volume 2: Beams, Plates and Shells.* CIMNE-Springer, 2009.

[OOB] Oñate, E., Oliver, J. and Bugeda, G., Finite element analysis of the non linear response of concrete dams subjected to internal loading. *Finite Element Methods for Non Linear Problems*, Europe–US Symposium, Trǫdheim, Norway, 1985, P. Bergan *et al.* (Eds.), Springer–Verlag, 1985.

[OPSE] Oñate, E., Pasenau, M., Suit, J. and Escolano, E., New paradigms and tools for interfacing geometrical models, analysis data, numerical methods and visualization software in computational science and engineering: The GiD project. Publication CIMNE, PI 262, May 2005.

[Or] O´Rourke, J., *Computational Geometry in C.* Cambridge University Press, 2nd Edition, 2001.

[OSCS] Owen, S.J., Staten, M.L., Canann, S.A. and Saigal, S., Q-morph: An indirect approach to advancing front quad meshing. *Int. J. Num. Meth. Engrg.*, **44**, 1317–1340, 1999.

[Pe] Peiró, J., *A finite element procedure for the solution of the Euler equations on unstructured meshes*. Ph.D. Thesis, Civil Eng. Dpt., Univ. College of Swansea, U.K., 1989.

[Pe2] Peiró, J., Surface grid generation. In *Handbook of Grid Generation*, Chapter 19, 19.1–19.20, CRC Press, 1999.

[PFGS] Pereda, J., Foces, A., Garrido, J.A. and Suárez, S., Multicellular box docks: An user friendly Internet based software for design in civil engineering. 2nd Conference on Advances and Applications of GiD, Publication CIMNE No. 210, Barcelona, 2004.

[PFTV] Pres, W.H., Flannery, B.P., Teukolsky, S.A. and Vetterling, W.T., *Numerical Recipes. The art of Scientific Computing*. Cambridge Univ. Press, 1986.

[PGH] Parthasarathy, V.N., Graichen, C.M. and Hathaway, A.F., A comparison of tetrahedron quality measures. *Finite Elements in Analysis and Design*, **15**, 255–261, 1993.

[Pi] Pissanetzky, S., Kubik: An automatic three dimensional mesh generator. *Int. J. Num. Meth. Engrg.*, **17**(2), 255–69, 1981.

[PP] Prinja, N.K. and Puri, A.K., *An Introduction to the Use of Material Models in FE*. NAFEMS, November 2005.

[PPFM+] Peraire, J., Peiró, J., Formaggio, L., Morgan, K.J. and Zienkiewicz, O.C., Finite element Euler computations in three dimensions. *Int. J. Num. Meth. Engrg.*, **26**, 2135–2159, 1988.

[PPM] Peraire, J., Peiró, J. and Morgan, K., Adaptive remeshing for 3–dimensional compressible flow computations. *Journal of Computational Physics*, **103**, 269–285, 1992.

[Pr] Przemienieck, J.S., *Theory of matrix structural analysis*. McGraw-Hill, New York, 1968.

[PS] Preparata, F.P. and Shamos, M.I., *Computational Geometry*. Springer-Verlag, New York, 1988.

[PT] Piegl, L. and Tiller, W., *The NURBS Book*. 2nd Edition, Springer-Verlag, New York, 646, 1997.

[PVMZ] Peraire, J., Vahdati, M., Morgan, K. and Zienkiewicz, O.C., Adaptive remeshing for compressible flow computation. *J. Comp. Phys.*, **72**, 449–66, 1987.

[Rad] Radau, *Journal of Math.*, **3**, p. 283, 1880.

[Ral] Ralston, A., *A First Course in Numerical Analysis*. McGraw-Hill, New York, 1965.

[RCMR+] Roca, P., Casarin, F., Modena, C., Rodríguez, I. and Garay, A., Damage monitoring of long-span historical structures. In *Damage of massive historical constructions*, L. Vinda (ed.), WIT Press, Southampton, UK, 2006.

[Re] Rebay, S., Efficient unstructured mesh generation by means of Delaunay triangulation and Bowyer-Watson algorithm. *Journal of Computational Physics*, **106**, 125–138, 1993.

[REMFI] REMFI: Rear fuselage and empennage flow investigation. Project No. AST3-CT-2004-502895, sponsored by the FP6 Programme of the European Commission (EC), 3/1/2004–2/28/2007, www.cimne.com/remfi/.

[RG] Rosanoff, R.A. and Ginisburg, T.A., Matrix error analysis for engineers. *Proc. First Conf. on Matrix Meth in Struct. Mech.*, Wright-Patterson, Air Force B., Ohio, 1965.

[RGAA] Roca, P., González, J.L., Aguerri, F. and Aguerri, J.I., Monitoring of long-term damage in Gothic Cathedrals. In *Structural Studies, Repairs and Maintenance of Heritage Architecture VII (STREMAH 7)*, C. Brebbia (Ed.), WIT Press, Southampton, pp. 109–115, Halkidiki, Grecia, 2003.

[RGL] Rosanoff, R.A., Gloudemann, J.F. and Levy, S., Numerical conditioning of stiffness matrix formulations for frame structures. *Proc. Second Conf. on Matrix Meth. in Struct. Mech.*, Wright–Pattersson, AFB, Ohio, 1968 (AFFDL-TR-68-150. Dec, 1969: AD-703-685 N.T.T.S) 1029–60.

[RGMO] Roca, P., González, J.L., Mari, A.R. and Oñate, E. (Eds.), *Structural Analysis of Historical Constructions I. Possibilities of Numerical and Experimental Techniques.* CIMNE, Barcelona, 1997.

[RGOL] Roca, P., González, J.L., Oñate, E. and Lourenço, P.B. (Eds.), *Structural Analysis of Historical Constructions II. Possibilities of Numerical and Experimental Techniques.* CIMNE, Barcelona, 1998.

[RM] Roca, P. and Molins, C., Structural assessment of the cimborios of two Gothic Cathedrals. In *IASS-MSU International Symposium. Bridging Large Spans: From Antiquity to the Present*, Sanayi-i-Nefise, pp. 145–154, Istambul, Turquía, 2000.

[Ro] Roache, P.J., *Verification and Validation in Computational Science and Engineering.* Hermosa, Albuquerque, 1998.

[Ro2] Roca, P., Structural Monitoring of three Gothic Cathedrals. In *On-site control and evaluation of masonry structures*, RILEM Publication, L. Vinda (Ed.), pp. 105–118, Bagneux, Francia, 2003.

[ROD] Rachowicz, W., Oden, J.T. and Demkowicz, L., Toward a universal $h-p$ adaptive finite element strategy. Part 3: Design of $h-p$ meshes. *Comput. Meth. in Appl. Mech. Engrg*, **77**, 181–211, 1989.

[RPOH] Roca, P., Pellegrini, L., Oñate, E. and Hanganu, A., Analysis of the structure of Gothic Cathedrals. Application to Barcelona Cathedral. In *Structural Analysis of Historical Constructions II. Possibilities of Numerical and Experimental Techniques*, Roca *et al.* (Eds.), CIMNE, Barcelona, 1998.

[RSS] Rank, E., Schweingeruber, M. and Sommer, M., Adaptive mesh generation and transformation of triangular to quadrilateral meshes. *Communications in Numerical Methods in Engineering*, **9**, 121–129, 1993.

[Sa] Samuelsson, A., The global constant strain condition and the patch test. *Energy Methods in Finite Element Analysis*, R. Glowinski, E.Y. Rodin and O.C. Zienkiewicz (Eds.), **3**, 46–68, J. Wiley, 1979.

[SC] Sunder, K.S. and Cookson, R.A., Integration points for triangles and tetrahedrons obtained from the Gaussian quadrature points for a line. *Computers & Structures*, **21**(5), 881–885, 1985.

[Sch] Schwer, L.E., Guide for verification and validation in computational solid mechanics. *IACM Expressions*, N. 20, 14-21, January 2007 (www.iacm.info).

[SDH] Stein, E., De Borst, R. and Hughes, T.J.R. (Eds.), *Encyclopedia of Computational Mechanics. 1 Fundamentals.* J. Wiley, 2004.

[SDH2] Stein, E., De Borst, R. and Hughes, T.J.R. (Eds.), *Encyclopedia of Computational Mechanics. 2 Solids and Structures.* J. Wiley, 2004.

[SF] Strang, S. and Fix, G.J., *An analysis of the finite element method.* Prentice Hall, 1973.

[SFPC] Suárez, S., Foces, A., Pereda, J. and Cacho, M., Treating the geometry of a floating caisson to obtain a structured mesh in GiD. *1st Conference on Advances and Applications of GiD*, Publication CIMNE No. 210, Barcelona, 2002.

[SG] Shephard, M.S. and Georges, M.K., Automatic three-dimensional mesh generation by the finite octree technique. *Int. J. Num. Meth. Engrg.*, **32**, pp. 709–749, 1991.

[SG2] Suárez, B. and González, J.M., Analysis of Melonares dam. CIMNE Report (in Spanish), Barcelona, 2008.

[SG3] Suárez, B. and González, J.M., Structural analysis and design of an arch in Jaen (Spain) accounting for the effect of the terrain. CIMNE Report (in Spanish), Barcelona, 2008.

[SHM] Suárez, B., Hanganu, A. and Miquel, J., Optimum design of multicellular reinforced concrete box docks. *Journal of Structural Engineering (ASCE)*, **128**(5), 603–611, 2002.

[SM] Suárez, B. and Martel, E., *Finite element analysis of Santa Coloma dam* (in spanish). Pub. IT-26, CIMNE, Barcelona, 1991.

[SM2] Suárez, B., and Martel, E., *Analysis of turbine component in a hydroelectrical power plant.* CIMNE, Report, Barcelona, 1991.

[SR] Stein, E. and Rüter, M., Finite element methods for elasticity with error-controlled discretization and model adaptivity. *Encyclopedia of Computational Mechanics*, Vol. 1, 5–58, E. Stein, R. de Borst and T.J.R. Hughes (Eds.), Wiley, 2004.

[SRO] Stein, E., Rüter, M. and Ohnimus, S., Adaptive finite element analysis and modelling of solids and structures. Findings, problems and trends. *Int. J. Num. Meth. Engrg.*, **60**, 103–138, 2004.

[SS] Schroeder, W. and Shephard, M.S., Geometry-based fully automatic mesh generation and the Delaunay triangulation. *Int. J. Num. Meth, Engrg.*, **26**, 2503–2524, 1988.

[SS2] Schroeder, W. and Shephard, M.S., Computational Visualization. In *Encyclopedia of Computational Mechanics.* Chapter 18, 1 Fundamentals, E. Stein, R. De Borst, J.R. Thomas, editors. J. Wiley, 2004.

[Str] Strang, G., *Linear Algebra and its Application.* Academic Press, New York, 1976.

[Ta] Taig, O.C., *Structural analysis by the matrix displacement method*. Electric Aviation Report, No. 5017, 1961.

[Tay] Taylor, R.L., Solution of linear equations by a profile solver. *Engineering Computations*, **2**, 344–350, 1985.

[TBM] Tautges, T.J., Blacker, T.D. and Mitchell, S.A., The Whisker-Weaving algorithm: A connectivity-based method for constructing all-hexahedral finite element meshes. *Int. J. Num. Meth, Engrg.*, **39**, 3327–3349, 1996.

[TBW] Taylor, R.L., Beresford, P.J. and Wilson, E.L., A non conforming element for stress analysis. *Int. J. Num. Meth. Engrg.*, **10**, 1211–20, 1976.

[TCMT] Turner, M.J., Clough, R.W., Martin, H. y Topp, L.J., Stiffness and deflection analysis of complex structures. *J. Aeron. Sci.*, **23**, 805–23, 1956.

[TG] Timoshenko, S.P. and Goodier, J.N., *Theory of elasticity*. McGraw-Hill, New York, 3rd Edition, 1969.

[Ti] Timoshenko, S.P., *Strength of Materials. Part 1: Elementary Theory and Problems*. Van Nostrand Company Inc., New York, 1958.

[Ti2] Timoshenko, S.P., *Strength of Materials. Part 2: Advanced Theory and Problems*. 3rd. Edition, Van Nostrand-Reinhold, New York, 1956.

[TK] Taig, I.C. and Kerr, R.I., Some problems in the discrete element representation of aircraft structures. In *Matrix Methods of Structural Analysis*, B.M. Fraeijs de Veubeke (ed.), Pergamon Press, 1964.

[TK2] Topping, B.H.V. and Khan, A.I., Parallel Finite Element Computations. Saxe-Coburg Publications, Edinburgh, UK, 1996.

[TMIP+] Topping, B.H.V., Muylle, J., Iványi, P., Putanowicz, R. and Cheng, B., *Finite Element Mesh Generation*. Saxe-Coburg Publications, Stirling, 2004.

[TSW] Thompson, J.F., Soni, B.K. and Weatherill, N.P. (Eds.), *Handbook of Grid Generation*. CRC Press, January 1999.

[TTM] Thompson, J.F., Thames, F.C. and Martin C.W., Automatic numerical generation of body-fitted curvilinear coordinates for a field containing any number of arbitrary two dimensional bodies. *J. Comput. Phys.*, **15**, 299–319, 1974.

[TW] Thompson, J.F. and Warsi, Z.U.A., Boundary-fitted coordinate systems for numerical solution of partial differential equations. *J. Comput. Phys.*, **47**, 1–108, 1982.

[TWM] Thompson, J.F., Warsi, Z.U.A. and Martin C.W., *Numerical Grid Generation: Foundations and Applications*. North-Holland, Dordrecht, 1987.

[TY] Timoshenko, S.P., and Young, D.M., *Theory of Structures*. 2d Edition. McGraw-Hill, New York, 1965.

[Ut] Utku, S., Explicit expressions for triangular torus element stiffness matrix. AIAA Journal, 1174–1176, 1968.

[Ve] Verfürth, R., *A review of a Posterior Error Construction and Adaptive Refinement Procedures*. Wiley-Teubner, 1996.

[Vo] Voronoi, G., Nouvelles applications des paramètres continus à la théorie des formes quadratiques. *J. Reine Angew. Math*, **133**, 97–178, 1907.

[vRBBB] van Rens, B.J.E., Brokken, D., Brekelmans, W.A.M. and Baaijens, F.P.T., A two dimensional paving mesh generator for triangles with controllable aspect ratio and quadrilaterals with high quality. *Engineering with Computers*, **14**(3), 248–259, 1998.

[Was] Washizu, K., *Variational methods in elasticity and plasticity*. Pergamon Press, 1975.

[Wat] Watson, D.F., Computing the n-dimensional Delaunay tessellation with application to Voronoi polytopes. *Comp. J.*, **24**, 167–172, 1981.

[We] Weatherill, N.P., A method for generating irregular computation grids in multiply connected planar domains. *Int. J. Num. Meth, Engrg.*, **8**, 181–197, 1988.

[We2] Weatherill, N.P., The integrity of geometrical boundaries in the two-dimensional Delaunay triangulation. *Commun. Appl. Numer. Methods*, **6**, 101–109, 1990.

[WH] Weatherill, N.P. and Hassan, O., Efficient 3-dimensional Delaunay triangulation with automatic point generation and imposed boundary constraints. *Int. J. Num. Meth, Engrg.*, **37**, 2005–2039, 1994.

[WJ] Weaver, W. (Jr.) and Johnston, P.R., *Finite elements for structural analysis*. Prentice Hall, 1984.

[WL] Wiberg, N.-E. and Li, X.D., Superconvergent patch recovery of finite element solutions and a posteriori l2 norm error estimate. *Commun. Numer. Methods Engrg.*, **10**, 313–320, 1994.

[WR] Wilkinson, J.H. and Reinsch, C., *Linear Algebra. Handbook for Automatic Computation.* Volume II. Springer-Verlag, Berlin, 1971.

[Ya] Yang, T.Y., *Finite Element Structural Analysis*. Prentice Hall, 1986.

[Yam] Yamaguchi, F., *Curves and Surfaces in Computer Aided Geometric Design.* Springer- Verlag, Berlin, 1988.

[YS] Yerry, M.A. and Shephard, M.S., Automatic three-dimensional mesh generation by the modified octree technique. *Int. J. Num. Meth. Engrg.*, **20**, 1965–1990, 1984.

[ZL] Zhang, Z. and Lin, R., Ultraconvergence of ZZ patch recovery at mesh symmetry points. *Numer. Math*, **95**, 781–801, 2003.

[ZP] Zienkiewicz, O.C. and Phillips, D.V., An automatic mesh generation scheme for plane and curved surfaces by isoparametric coordinates. *Int. J. Num. Meth, Engrg.*, **3**, 519–528, 1971.

[ZS] Zienkiewicz, O.C. and Scott, F.C., On the principle of repeability and its application in the analysis of turbine and pump impellers. *Int. J. Num. Meth. Engrg.*, **4**(3), 445–448, 1972.

[ZT] Zienkiewicz, O.C. and Taylor, R.L. *The Finite Element Method for Solid and Structural Mechanics*. Sixth Edition, Elsevier, 2005.

[ZTZ] Zienkiewicz, O.C., Taylor, R.L. and Zhu, J.Z., *The Finite Element Method. Its Basis and Fundamentals*. Sixth Edition, Elsevier, 2005.

[ZVK] Zienkiewicz, O.C., Valliappan, S. and King, I.P., Stress analysis of rock as a no tension material. *Geotechnique*, **18**, 56, 1968.

[ZZ] Zienkiewicz, O.C. and Zhu, J.Z. The superconvergent patch recovery (SPR) and a posteriori error estimations. Part I: The recovery technique. *Int. J. Num. Meth. Engrg.*, **33**(7), 1331–1364, 1992.

[ZZ2] Zienkiewicz, O.C. and Zhu, J.Z., The superconvergent patch recovery (SPR) and a posteriori error estimations. Part II: Error estimates and adaptivity. *Int. J. Num. Meth. Engng.*, **33**(7), 1365–1382, 1992.

[ZZ3] Zienkiewicz, O.C. and Zhu, J.Z., The superconvergent patch recovery (SPR) and and adaptive finite element refinement. *Comput. Meth. in Appl. Mech. Engrg*, **101**, 207–224, 1992.

[ZZHW] Zhu, J.Z., Zienkiewicz, O.C., Hinton, E. and Wu, J., A new approach to the development of automatic quadrilateral mesh generation. *Int. J. Num. Meth. Engrg.*, **32**, 849–866. 1991.

Author index

Subject index